Methods in Molecular Biology™

Series Editor
John M. Walker
School of Life Sciences
University of Hertfordshire
Hatfield, Hertfordshire, AL10 9AB, UK

For further volumes:
http://www.springer.com/series/7651

Neisseria meningitidis

Advanced Methods and Protocols

Edited by

Myron Christodoulides

Division of Infection, Inflammation, and Immunity, Sir Henry Wellcome Laboratories,
University of Southampton Medical School, Southampton, UK

Humana Press

Editor
Myron Christodoulides, Ph.D
Division of Infection, Inflammation, and Immunity
Sir Henry Wellcome Laboratories
University of Southampton Medical School
Southampton, UK
M.Christodoulides@soton.ac.uk

ISSN 1064-3745 e-ISSN 1940-6029
ISBN 978-1-61779-345-5 e-ISBN 978-1-61779-346-2
DOI 10.1007/978-1-61779-346-2
Springer New York Dordrecht Heidelberg London

Library of Congress Control Number: 2011937568

Printed on acid-free paper

Cover illustration: The cover picture is a confocal microscopy image of human meningeal cells (stained red with Evan's blue) infected with *Neisseria meningitidis* (bacteria reacted with specific antibody to outer membranes and stained green with FITC-labelling). Image copyright Myron Christodoulides.

Humana Press is part of Springer Science+Business Media (www.springer.com)

Preface

Neisseria meningitidis (the meningococcus, family *Neisseriaceae*) is a major causative agent, worldwide, of potentially life-threatening meningitis and septicemia. The organism is a Gram-negative β proteobacterium that resides normally as a commensal on the mucosal epithelium of the human nasopharynx. However, a combination of the expression of key microbial virulence factors, host susceptibility, inadequate innate immune recognition, and changes in niche environmental conditions can favor bacterial invasion of the host.

The pathogenesis of meningococcal infection involves initial penetration of the nasopharyngeal mucosal epithelium, entry into the blood, and the development of a bacteremia, which occurs in the absence of host humoral immunity. The meningococcus is a classical meningeal pathogen, i.e., it is capable of passing across the blood-cerebrospinal fluid (CSF) barrier to enter the CSF-filled subarachnoid space (SAS) to colonize the leptomeninges [1–3]. The consequences to the host are a compartmentalized intravascular inflammatory response, septicemia, and/or a compartmentalized intracranial inflammatory response, leptomeningitis [4]. Meningococcal infection can be sporadic, hyper-sporadic, or epidemic in nature, and global estimates for infection are ~1.2 million cases each year with a toll of ~135,000 deaths [5]. Despite successful antibiotic treatment and advances in intensive care management of patients with meningococcal infection, the mortality rate is still high at ~15% in industrialized countries, and the disease is rightly feared by the public, parents, and physicians for the rapidity of onset of both clinical symptoms and the decline of patients' health. Moreover, survivors are often faced with permanent physical and neurological sequelae, including loss of limbs; auditory and visual impairment; cognitive dysfunction; educational, behavioral, and developmental problems; seizures and persistent headaches; and motor nerve deficits, hydrocephalus, and permanent brain damage [6]. These outcomes impact the quality of life of the affected individuals and their families and on the providers of health, welfare, and social services.

Since Weichselbaum first identified the meningococcus from the CSF of a patient with meningitis in 1887 [7], key discoveries about the nature of the organism, the pathogenesis of infection, and the mechanisms of natural immunity have led, finally, to the development and introduction of safe and effective vaccines. The classical studies by Goldschneider and colleagues in the late 1960s identified the correlation between the development of serum bactericidal antibodies and protection from meningococcal infection [8, 9]. Combined with our increased understanding of meningococcal surface antigen structure and function, capsule polysaccharide-protein conjugate vaccines against meningococci that express the serogroup A, C, Y, and W-135 capsules have been developed and introduced into adolescent and adult immunization schedules [10, 11] and trialed in young children [12]. Conjugate vaccines have already significantly reduced meningococcal disease in industrialized countries, e.g., for serogroup C infection [13], and the long-term expectation is that the introduction of new conjugate vaccines against serogroup A will lead to a decline in epidemic disease reported in the "African Meningitis Belt" [14]. Today, a concerted effort is aimed at developing vaccines to serogroup B meningococci, which present a greater technical challenge, due to the poor immunogenicity of its capsule and the molecular mimicry of foetal NCAM, thereby necessitating a search for subcapsular antigens capable of inducing protective immune responses. The past decade has seen the publication of many *Neisseria* spp. genomes [15–22], and these have under-pinned advances in molecular methods and techniques with applications to vaccine design. The result is new vaccines for serogroup B that are showing promise in clinical trials [23–25] and a plethora of experimental vaccines that have shown success in preclinical, laboratory studies.

The key discoveries about the meningococcus would not have been possible without significant developments in laboratory methods for studying the pathogen at the molecular and cellular levels. Many of these laboratory methods have been described in the landmark books on meningococcal vaccines and meningococcal disease, edited by Andrew Pollard and Martin Maiden for the series *Methods in Molecular Medicine* [26, 27]. This new book, *Neisseria meningitidis: Advanced Methods and Protocols*, does not simply revisit and update the methods and protocols described in the books from Pollard and Maiden, a task of itself unnecessary due to the comprehensive nature of these previous volumes, but it offers a collection of advanced methods and protocols that in many ways reflect the development and refinement of several new technologies applied to the meningococcus. Several of the chapters in this book describe methods that rely on the collection of complete sets of biological data, for example, using the genome to generate transcriptomes, proteomes, and metabolomes. However, there are many new *-omics* that are being developed both theoretically and practically (e.g., the interactome, moleculome, cytome, and regulome, to name but a few), and these are now beginning to be applied to the study of *Neisseria* and many other human pathogens. Laboratory methods and protocols for these new *-omics* could be the subjects of future volumes.

Neisseria meningitidis: Advanced Methods and Protocols begins with a review of the biology, microbiology, and epidemiology of the meningococcus, which is followed by two chapters that provide a clinical context, namely, in the classification and pathogenesis of meningococcal infections and a technique for detecting the pathogen in CSF samples from patients. In cases of undefined meningeal irritation, the latter method provides a means for identifying between the important bacterial and viral causes of meningitis.

A number of chapters then follow that provide methods and protocols for investigating the molecular biology and biochemistry of the meningococcus. These techniques can provide useful tools for vaccine and pathogen–host interaction studies and include methods for generating knock-out and complementation strains of the meningococcus, for identifying and characterizing small RNA molecules and for the expression of purified meningococcal proteins for crystallization. A particular area of *Neisseria* research that, if not exactly neglected, is not always appreciated relates to the metabolism of the meningococcus. Hence, a chapter is provided that explains how genome-scale metabolic networks can be constructed using a constraint-based modeling approach, using available genome sequence databases and high-throughput bioinformatics. This is followed by a complementary method for studying adaptations in meningococcal/microbial proteomes to changing environmental conditions.

The next collection of chapters broadly covers pathogen–host cell interactions and is prefaced with an introductory review of our current knowledge of meningococcal surface ligands and their respective host cell receptors. A protocol for studying meningococcal interactions with an animal model is presented, followed by methods used for culturing and investigating biofilms in vitro. Interactions of bacteria with host cells are subject to environmental stress and physical forces applied to bacterial ligand–host cell receptor binding events, and methods are provided for investigating bacterial adhesion under shear stress and the forces exerted by the meningococcal pilus adhesin. The literature is replete with in vitro cell culture models used for studying bacterial interactions, from human explant models to monocultures of primary cells and transformed cells of myeloid and nonmyeloid origins: for this book, protocols are described for isolating human dendritic cells and using them to study host–*Neisseria* interactions, acknowledging the important role that these cells have in sentinel immune recognition during *Neisseria* infection and in driving polarization of naïve T-cell helper responses. Finally, the events that follow bacterial interaction with host cell receptors are considered in two chapters that present methods for investigating ligand–receptor interactions, by using hydrogen/deuterium exchange coupled to mass spectrometry and nanoscale imaging techniques to visualize the interactions between pathogen-associated molecular patterns (PAMP) and host pattern recognition receptors (PRR).

The next chapters consider the consequences of meningococcal interaction with host cells; in-depth protocols are provided for analyzing the transcriptome of the pathogen and host epithelial and endothelial cell models, followed by a detailed technical review on the experimental design that allows the researcher to generate valuable and reliable data from using the pan-*Neisseria* microarray. A major consequence of meningococcal infection is host cell damage, clearly seen in patients with sepsis and meningitis, and in vitro methods are provided for investigating host cellular apoptosis/necrosis induced by the pathogenic *Neisseriae*.

The final part of this book focuses on methods and protocols for vaccine antigen discovery and vaccine design. Methods are provided for two different approaches, one using proteomics to analyze the human immune response to *Neisseria meningitidis* and the other, "reverse vaccinology." In particular, the latter chapter provides detailed methods for in silico identification and selection of antigens, through production of recombinant proteins for immunization and analyses of the immune response. The final chapter provides a protocol for preparing experimental DNA vaccines to bacterial polypeptides.

Many of the techniques described herein can be readily used to study other pathogens and diseases and should have broad appeal to clinical and nonclinical scientists alike. I do accept that some of these methods can seem daunting or require specialized equipment, but I do hope that they stimulate collaboration between readers and authors. This book could not have been possible without the contributions of many, and I would like to express my gratitude toward all authors, all of whom enthusiastically contributed their articles and showed patience with my editing; to the staff at Humana Press for commissioning this volume and especially to the series editor, John Walker, who has provided support and advice when needed. Finally, although this past decade has seen tremendous advances in the fight against meningococcal infection, there is still much to learn about the meningococcus and not only does it continue to surprise us with its complex nature, but also its relationship with its host reveals a great deal about human biology.

Southampton, UK ***Myron Christodoulides***

References

1. Carbonnelle E, Hill DJ, Morand P et al (2009) Meningococcal interactions with the host. Vaccine 27: B78–B89.
2. Christodoulides M, Heckels JE, Weller RO (2002) The role of the leptomeninges in meningococcal meningitis In: Ferreiros C, Criado MT, Vazquez J (ed) Emerging strategies in the fight against meningitis. Horizon Press.
3. Join-Lambert O, Morand PC, Carbonnelle E et al (2010) Mechanisms of meningeal invasion by a bacterial extracellular pathogen, the example of *Neisseria meningitidis.* Prog Neurobiol 91: 130–139.
4. Brandtzaeg P (1995) Pathogenesis of meningococcal infections. In: Cartwright KAV (ed) Meningococcal Disease. Wiley, New York.

5. Stephens DS, Greenwood B, Brandtzaeg P (2007) Epidemic meningitis, meningococcaemia, and *Neisseria meningitidis*. Lancet 369: 2196–2210.
6. Steven N, Wood M (1995) The clinical spectrum of meningococcal disease. In: Cartwright KAV (ed) Meningococcal Disease. Wiley, New York.
7. Weichselbaum A (1887) Ueber die aetiologie der akuten meningitis cerebrospinalis. Fortschr Med 5: 573.
8. Goldschneider I, Gotschlich EC, Artenstein MS (1969) Human immunity to the meningococcus. I. The role of humoral antibodies. J Exp Med 129: 1307–1326.
9. Goldschneider I, Gotschlich EC, Artenstein MS (1969) Human immunity to the meningococcus. II. Development of natural immunity. J Exp Med 129: 1327–1348.
10. Pace D (2010) Novel quadrivalent meningococcal A, C, W-135 and Y glycoconjugate vaccine for the broader protection of adolescents and adults. Future Microbiol 5: 1629–1640.
11. Keyserling HL, Pollard AJ, Detora LM et al (2006) Experience with MCV-4, a meningococcal, diphtheria toxoid conjugate vaccine against serogroups A, C, Y and W-135. Expert Rev Vaccine 5: 445–459.
12. Halperin SA, Gupta A, Jeanfreau R et al (2010) Comparison of the safety and immunogenicity of an investigational and a licensed quadrivalent meningococcal conjugate vaccine in children 2-10 years of age. Vaccine 28: 7865–7872.
13. Borrow R, Miller E (2006) Long-term protection in children with meningococcal C conjugate vaccination: lessons learned. Expert Rev Vaccine 5: 851–857.
14. Okoko BJ, Idoko OT, Adegbola RA (2009) Prospects and challenges with introduction of a mono-valent meningococcal conjugate vaccine in Africa. Vaccine 27: 2023–2029.
15. Tettelin H, Saunders NJ, Heidelberg J et al (2000) Complete genome sequence of *Neisseria meningitidis* serogroup B strain MC58. Science 287: 1809–1815.
16. Parkhill J, Achtman M, James KD et al (2000) Complete DNA sequence of a serogroup A strain of *Neisseria meningitidis* Z2491. Nature 404: 502–506.
17. Bentley SD, Vernikos GS, Snyder LAS et al (2007) Meningococcal genetic variation mechanisms viewed through comparative analysis of serogroup C strain FAM18. PloS Genetics 3: 230–240.
18. Peng JP, Zhang XB, Yang E et al (2007) Characterization of serogroup C meningococci isolated from 14 provinces of China during 1966-2005 using comparative genomic hybridization. Sci China Ser C-Life Sciences 50: 1–6.
19. Schoen C, Blom J, Claus H et al (2008) Whole-genome comparison of disease and carriage strains provides insights into virulence evolution in *Neisseria meningitidis*. Proc Nat Acad Sci USA 105: 3473–3478.
20. Bennett JS, Bentley SD, Vernikos GS et al (2010) Independent evolution of the core and accessory gene sets in the genus *Neisseria*: insights gained from the genome of *Neisseria lactamica* isolate 020-06. Bmc Genomics 11:652.
21. Lavezzo E, Toppo S, Barzon L et al (2010) Draft genome sequences of two *Neisseria meningitidis* serogroup C clinical isolates. J Bact 192: 5270–5271.
22. Marri PR, Paniscus M, Weyand NJ et al (2010) Genome sequencing reveals widespread virulence gene exchange among human *Neisseria* species. PlosSOne 5:e11835.
23. Rinaudo CD, Telford JL, Rappuoli R et al (2009) Vaccinology in the genome era. J Clin Invest 119: 2515–2525.
24. Findlow J, Borrow R, Snape MD et al (2010) Multicenter, open-label, randomized Phase II controlled trial of an investigational recombinant meningococcal serogroup B vaccine with and without outer membrane vesicles, administered in infancy. Clin Infect Dis 51: 1127–1137.
25. Richmond P, Marshall H, Sheldon E et al (2010) Safety and immunogenicity of serogroup B *Neisseria meningitidis* (MnB) rLP2086 vaccine in adults and adolescent subjects: overview of 3 clinical trials, in: 17th International Pathogenic *Neisseria* Conference.
26. Pollard AJ, Maiden MCJ (2001) Meningococcal vaccines: methods and protocols. Humana Press, Totowa.
27. Pollard AJ, Maiden MCJ (2001) Meningococcal disease: methods and protocols. Humana Press, Totowa.

Contents

Contributors

MICHAEL A. APICELLA • *The Department of Microbiology, Carver College of Medicine, The University of Iowa, Iowa City, IA, USA*
GINO J.E. BAART • *VIB Department of Plant Systems Biology/Department of Biology, Protistology and Aquatic Ecology, Ghent University, Ghent, Belgium*
CHRISTIAN BAUMANN • *Proteome Sciences R&D, Frankfurt am Main, Germany*
NICOLAS BIAIS • *Department of Biological Sciences, Columbia University, New York, NY, USA*
MARTINE P. BOS • *Department of Molecular Microbiology, Utrecht University, Utrecht, The Netherlands*
PETTER BRANDTZAEG • *Departments of Pediatrics and Medical Biochemistry, University of Oslo, Oslo, Norway*
BRUNELLA BRUNELLI • *Novartis Vaccines and Diagnostics, Siena, Italy*
MYRON CHRISTODOULIDES • *Division of Infection, Inflammation, and Immunity, Sir Henry Wellcome Laboratories, University of Southampton Medical School, Southampton, UK*
VINCENT VAN DAM • *Department of Molecular Microbiology, Utrecht University, Utrecht, The Netherlands*
JOHN K. DAVIES • *Nursing and Health Sciences, Monash University, VIC, Australia*
JEREMY P. DERRICK • *University of Manchester, Manchester, UK*
MARCEL VAN DEUREN • *Department of Internal Medicine and Nijmegen Institute for Infection, Inflammation and Immunity, Radboud University Nijmegen Medical Centre, Nijmegen, The Netherlands*
GARTH L.J. DIXON • *Department of Microbiology, Camelia Botnar Laboratories, Great Ormond Street Hospital, London, UK*
GUILLAUME DUMÉNIL • *INSERM, U970, Paris Cardiovascular Research Center PARCC, Paris, France; Université Paris Descartes, UMR-S970, Paris, France*
ARIE VAN DER ENDE • *Department of Medical Microbiology, Academic Medical Center, Center for Infection and Immunity, Amsterdam, The Netherlands*
LUIGI FIASCHI • *Novartis Vaccines and Diagnostics, Siena, Italy*
MATTHIAS FROSCH • *Institute for Hygiene and Microbiology, Würzburg, Germany*
JOHN E. HECKELS • *Division of Infection, Inflammation, and Immunity, Sir Henry Wellcome Laboratories, University of Southampton Medical School, Southampton, UK*
DUSTIN HIGASHI • *Department of Immunobiology and the BIO5 Institute, University of Arizona, Tucson, AZ, USA*
DARRYL J. HILL • *School of Cellular and Molecular Medicine, University of Bristol, Bristol, UK*
HANNAH E. JONES • *Infectious Diseases and Microbiology Unit, Institute of Child Health, University College London, London, UK*

ANN-BETH JONSSON • *Department of Genetics, Microbiology, and Toxicology (GMT), Stockholm University, Stockholm, Sweden*
BIJU JOSEPH • *Institute for Hygiene and Microbiology, Würzburg, Germany*
NIGEL KLEIN • *Infectious Diseases and Microbiology Unit, Institute of Child Health, University College London, London, UK*
KARSTEN KUHN • *Proteome Sciences R&D, Frankfurt am Main, Germany*
BENOIT LADOUX • *Matières et Systèmes Complexes, CNRS UNR7057/Université Paris 7, Paris, France*
JESSMI M.L. LING • *Department of Microbiology and Infectious Diseases, University of Calgary, Calgary, AB, Canada*
SARA MARCHI • *Novartis Vaccines and Diagnostics, Siena, Italy*
DIRK E. MARTENS • *Bioprocess Engineering Group, Wageningen University, Wageningen, The Netherlands*
PAOLA MASSARI • *Department of Medicine, Evans BioMedical Research Center, Section of Infectious Diseases, Boston University School of Medicine, Boston, MA, USA*
JENS KJØLSETH MØLLER • *Department of Clinical Microbiology, Vejle Hospital, Vejle, Denmark*
JEREMY MOORE • *Centre for Structural Biology, Imperial College London, London, UK*
R. BROCK NEIL • *The Department of Microbiology, Carver College of Medicine, The University of Iowa, Iowa City, IA, USA*
EMMANUELLE PALUMBO • *Novartis Vaccines and Diagnostics, Siena, Italy*
YVONNE PANNEKOEK • *Department of Medical Microbiology, Academic Medical Center, Center for Infection and Immunity, Amsterdam, The Netherlands*
MARIAGRAZIA PIZZA • *Novartis Vaccines and Diagnostics, Siena, Italy*
THORSTEN PRINZ • *Proteome Sciences R&D, Frankfurt am Main, Germany*
JASON RICE • *Division of Cancer Sciences, Genetic Vaccines Group, University of Southampton Medical School, Southampton, UK*
NADINE G. ROUPHAEL • *Department of Medicine, Division of Infectious Diseases, Emory University School of Medicine, Atlanta, GA, USA*
MUHAMMAD SALEEM • *University of Manchester, Manchester, UK*
NIGEL J. SAUNDERS • *Sir William Dunn School of Pathology, University of Oxford, Oxford, UK*
SILVANA SAVINO • *Novartis Vaccines and Diagnostics, Siena, Italy*
CHRISTOPH SCHOEN • *Institute for Hygiene and Microbiology, Würzburg, Germany*
DAVID C. SCHRIEMER • *Department of Biochemistry and Molecular Biology, University of Calgary, Calgary, AB, Canada*
ANTHONY B. SCHRYVERS • *Department of Microbiology and Infectious Diseases, University of Calgary, Calgary, AB, Canada*
ALEXANDRA SCHUBERT-UNKMEIR • *Institute for Hygiene and Microbiology, University of Würzburg, Würzburg, Germany*
JIANQIANG SHAO • *The Department of Microbiology, Carver College of Medicine, The University of Iowa, Iowa City, IA, USA*
LESLIE SILVA • *Department of Biochemistry and Molecular Biology, University of Calgary, Calgary, AB, Canada*
HONG SJÖLINDER • *Department of Genetics, Microbiology, and Toxicology (GMT), Stockholm University, Stockholm, Sweden*

MAGDALENE SO • *Department of Immunobiology and the BIO5 Institute, University of Arizona, Tucson, AZ, USA*
MAGALI SOYER • *INSERM, U970, Paris Cardiovascular Research Center PARCC, Paris, France; Université Paris Descartes, UMR-S970, Paris, France*
DAVID S. STEPHENS • *Department of Medicine, Division of Infectious Diseases, Emory University School of Medicine, Atlanta, GA, USA; Laboratories of Microbial Pathogenesis, Atlanta, VA Medical Center, Decatur, GA, USA*
JAN TOMMASSEN • *Department of Molecular Microbiology, Utrecht University, Utrecht, The Netherlands*
KATHY TRIANTAFILOU • *Department of Child Health, School of Medicine, University Hospital of Wales, Cardiff University, Cardiff, UK*
MARTHA TRIANTAFILOU • *Department of Child Health, School of Medicine, University Hospital of Wales, Cardiff University, Cardiff, UK*
MUMTAZ VIRJI • *School of Cellular and Molecular Medicine, University of Bristol, Bristol, UK*
LEE M. WETZLER • *Department of Medicine, Evans BioMedical Research Center, Section of Infectious Diseases, Boston University School of Medicine, Boston, MA, USA*
JEANNETTE N. WILLIAMS • *Division of Infection, Inflammation, and Immunity, Sir Henry Wellcome Laboratories, University of Southampton Medical School, Southampton, UK*

Chapter 1

Neisseria meningitidis: Biology, Microbiology, and Epidemiology

Nadine G. Rouphael and David S. Stephens

Abstract

Neisseria meningitidis (the meningococcus) causes significant morbidity and mortality in children and young adults worldwide through epidemic or sporadic meningitis and/or septicemia. In this review, we describe the biology, microbiology, and epidemiology of this exclusive human pathogen. *N.meningitidis* is a fastidious, encapsulated, aerobic gram-negative diplococcus. Colonies are positive by the oxidase test and most strains utilize maltose. The phenotypic classification of meningococci, based on structural differences in capsular polysaccharide, lipooligosaccharide (LOS) and outer membrane proteins, is now complemented by genome sequence typing (ST). The epidemiological profile of *N. meningitidis* is variable in different populations and over time and virulence of the meningococcus is based on a transformable/plastic genome and expression of certain capsular polysaccharides (serogroups A, B, C, W-135, Y and X) and non-capsular antigens. *N. meningitidis* colonizes mucosal surfaces using a multifactorial process involving pili, twitching motility, LOS, opacity associated, and other surface proteins. Certain clonal groups have an increased capacity to gain access to the blood, evade innate immune responses, multiply, and cause systemic disease. Although new vaccines hold great promise, meningococcal infection continues to be reported in both developed and developing countries, where universal vaccine coverage is absent and antibiotic resistance increasingly more common.

Key words: *Neisseria meningitidis*, Pathogenesis, Bacterial infections, Microbiology, Epidemiology

1. Introduction

In 1887, Weichselbaum (1) was the first to identify the meningococcus from the cerebrospinal fluid (CSF) of a patient with meningitis. Epidemics of meningococcal meningitis were first described during the early nineteenth century, in 1805 in Geneva, Switzerland by Vieusseux (2), in 1806 in New Bedford, Massachusetts by Danielson and Mann (3) and in the early 1900s in the African meningitis belt (4). The meningococcus was recognized as a habitant

Myron Christodoulides (ed.), *Neisseria meningitidis: Advanced Methods and Protocols*, Methods in Molecular Biology, vol. 799,
DOI 10.1007/978-1-61779-346-2_1, © Springer Science+Business Media, LLC 2012

of the nasopharynx of healthy individuals (5), especially seen in the setting of military recruits camps (6) at the beginning of the twentieth century. Treatment for meningococcal disease included serum therapy introduced in 1913 by Flexner (7) and sulfonamides introduced in 1937 (8). The emergence of resistance to sulfonamides (9) in the 1960s prompted the development of the first vaccines against meningococci (10).

Despite an understanding of the pathogenesis, the availability of therapeutic and prophylactic antibiotics and immunizations against important serogroups, the meningococcus remains a leading cause worldwide of bacterial meningitis (11). Invasive meningococcal disease results from the interplay of: (1) microbial factors influencing the virulence of the organism, (2) environmental conditions facilitating exposure and acquisition, and (3) host susceptibility factors favoring bacterial acquisition, colonization, invasion, and survival. In the pre-serum therapy and pre-antibiotic eras, 70–85% of meningococcal disease cases were fatal; today, the overall mortality rate in invasive meningococcal disease still remains high, at between 10 and 15% (12). Meningococcal disease is also associated with marked morbidity including limb loss, hearing loss, cognitive dysfunction, visual impairment, educational difficulties, developmental delays, motor nerve deficits, seizure disorders, and behavioral problems (13). In this chapter, we review the biology, microbiology, and epidemiology of the meningococcus.

2. Biology of the Meningococcus

The virulence (14) of *N. meningitidis* is influenced by multiple factors: capsule polysaccharide expression, expression of surface adhesive proteins (outer membrane proteins including pili, porins PorA and B, adhesion molecules Opa and Opc), iron sequestration mechanisms, and endotoxin (lipooligosaccharide, LOS). *N. meningitidis* also has evolved genetic mechanisms resulting in a horizontal genetic exchange, high frequency phase, antigenic variation, and molecular mimicry, allowing the organism to successfully adapt at mucosal surfaces and invade the host (14).

2.1. Genetics

Genome sequences for a number of *N. meningitidis* strains including MC58 (serogroup B, ST-32) (15), Z2491 (serogroup A, ST-4) (16), FAM18 (serogroup C, ST-11), and NMB-CDC (serogroup B, ST-8) have been reported. Based on the sequencing of several genomes, the chromosome is between 2.0 and 2.2 megabases in size and contains about 2,000 genes (17). Except for the IHT-A1 capsule locus, no specific core pathogenome has been identified, suggesting that virulence may be clonal group-dependent. The core meningococcal genome that encodes for essential metabolic functions represents about 70% of the genome. Large genetic islands are present in different strains and are predicted to encode hypothetical surface proteins and virulence factors. The IHT-A1 locus contains the genes for capsule biosynthesis and transport, the IHT-A2 locus is predicted to encode an ABC transporter and a secreted protein, and the IHT-C locus is predicted to encode 30 open reading frames including toxin homologues, a bacteriophage, and potential virulence proteins (18).

The meningococcus shares about 90% homology at the nucleotide level with either *N. gonorrhoeae* or *N. lactamica*. Mobile genetic elements including IS elements and prophage sequences make up to ~10% of the genome. Transfer of DNA occurs between meningococci, gonococci, and commensal *Neisseria* spp. as well as other bacteria (such as *Haemophilus*) (19). Several repetitive sequence and polymorphic regions are present, usually in large heterogeneous arrays, suggesting active areas of genetic recombination. Another characteristic of the meningococcal genome is the presence of multiple genetic switches (e.g., slipped-strand misparing, IS element movement), contributing to the expression of pathogen-associated genes (20). In summary, a central characteristic in the evolution of the meningococcus is the plasticity of the genome and the capacity created by this plasticity for diversity of phenotype.

2.2. Capsule

N. meningitidis can be either encapsulated or not. However, *N. meningitidis* strains causing invasive disease and isolated from sterile sites such as the blood or the CSF are almost always encapsulated. The capsule is essential for the survival of the organism in the blood as it provides resistance to antibody/complement-mediated killing and inhibits phagocytosis (21). Antibodies directed at capsule play a major part in protection against meningococcal disease and capsule forms the basis for licensed polysaccharide (22, 23) and new conjugate-polysaccharide meningococcal vaccines (24) (except for serogroup B) and for the classification of meningococci into serogroups.

The main meningococcal capsular polysaccharides associated with invasive disease are composed of sialic acid derivatives, except for the serogroup A capsule, which consists of repeating units of N-acetyl-mannosamine-1-phosphate. In *N. meningitidis*, N-acetylneuraminic acid (Neu5Ac), unlike in mammalian cells, is synthesized from N-acetylmannosamine (ManNAc) and phosphoenolpyruvate without phosphorylated intermediates (25). Neu5Ac is the most common form of sialic acid in humans and plays an important role in intercellular and/or intermolecular recognition (26). The incorporation of Neu5Ac into meningococcal capsules allows the meningococcus to become less visible to the host immune system (27, 28) because of molecular mimicry. The most striking example is observed in the serogroup B capsule (29), an α(2–8)-linked sialic acid homopolymer identical in structure to the human fetal neural cell-adhesion molecule (NCAM). Such identity is responsible for the particularly poor immune response generated against serogroup B capsule by humans (30).

Capsular genes are located in a single locus *cps* within the IHT-A1 24 kb virulence island that is divided in three regions A, B, and C (31). Region A contains genes responsible for the synthesis and the polymerization of the polysaccharide. Regions B and C contain genes responsible for translocation of the polysaccharide from the cytoplasm to the surface. "Capsule switching" occurs due to genetic identity of parts of the capsule loci and is the result of horizontal exchange by transformation and recombination in the locus of serogroup specific capsule biosynthesis genes (32). Capsule switching is another mechanism of escape from vaccine-induced or natural protective immunity and a virulence mechanism shown by other encapsulated bacterial pathogens (e.g., *Streptococcus pneumoniae*). For example, the emergence of W-135 as a cause of outbreaks in 2000–2002 was due to the spread of W-135 ET-37 (ST-11) strains closely related to ET-37 (ST-11) serogroup C strains (33, 34). The phenomenon of capsule switching (32, 35) has raised concerns about the immune pressure that capsule-based vaccination programs may apply. However, so far, capsule switching has not caused significant problems after meningococcal conjugate vaccine introduction (36, 37).

2.3. Cell Envelope

In gram-negative bacteria, such as *N. meningitidis*, the subcapsular cell envelope consists of an outer membrane (OM), a peptidoglycan layer, and a cytoplasmic or inner membrane (Fig. 1). The OM has an outside layer primarily composed of lipopolysaccharide (LPS), actually a lipooligosaccharide, and proteins and an inside layer composed of phospholipids (38) that contains proteins primarily responsible for regulating the flow of nutrients and metabolic products. The major phospholipid component of *Neisseria* membranes consists largely of phosphatidylethanolamine (PE), with varying amounts of phosphatidylglycerol (PG), cardiolipin (CL), and phosphatidate (PA) (39). The structure of peptidoglycan of different *N. meningitidis* strains consists of a maximum of two layers (40) with different variations in the degree of cross-linking and O-acetylation. Typically, the percentage of cross-linking of the meningococcal peptidoglycan is around 40%, similar to other gram-negative bacteria (41). The O-acetylation of peptidoglycan results in resistance to lysozyme and to other muramidases (42). Also, peptidoglycan structures are recognized by components of the innate immune system (43).

2.4. Lipopoly saccharide (LPS)

Meningococcal LPS or LOS (endotoxin) (44, 45) plays a role in the adherence of the meningococcus (46) and in activation of the innate immune system (47). Meningococcal LOS lacks a repeating O-side chain of LPS found in enteric gram-negative bacilli (46). Meningococcal LPS consists of three parts (48): lipid A containing

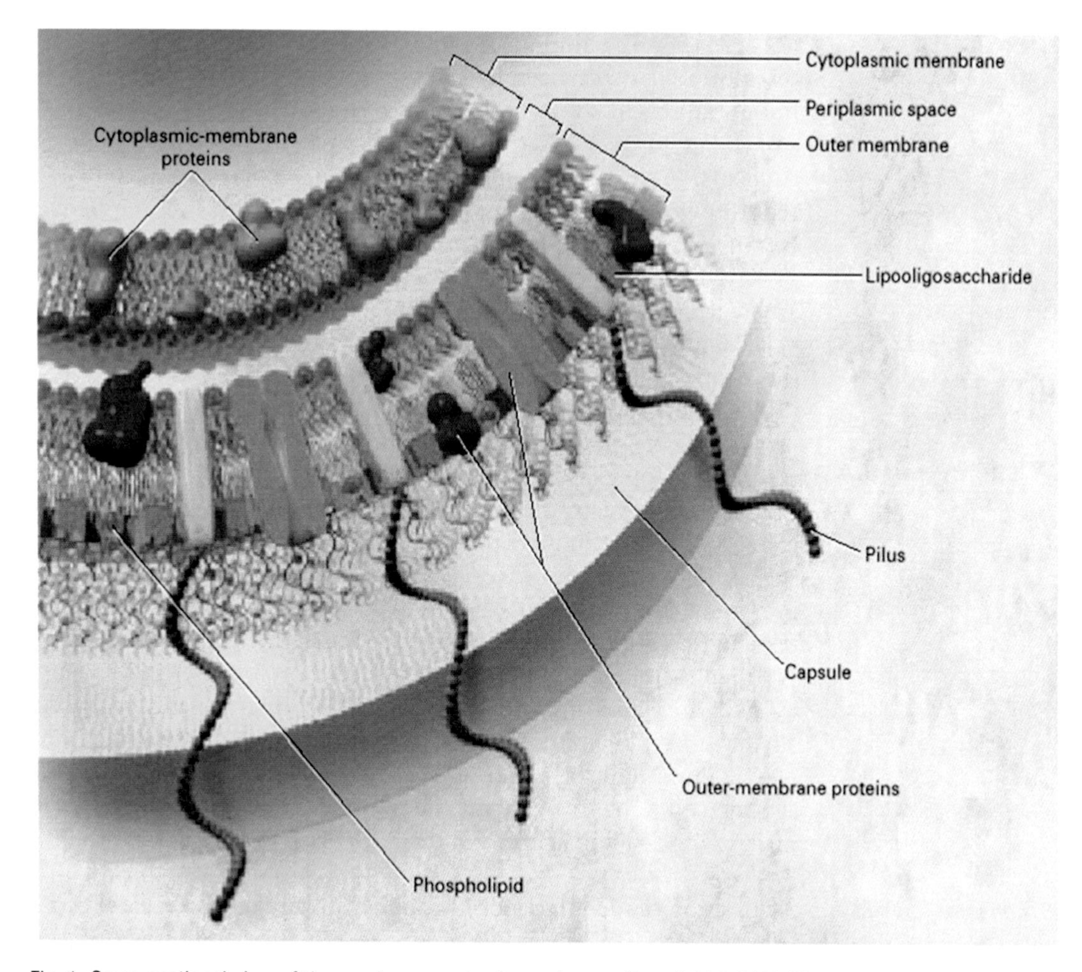

Fig. 1. Cross-sectional view of the meningococcal cell membrane.

hydroxy fatty-acid chains and phosphoethanolamine, a core oligosaccharide containing 3-deoxy-D-manno-oct-2-ulosonic acid (KDO) and heptose residues, and highly variable short oligosaccharides. The heptose residues provide linkage to the short oligosaccharide residues of the α-, β-, and γ-chains. Meningococcal lipid A, a disaccharide of pyranasol N-acetyl glucosamine residues, is responsible for much of the biological activity and toxicity of meningococcal endotoxin. Sialic acid can be a terminal component of the α-chain lacto-*N*-neotetraose and is available from endogenous (sialic acid production) or exogenous (host sialic acid at the cell surfaces) sources. The α-chain structures of meningococcal LOS can mimic the human I and i antigens, an example of host molecular mimicry and immune escape mechanisms (49). Phase and antigenic variations (50), leading to different chain oligosaccharide and inner core composition, dramatically alter the antigenic properties of LOS and form the basis of the classification into different immunotypes

(L1-12) (35, 51). Immunotyping screenings of strains from invasive disease and from carriage have indicated that carriage isolates commonly expressed shorter structures termed L1 and L8, while invasive isolates were characterized by the expression of LOS with long α-chains. Also, ~97% of isolates from a serogroup B epidemic reported in England expressed the L3, 7, and 9 (36) immunotype, but the immunotypes of carriers were more heterogeneous.

Meningococcal LOS binds to a series of host transfer molecules and receptors on monocytic and dendritic cells of the innate immune system, including LPS-binding protein (LBP), CD14, and myeloid differentiation protein 2 (MD2), part of the Toll-like receptor 4 (TLR4) (52, 53) complex. This triggers the secretion of various cytokines (54) (including IL-6 and TNF-α) that at high levels can result in endothelial damage and capillary leakage. There is a direct correlation between LPS levels and severity of meningococcal disease (55, 56). LPS also induces the release of chemokines, reactive oxygen species (ROS), and nitric oxide (NO). In addition, LOS also plays an important role in resistance to other host defenses. Meningococci are resistant to cationic antimicrobial peptides (CAMPs) due to the lipid A phosphoethanolamine structures present on lipid A head groups. CAMPs are present in macrophages and neutrophils, and occasionally produced by epithelial cells at mucosal surfaces. These peptides play an important role in host defense against microbial infection and are key components of the innate immune responses through their nonoxidative killing action and their signaling functions.

2.5. Adhesins

Acquisition of meningococci through exposure to respiratory secretions and attachment on human upper respiratory mucosal surfaces by *N. meningitidis* are the first steps in establishing a human carrier state and invasive meningococcal disease. Meningococcal carriage occurs in 8–25% (57–59) of the human population with adolescents being the major reservoir. Duration of carriage can vary from days to several months. Meningococcal transmission among humans occurs largely through large respiratory droplets; the acquisition may be asymptomatic or may result in local inflammation. Invasive disease usually occurs 1–14 days after acquisition. Multiple structures, termed major and minor adhesions, facilitate meningococcal adherence.

2.5.1. Major Adhesins

Pili: The adhesive properties of capsulate *N. meningitidis* are mediated by pili (60), which extend several thousand nm beyond the capsule and initiate binding to epithelial cells (61). Twitching motility generated by pilus retraction is important for passage through the epithelial mucus layer, movement over epithelial surfaces, and microcolony formation (62). Piliated meningococci attach to human nasopharyngeal cells (63) in greater numbers when compared to meningococci devoid of pili. In addition, pili are involved in facilitating the uptake of DNA by meningococci (64) and also enable adherence to endothelial cells and erythrocytes. Meningococcal pili (60) are composed of two major pilin families and undergo both phase and antigenic variation. Neisserial pili can undergo posttranscriptional modifications such as pilin glycosylation (65). Glycosylation (66) may promote secretion of the soluble pilin units that compete for both anti-pili antibodies and host cell receptors allowing protection of the organism.

Opacity proteins: *N. meningitidis* strains commonly express two types of OM opacity proteins, Opa and Opc. While Opc is only expressed by *N. meningitidis* and encoded by a single gene, Opa proteins are expressed by both meningococci and gonococci, and encoded by multiple genes. Opa expression is subject to antigenic and phase variation and certain Opa types may predominate in clinical isolates due to their adhesion/virulence properties (67). Opa interacts with multiple members of the CEACAM (carcino-embryonic antigen-related cell-adhesion molecule) family (68); during inflammation high levels of CEACAM are expressed, facilitating Opa interactions and therefore cellular attachment and invasion. Both Opa and Opc (69) also interact with cell-surface associated HSPGs (heparan sulfate proteoglycans) (70).

2.5.2. Minor Adhesins

These molecules are often expressed at low levels in vitro, but may be upregulated in vivo; their potential roles in pathogenesis are not fully defined. Examples of minor adhesins include NadA (neisserial adhesinA), NhhA (*Neisseria* hia homologue A), App (adhesion and penetration protein), and MspA (meningococcal serine protease A) (71).

2.6. Other Molecules

Iron-binding proteins enable meningococci to acquire iron, a crucial growth factor during colonization and disease (72, 73). Meningococcal iron-acquiring proteins include HmbR (hemoglobin), TbpA and TbpB (transferrin), HbpA and HbpB (lactoferrin), HpnA and HpnB (hemoglobin-haptoglobin complex), and possible siderophore homologues.

N. meningitidis expresses two distinct porins, PorA and PorB, through which small hydrophilic nutrients diffuse into the bacterium via cation or anion selection. OM porins are also involved in host cell interactions and as targets for bactericidal antibodies (74). PorB is the major OM porin that inserts in membranes, induces Ca^{2+} influx and activates TLR2 and cell apoptosis (75). PorA is a major component of OM vesicle-based vaccines and a target for bactericidal antibodies (76). Of note, with no universal vaccine for serogroup such as PorA and in other proteins B, there is a particular interest in non-capsular OM antigens identified as vaccine targets through several experimental approaches including "reverse vaccinology" (77).

3. Microbiology of the Meningococcus

N. meningitidis is a gram-negative β proteobacterium and member of the bacterial family of Neisseriaceae. *N. meningitidis* is a fastidious bacterium, dying within hours on inanimate surfaces, and is either an encapsulated or unencapsulated, aerobic diplococcus with a "kidney" or "coffee-bean" shape (Fig. 2). Optimal growth for the organism occurs at 35–37 °C with 5–10% (v/v) carbon dioxide. The organism grows on different media such as blood agar, trypticase soy agar, supplemented chocolate agar, and Mueller-Hinton agar. *N. meningitidis* colonies on blood agar are grayish, nonhemolytic, round, convex, smooth, moist, and glistening with a clearly defined edge. *N. meningitidis,* however, tends to undergo rapid autolysis in stationary phase. Colonies are positive by the oxidase test and the result is confirmed with carbohydrate reactions (meningococci oxidize glucose and usually maltose, but not sucrose and lactose).

3.1. Classification

Meningococci are classified according to serological typing (78, 79) and serogrouping is the traditional approach. Further classification into serosubtype, serotype, and immunotype is based on PorA, PorB, and LOS structure, respectively (13). At least 13 distinct

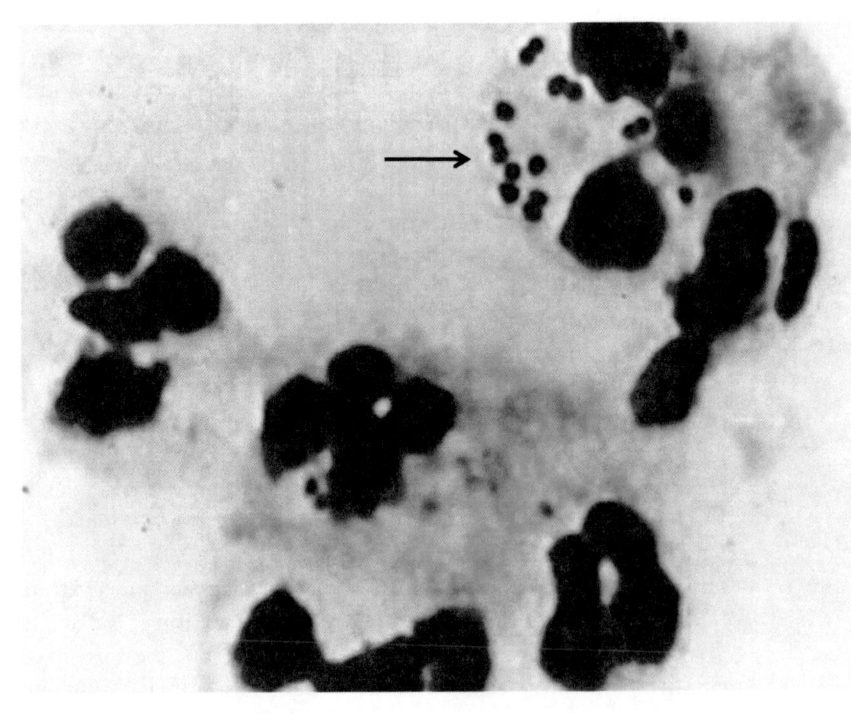

Fig. 2. Intracellular gram-negative diplococci and leukocytes in the CSF from a patient with meningococcal meningitis. The arrow denotes diplococci in proximity and within leukocytes.

meningococcal groups have been defined on the basis of their immunological reactivity and structure of the capsule's polysaccharide (80). These serogroups are the following: A, B, C, E-29, H, I, K, L, W-135, X, Y, Z, and Z' (29E). Only six serogroups (A, B, C, W-135, X, Y) cause life-threatening disease. Serogroup identification is done by slide agglutination or polymerase chain reaction (PCR) assays, while other meningococcal typing is performed using monoclonal antibodies (mAbs), PCR, and DNA sequencing (81).

3.2. Molecular Typing

Molecular typing is now the preferred approach for identifying clonal groups, closely related strains, strains with the potential to cause outbreaks, and in predicting vaccine coverage and understanding the genome of *N. meningitidis*. Molecular typing has used multiple techniques (82) including pulsed-field gel electrophoresis (PFGE) (83), multilocus enzyme electrophoresis (MLEE) (84), multilocus sequence typing (MLST) (85), and PCR (86). Currently, MLST is the gold standard for molecular typing (85) and classifies meningococcal strains into different STs (sequence types) based upon polymorphisms in seven housekeeping genes. Many meningococcal STs have been identified, which are independent of the serogroup. Of these, some are disproportionately associated with disease relative to carriage levels and so have been termed hyperinvasive lineages (87). Multilocus and antigen sequence typing data are assembled in large databases accessible via the internet (81).

3.3. Antimicrobial Susceptibility

Antimicrobial susceptibility testing of *N. meningitidis* should not be performed by disk diffusion, but by either minimal inhibitory concentration (MIC) determination by broth microdilution, or by use of the Etest® strip. To date, antibiotic resistance except for sulfonamides is relatively uncommon in the meningococcus. However, emergence of ciprofloxacin resistant strains (88) was recently observed and a decrease in penicillin susceptibility has been seen with some strains. Laboratory personnel at risk for exposure to aerosolized *N. meningitidis* should ensure their protective vaccination status is current, and they should work in a biological safety cabinet. Researchers who manipulate invasive *N. meningitidis* isolates in a manner that could induce aerosolization or droplet formation (i.e., plating, subculturing and serogrouping) on an open bench top and in the absence of effective protection from droplets or aerosols should consider antimicrobial chemoprophylaxis.

3.4. Metabolism

Very little direct work on the metabolism of the meningococcus has been carried out in the past 25 years. In 2007, Bart et al (89) screened the genome of MenB for open reading frames (ORFs) that code for enzymes present in the primary metabolism, yielding a genome-scale metabolic network (90, 91). Their genome-scale flux model was verified using flux balance analysis. According to the model, glucose can be completely catabolized mainly through the Entner-Douderoff pathway (ED), but also through the pentose phosphate pathway (PP), but not the Embden-Meyerhof-Parnas glycolytic pathway (EMP). The ED cleavage synthesizes the major part of pyruvate (67–87%) (92) and the PP pathway accounts for the remainder. N. *meningitidis* requires glucose, pyruvate, or lactate as sole carbon source (93) with a certain level of environmental CO_2 tension to initiate growth (94). During cultivation on any of the carbon sources, secretion of acetate into the medium occurs (95). Studies on lactate utilization showed that lactate can be utilized by different meningococcal lactate dehydrogenases (LDH) (96). All biochemical pathways for amino acid synthesis in *N. meningitidis* are available (97). Reduced sulfur in the form of cysteine, cystine, or thiosulfate is required by the meningococcus for growth (98).

4. Epidemiology of the Meningococcus

Meningococcal infection is a global but not uniform problem occurring as sporadic, hyper-sporadic, and epidemic disease (Fig. 3). There are an estimated 1.2 million cases of meningococcal infection per year, with a death toll of ~135,000 worldwide. Disease patterns vary widely over time and between geographical areas, age groups, and bacterial serogroups. Most disease is caused by a few genetically defined clonal complexes of *N. meningitidis* that can emerge and spread worldwide (85).

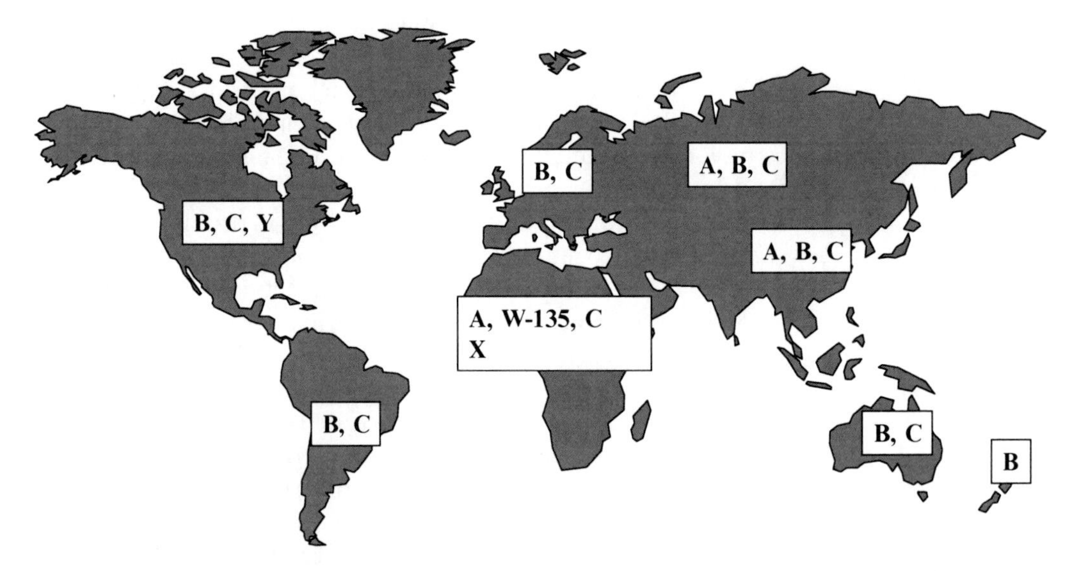

Fig. 3. Worldwide serogroup distribution of invasive meningococcal disease. (FEMS Microbiol Rev; used with permission).

4.1. Geographical Areas

1. United States: large serogroup A outbreaks occurred in the US during the first part of the twentieth century, but since the 1950s serogroup A meningococcal disease outbreaks disappeared in the USA as well as other industrialized countries (13) for unknown reasons. In the USA, the attack rate is now less than one case per 100,000 per year. During the 10-year period of 1998–2007 and according to the Active Bacterial Core surveillance (ABCs), the annual incidence decreased by 64.1%, from 0.92 cases per 100,000 in 1998 to 0.33 cases per 100,000 in 2007 with an average of 0.53 cases per 100,000 per year (99). A decrease in racial disparities was observed as well. The decrease in rates could not be readily explained by a change in environmental factors, such as smoking (100) and crowding, which are known to be risk factors for meningococcal disease. The decrease in rates was observed prior to implementation of the quadrivalent (serogroups A, C, Y, and W-135) meningococcal conjugate vaccine in 2005 in the USA, which for several reasons had low vaccine uptake in the early years, 11.4% uptake in 2006 and 32.4% uptake in 2007 (99). Infants aged <1 year have the highest incidence of meningococcal disease (5.38 cases per 100,000) (99), and a quadrivalent meningococcal vaccine is now recommended in the US for children 9 months to two years at increased risk of meningococcal disease. Today, serogroups C (101), Y (since the mid 1990s) and B cause most disease in the USA (102).
2. Europe: the attack rates (≥2 per 100,000 per year) in Europe have been higher than those observed in the USA. In the United Kingdom, rates of 5 per 100,000 per year prompted universal vaccination against serogroup C (103). In 1999, the UK was the first country to introduce serogroup C meningococcal conjugate vaccines to control the growing burden of serogroup C disease. This program has been very successful in reducing the incidence of serogroup C meningococcal disease in vaccinated persons (104). A major attribute of these vaccines was their reduction of serogroup C meningococcal carriage, leading to a decrease in the incidence of serogroup C meningococcal disease in the unvaccinated population as a result of herd immunity (105). With the success of the serogroup C immunization campaign in the UK, around 90% of remaining cases of invasive meningococcal disease in that country are now caused by serogroup B. The success of this program led to the subsequent introduction of meningococcal conjugate vaccines into routine immunization programs in other European countries (106), as well as Canada and Australia.
3. Africa: the term "meningitis belt" was defined by Lapeyssonnie (107) in 1963 when he described the sub-Saharan region from Ethiopia to Senegal, which includes 18 countries with more than 270 million people. The "meningitis belt" is characterized by periodic large epidemics of predominantly meningococcal

meningitis. Epidemics occur every 8–10 years and began around 1905. The reasons for development and persistence of these outbreaks are not well understood, but environmental factors such as humidity and dust contribute (108, 109). Meteorological data may provide early warning for an impending epidemic (110) in the belt. While the peak of meningococcal disease coincides with respiratory viral illnesses (111, 112) during winter months in developed countries, meningococcal disease in Africa occurs in the dry season (109). However, mycoplasma infections have been associated with African outbreaks. Rates vary from 20 to 1,000 per 100,000 depending on the year and the presence or absence of epidemics.

4. Latin America: the epidemiology of meningococcal disease in Latin America is characterized by marked differences between countries (113). The overall incidence of meningococcal disease per year varies from less than 0.1 case per 100,000 in countries like Mexico to 2 cases per 100,000 in Brazil and Chile with major outbreaks. Serogroup A is rare in Latin America and most cases are due to serogroups B and C, with serogroups W-135 and Y now emerging in certain countries (e.g., serogroup Y was the most prevalent serogroup causing disease in Columbia and Venezuela in 2006).
5. Asia: large outbreaks of serogroup A have historically occurred in China (114), Nepal, India, and Russia but have recently been replaced by localized disease due to serogroups B and C. The incidence in Japan since WWII has fallen to very low levels, ~0.1 case per 100,000. Serogroups B and C are dominant now in Australia with a prolonged serogroup B epidemic in New Zealand occurring in the 1990s.

4.2. Meningococcal Serogroups

Most of the cases of meningococcal disease worldwide are caused by six serogroups (A, B, C, Y, W-135 and recently, in sub-Saharan Africa, X). Serogroup A has been responsible for the largest and most devastating meningococcal outbreaks in sub-Saharan Africa (115). In the largest meningococcal epidemic outbreak recorded in 1996–1997 in Africa, an estimated 300,000 cases and 30,000 deaths occurred in the meningitis belt due to serogroup A infection. Serogroup A was the cause of most meningococcal disease in the first part of the twentieth century in developed countries, but is now rare in the US and Europe.

Serogroup B is typically associated with a lower incidence of disease when compared to serogroup A or C, but prolonged outbreaks of serogroup B disease cause significant morbidity and mortality. Currently, serogroup B is the most important cause of endemic disease in developed countries, causing 30–40% of disease in the US and up to 80% in Europe. Serogroup B polysaccharide is poorly immunogenic (116), but serosubtype-specific vaccines have been developed for countries such as Cuba, New Zealand, and Norway that have experienced prolonged epidemic serogroup B disease. These vaccines are based on strain-specific OM vesicle (OMV) preparations (PorA major target) and were successful in reducing the incidence of local serogroup B outbreaks (117, 118). In New Zealand, the incidence of meningococcal disease increased from 1.6 cases per 100,000 population in 1990 to a peak of 17.4 cases per 100,000 population in 2001, with 85% of cases due to serogroup B. An OMV vaccine against the epidemic strain was introduced in 2004. The incidence decreased to 2.6 cases per 100,000 population by 2007, and the estimated effectiveness of the vaccine was 80% in fully immunized children aged 6 months to 15 years (119). Broadly protective vaccines that prevent both endemic and epidemic serogroup B disease, however, remain a major need worldwide.

Serogroup C is responsible for part of the reported endemic disease and localized epidemic outbreaks in developed countries, accounting for 30% of disease in the US (101) and Europe. Serogroup C has occasionally caused large epidemics. Serogroup Y has emerged in the US and caused more than a quarter of the disease due to meningococci in the US in the last decade (102). Compared with the early-1990s, when the proportion of serogroup Y cases was 2% during 1989–1991, the rates increased to 32.6% in 1996 (102) and serogroup Y still caused 26% of meningococcal disease cases in 2007 (99). Serogroup Y causes meningococcal pneumonia in older adults, but is also responsible for a large proportion of meningococcaemia and meningitis among infants less than 6 months of age (99). Serogroup Y has also been seen recently in South Africa, South America, and Israel.

Serogroup W-135 has emerged in the last 20 years as a cause of epidemic disease and like serogroup A previously, it has been also important especially in relationship to the Hajj pilgrimage (120). Since the 1990s, W-135 meningococcal infection has affected the health of these travelers to Saudi Arabia and their contacts in countries throughout the world (121–123). Saudi authorities require Hajj pilgrims to show evidence of immunization with the tetravalent meningococcal vaccines (A, C, W-135 and Y) and not just with mono- or bivalent meningococcal vaccines. Serogroup W-135 has also emerged in South America (124) and in Africa; Burkina Faso (125–128) in particular witnessed a large outbreak of serogroup W-135 infection in 2002. Serogroup X has recently been found responsible for meningococcal cases and outbreaks in certain African countries such as Kenya (129), Niger (130), and Ghana (131).

4.3. Age Groups

The meningococcus remains a common cause of bacterial meningitis in children and young adults in the USA (132), now mostly affecting children less than 2 years of age (102, 133). Two-thirds of meningococcal disease in the first year of life in the US occurs in infants less than 6 months of age (134). Worldwide, the rates of meningococcal disease are also highest for young children due to waning protective maternal antibody, but in epidemic outbreaks, older children and adolescents can have high rates of disease. Fifty percent of cases in infants in the US are due to serogroup B; serogroup C is mostly seen in adolescents and serogroups B and Y in older adults. Even though peak incidence occurs among infants and adolescents; one-third to one-half of sporadic cases are seen in adults older than 18 years.

4.4. Clonal Complexes

The advances in molecular typing have created a better understanding of the epidemiology and population biology of the meningococcus. The genetic diversity of the meningococcus, even though extensive, is highly structured. Groups of genetically closely related meningococci are grouped into clonal complexes. A minority of clonal complexes, the so-called hyperinvasive lineages, are relatively stable with life spans of many decades and with global geographic spread (135). The hyperinvasive lineages are responsible for a disproportionate number of cases of disease with specific phenotype features (Table 1). While the ST-1, ST-4, and ST-5 complexes are restricted nearly exclusively to strains of serogroup A, other clonal complexes, like the ST-11 complex, may be associated with various serogroups. For example, the emergence of W-135 as a cause of outbreaks in 2000–2002 was

Table 1
Epidemiology and genotypic features of meningococcal clonal complexes

Epidemiology					
Clonal complex	ST-32	ST-8	ST-4	ST-11	ST-23
Protype Strain	MC58	NMB-CDC	Z2491	FAM18	M6049
Disease/carriage ratio (136)	3.5	24.5	19.5	6.6	0.8
Feature/Genotype					
IHTA Capsule synthesis	Yes (B, C)	Yes (B)	Yes (A)	Yes (C, W-135)	Yes (Y, W-135, NG)
IHTB/C/D	Yes	No	No	No	No
IHTE Lamboid bacteriophage	No	Yes	No	Yes	No
Pnm1 or 2 Mu-like prophage	Yes	No	Yes	No	Variable
MDA Disease associated phage	Yes	Yes	Yes	Yes	Variable
hmbR locus	Yes	Yes	Yes	Yes	No

ST sequence type, *IHT* islands of horizontal transfer

the result of the spread of W-135, ST-11 strains closely related to the ST-11 serogroup C strains. Other "virulent" clonal complexes include the ST-269 complex, a significant cause of serogroup B disease since the 1990s in the UK and now recognized worldwide; the ST-8, ST-32, and ST41/44 complexes associated with serogroup B worldwide; the ST-23 complex associated with serogroup Y disease in the US and now in other countries.

5. Conclusions

The human species is the only natural host for the meningococcus. The meningococcus has evolved multiple mechanisms to be able to transmit from, adapt to, and colonize predominantly human upper respiratory tract mucosal surfaces. Certain clonal groups of meningococci have also evolved the capacity (e.g., expression of certain capsular polysaccharides) to cause invasive disease. N. *meningitidis* remains a global threat causing sporadic cases, case clusters, epidemics, and pandemics, although new conjugate and protein-based vaccines hold great promise and are already influencing the incidence of meningococcal disease.

Acknowledgements

We would like to thank Lane Pucko for her help in preparing this review. Research is supported by NIH/NIAID grants (R01 AI33517 and R01 AI40247) to D.S.S. and "Atlanta Clinical and Translational Science Institute" (UL1RR025008; KL2FF025009; TL1RR025010) and Georgia Research Alliance (GRA.VAC.09.K) grants to N.G.R. and D.S.S.

References

1. Weichselbaum A (1887) Ueber die Aetiologie der akuten meningitis cerebrospinalis. Fortschr Med 5:573.
2. Vieusseux M (1805) Mémoire sur la maladie qui a régné à Genéve au printemps de 1805. J Med Clin Pharm 11:163–82.
3. Danielson L, Mann E (1806) A history of a singular and very noted disease, which lately made its appearance in Medfield. Med Agricultural Reg 1:65–9.
4. Greenwood B (1999) Manson Lecture. Meningococcal meningitis in Africa. Trans R Soc Trop Med Hyg 93:341–53.
5. Kiefer F (1896) Zur differential Diagnose des Erregers der epidemischen Cerebrospinal meningitis und der Gonorrhoea Berl Klin Wochenschr 33:628.
6. Glover J (1918) The cerebrospinal fever epidemic of 1917 at "X" depot. J R Army Med Corps 30:23.
7. Flexner S (1913) The results if the serum treatment in thirteen hundred cases of epidemic meningitis. J Exp Med 17:553.
8. Schwentker F, Gelman S, Long P (1937) The treatment of meningococcic meningitis with sulfonamide. Preliminary report. JAMA 108:1407.
9. Schoenback E, Phair J (1948) The sensitivity of meningococci to sulfadiazine. Am J Hyg 47:177–86.

10. Artenstein MS, Gold R, Zimmerly JG et al (1970) Prevention of meningococcal disease by group C polysaccharide vaccine. N Engl J Med 282:417–20.
11. Stephens DS, Greenwood B, Brandtzaeg P (2007) Epidemic meningitis, meningococcaemia, and *Neisseria meningitidis*. Lancet 369:2196–210.
12. Sharip A, Sorvillo F, Redelings MD et al (2006) Population-based analysis of meningococcal disease mortality in the United States: 1990-2002. Pediatr Infect Dis J 25:191–4.
13. Rosenstein NE, Perkins BA, Stephens DS et al (2001) Meningococcal disease. N Engl J Med 344:1378–88.
14. Stephens DS (2009) Biology and pathogenesis of the evolutionarily successful, obligate human bacterium *Neisseria meningitidis*. Vaccine 27 Suppl 2:B71–7.
15. Tettelin H, Saunders NJ, Heidelberg J et al (2000) Complete genome sequence of *Neisseria meningitidis* serogroup B strain MC58. Science 287:1809–15.
16. Parkhill J, Achtman M, James KD et al (2000) Complete DNA sequence of a serogroup A strain of *Neisseria meningitidis* Z2491. Nature 404:502–6.
17. Schoen C, Blom J, Claus H et al (2008) Whole-genome comparison of disease and carriage strains provides insights into virulence evolution in *Neisseria meningitidis*. Proc Natl Acad Sci USA 105:3473–8.
18. Hotopp JC, Grifantini R, Kumar N et al (2006) Comparative genomics of *Neisseria meningitidis*: core genome, islands of horizontal transfer and pathogen-specific genes. Microbiology 152:3733–49.
19. Davidsen T, Tonjum T (2006) Meningococcal genome dynamics. Nat Rev Microbiol 4:11–22.
20. Hilse R, Hammerschmidt S, Bautsch W et al (1996) Site-specific insertion of IS1301 and distribution in *Neisseria meningitidis* strains. J Bacteriol 178:2527–32.
21. Uria MJ, Zhang Q, Li Y et al (2008) A generic mechanism in *Neisseria meningitidis* for enhanced resistance against bactericidal antibodies. J Exp Med 205:1423–34.
22. Goldschneider I, Gotschlich EC, Artenstein MS (1969) Human immunity to the meningococcus. I. The role of humoral antibodies. J Exp Med 129:1307–26.
23. Goldschneider I, Gotschlich EC, Artenstein MS (1969) Human immunity to the meningococcus. II. Development of natural immunity. J Exp Med 129:1327–48.
24. Snape MD, Pollard AJ (2005) Meningococcal polysaccharide-protein conjugate vaccines. Lancet Infect Dis 5:21–30.
25. Blacklow RS, Warren L (1962) Biosynthesis of sialic acids by *Neisseria meningitidis*. J Biol Chem 237:3520–6.
26. Varki A (1997) Sialic acids as ligands in recognition phenomena. FASEB J 11:248–55.
27. Estabrook MM, Griffiss JM, Jarvis GA (1997) Sialylation of *Neisseria meningitidis* lipooligosaccharide inhibits serum bactericidal activity by masking lacto-N-neotetraose. Infect Immun 65:4436–44.
28. Kahler CM, Martin LE, Shih GC et al (1998) The (alpha2-->8)-linked polysialic acid capsule and lipooligosaccharide structure both contribute to the ability of serogroup B *Neisseria meningitidis* to resist the bactericidal activity of normal human serum. Infect Immun 66:5939–47.
29. Hobb RI, Tzeng YL, Choudhury BP et al (2010) Requirement of NMB0065 for connecting assembly and export of sialic acid capsular polysaccharides in *Neisseria meningitidis*. Microbes Infect 12:476–87.
30. Zimmer SM, Stephens DS (2006) Serogroup B meningococcal vaccines. Curr Opin Investig Drugs 7:733–9.
31. Frosch M, Weisgerber C, Meyer TF (1989) Molecular characterization and expression in *Escherichia coli* of the gene complex encoding the polysaccharide capsule of *Neisseria meningitidis* group B. Proc Natl Acad Sci USA 86:1669–73.
32. Swartley JS, Marfin AA, Edupuganti S, et al (1997) Capsule switching of *Neisseria meningitidis*. Proc Natl Acad Sci USA 94:271–6.
33. Aguilera JF, Perrocheau A, Meffre C et al (2002) Outbreak of serogroup W135 meningococcal disease after the Hajj pilgrimage, Europe, 2000. Emerg Infect Dis 8:761–7.
34. Raghunathan PL, Jones JD, Tiendrebeogo SR et al (2006) Predictors of immunity after a major serogroup W-135 meningococcal disease epidemic, Burkina Faso, 2002. J Infect Dis 193:607–16.
35. Harrison LH, Shutt KA, Schmink SE et al (2010) Population structure and capsular switching of invasive *Neisseria meningitidis* isolates in the pre-meningococcal conjugate vaccine era-United States, 2000-2005. J Infect Dis 201:1208–24.
36. Alonso JM, Gilmet G, Rouzic EM et al (2007) Workshop on vaccine pressure and *Neisseria meningitidis*, Annecy, France, 9–11 March 2005. Vaccine 25:4125–9.

37. Balmer P, Borrow R, Miller E (2002) Impact of meningococcal C conjugate vaccine in the UK. J Med Microbiol 51:717–22.
38. Nikaido H (1999) Microdermatology: cell surface in the interaction of microbes with the external world. J Bacteriol 181:4–8.
39. Rahman MM, Kolli VS, Kahler CM et al (2000) The membrane phospholipids of *Neisseria meningitidis* and *Neisseria gonorrhoeae* as characterized by fast atom bombardment mass spectrometry. Microbiology 146 (Pt 8):1901–11.
40. Antignac A, Rousselle JC, Namane A et al (2003) Detailed structural analysis of the peptidoglycan of the human pathogen *Neisseria meningitidis.* J Biol Chem 278:31521–8.
41. Quintela JC, Caparros M, de Pedro MA (1995) Variability of peptidoglycan structural parameters in gram-negative bacteria. FEMS Microbiol Lett 125:95–100.
42. Clarke AJ, Dupont C (1992) O-acetylated peptidoglycan: its occurrence, pathobiological significance, and biosynthesis. Can J Microbiol 38:85–91.
43. Girardin SE, Boneca IG, Viala J et al (2003) Nod2 is a general sensor of peptidoglycan through muramyl dipeptide (MDP) detection. J Biol Chem 278:8869–72.
44. Jennings HJ, Johnson KG, Kenne L (1983) The structure of an R-type oligosaccharide core obtained from some lipopolysaccharides of *Neisseria meningitidis.* Carbohydr Res 121:233–41.
45. Gamian A, Beurret M, Michon F et al (1992) Structure of the L2 lipopolysaccharide core oligosaccharides of *Neisseria meningitidis.* J Biol Chem 267:922–5.
46. Kahler CM, Stephens DS (1998) Genetic basis for biosynthesis, structure, and function of meningococcal lipooligosaccharide (endotoxin). Crit Rev Microbiol 24:281–334.
47. Plant L, Sundqvist J, Zughaier S et al (2006) Lipooligosaccharide structure contributes to multiple steps in the virulence of *Neisseria meningitidis.* Infect Immun 74:1360–7.
48. Zughaier SM, Lindner B, Howe J et al (2007) Physicochemical characterization and biological activity of lipooligosaccharides and lipid A from *Neisseria meningitidis.* J Endotoxin Res 13:343–57.
49. Mandrell RE, Griffiss JM, Macher BA (1988) Lipooligosaccharides (LOS) of *Neisseria gonorrhoeae* and *Neisseria meningitidis* have components that are immunochemically similar to precursors of human blood group antigens. Carbohydrate sequence specificity of the mouse monoclonal antibodies that recognize crossreacting antigens on LOS and human erythrocytes. J Exp Med 168:107–26.
50. Jennings MP, Srikhanta YN, Moxon ER et al (1999) The genetic basis of the phase variation repertoire of lipopolysaccharide immunotypes in *Neisseria meningitidis.* Microbiology 145:3013–21.
51. Mandrell RE, Zollinger WD (1977) Lipopolysaccharide serotyping of *Neisseria meningitidis* by hemagglutination inhibition. Infect Immun 16:471–5.
52. Zughaier SM, Tzeng YL, Zimmer SM et al (2004) *Neisseria meningitidis* lipooligosaccharide structure-dependent activation of the macrophage CD14/Toll-like receptor 4 pathway. Infect Immun 72:371–80.
53. Zughaier S, Steeghs L, van der Ley P et al (2007) TLR4-dependent adjuvant activity of *Neisseria meningitidis* lipid A. Vaccine 25:4401–9.
54. Braun JM, Blackwell CC, Poxton IR et al (2002) Proinflammatory responses to lipooligosaccharide of *Neisseria meningitidis* immunotype strains in relation to virulence and disease. J Infect Dis 185:1431–8.
55. Brandtzaeg P, Kierulf P, Gaustad P et al (1989) Plasma endotoxin as a predictor of multiple organ failure and death in systemic meningococcal disease. J Infect Dis 159:195–204.
56. Brandtzaeg P, Bryn K, Kierulf P et al (1992) Meningococcal endotoxin in lethal septic shock plasma studied by gas chromatography, mass-spectrometry, ultracentrifugation, and electron microscopy. J Clin Invest 89: 816–23.
57. Greenfield S, Sheehe PR, Feldman HA (1971) Meningococcal carriage in a population of "normal" families. J Infect Dis 123:67–73.
58. Stephens DS (1999) Uncloaking the meningococcus: dynamics of carriage and disease. Lancet 353:941–2.
59. Caugant DA, Hoiby EA, Magnus P et al (1994) Asymptomatic carriage of *Neisseria meningitidis* in a randomly sampled population. J Clin Microbiol 32:323–30.
60. Pinner RW, Spellman PA, Stephens DS (1991) Evidence for functionally distinct pili expressed by *Neisseria meningitidis.* Infect Immun 59:3169–75.
61. Virji M, Alexandrescu C, Ferguson DJ et al (1992)Variations in the expression of pili: the effect on adherence of *Neisseria meningitidis* to human epithelial and endothelial cells. Mol Microbiol 6:1271–9.
62. Merz AJ, So M (2000) Interactions of pathogenic neisseriae with epithelial cell membranes. Annu Rev Cell Dev Biol 16:423–57.

63. Stephens DS, McGee ZA (1981) Attachment of *Neisseria meningitidis* to human mucosal surfaces: influence of pili and type of receptor cell. J Infect Dis 143:525–32.
64. Proft T, Baker EN (2009) Pili in Gram-negative and Gram-positive bacteria - structure, assembly and their role in disease. Cell Mol Life Sci 66:613–35.
65. Kahler CM, Martin LE, Tzeng YL et al (2001) Polymorphisms in pilin glycosylation Locus of *Neisseria meningitidis* expressing class II pili. Infect Immun 69:3597–604.
66. Marceau M, Forest K, Beretti JL et al (1998) Consequences of the loss of O-linked glycosylation of meningococcal type IV pilin on piliation and pilus-mediated adhesion. Mol Microbiol 27:705–15.
67. Callaghan MJ, Jolley KA, Maiden MC (2006) Opacity-associated adhesin repertoire in hyperinvasive *Neisseria meningitidis*. Infect Immun 74:5085–94.
68. Virji M, Watt SM, Barker S et al (1996) The N-domain of the human CD66a adhesion molecule is a target for Opa proteins of *Neisseria meningitidis* and *Neisseria gonorrhoeae*. Mol Microbiol 22:929–39.
69. Virji M, Makepeace K, Ferguson DJ et al (1992) Expression of the Opc protein correlates with invasion of epithelial and endothelial cells by *Neisseria meningitidis*. Mol Microbiol 6:2785–95.
70. Virji M, Evans D, Hadfield A et al (1999) Critical determinants of host receptor targeting by *Neisseria meningitidis* and *Neisseria gonorrhoeae*: identification of Opa adhesiotopes on the N-domain of CD66 molecules. Mol Microbiol 34:538–51.
71. Hill DJ, Griffiths NJ, Borodina E et al (2010) Cellular and molecular biology of *Neisseria meningitidis* colonization and invasive disease. Clin Sci (Lond) 118:547–64.
72. Perkins-Balding D, Ratliff-Griffin M, Stojiljkovic I (2004) Iron transport systems in *Neisseria meningitidis*. Microbiol Mol Biol Rev 68:154–71.
73. Schryvers AB, Stojiljkovic I (1999) Iron acquisition systems in the pathogenic Neisseria. Mol Microbiol 32:1117–23.
74. Tzeng YL, Stephens DS (2000) Epidemiology and pathogenesis of *Neisseria meningitidis*. Microbes Infect 2:687–700.
75. Massari P, Ram S, Macleod H et al (2003) The role of porins in neisserial pathogenesis and immunity. Trends Microbiol 11:87–93.
76. Vermont CL, van Dijken HH, Kuipers AJ et al (2003) Cross-reactivity of antibodies against PorA after vaccination with a meningococcal B outer membrane vesicle vaccine. Infect Immun 71:1650–5.
77. Rappuoli R (2001) Reverse vaccinology, a genome-based approach to vaccine development. Vaccine 19:2688–91.
78. Frasch CE, Zollinger WD, Poolman JT (1985) Serotype antigens of *Neisseria meningitidis* and a proposed scheme for designation of serotypes. Rev Infect Dis 7:504–10.
79. Slaterus KW (1961) Serological typing of meningococci by means of micro-precipitation. Antonie Van Leeuwenhoek 27:305–15.
80. Branham S (1953) Serological relationships among meningococci. Bact Rev 17:175–88.
81. Vogel U (2010) Molecular epidemiology of meningococci: Application of DNA sequence typing. Int J Med Microbiol 300(7):415–20.
82. Caugant DA, Froholm LO, Bovre K et al (1986) Intercontinental spread of a genetically distinctive complex of clones of *Neisseria meningitidis* causing epidemic disease. Proc Natl Acad Sci USA 83:4927–31.
83. Bevanger L, Bergh K, Gisnas G et al (1998) Identification of nasopharyngeal carriage of an outbreak strain of *Neisseria meningitidis* by pulsed-field gel electrophoresis versus phenotypic methods. J Med Microbiol 47:993–8.
84. Weis N, Lind I (1998) Epidemiological markers in *Neisseria meningitidis*: an estimate of the performance of genotyping vs phenotyping. Scand J Infect Dis 30:69–75.
85. Maiden MC, Bygraves JA, Feil E et al (1998) Multilocus sequence typing: a portable approach to the identification of clones within populations of pathogenic microorganisms. Proc Natl Acad Sci USA 95:3140–5.
86. Mothershed EA, Sacchi CT, Whitney AM et al (2004) Use of real-time PCR to resolve slide agglutination discrepancies in serogroup identification of *Neisseria meningitidis*. J Clin Microbiol 42:320–8.
87. Yazdankhah SP, Kriz P, Tzanakaki G et al (2004) Distribution of serogroups and genotypes among disease-associated and carried isolates of *Neisseria meningitidis* from the Czech Republic, Greece, and Norway. J Clin Microbiol 42:5146–53.
88. Wu HM, Harcourt BH, Hatcher CP et al (2009) Emergence of ciprofloxacin-resistant *Neisseria meningitidis* in North America. N Engl J Med 360:886–92.
89. Baart GJ, Zomer B, de Haan A et al (2007) Modeling *Neisseria meningitidis* metabolism: from genome to metabolic fluxes. Genome Biol 8:R136.
90. Price ND, Papin JA, Schilling CH et al (2003) Genome-scale microbial in silico models: the

constraints-based approach. Trends Biotechnol 21:162–9.

91. Papin JA, Price ND, Wiback SJ et al (2003) Metabolic pathways in the post-genome era. Trends Biochem Sci 28:250–8.
92. Jyssum K, Borchgrevink B, Jyssum S (1961) Glucose catabolism in *Neisseria meningitidis*. 1. Glucose oxidation and intermediate reactions of the Embden-Meyerhof pathway. Acta Pathol Microbiol Scand 53:71–83.
93. Frantz ID (1942) Growth requirements of the meningococcus. J Bacteriol 1942;43:757–61.
94. Chapin (1918) Carbon dioxide in the primary cultivation of the gonococcus. J Infect Dis 19:558–61.
95. Grossowicz N (1945) Growth requirements and metabolism of *Neisseria intracellularis*. J Bacteriol 50:109–15.
96. Erwin AL, Gotschlich EC (1996) Cloning of a *Neisseria meningitidis* gene for L-lactate dehydrogenase (L-LDH): evidence for a second meningococcal L-LDH with different regulation. J Bacteriol 178:4807–13.
97. Leighton MP, Kelly DJ, Williamson MP et al (2001) An NMR and enzyme study of the carbon metabolism of *Neisseria meningitidis*. Microbiology 147:1473–82.
98. Port JL, DeVoe IW, Archibald FS (1984) Sulphur acquisition by *Neisseria meningitidis*. Can J Microbiol 30:1453–7.
99. Cohn AC, MacNeil JR, Harrison LH et al (2010) Changes in *Neisseria meningitidis* disease epidemiology in the United States, 1998–2007: implications for prevention of meningococcal disease. Clin Infect Dis 50:184–91.
100. Fischer M, Hedberg K, Cardosi P et al (1997) Tobacco smoke as a risk factor for meningococcal disease. Pediatr Infect Dis J 16:979–83.
101. Jackson LA, Schuchat A, Reeves MW et al (1995) Serogroup C meningococcal outbreaks in the United States. An emerging threat. JAMA 273:383–9.
102. Rosenstein NE, Perkins BA, Stephens DS et al (1999) The changing epidemiology of meningococcal disease in the United States, 1992–1996. J Infect Dis 180:1894–901.
103. Cartwright K, Noah N, Peltola H (2001) Meningococcal disease in Europe: epidemiology, mortality, and prevention with conjugate vaccines. Report of a European advisory board meeting Vienna, Austria, 6–8 October, 2000. Vaccine 19:4347–56.
104. Trotter CL, Andrews NJ, Kaczmarski EB et al (2004) Effectiveness of meningococcal serogroup C conjugate vaccine 4 years after introduction. Lancet 364:365–7.
105. Maiden MC, Ibarz-Pavon AB, Urwin R et al (2008) Impact of meningococcal serogroup C conjugate vaccines on carriage and herd immunity. J Infect Dis 197:737–43.
106. Trotter CL, Ramsay ME (2007) Vaccination against meningococcal disease in Europe: review and recommendations for the use of conjugate vaccines. FEMS Microbiol Rev 31:101–7.
107. Lepeyssonnie (1963) La méningite cérébrospinale en Afrique. Bull WHO 28:53–114.
108. Molesworth AM, Cuevas LE, Connor SJ et al (2003) Environmental risk and meningitis epidemics in Africa. Emerg Infect Dis 9:1287–93.
109. Greenwood BM, Bradley AK, Wall RA et al (1985) Meningococcal disease and season in sub-Saharan Africa. Lancet 2:829–30.
110. Lewis R, Nathan N, Diarra L et al (2001) Timely detection of meningococcal meningitis epidemics in Africa. Lancet 358:287–93.
111. Artenstein MS, Rust JH Jr., Hunter DH et al (1967) Acute respiratory disease and meningococcal infection in army recruits. JAMA 201:1004–7.
112. Young LS, LaForce FM, Head JJ et al (1972) A simultaneous outbreak of meningococcal and influenza infections. N Engl J Med 287:5–9.
113. Safadi MA, Cintra OA (2010) Epidemiology of meningococcal disease in Latin America: current situation and opportunities for prevention. Neurol Res32:263–71.
114. Wang JF, Caugant DA, Li X et al (1992) Clonal and antigenic analysis of serogroup A *Neisseria meningitidis* with particular reference to epidemiological features of epidemic meningitis in the People's Republic of China. Infect Immun 60:5267–82.
115. Hart CA, Cuevas LE (1997) Meningococcal disease in Africa. Ann Trop Med Parasitol 91:777–85.
116. Wyle FA, Artenstein MS, Brandt BL et al (1972) Immunologic response of man to group B meningococcal polysaccharide vaccines. J Infect Dis 126:514–21.
117. Oster P, Lennon D, O'Hallahan J et al (2005) MeNZB: a safe and highly immunogenic tailor-made vaccine against the New Zealand *Neisseria meningitidis* serogroup B disease epidemic strain. Vaccine 23:2191–6.
118. Rodriguez AP, Dickinson F, Baly A et al (1999) The epidemiological impact of antimeningococcal B vaccination in Cuba. Mem Inst Oswaldo Cruz 94:433–40.

119. Galloway Y, Stehr-Green P, McNicholas A et al (2009) Use of an observational cohort study to estimate the effectiveness of the New Zealand group B meningococcal vaccine in children aged under 5 years. Int J Epidemiol 38:413–8.
120. Dull PM, Abdelwahab J, Sacchi CT et al (2005) *Neisseria meningitidis* serogroup W-135 carriage among US travelers to the 2001 Hajj. J Infect Dis 191:33–9.
121. Lingappa JR, Al-Rabeah AM, Hajjeh R et al (2003) Serogroup W-135 meningococcal disease during the Hajj, 2000. Emerg Infect Dis 9:665–71.
122. Wilder-Smith A, Goh KT, Barkham T et al (2003) Hajj-associated outbreak strain of *Neisseria meningitidis* serogroup W135: estimates of the attack rate in a defined population and the risk of invasive disease developing in carriers. Clin Infect Dis 36:679–83.
123. Hahne SJ, Gray SJ, Jean F et al (2002) W135 meningococcal disease in England and Wales associated with Hajj 2000 and 2001. Lancet 359:582–3.
124. Efron AM, Sorhouet C, Salcedo C et al (2009) W135 invasive meningococcal strains spreading in South America: significant increase in incidence rate in Argentina. J Clin Microbiol 47:1979–80.
125. Koumare B, Ouedraogo-Traore R, Sanou I et al (2007) The first large epidemic of meningococcal disease caused by serogroup W135, Burkina Faso, 2002. Vaccine 25 Suppl 1:A37–41.
126. Traore Y, Njanpop-Lafourcade BM, Adjogble KL et al (2006) The rise and fall of epidemic *Neisseria meningitidis* serogroup W135 meningitis in Burkina Faso, 2002-2005. Clin Infect Dis 43:817–22.
127. Decosas J, Koama JB (2002) Chronicle of an outbreak foretold: meningococcal meningitis W135 in Burkina Faso. Lancet Infect Dis 2:763–5.
128. Ouedraogo-Traore R, Hoiby EA, Sanou I et al (2002) Molecular characteristics of *Neisseria meningitidis* strains isolated in Burkina Faso in 2001. Scand J Infect Dis 34:804–7.
129. Materu S, Cox HS, Isaakidis P et al (2007) Serogroup X in meningococcal disease, Western Kenya. Emerg Infect Dis 13:944–5.
130. Boisier P, Nicolas P, Djibo S et al (2007) Meningococcal meningitis: unprecedented incidence of serogroup X-related cases in 2006 in Niger. Clin Infect Dis 44:657–63.
131. Gagneux SP, Hodgson A, Smith TA et al (2002) Prospective study of a serogroup X *Neisseria meningitidis* outbreak in northern Ghana. J Infect Dis 185:618–26.
132. Harrison LH, Dwyer DM, Maples CT (1999) Risk of meningococcal infection in college students. JAMA 281:1906–10.
133. Kaplan SL, Schutze GE, Leake JA et al (2006) Multicenter surveillance of invasive meningococcal infections in children. Pediatrics 118:e979–84.
134. Shepard CW, Rosenstein NE, Fischer M (2003) Neonatal meningococcal disease in the United States, 1990 to 1999. Pediatr Infect Dis J 22:418–22.
135. Caugant DA (2008) Genetics and evolution of *Neisseria meningitidis*: importance for the epidemiology of meningococcal disease. Infect Genet Evol 8:558–65.
136. Caugant DA, Maiden MCJ (2009) Meningococcal carriage and disease – population biology and evolution Vaccine 27:B64–B70.

Chapter 2

Classification and Pathogenesis of Meningococcal Infections

Petter Brandtzaeg and Marcel van Deuren

Abstract

The clinical symptoms induced by *Neisseria meningitidis* reflect compartmentalized intravascular and intracranial bacterial growth and inflammation. In this chapter, we describe a classification system for meningococcal disease based on the nature of the clinical symptoms. Meningococci invade the subarachnoid space and cause meningitis in as many as 50–70% of patients. The bacteremic phase is moderate in patients with meningitis and mild systemic meningococcemia but graded high in patients with septic shock. Three landmark studies using this classification system and comprising 862 patients showed that 37–49% developed meningitis without shock, 10–18% shock without meningitis, 7–12% shock and meningitis, and 18–33% had mild meningococcemia without shock or meningitis. *N. meningitidis* lipopolysaccharide (LPS) is the principal trigger of the innate immune system via activation of the Toll-like receptor 4-MD2 cell surface receptor complex on myeloid and nonmyeloid human cells. The intracellular signals are conveyed via MyD88-dependent and -independent pathways altering the expression of >4,600 genes in target cells such as monocytes. However, non-LPS molecules contribute to inflammation, but 10–100-fold higher concentrations are required to reach the same responses as induced by LPS. Activation of the complement and coagulation systems is related to the bacterial load in the circulation and contributes to the development of shock, organ dysfunction, thrombus formation, bleeding, and long-term complications in patients. Despite rapid intervention and advances in patient intensive care, why as many as 30% of patients with systemic meningococcal disease develop massive meningococcemia leading to shock and death is still not understood.

Key words: Meningococcal meningitis, Septicemia, Classification, Lipopolysaccharide, Shock

1. Introduction

Neisseria meningitidis is an obligate human Gram-negative diplococcus residing asymptomatically in the upper respiratory tract (1). The highest carriage rate is found among adolescents and young adults representing the main reservoir of the bacterium. Few young children carry meningococci (1). Most carried strains never cause

Myron Christodoulides (ed.), *Neisseria meningitidis: Advanced Methods and Protocols*, Methods in Molecular Biology, vol. 799, DOI 10.1007/978-1-61779-346-2_2, © Springer Science+Business Media, LLC 2012

invasive disease. A limited number of meningococci belonging to specific clones or clonal complexes with serogroups A, B, C, Y, and W135 cause >95% of systemic meningococcal disease (SMD) (1). Encapsulated meningococci expressing pili, opacity proteins, and other subcapsular adhesion molecules are transferred by droplets or by direct contact (kissing) from an asymptomatic carrier to a nonimmune person.

After attachment to nonciliated columnar epithelial cells in the nasopharynx or the epithelium covering the tonsils, meningococci adapt and start to proliferate. Type 4 pili and various outer membrane proteins including opacity proteins undergo phase variation (2). Studies in vitro suggest that *N. meningitidis* may transverse the epithelial cells through "parasite directed endocytosis" (3). It is assumed that meningococci enter the blood stream through the capillaries and small veins in the underlying submucosal tissue. Survival and growth of encapsulated *N. meningitidis* in the circulation is a fundamental requirement for developing SMD.

Clinical disease usually develops within a week following breach of the mucosal barrier. A century ago, before any effective treatment existed, 70–90% of the patients died during the natural course of infection (4). After the introduction of antimeningococcal serum therapy in 1905, the case fatality rate (CFR) declined to approximately 40% (4). Treatment with sulfonamides in 1937 and with penicillin in 1945 resulted in a further decline of the CFR to approximately 10%. In the last decade, the CFR has ranged from 7 to 11% in Europe and the USA for sporadic cases, increasing to above 20% in outbreak situations (5–7).

Two characteristics make *N. meningitidis* unique among human pathogens: (1) The propensity to invade the meninges and (2) the ability to proliferate rapidly in the blood leading to shock and multiple organ failure in as many as 30% of the patients contracting serogroup B and C strains (6, 8, 9). Meningococci are known among physicians, epidemiologists, and lay people as the bacterium causing outbreaks of meningitis. In industrialized countries, physicians should rightly fear *N. meningitidis* as the bacterium that may cause overwhelming septicemia, shock, and death. Meningococci may kill previously healthy children or adults within 12–24 h (6, 10, 11). Septic shock and multiple organ failure is the direct cause of death in nine out of ten patients with lethal meningococcal infections in Western countries (Table 1) (6, 8–11). A minority dies of meningitis leading to brain edema, herniation, and arrest of brain circulation (6, 9–11). In Third World countries, particularly in sub-Saharan Africa, *N. meningitidis* serogroup A still causes large-scale outbreaks involving tens of thousands of people. Meningitis without shock appears to be the dominant clinical presentation (12).

A lack of bactericidal antibodies is the single most important predisposing factor for developing SMD (11, 13, 14). Defects in

Table 1
Classification of systemic meningococcal disease presentations

References	Year	n	Shock without meningitis (%)	CFR (%)	Shock + meningitis (%)	CFR (%)	Meningitis without shock (%)	CFR (%)	Mild meningo-coccemia (%)	CFR (%)
(8)	1983	115	18	52	15	12	49	0	18	0
(9)	1987	206	17	29	13	7	37	1	33	0
(6)	2008	541	10	16	20	11	49	1	21	5

n denotes the total number of patients included in the study. CFR denotes case fatality rate of the previous column (6, 8, 9). The clinical presentations are defined as follows:
Shock without meningitis: (1) severe septic shock lasting for 24 h or until death requiring fluid and pressor drug therapy and (2) $<100 \times 10^6$/L leukocytes in CSF. Spinal puncture is presently not recommended for these shock patients. The leukocyte count is therefore substituted by clinical signs of distinct meningism
Meningitis without shock: (1) clinical distinct symptoms and signs of meningism, $\geq 100 \times 10^6$/L leukocytes in CSF (or if the results of CSF are not obtainable, signs of distinct meningism) and (2) lack of persistent septic shock lasting >24 h requiring fluid and pressor drug therapy
Shock and meningitis: (1) severe septic shock lasting for 24 h or until death requiring fluid and pressor drug therapy and (2) $\geq 100 \times 10^6$/L leukocytes (or if the results of CSF are not obtainable, signs of distinct meningism)
Meningococcemia without shock or meningitis (mild SMD): Meningococcemia detected by blood culture, PCR or by other methods without (1) severe septic shock requiring treatment for at least 24 h and (2) $<100 \times 10^6$/L leukocytes (or if the results of CSF are not obtainable, lack of marked meningism)

the alternative and terminal pathways of complement increase the risk of contracting invasive infection (11, 13, 14). Cohort and genome-wide association studies suggest that complement factor H and related proteins determine the susceptibility to meningococcemia, whereas the role of mannose binding lectin is uncertain as a predisposing factor (15, 16). Meta-analysis suggests that genes related to fibrinolysis and interleukin-1 influence the severity of outcome (17).

2. Clinical Classification of Systemic Meningococcal Disease for Research Purposes

In the 1960s, it was recognized that patients with meningococcal infections presenting with the classical symptoms of meningitis (nuchal and back rigidity) and marked pleocytosis, i.e., $\geq 100 \times 10^6$/L leukocytes in the cerebrospinal fluid (CSF) had a much better prognosis than patients with shock and minimal pleocytosis, i.e., $<100 \times 10^6$/L leukocytes in the CSF (18). Development of shock was the decisive factor determining survival or death and this conclusion was later confirmed by others (6, 8–11, 14, 19, 20). The degree of pleocytosis, i.e., the number of leukocytes in CSF $\geq 100 \times 10^6$/L leukocytes became a cut-off marker for defining distinct meningitis in meningococcal research. Based on two easy recognizable criteria: (1) shock and (2) meningitis ($\geq 100 \times 10^6$/L leukocytes in CSF or distinct signs of meningitis if CSF is not available) a classification system was established for research purposes (8). By applying this classification system in subsequent clinical studies it was possible to delineate the inflammatory responses induced by *N. meningitidis* and to describe the underlying pathophysiology in great detail (11, 14, 19, 20).

Three landmark studies used this clinical classification and comprised 862 patients, of whom 656 were studied prospectively, infected with serogroup B or C strains (6, 8, 9) (Table 1). Seventy percent of the 862 patients did not develop septic shock. They presented with distinct meningitis or mild meningococcemia. The CFR was low (Table 1). Thirty percent of the patients developed septic shock with a much higher CFR than the nonshock group (Table 1). The majority of the shock patients lacked distinct signs of meningitis. Persistent shock was the major cause of death (Table 1).

2.1. Shock Without Meningitis (Fulminant Septicemia, Waterhouse–Friderichsen Syndrome)

The symptoms develop rapidly. The median time between the onset of symptoms to hospital admission was 12–13 h in five Western European studies (6, 11, 19). The clinical picture is dominated by septic shock requiring large volumes of fluids and vasoactive drugs, artificial ventilation due to acute respiratory distress syndrome (ARDS), renal failure, large hemorrhagic skin lesions, and adrenal hemorrhage reflecting severe disseminated intravascular

Table 2
Plasma levels of *Neisseria meningitidis* DNA copy numbers and LPS levels in 65 Norwegian patients with confirmed SMD

	Shock ($n=21$)	Shock + meningitis ($n=2$)	Meningitis without shock ($n=28$)	Mild meningococcemia ($n=14$)
Nm DNA copies	2×10^7	1×10^6	$<10^3$	7.7×10^3
LPS EU/mL	43	9.4	<0.5	<0.5

n denotes the number of patients analyzed. All levels are given as median values (23). SMD denotes systemic meningococcal disease

Table 3
Plasma LPS levels vs. clinical presentation in 150 Norwegian patients with confirmed SMD

LPS (EU/mL)	Number of patients	Number dead	Case fatality rate (%)
>250	7	7	100
250–50	20	17	85
50–10	24	6	25
10–0.5	31	1	3
<0.5	68	0	0

EU/mL denotes endotoxin units per mL as determined by the limulus amebocyte lysate (LAL) assay. SMD denotes systemic meningococcal disease. The *dashed line* indicates the plasma shock level (10 EU/mL) (10, 19)

coagulation (DIC) (11, 14, 19). Patients lack marked symptoms and signs of meningitis, reveal minimal pleocytosis, and few meningococci are present in the CSF. This clinical presentation occurred in 10–18% of the 862 patients studied. The CFR varied from 16 to 52% depending on the pathogenicity of the strain and between the endemic vs. epidemic situation (5, 6, 8–11, 14, 19, 20). The severity, development of shock and death are closely associated with the level of meningococci in the circulation as measured by the number of *N. meningitidis* DNA copies reaching levels as high as 10^8/mL (21–23) (Table 2). The plasma level of lipopolysaccharide (LPS, endotoxin) as quantified by the Limulus Amebocyte Lysate (LAL) assay is as high as 2,150 Endotoxin Units (EU)/mL (23). If the LPS level in plasma passes 8–10 EU/mL, 95% of the patients will develop septic shock (10, 19) (Tables 2 and 3).

The exponentially escalating numbers of meningococci induce an exaggerated and destructive inflammatory response in the vasculature, heart, kidneys, and lungs with high levels of pro- and anti-inflammatory cytokines (11, 19). Fifty percent of nonsurviving patients die within 12 h after hospital admission (6, 10, 11, 19). Surviving patients may require treatment for weeks with prolonged ventilation for ARDS, dialysis for renal failure, amputation of peripheral part of the extremities, and extensive skin grafting. The late complications are a consequence of DIC leading to thrombosis in various organs. The CFR varied from 16 to 52% in the three studies comprising these 862 patients (Table 1).

Petechial rash, i.e., small hemorrhagic skin lesions (Ø= 1–4 mm) is common and one of the most characteristic symptoms of SMD. The rash combined with fever is considered the hallmark of this infection. Petechiae are, however, not universally present. Depending on meningococcal strains and host factors, typical hemorrhagic lesions are observed in 28–78% of the patients (6, 14). The petechial lesion is a consequence of a preceding meningococcemia, local attachment of meningococci to the endothelial cells of capillaries and small veins in the skin. The meningococci express capsule polysaccharide, type IV pili, and PorA outer membrane protein (24). The organisms "dock" at and migrate through, or between, endothelial cells, altering the antithrombotic surface of the endothelium. The result is formation of thrombi and extravasation of erythrocytes, which is seen as skin hemorrhages (11, 14, 19). The meningococci may divide locally and can be cultured 13 h after initiation of antibiotic treatment (25). Up to 100 meningococci have been visualized in a small area in biopsies of skin lesions (24). Hemorrhagic lesions with a diameter >1 cm (ecchymoses) are primarily observed in patients developing shock with high levels (>10 EU/mL) of LPS. The presence of these lesions indicates severe DIC (11, 14, 19).

The hemorrhagic skin lesions and thrombosis in different organs are associated with massive up-regulation of tissue factor (TF) in circulating monocytes (11, 14, 19). TF activates the extrinsic pathway of the coagulation system. Concomitantly, the fibrinolysis is initially activated by tissue plasminogen activactor (tPA). Subsequently, tPA is blocked by plasminogen activator inhibitor-1 (PAI-1) facilitating thrombi formation. Plasma fibrinogen is consumed with declining plasma levels (11, 14, 19). Activation of thrombin and plasmin is reflected in increased levels of thrombin–antithrombin (TAT) and plasmin–antiplasmin (PAP) complexes in plasma. The natural coagulation inhibitor protein C is reduced to 10–20% of normal levels (19). Reduction to such low levels is associated with purpura fulminans, i.e., diffuse thromboses of skin vessels (19). Antithrombin is often reduced to 50% of normal levels (11, 14, 19). The functional levels of tissue factor pathway inhibitor (TFPI) increases (19). The platelets are activated, adhere to surface structures on the vessel walls, and contribute to the formation of thrombi. The coagulopathy, which is more pronounced than in most other cases of severe sepsis, is a consequence of the very high number of meningococci and the accompanying endotoxinemia in the blood.

Shock, i.e., persistent hypotension and hypoperfusion of different organs, is the single most important indicator of survival or death in SMD (6, 8–11, 14, 18–20, 26). Persistent shock after fluid resuscitation is a consequence of vasodilation, vascular leakage, and gradually reduced contractility of the myocardium (11, 14, 19, 26). After volume treatment, the circulation is initially hyperdynamic with increased cardiac output and tachycardia compensating for low peripheral resistance. The myocardial contractility is gradually reduced, which is not compensated by massive fluid treatment and vasopressors (dopamine, norepinephrine, epinephrine, and dobutamine). The declining myocardial performance can be visualized by serially ultrasonographic examinations. At the terminal shock stage, the myocardial contractility and ejection fraction are severely reduced and the peripheral vascular resistance low. The patients often die of arrhythmia after a period of tissue hypoxia and acidosis (10). A myriad of factors contribute to the reduced cardiovascular performance, including nitric oxide (NO) produced locally in the endothelium and myocytes after up-regulating nitric oxide synthase 2 (NOS 2) (27).

2.2. Shock and Meningitis

The patients present with shock in combination with a marked pleocytosis or clinical symptoms of meningitis. This combination occurred in 13–20% of the 862 patients. The CFR varied from 7 to 12% and was lower than for patients with shock alone, but significantly higher than for patients with meningitis alone (Table 1) (6, 8, 9).

2.3. Meningitis Without Shock

This is the most common clinical presentation (6, 8–11) (Table 1). The symptoms develop more gradually than patients with shock with a median onset – admission time of 23–29 h in five European studies (6, 11, 19). After a comparatively low graded meningococcemia, the bacteria may transverse the blood–CSF barrier of meningeal blood vessels and enter the subarachnoid space and possibly also enter the lateral ventricles of the brain by passing through the choroid plexus (19). In general, meningococci proliferate in the CSF to levels that are 3–5 log_{10}-fold higher than the plasma levels. Approximately 70% of these patients have *N. meningitidis* DNA copy number $<10^3$/mL in plasma. The median LPS level in plasma is <0.5 EU/mL (23). CSF may contain up to 10^9/mL meningococcal DNA copies (23). The CSF concentration of LPS may reach 4,000 EU/mL (19). Concomitantly, bacteria trigger the release of a variety of inflammatory mediators including cytokines and chemokines, whose levels are several log_{10}-fold higher in CSF than the levels measured in plasma. The patients present with distinct clinical symptoms and signs of meningism, i.e., neck and back rigidity, headache, vomiting, photophobia, and positive Kernig's and Brudzinski's signs. The CSF contains $\geq 100 \times 10^6$/L leukocytes, an elevated level of proteins, and a lower than normal content of glucose. In the 862-patient cohort, 37–49% were classified as meningitis without shock. The CFR was ≤1% (Table 1) and the major sequel was deafness (14).

2.4. Meningococcemia Without Shock or Meningitis (Mild SMD)

The patients usually present with fever. They may have a petechial rash but lack signs of septic shock or meningitis. Blood cultures are positive in 70–80% of the cases (19). The meningococcemia is occasionally transient and only present in the prehospital phase. *N. meningitidis* is detected in cultures from blood or biopsies(s) and PCR or ELISA (for research purpose). The median number of meningococcal DNA copies was 7.7×10^3/mL and LPS level <0.5 EU/mL in one study (23). If a spinal puncture is performed, the CSF is normal or shows signs of slight pleocytosis ($<100 \times 10^6$/L leukocytes). Given the fact that these patients do not develop long-lasting circulatory impairment or a massive inflammation in the subarachnoid space the CFR is low. When death does occur, it is related to the development of shock or brain edema after admission. This clinical presentation occurred in 18–33% of the 862 patients studied prospectively with a CFR of 0–5% (Table 1). It is a composite clinical group. Untreated, some of these patients might have developed shock or meningitis later, whereas others are transient meningococcemia possibly seeding joints, pericardium, or eyes (4, 14). Very few develop subchronic or chronic meningococcemia with intermittent fever and rash (4, 11, 14). In the Netherlands, the CFR among 752 prospectively studied patients from 2003 to 2005 was inversely related to the duration of the disease before admission (Table 4) (6). The longer the duration of the symptoms the lower the proliferation rate of meningococci in the circulation.

3. *N. meningitidis* Interacting with the Human Innate Immune System

The disease severity and inflammatory response are closely associated with the level of LPS in plasma and CSF, reflecting the growth of meningococci in different compartments (10, 19, 23, 28, 29). In all four clinical presentations described earlier, meningococci can be grown in both the blood and CSF and the levels of bacteria in the two compartments determine the clinical symptoms (19, 23) (Table 5).

Lipopolysaccharides (LPS, endotoxin) are the principal, but not the only molecules, in the outer membrane of the meningococcus leading to a generalized activation of the innate immune system (11, 19, 23, 28–32). The immunological response is reflected

Table 4
Inverse relationship between case fatality rate (CFR) and duration of the disease before admission (6)

Duration	<12 h (%)	12–18 h (%)	18–36 h (%)	>36 h (%)
Case fatality rate	10.2	7.8	3.5	2.2

Table 5
The percentage of positive cultures of blood (157 patients) and CSF (119 patients) with microbiologically confirmed diagnosis of SMD before receiving antibiotics (19)

	Shock (%)	Shock + meningitis (%)	Meningitis without shock (%)	Mild meningococcemia (%)
Blood	93	87	50	77
CSF	59	83	84	47

by cytokine levels, release of other inflammatory mediators, activation of complement, and coagulation and inhibition of the fibrinolytic systems (11, 19). Non-LPS molecules do activate the innate immune system, albeit less strongly than LPS. Porins (PorB) and presumably other lipoproteins have been reported to activate Toll Like Receptor (TLR) 2, meningococcal DNA activates TLR9 through CpG repeats, and fragments of peptidoglycan may activate NOD1 and NOD2 (33–37). The role of the non-LPS molecules in human disease is still unknown.

3.1. Structure of N. meningitidis LPS

Meningococcal LPS is often referred to as lipooligosaccharide (LOS) owing to a short side chain typical for Gram-negative bacteria found in the upper respiratory tract. It consists of lipid A, a core structure comprising two 2-keto-3-deoxy-octulosonic acid (KDO) and two heptoses (L-glycero-D-manno-heptopyranoside) substituted with short polysaccharide side chain (29, 38, 39). Lipid A and KDO represent the toxic moiety, i.e., the immune stimulatory part activating TLR4 when in complex with myeloid differentiation factor (MD)2 (40). The full length side chain comprises lacto-N-neotetraose (Galβ1→4GluNAcβ1→3Galβ1→4Glc) and a terminal sialic acid. Lacto-N-neotetraose and sialic acid play a role in avoidance of immune recognition (molecular mimicry), serum resistance, and cell adherence. Based on antibody specificity, 12 different LPS immunotypes have been described. Immunotypes 3, 7, and 9 are found in approximately 80% of isolates from Dutch patients and represent LPS with a maximum number of sugars including a terminal sialic acid (41). Carrier isolates often harbor a shorter LPS (L8) making them more adhesive, but vulnerable to bactericidal antibodies and less virulent (29).

3.2. Variation in the N. meningitidis LPS Lipid A Structure Is Reflected by a Reduced Capacity to Induce Cytokines

Mutated strains of *N. meningitidis* comprising four or five fatty acids in the lipid A moiety, as opposed to hexavalent lipid A, are less biological active than the wild-type parent strains (42). Wild-type strains with penta-acylated lipid A may infect patients (43). They apparently cause infections with less activation of the coagulation system than the hexa-acylated wild-type strains usually isolated from patients. The ability of Neisseria isolates to activate the LAL assay differs significantly and is presumably related to certain variations in the lipid A structure, including variable phosphorylation of the backbone structure (44).

3.3. Minimal Chain Length Requirement of N. meningitidis LPS for Optimal TLR4-MD2 Activation

Based on studies of *E. coli* lipid A it was assumed for a long time that lipid A derived from meningococci represented the toxic moiety of the molecule. However, by systematically testing the biological activity of *N. meningitidis* mutants expressing variable lengths of the polysaccharide side chain and core region of LPS, it has been documented that lipid A per se is less active than lipid A to which two KDO molecules are attached (40). Synthetic lipid A containing one KDO is markedly more potent than lipid A alone (45). MD2 in complex with lipid A is crucial for TLR4 activation (46).

3.4. Studying the Activation of Complement and Different Cell Types by Wild Type and LPS-Deficient N. meningitidis

Two international reference strains (H44/76 and FAM20) have been transformed into viable mutants completely lacking LPS (47, 48). By systematically comparing the effects of the LPS-containing wild-type strain with the LPS-deficient mutant (H44/76 *lpxA*-) in various cell lines, animal models, and in whole human blood models, the contribution of LPS in the inflammatory response has been clarified (49). A general conclusion is that LPS in *N. meningitidis* is the most potent group of molecules for inducing inflammation in the host. Ten to 100 times higher numbers of mutant bacteria are required to induce inflammatory responses of the same magnitude when assayed in human whole blood models, different cell lines, and mice (31, 32, 49).

3.4.1. Complement Activation vs. Load of N. meningitidis

Studies of complement activation in patients revealed a dose-dependent association between the levels of LPS and formation of the terminal complement complex (C5b-polyC9, TCC) in plasma, the clinical presentation and lethal outcome (50). Subsequent studies documented that complement activation induced by wild-type meningococci and the LPS-deficient mutant was independent of surface exposed LPS (51). LPS in patient plasmas served merely as a marker of other meningococcal molecules activating complement. In vitro experiments showed that 3×10^7/mL *N. meningitidis* in serum was required to generate significantly increased levels of

TCC, indicating terminal pathway activation (31). Looking back at the patient data, the marked elevation of TCC in EDTA-plasma was associated with high-grade meningococcemia and LPS levels ≥100 EU/mL, which is roughly equivalent to $\geq 10^7$/mL copies of *N. meningitidis* DNA and an 85% CFR (23, 50).

3.4.2. Interaction of Toll-Like Receptors and Complement in Human Whole Blood, Simulating Sepsis

By simulating the log phase growth of *N. meningitidis* to fulminant sepsis it has been documented that TLR4 is activated with much lower levels of meningococci than are required to activate the complement system with significant increase of TCC. The difference was in the magnitude of 3–4 $\log_{10}$ (31). However, blocking experiments using whole blood triggered by increasing doses of meningococci documented a synergy between TLR4 and complement activation. The cytokine production was augmented with much lower levels of meningococci than were required to induce TCC as measured by ELISA, which indicated low-grade complement activation (31).

3.4.3. N. meningitidis Activating Human Monocytes

Human monocytes are precursors of tissue macrophages in liver, spleen, lungs, peritoneum, and other tissues. They are key cells in the mononuclear phagocyte system. In a comparative study using microarray to evaluate the influence of meningococcal LPS vs. non-LPS structures on gene regulation, 4,689 monocyte genes were either twofold up-regulated or down-regulated by 10^6/mL of wild-type *N. meningitidis* strain H44/76 (52). Only 72 genes were differentially regulated by 10^6/mL of LPS-deficient mutant bacteria. By increasing the numbers of LPS-deficient mutant bacteria to 10^8/mL, the expression of 3,905 genes was altered, indicating a dose–response activation of the monocytes induced by a 100-fold higher number of non-LPS molecules. Further analysis revealed that 2,288 genes were particularly LPS sensitive: these genes were up- or down-regulated by 10^6/mL wild-type meningococci but not by 10^8/mL LPS-deficient mutant bacteria (52). The same quantitative pattern of cytokine production is present in human whole blood models where increasing numbers of wild-type *N. meningitidis* H44/76 are compared with the LPS-deficient mutant (31).

3.4.4. N. meningitidis Activates Endothelial Cells

Current hypotheses suggest that meningococci are most likely to enter the subarachnoid space between or through endothelial cells of blood vessels located within the meninges (53–55). Using different tissue culture models simulating the interaction between *N. meningitidis* and human brain microvascular endothelial cells, a significant reprogramming of gene expression occurred in the host cells (53, 56). Key mRNA transcripts for cytokines and chemokines including TNF-α, IL-6, and IL-8 were up-regulated in the cells. ICAM-1 mRNA was likewise up-regulated, facilitating the transition of activated neutrophils through the capillary wall. IL-1β was unchanged. Expression of pili increased gene expression, whereas the presence of capsule reduced expression (53). The cells also showed signs of antibacterial resistance by activating antiapoptotic mechanisms (56).

3.4.5. Intracellular Signaling Related to LPS and Non-LPS Meningococcal Molecules

Transmembrane signals generated by wild-type meningococci are conveyed via myeloid differentiation primary response protein (MyD)88-dependent and -independent pathways (52, 57). Genes for cardinal cytokines including TNF-α, IL-1β, IL-6, IL-8, and MCP-1 are activated via the MyD88 pathway primarily by LPS, but also with very high levels of the mutant, by non-LPS molecules (31). The large numbers of particular LPS-responsive genes are primarily activated by MyD88-independent mechanisms via increased levels of interferon-β, which activates interferon-β inducible genes (type I interferon-β signaling pathway) (52). NO, a powerful vasodilator contributing to the shock, is induced via the MyD88-independent pathway (57).

4. A New Porcine Model Simulating Fulminant Meningococcemia with Profound Shock

Stepwise increasing doses of heat-killed wild-type serogroup B meningococci (H44/76) or LPS-deficient (H44/76 *lpxA*-) strains were infused into anesthetized healthy landrace pigs, simulating the logarithmic growth of meningococci in patients with shock (58, 59). Pulmonary hypertension induced by thromboxane A2, endothelin, and other mediators were the first significant changes in the cardiovascular system. They were induced earlier and more intensively by the wild-type strain than the LPS-deficient mutant but finally reached the same level. In parallel, a general vascular leakage occurred. Subsequently, reduced vascular resistance and decreasing mean arterial pressure followed in the animals given wild-type meningococci but not in the pigs receiving the LPS-deficient mutant (59). Animals challenged by the wild-type strain but not by the mutant lacking LPS became profoundly leukopenic, indicating "margination" of leukocytes in the vasculature. The leukopenia indicated up-regulation of various adhesion molecules on the leukocytes and the endothelial cells by LPS. The wild-type strain activated the coagulation system more profoundly than the mutant. This model confirmed the potent immunostimulatory role of LPS as the major sepsis-inducing molecule of meningococci. However, the non-LPS molecules of the mutant were fully capable of inducing TNF-α, IL-6, and IL-10 but not IL-1β, IL-8, and IL-12. The wild-type strain was approximately ten times more potent than the LPS-deficient mutant (59). Massive complement activation resulting in increased TCC in plasma did not occur since the levels of meningococci in the blood were below 10^7 *N. meningitidis* DNA copies/mL (31, 59).

5. Conclusion

The inflammation induced by *N. meningitidis* has been studied in patients and in various experimental models. LPS is crucial, but not the only molecule capable of activating the innate immune system. Rapid intravascular proliferation of meningococci leading to massive endotoxinemia and shock is the greatest threat to the patient. Proliferation in the subarachnoid space leading to distinct meningitis has a much better prognosis if the patient is treated with appropriate antibiotics. Why as many as 30% of patients with SMD develop massive meningococcemia leading to septic shock and multiple organ failure is presently not understood and not observed with any other human invasive bacterium.

Acknowledgments

Peter Kierulf, Reidun Övstebö, Berit Brusletto, Tom Eirik Mollnes, Bernt Christian Hellerud, Arne Höiby, Anne Marie Siebke Tröseid, Tom Sprong, Chris Neeleman, and Sabine de Greeff have all contributed extensively to the studies.

The studies have been financed by Ullevål University Hospital, Oslo, the Reginal Health Authority Helse Söröst, Norway, and the Dutch Organisation for Scientific Research, the Netherlands. A particular thank you to all patients, parents, and relatives who consented to participate in the studies.

References

1. Caugant DA, Maiden MC (2009) Meningococcal carriage and disease-population biology and evolution. Vaccine 27 Suppl 2:B64–B70
2. Virji M (2009) Pathogenic neisseriae: surface modulation, pathogenesis and infection control. Nat Rev Microbiol 7:274–286
3. Stephens DS, Farley MM (1991) Pathogenic events during infection of the human nasopharynx with *Neisseria meningitidis* and *Haemophilus influenzae*. Rev Infect Dis 13:22–33
4. Flexner S (1913) The results of serum treatment in thirteen hundred cases of epidemic meningitis. J Exp Med 17:553–576
5. Brooks R, Woods CW, Benjamin DK Jr et al (2006) Increased case-fatality rate associated with outbreaks of *Neisseria meningitidis* infection, compared with sporadic meningococcal disease, in the United States, 1994-2002. Clin Infect Dis 43:49–54
6. de Greeff SC, de Melker HE, Schouls LM et al (2008) Pre-admission clinical course of meningococcal disease and opportunities for the earlier start of appropriate intervention: a prospective epidemiological study on 752 patients in the Netherlands, 2003-2005. Eur J Clin Microbiol Infect Dis 27:985–992
7. Rosenstein NE, Perkins BA, Stephens DS et al (1999) The changing epidemiology of meningococcal disease in the United States, 1992-1996. J Infect Dis 180:1894–1901
8. Gedde-Dahl TW, Hoiby EA, Schillinger A et al (1983) An epidemiological, clinical and microbiological follow-up study of incident meningococcal disease cases in Norway, winter 1981-1982. Material and epidemiology in the MenOPP project. NIPH Ann 6:155–168
9. Halstensen A, Pedersen SH, Haneberg B et al (1987) Case fatality of meningococcal disease in western Norway. Scand J Infect Dis 19:35–42

10. Brandtzaeg P, Kierulf P, Gaustad P et al (1989) Plasma endotoxin as a predictor of multiple organ failure and death in systemic meningococcal disease. J Infect Dis 159:195–204
11. van Deuren M, Brandtzaeg P, van der Meer JW (2000) Update on meningococcal disease with emphasis on pathogenesis and clinical management. Clin Microbiol Rev 13:144–66
12. Boisier P, Mainassara HB, Sidikou F et al (2007) Case-fatality ratio of bacterial meningitis in the African meningitis belt: we can do better. Vaccine 25 Suppl 1:A24–A29
13. Rosenstein NE, Perkins BA, Stephens DS et al (2001) Meningococcal disease. N Engl J Med 344:1378–1388
14. Stephens DS, Greenwood B, Brandtzaeg P (2007) Epidemic meningitis, meningococcaemia, and *Neisseria meningitidis.* Lancet 369:2196–2210
15. Davila S, Wright VJ, Khor CC et al (2010) Genome-wide association study identifies variants in the CFH region associated with host susceptibility to meningococcal disease. Nat Genet 42:772–776
16. Haralambous E, Dolly SO, Hibberd ML et al (2006) Factor H, a regulator of complement activity, is a major determinant of meningococcal disease susceptibility in UK Caucasian patients. Scand J Infect Dis 38:764–771
17. Brouwer MC, Read RC, van de Beek D (2010) Host genetics and outcome in meningococcal disease: a systematic review and meta-analysis. Lancet Infect Dis 10:262–274
18. Stiehm ER, Damrosch DS (1966) Factors in the prognosis of meningococcal infection. Review of 63 cases with emphasis on recognition and management of the severely ill patient. J Pediatr 68:457–467
19. Brandtzaeg P (2006) Pathogenesis and pathophysiology of invasive meningococcal disease. In: Frosch M, Maiden CJ (eds) Handbook of meningococcal disease. Wiley-VCH, Weinheim, pp 427–480
20. Brandtzaeg P (2010) Meningococcal Infections. In: Warrell DA, Cox TM, Firth JD (eds) Oxford Textbook of Medicine, 5th edn. Oxford University Press, Oxford, pp 709–722
21. Hackett SJ, Guiver M, Marsh J et al (2002) Meningococcal bacterial DNA load at presentation correlates with disease severity. Arch Dis Child 86:44–46
22. Darton T, Guiver M, Naylor S et al (2009) Severity of meningococcal disease associated with genomic bacterial load. Clin Infect Dis 48:587–594
23. Ovstebo R, Brandtzaeg P, Brusletto B et al (2004) Use of robotized DNA isolation and real-time PCR to quantify and identify close correlation between levels of *Neisseria meningitidis* DNA and lipopolysaccharides in plasma and cerebrospinal fluid from patients with systemic meningococcal disease. J Clin Microbiol 42:2980–2987
24. Harrison OB, Robertson BD, Faust SN et al (2002) Analysis of pathogen-host cell interactions in purpura fulminans: expression of capsule, type IV pili, and PorA by *Neisseria meningitidis* in vivo. Infect Immun 70:5193–5201
25. van Deuren M, van Dijke BJ, Koopman RJ et al (1993) Rapid diagnosis of acute meningococcal infections by needle aspiration or biopsy of skin lesions. BMJ 306:1229–1232
26. van Deuren M, Brandtzaeg P (2005) Myocardial dysfunction in meningococcal septic shock: no clear answer yet. Crit Care Med 33:1884–1886
27. Baines PB, Stanford S, Bishop-Bailey D et al (1999) Nitric oxide production in meningococcal disease is directly related to disease severity. Crit Care Med 27:1187–1190
28. Bjerre A, Brusletto B, Rosenqvist E et al (2000) Cellular activating properties and morphology of membrane-bound and purified meningococcal lipopolysaccharide. J Endotoxin Res 6:437–445
29. Brandtzaeg P, Bjerre A, Ovstebo R et al (2001) *Neisseria meningitidis* lipopolysaccharides in human pathology. J Endotoxin Res 7:401–420
30. Bjerre A, Brusletto B, Ovstebo R et al (2003) Identification of meningococcal LPS as a major monocyte activator in IL-10 depleted shock plasmas and CSF by blocking the CD14-TLR4 receptor complex. J Endotoxin Res 9:155–163
31. Hellerud BC, Stenvik J, Espevik T et al (2008) Stages of meningococcal sepsis simulated *in vitro*, with emphasis on complement and Toll-like receptor activation. Infect Immun 76:4183–4189
32. Sprong T, Stikkelbroeck N, van der Ley P et al (2001) Contributions of *Neisseria meningitidis* LPS and non-LPS to proinflammatory cytokine response. J Leukoc Biol 70:283–288
33. Girardin SE, Travassos LH, Herve M et al (2003) Peptidoglycan molecular requirements allowing detection by Nod1 and Nod2. J Biol Chem 278:41702–41708
34. Ingalls RR, Lien E, Golenbock DT (2001) Membrane-associated proteins of a lipopolysaccharide-deficient mutant of *Neisseria meningitidis* activate the inflammatory response through toll-like receptor 2. Infect Immun 69:2230–2236
35. Massari P, Ram S, Macleod H et al (2003) The role of porins in neisserial pathogenesis and immunity. Trends Microbiol 11:87–93

36. Mogensen TH, Paludan SR, Kilian M et al (2006) Live *Streptococcus pneumoniae*, *Haemophilus influenzae*, and *Neisseria meningitidis* activate the inflammatory response through Toll-like receptors 2, 4, and 9 in species-specific patterns. J Leukoc Biol 80:267–277
37. Pridmore AC, Wyllie DH, Abdillahi F et al (2001) A lipopolysaccharide-deficient mutant of *Neisseria meningitidis* elicits attenuated cytokine release by human macrophages and signals via toll-like receptor (TLR) 2 but not via TLR4/MD2. J Infect Dis 183:89–96
38. Kahler CM, Stephens DS (1998) Genetic basis for biosynthesis, structure, and function of meningococcal lipooligosaccharide (endotoxin). Crit Rev Microbiol 24:281–334
39. Kulshin VA, Zahringer U, Lindner B et al (1992) Structural characterization of the lipid A component of pathogenic *Neisseria meningitidis*. J Bacteriol 174:1793–1800
40. Zughaier S, Agrawal S, Stephens DS et al (2006) Hexa-acylation and KDO(2)-glycosylation determine the specific immunostimulatory activity of *Neisseria meningitidis* lipid A for human monocyte derived dendritic cells. Vaccine 24:1291–1297
41. Scholten RJ, Kuipers B, Valkenburg HA et al (1994) Lipo-oligosaccharide immunotyping of *Neisseria meningitidis* by a whole-cell ELISA with monoclonal antibodies. J Med Microbiol 41:236–243
42. van der Ley P, Steeghs L, Hamstra HJ et al (2001) Modification of lipid A biosynthesis in *Neisseria meningitidis* lpxL mutants: influence on lipopolysaccharide structure, toxicity, and adjuvant activity. Infect Immun 69:5981–5990
43. Fransen F, Heckenberg SG, Hamstra HJ et al (2009) Naturally occurring lipid A mutants in *Neisseria meningitidis* from patients with invasive meningococcal disease are associated with reduced coagulopathy. PLoS Pathog 5:e1000396
44. John CM, Liu M, Jarvis GA (2009) Natural phosphoryl and acyl variants of lipid A from *Neisseria meningitidis* strain 89I differentially induce tumor necrosis factor-alpha in human monocytes. J Biol Chem 284:21515–21525
45. Zhang Y, Gaekwad J, Wolfert MA et al (2008) Innate immune responses of synthetic lipid A derivatives of *Neisseria meningitidis*. Chemistry 14:558–569
46. Zimmer SM, Zughaier SM, Tzeng YL et al (2007) Human MD-2 discrimination of meningococcal lipid A structures and activation of TLR4. Glycobiology 17:847–856
47. Steeghs L, den Hartog R, den Boer A et al (1998) Meningitis bacterium is viable without endotoxin. Nature 392:449–450
48. Albiger B, Johansson L, Jonsson AB (2003) Lipooligosaccharide-deficient *Neisseria meningitidis* shows altered pilus-associated characteristics. Infect Immun 71:155–162
49. Brandtzaeg P (2003) Host response to *Neisseria meningitidis* lacking lipopolysaccharides. Expert Rev Anti Infect Ther 1:589–596
50. Brandtzaeg P, Mollnes TE, Kierulf P (1989) Complement activation and endotoxin levels in systemic meningococcal disease. J Infect Dis 160:58–65
51. Bjerre A, Brusletto B, Mollnes TE et al (2002) Complement activation induced by purified *Neisseria meningitidis* lipopolysaccharide (LPS), outer membrane vesicles, whole bacteria, and an LPS-free mutant. J Infect Dis 185:220–228
52. Ovstebo R, Olstad OK, Brusletto B et al (2008) Identification of genes particularly sensitive to lipopolysaccharide (LPS) in human monocytes induced by wild-type versus LPS-deficient *Neisseria meningitidis* strains. Infect Immun 76:2685–2695
53. Schubert-Unkmeir A, Sokolova O, Panzner U et al (2007) Gene expression pattern in human brain endothelial cells in response to *Neisseria meningitidis*. Infect Immun 75:899–914
54. Christodoulides M, Heckels JE, Weller RO (2002). The role of the leptomeninges in meningococcal meningitis. In: Ferreiros C, Criado MT, and Vavquez J (eds) Emerging strategies in the fight against meningitis: molecular and cellular aspects. Horizon Scientific Press, Norfolk: pp1–34
55. Join-Lambert O, Morand PC, Carbonnelle E et al (2010) Mechanisms of meningeal invasion by a bacterial extracellular pathogen, the example of *Neisseria meningitidis*. Prog Neurobiol 91:130–139
56. Wells DB, Tighe PJ, Wooldridge KG et al (2001) Differential gene expression during meningeal-meningococcal interaction: evidence for self-defense and early release of cytokines and chemokines. Infect Immun 69:2718–2722
57. Zughaier SM, Zimmer SM, Datta A et al (2005) Differential induction of the toll-like receptor 4-MyD88-dependent and -independent signaling pathways by endotoxins. Infect Immun 73: 2940–2950
58. Nielsen EW, Hellerud BC, Thorgersen EB et al (2009) A new dynamic porcine model of meningococcal shock. Shock 32:302–309
59. Hellerud BC, Nielsen EW, Thorgersen EB et al (2010) Dissecting the effects of lipopolysaccharides from nonlipopolysaccharide molecules in experimental porcine meningococcal sepsis. Crit Care Med 38:1467–1474

Chapter 3

Detection of *Neisseria meningitidis* in Cerebrospinal Fluid Using a Multiplex PCR and the Luminex Detection Technology

Jens Kjølseth Møller

Abstract

Rapid clinical and laboratory diagnoses are the foundation for a successful management of serious infections with *Neisseria meningitidis*. A species-specific multiplex polymerase chain reaction (PCR) coupled with fluidic microarrays using microbeads (the Luminex xMAP™ Technology) can detect pathogens most frequently found in the cerebrospinal fluid of patients. The Luminex suspension array system uniquely combines flow cytometry, microspheres, laser technology, digital signal processing, and traditional chemistry. In this method, the reaction is carried out in one vessel, in which distinctly color-coded bead sets, each conjugated with a different specific nucleic acid reactant, are hybridized with the PCR products, and a reporter molecule is used to quantify the interaction. The flow-based Luminex array reader identifies each reaction (bead set) after excitation by a red classification laser. Reporter signals from each reaction are simultaneously quantified by fluorescence generated by a green reporter laser. This nonculture, multiplex assay may prove to be an important tool for optimal laboratory diagnosis, not only of meningococcal meningitis, but also of meningitis caused by other bacterial or viral pathogens.

Key words: *Neisseria meningitidis*, Multiplex PCR, Cerebrospinal fluid, Luminex, Suspension array, Microspheres

1. Introduction

Bacterial meningitis is a serious infection affecting the central nervous system (CNS), with high morbidity and mortality. *Neisseria meningitidis* is one of the major causes of meningitis. Meningococcal meningitis may rapidly progress and result in permanent damage of the CNS. *N. meningitidis* is also known to cause epidemic outbreaks in crowded surroundings such as schools, refugee camps, and military recruit training centers. Rapid clinical and laboratory diagnoses

Myron Christodoulides (ed.), *Neisseria meningitidis: Advanced Methods and Protocols*, Methods in Molecular Biology, vol. 799, DOI 10.1007/978-1-61779-346-2_3, © Springer Science+Business Media, LLC 2012

are therefore mandatory for the successful management of meningococcal meningitis patients, as well as for interventions preventing the spread of meningococcal disease.

The traditional laboratory diagnostic investigations comprise Gram stain and culture of cerebrospinal fluid (CSF), a blood culture, a naso-pharyngeal culture and if relevant, a sample taken from a petechial hemorrhage in the skin. However, culture for the identification of bacterial pathogens takes 24 h or more. Furthermore, the practice of starting antimicrobial therapy before clinical sample collection decreases the ability to confirm the pathogenic microorganisms of bacterial meningitis by culture (1). Nonculture methods have therefore been recognized as an important tool for optimal laboratory diagnosis of meningococcal infections. Various Nucleic Acid Amplification Tests (NAATs) from the early 1990s were employed in detection of *N. meningitidis* in CSF (2). Later on, detection of nucleic acids of the 16S rRNA gene by polymerase chain reaction (PCR) was introduced for the diagnosis of meningococcal disease (3). Recently, real-time PCR assays have included other targets, e.g., the *porA* gene encoding the outer membrane PorA porin (4), the conserved regulatory gene, *crgA* (5), and the *ctrA* gene (6) involved in the transport of the capsular polysaccharide in meningococci (7). A species-specific multiplex PCR coupled with fluidic microarrays using microbeads (Luminex xMAP™ Technology) for the detection of bacterial pathogens most frequently found in patient's CSF has recently been described (8). The eight-plex PCR described includes primers targeting the *ctrA* gene in *N. meningitidis.* Compared to real-time PCR, the advantage of the Luminex microsphere-based suspension array platform is that up to 100 different reactions in a single reaction vessel may be analyzed and reported at a time.

The Luminex system based on MicroPlex™ Microspheres (formerly named xMAP Multianalyte Carboxylated Microspheres) is based on a unique combination of existing technologies (flow cytometry, microspheres, lasers, digital signal processing, and traditional chemistry) that use distinctly color-coded bead sets, each of which can be conjugated with a different specific analyte, e.g., a protein or nucleic acid reactant (9). A reporter molecule, specific for the analyte, is used to quantify the interaction. A sample from the reaction vessel is drawn up into the flow-based Luminex array reader (Luminex 100/200), where each reaction (bead set) is identified by its spectral signature after excitation by the red classification laser. The attendant reporter signal from each reaction is simultaneously quantified by fluorescence generated by the green reporter laser (Fig. 1). Automated data analysis and detailed summary reports are provided by the Luminex software. The Luminex xMAP™ Technology is a flexible, open-architecture design, which can be configured to perform a wide variety of bioassays quickly, accurately, and cost-effectively.

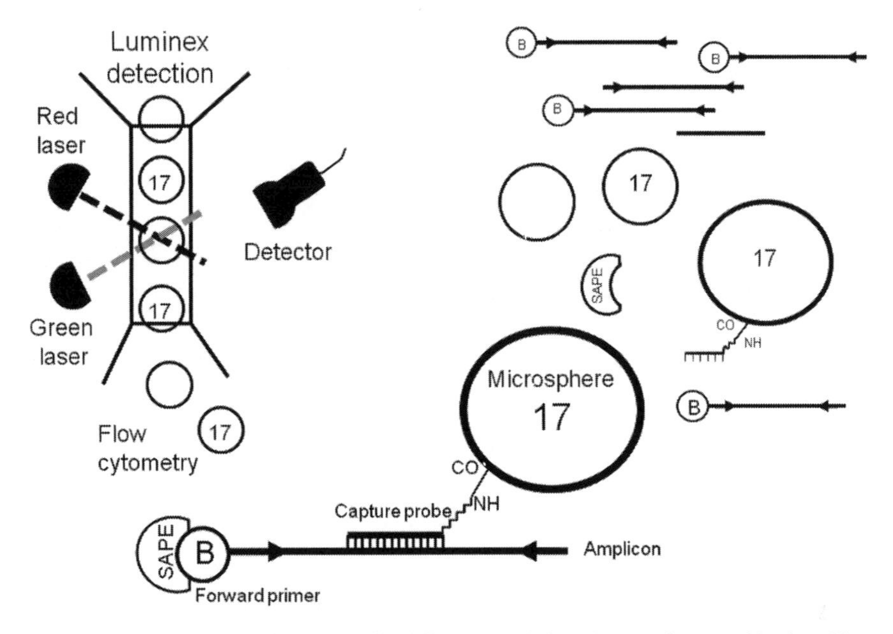

Fig. 1. The LUMINEX detection system (xMAP Technology). The system is based on a unique combination of flow cytometry, microspheres (beads), lasers, digital signal processing, and traditional chemistry that uses distinctly color-coded bead sets, conjugated with a different specific analyte, e.g., a nucleic acid reactant (capture probe).

2. Materials

2.1. Coupling of Microspheres

The purpose of the carbodiimide coupling of bead sets with specific oligonucleotides is to create the suspension microarray capable of capturing PCR amplicons with a corresponding nucleotide sequence. The optimal amount (see Note 1) of oligonucleotide probe for a coupling reaction will depend on the length of the probe to be coupled and the size of the target to be detected (see Note 2).

1. EDC (1-ethyl-3-[3dimethylaminopropyl] carbodiimide hydrochloride) (Pierce) (see Note 3). Prepare a fresh solution of 10 mg/mL of EDC in distilled water.
2. Stock uncoupled microspheres (Luminex) (see Note 4). Store the microspheres at 2–8°C in the dark. Before use, remove from cold storage and allow equilibration to room temperature (see Note 5).
3. Capture probe: suspend the amine-substituted oligonucleotide (Pas in Table 1 and 2) to 1 mM (1 nmol/μL) in distilled water. Prepare a 1/10 dilution of the 1 mM stock capture oligonucleotide in distilled water (i.e., to a working concentration of 0.1 nmol/μL).
4. Coupling buffer, 0.1 M MES buffer, pH 4.5: to prepare this buffer, suspend 4.88 g of MES (2-[N-Morpholino]ethanesulfonic

Table 1
Primers and probe for the 120 bp target sequence of *N. meningitidis*

Oligo name	Sequence (5′-3′)	Position[a]
Nme-s	*Biotin*-GTGATGGTGCGTTTGGTGCAGAATA	504–528
Nme-as	CACATTTGCCGTTGAACCACCTACC	644–620
Nme-Pas	*C12*-CAACACACGCTCACCGGCTGCCGTCAGCGGCATAC	605–571

s sense; *as* antisense; *Pas* antisense probe; *C12* 5′-amino modifier
[a]According to the *ctrA* nucleotide sequence of *N. meningitidis* (GenBank reference sequence GI: 21902494)

acid hydrate) (Sigma-Aldrich) in 250 mL of distilled water and add five drops of 5 N NaOH. Filter sterilize (filter pore size 0.2 μm) the solution and store at it room temperature.

5. Wash buffer I, 0.02% (v/v) Tween-20: add 50 μL of TWEEN® 20 (Polyoxyethylenesorbitan monolaurate; Sigma-Aldrich) to 250 mL of distilled water. Filter sterilize (filter pore size 0.2 μm) the buffer and store it at room temperature.
6. Wash buffer II, 0.1% (w/v) sodium dodecyl sulfate (SDS): add 2.5 mL of a 10% (w/v) solution of SDS into 250 mL of distilled water. Filter sterilize (filter pore size 0.2 μm) the buffer and store it at room temperature.
7. Sample diluent, TE buffer, pH 8.0: dilute 100× Tris-EDTA Buffer, pH 8.0 (Sigma-Aldrich) to a final concentration of 1× by adding 2.5 mL of stock solution to 250 mL of distilled water. Filter sterilize (filter pore size 0.2 μm) the buffer and store it at room temperature.
8. Centrifuge: Eppendorf 5415R.
9. Bransonic water bath sonicator (Branson 200 Ultrasonic Cleaner).
10. Bürker-Türk hemocytometer.

2.2. Multiplex PCR

1. Primers. Primers for the PCR and subsequent Luminex detection should be designed to yield amplicons in the 100–300 bp range. Table 1 shows the nucleotide sequence of the primers for the amplification of a 140-bp target of the *ctrA* gene in *N. meningitidis*. One PCR primer (Nme-s) must be labeled with a 5′-biotin (for the streptavidin–phycoerythrin reporter). Biotin labeling is performed by the manufacturer of the primers. The labeled strand of the amplicons must comprise a nucleotide sequence complementary to the capture probe (Nme-Pas) on the microspheres. The capture probe must have a primary amino group for coupling to the carboxyl group on the microsphere.

Table 2
Primers and probes for the seven other pathogens in the eight-plex PCR

Microorganism and length of amplicon	Oligo name	Sequence (5′-3′)
S. pneumoniae 124 bp	Spn-s	Biotin-CGCAATCTAGCAGATGAAGCAGGTT
	Spn-as	AAGGGTCAACGTGGTCTGAGTGGTT
	Spn-Pas	C12-ACTCGTGCGTTTTAATTCCAGCTAAACTCCC TGTA
S. aureus 150 bp	Sau-s	Biotin-GAATGTGAATGGTGGCGCTATTGCT
	Sau-as	AGCTGCACCCATGCCGACAC
	Sau-Pas	C12-CAATACACATCGTAACCATGCCGTAACGGCTAT
E. coli 106 bp	Eco-s	Biotin-TCCACTTTGCTGCTCACACTTGCTC
	Eco-as	CGTGGTGGTCGCTTTTACCACAGAT
	Eco-Pas	C12-GCGTTTATGCCAGTATGGTTTGTTGAATTTT TATT
L. monocytogenes 119 bp	Lmo-s	Biotin-GCTTTGCCGAAAAATCTGGAAGGTC
	Lmo-as	TGTAAACTTCGGCGCAATCAGTGAA
	Lmo-Pas	C12-GGGAAAATGCAAGAAGAAGTCATTAGTTTTA AACA
S. agalactiae 148 bp	Sag-s	Biotin-TCAGGGTTGGCACGCAATGAA
	Sag-as	GCCCAGCAAATGGCTCAAAAGC
	Sag-Pas1[a]	C12-TATCAAAGATAATGTTCAGGGAACAGATTAT GAAAAA<u>A</u>CGG
	Sag-Pas2	C12-TATCAAAGATAATGTTCAGGGAACAGATTAT GAAAAA<u>C</u>CGG
HSV-1 and HSV-2 130 bp	HSV-s	Biotin-CCACCTCGATCTCCAGGTAGTCC
	HSV-as	GGTGTTCGACTTTGCCAGCCTGTAC
	HSV-Pas	C12-CCCAGCATCATCCAGGCCCACAACCTGTGC TTCAG
VZV 136 bp	VZV-s	Biotin- GATGGTGCATACAGAGAACATTCC
	VZV-as	CCGTTAAATGAGGCGTGACTAA
	VZV-Pas	C12-AAAGTTCCGCGCTGCAGGTTCCAGTAATGC TCTA

s sense; *as* antisense; *Pas* antisense probe; *C12* 5′-amino modifier
[a]The difference between Sag-Pas1 and Pas2 is A/C at base pair position 38 (*underlined*)

Probes must also have a spacer between the reacting amine and the hybridizing sequence. As recommended by Luminex the capture probe (Nme-Pas) is synthesized with the 5′ Amino Modifier C12. The C12 amino modifier is added to the capture probe by the manufacturer of the probe.

For a successful multiplex PCR assay, the relative concentration of the primers, balance between the magnesium chloride and deoxynucleotide concentrations, cycling temperatures, and amount of template DNA and Taq DNA polymerase are important. Use of a commercially available preoptimized master mix greatly simplifies the PCR assay. Good quality control of the primers will help maintain consistent and reliable results over time. The supplier and what purification methods are used and how the concentration of primers is determined should carefully be considered. Primers synthesized may be controlled for quality by using HPLC (High Performance Liquid Chromatography), MALDI-TOF mass spectrometry, or polyacrylamide gel electrophoresis (PAGE) analysis. Mass spectrum analysis helps verify that the primer has the correct molecular weight (nucleotide sequence). A new lot of a particular primer should not be included in the routine Reaction Master Mix before being compared with the corresponding old primer and cleared in a PCR with a known amount of target DNA.

(a) Ensure that the primers are delivered quantified and fully documented. Data on molecular weight, amount and methods of purification, and quality control should appear on an analytical report and on the vial label.

(b) If the primers are delivered as lyophilized powders, add an appropriate volume of sterile distilled water to give a concentration of 100 μmol. Store the primers as recommended by the manufacturer.

2. MagNAPure Compact automatic system and the Total Nucleic Acid Isolation Kit I (Roche) or equivalent.
3. Multiplex PCR Kit (QIAGEN). This contains a ready-to-use preoptimized master mix, to minimize pipetting steps and eliminate tedious calculations. Only primers and template need to be added to prepare the final amplification mix. Prepare a 1× Multiplex Master Mix according to the manufacturer's instructions.
4. MicroAmp Optical 96-well Reaction plate (Applied Biosystems).
5. MWG thermo-cycler (AVISO GmbH) or equivalent.
6. Flat topped, individually capped, thin-walled PCR tubes, e.g., MicroAmp eight-tubes Strips (Applied Biosystems).
7. PCR cooler (cold block).

2.3. Hybridization and Addition of SAPE

1. Microsphere diluent, 1.5× TMAC Hybridization Solution: 250 mL: mix 225 mL of 5 M tetra-methyl ammonium chloride (TMAC) Hybridization Solution (Sigma-Aldrich), 18.75 mL of 1 M Tris–HCl buffer, pH 8.0, 3.0 mL of 0.5 M EDTA, pH 8.0, 1.88 mL of 20% (w/v) Sarkosyl solution (Sigma-Aldrich), and 1.37 mL of distilled water to give a final volume of 250 mL of diluent. Store the final solution at room temperature.
2. Detection buffer, 1× TMAC Hybridization Solution: mix 150 mL of 5 M TMAC Hybridization Solution, 12.5 mL of

1 M Tris–HCl buffer, pH 8.0, 2.0 mL of 0.5 M EDTA, pH 8.0, 1.25 mL of 20% (w/v) Sarkosyl solution, and add 84.25 mL of distilled water to give a final volume of 250 mL of buffer. Store the final buffer at room temperature.

3. Reporter mix: prepare a fresh solution of streptavidin-R-phycoerythrin (SAPE, 1 mg/mL, Molecular Probes, Invitrogen) by adding 0.1–25 μL of the detection buffer per PCR reaction. The final concentration in the reaction should be between 2 and 4 μg/mL of SAPE.
4. Working microsphere mixture: vortex and sonicate in the water bath sonicator the coupled microsphere stocks for 20 s and prepare a solution by diluting the stocks to 150 microspheres of each set/μL in microsphere diluent.
5. TE buffer, pH 8.0 (sample diluent, see Subheading 2.1).
6. Six MWG thermo-cycler (AVISO GmbH) or equivalent.
7. Costar Thermowell™ 96-well microplate (Fisher).

2.4. Detection of Microspheres and Amplicons

1. Luminex Analyzer.
2. 70% (v/v) isopropanol or 70% (v/v) ethanol.
3. Luminex xMAP Sheath fluid, LX100 (Luminex).
4. xMAP Classification Calibrator: microspheres, LX100, CL1 CL2 Calibration (Luminex).
5. xMAP Reporter Calibrator: microspheres, LX100, CL1 CL2 Calibration (Luminex).
6. xMAP Classification Control: microspheres, LX100, CL1 CL2 Control (Luminex).
7. xMAP Reporter Control: microspheres, LX100, RP1 Control (Luminex).

2.5. Analysis of Results

1. Luminex 100 IS software version 2.1.26 or a newer version of the Luminex software (manufacturer updates).

2.6. Sanitation

1. Household bleach, 10–20% (v/v).

3. Methods

The principle of the Luminex suspension array system has been comprehensively reviewed by Dunbar (9). In brief, species-specific PCR products labeled with biotin are mixed with microspheres (bead sets) that have been coupled to gene-specific probes. A universal array made of up to 100 different bead populations can be constructed for one particular assay (reaction vessel). PCR products

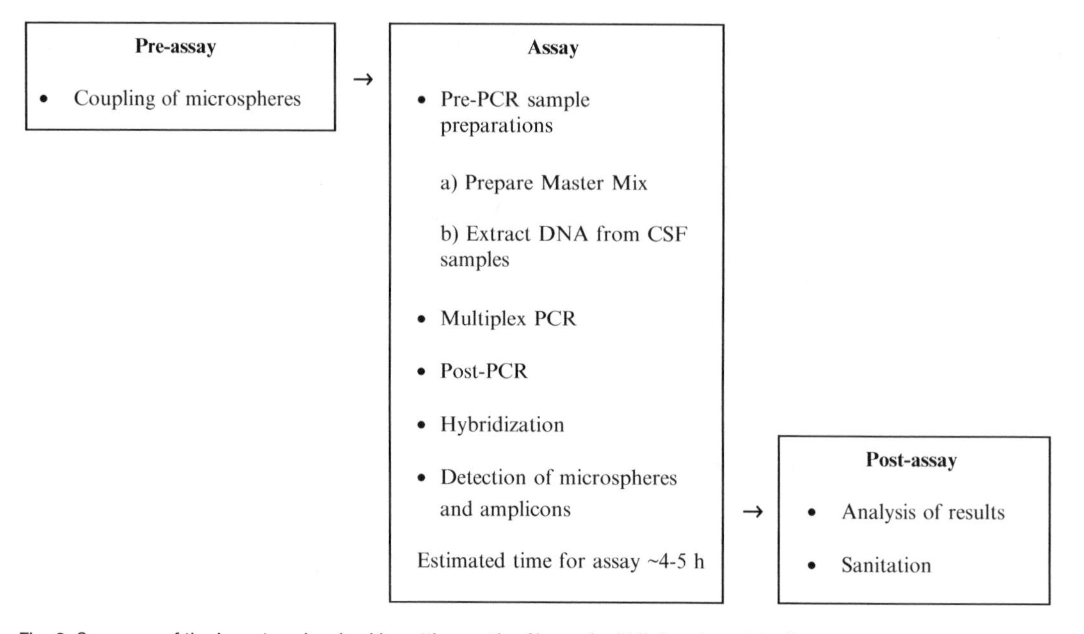

Fig. 2. Summary of the key steps involved in setting up the *N. meningitidis* Luminex detection assay.

are subsequently marked with streptavidin-R-phycoerythrin (SAPE) that binds to the biotin on the PCR products. Each set of microspheres coupled with probes will capture its specific PCR product if represented. Subsequently, each microsphere is analyzed by a red laser identifying the unique color of the microsphere and a green laser that analyzes its surface content of SAPE bound to hybridized PCR products (Fig. 1). Detection tests are run in a 96-well plate microplate and results of the overall meningitis assay are available within 4–5 h.

The major steps involved in setting up a Luminex detection assay are shown in Fig. 2. In a previously published study (8), the *N. meningitidis* assay was part of an eight-plex PCR assay for the detection of six bacterial and two viral pathogens frequently found in the CSF of meningitis patients. Coupling of eight sets of probes to Luminex MicroPlex™ Microspheres was performed according to the manufacturer's protocol. The primers and probes used for the other seven pathogens in the eight-plex PCR are given in Table 2. The individual steps necessary to carry out the assay are explained in detail below.

3.1. Coupling of Microspheres

The microbead sets made by carbodiimide coupling of an amine-modified oligonucleotide (capture probe) to the carboxylated microspheres can be stored long term at 2–8°C. Microspheres should be protected from prolonged exposure to light throughout the coupling procedure and subsequent storage.

1. Bring a fresh aliquot of desiccated EDC powder from −20°C storage to room temperature and prepare a fresh solution of 10 mg/mL EDC in distilled water (see Note 6).
2. Vortex and sonicate the stock uncoupled microspheres for 20 s.
3. Transfer 5.0×10^6 of the stock microspheres to an ordinary Eppendorf tube.
4. Pellet the stock microspheres by centrifugation at 14,000 × *g* for 2 min in the Eppendorf centrifuge.
5. Remove the supernatant and suspend the pelleted microspheres in 50 μL of the coupling buffer by vortex and sonication in the water bath sonicator for 20 s.
6. Add 2 μL (0.2 nmol) of a 1/10 diluted capture oligo (Nme-pas (Table 2) or one of the other capture oligos given in Table 3) to the chosen set of suspended microspheres and mix by vortex. Recommendations for scaling oligonucleotide–microsphere coupling reactions are given in Table 3.
7. One by one for each coupling reaction, add 2.5 μL of fresh EDC (10 mg/mL) to the microspheres and mix by vortex for 15 s.
8. Incubate for 30 min at room temperature in the dark.
9. Prepare a second fresh solution of 10 mg/mL EDC (10 mg/mL) (see Note 6) and repeat steps 8 and 9.
10. Next, add 1.0 mL of Wash buffer I to the coupled microspheres.
11. Pellet the coupled microspheres by microcentrifugation at 14,000 × *g* for 2 min.
12. Remove the supernatant and suspend the coupled microspheres in 1.0 mL of Wash buffer II by vortexing for 15 s.
13. Pellet the coupled microspheres by microcentrifugation at 14,000 × *g* for 2 min.
14. Remove the supernatant and suspend the coupled microspheres in 100 μL of sample diluent by vortex and sonication for 20 s.
15. In order to count the numbers of coupled microspheres, dilute the suspended, coupled microspheres 1/100 in distilled water, mix thoroughly by vortexing for 15 s, and transfer 10 μL to a hemocytometer. Count the microspheres within the four large corners of the hemocytometer grid and calculate the number using the formula:

    Microspheres/μL = (Sum of microspheres in four large corners) × 2.5 × 100 (dilution factor).
16. Store the coupled microspheres at 2–8°C in the dark (see Note 7).

Table 3
Recommendations for scaling the oligonucleotide–microsphere coupling reaction

Number of micro-spheres ($\times 10^6$)	Reaction volume (μL)	Probe input[a] (nM)	EDC concentration[b] (mg/mL)	Tween-20 wash volume (mL)	SDS wash volume (mL)	Final volume[c] (μL)
1	10	0.04–0.1	0.5–2.5	0.5	0.5	20
2.5	25	0.1–0.2	0.5–1	0.5	0.5	50
5	50	0.2–1	0.5–1	1.0	1.0	100
10	50	0.5–1	0.5–1	1.0	1.0	200
50	50–100	1–4	0.5–1	1.0	1.0	1,000
100	100	1–4	0.5–1	1.0	1.0	2,000

[a] see Note 1
[b] EDC input was not adjusted for reactions containing less than 5×10^6 microspheres
[c] Suspension volume of TE buffer, pH 8.0 for 50,000 microspheres/μL, assuming 100% recovery

3.2. Multiplex PCR

The multiplex PCR coupled with the Luminex fluidic microarrays using microbeads for detection involves opening of the amplification vessels and post-PCR sample handling. Even trace amounts of contamination can have deleterious effects on the final results of any PCR-based assay but in particular an open PCR system. Maintaining a contamination-free laboratory working environment involves following critical recommendations for all laboratory facilities, equipments, and procedures (see Notes 8–11). The main point is to prevent carryover contamination with amplicons from post-PCR procedures. Ultimately, clean laboratory practices and physical separation remain the most important anticontamination measures.

1. Extract bacterial DNA using the MagNAPure Compact automatic system and the Total Nucleic Acid Isolation Kit I, following the manufacturer's recommendations. The starting volume may vary between 100 and 400 μL of CSF or CSF supernatant and keep the extractions cold during setup. Elute the DNA in 50 μL of elution buffer, and subsequently store at −20°C.
2. Prepare a Reaction Master Mix in a total volume of 6 μL consisting of 5 μL of QIAGEN Multiplex Master Mix, 0.4 μL of primers (200 nM of each primer), and 0.6 μL of distilled water.
3. Setup the PCR reactions in 8-tube strips or a 96-well reaction plate on a PCR cooler (cold block). Add samples in a timely fashion.
4. Include a positive control (see Note 12) for each target and one or more negative controls (distilled water) in a 96-well plate together with the clinical samples to run.
5. A touchdown PCR reaction (see Note 13) was performed on a MWG thermo-cycler. Thermo-cycler Ramp ≤2°C per second. Perform the multiplex PCR amplification in a total volume of 10 μL, i.e., 4 μL of DNA template plus 6 μL of Reaction Master Mix using the following conditions (steps):

   95°C/15 min; 14 cycles of 94°C/30 s; 65°C −½°C/3 min, 60°C/30 s; 36 cycles of 94°C/10 s, 58°C/30 s, 60°C/30 s; 1 cycle of 72°C/5 min; 1 cycle of 4°C (final hold).

3.3. Hybridization and Addition of SAPE

Bead hybridization and the addition of SAPE are performed simultaneously. In a typical reaction, 5 μL of the PCR product produced in step 5 of Subheading 3.2 is analyzed in a 50-μL hybridization assay containing TMAC with 5,000 of each set of probe-coupled microspheres.

1. Preheat the MWG thermo-cycler to 95°C for the 20-min hybridization/detection step.
2. Select the appropriate oligonucleotide-coupled microsphere sets.

3. Mix the working microsphere mixture by vortex and sonication for 20 s.
4. To each sample or background well in a 96-well reaction microplate, add 33 μL of working microsphere mixture.
5. To each sample or background well, add 12 μL of TE buffer, pH 8.0 (sample diluent).
6. To each sample well, add 5 μL of amplified DNA incorporating the biotinylated primer.
7. Mix the reaction wells gently by pipetting up and down several times.
8. Cover the reaction plate to prevent evaporation and incubate at 95–100°C for 5 min to denature the amplified biotinylated DNA.
9. Incubate the reaction plate for 15 min at the hybridization temperature of 55°C.
10. With the reaction plate still in the thermo-cycler, add 25 μL of Reporter mix containing SAPE to each well and mix gently by pipetting up and down several times (see Note 11).
11. Incubate the reaction plate for 5 min at the hybridization temperature of 55°C (see Note 14).

3.4. Detection of Microspheres and Amplicons

In the Luminex Analyzer, samples are injected into the bottom of the cuvette at a flow rate of 1 μL/s. The sheath fluid is injected around the sample at a flow rate of 90 μL/s and the microspheres aspirated through a narrow column, which forces alignment of the beads. Every object that passes through the cuvette is interrogated. The red laser excites internal dyes on the microspheres colored by the unique combination of red and infra-red dyes, and the internal dye fluorescence is measured. The green laser excites the external fluorophores bound to a microsphere, and the reporter dye fluorescence is measured (Fig. 1) and expressed in Median Fluorescence Intensity (MFI) units.

1. Before the hybridization step, follow the Luminex daily start up and maintenance procedures as described in the user manual (see Note 15).
   (a) Turn on the Luminex System and the computer attached. Wait 30 min for the Luminex Analyzer and the optics system to warm up.
   (b) Verify the levels of sheath fluid and waste fluid.
   (c) Prime the analyzer.
   (d) Perform a wash step. Perform an alcohol flush using at least 1.2 mL of 70% (v/v) isopropanol or 70% (v/v) ethanol in the reservoir.
   (e) Run two wash commands using distilled water.

(f) Prime with alcohol and sheath fluid, two times for each solution.

(g) Check that the Luminex XYP instrument sample probe has been vertically aligned for the plate to be used. Refer to the Luminex user manual for the adjustment procedure.

(h) Ensure that the XYP heater block has been set to 55°C at least 10 min before the end of the hybridization/detection step.

2. After the 20-min hybridization, transfer within 10 min the 96-well plate to the plate holder in the Luminex instrument and retract it.
3. Activate the Luminex program (double click on the Luminex icon on the screen).
4. Press Start Plate.
5. At the end of the run, discard the reaction plate wrapped in a closed plastic bag.

3.5. Analysis of results

1. Analyze the reactions using the Luminex analyzer software. One hundred microspheres are analyzed per sample for each target (species) (see Note 16).
2. Export data files (.csv) containing the results from batches of samples run to an Excel spreadsheet.
3. Determine whether the sample is positive or negative for the microorganisms in the eightplex PCR, based on the cut-off value for each set of microspheres.

3.6. Sanitation

Clean the Luminex system after running samples, following the procedures described in the manufacturer's user manual.

1. Sanitize with household bleach.
2. Run two "Wash Cycles" with distilled water.
3. Soak with distilled water. Soaking the probe replaces sheath fluid in the probe with water and takes approximately 30 s.
4. Turn off the Luminex analyzer.

4. Notes

1. The optimal amount of probe for the particular application must be determined by titration.
2. It should be noted that Luminex offers another technology known as xTAG® technology. This technology uses a proprietary universal tag system that allows easy optimization, product

development, and expansion of molecular diagnostic assays. In the xTAG system, the PCR reaction is subjected to a primer extension step (TSPE = Target Specific Primer Extension) that is specific for the infectious agent that is being analyzed. The 5′ end of the TSPE primers is attached to an xTAG® universal tag sequence. The 5′ universal tag sequence is hybridized to the complementary antitag sequence coupled to a particular xMAP bead set by the manufacturer (Luminex). Thus, the xTAG bead set may be used for any assay where a complementary 5′ universal tag sequence has been employed in the TSPE of a given microorganism specific primer. This xTAG® technology is used in the xTAG® Respiratory Viral Panel (RVP kit) which is a recently described commercial multiplex PCR assay involving 21 primer sets for the detection of multiple specific respiratory virus types and subtypes (10).

3. Pierce EDC powder is very reactive to moisture and should be kept dry at all times. Make aliquots of the EDC powder corresponding to a single coupling reaction and store at −20°C. Discard the unused material after coupling a capture probe to a set of beads.
4. The particular bead sets for a multiplex assay may be picked among any of the 100 available sets and used together with other sets in the same vessel (assay).
5. The xMAP beads will settle if left undisturbed. Always ensure that the xMAP beads are suspended homogeneously prior to dispensing. The uncoupled xMAP beads are not monodispersed and tend to aggregate until coated.
6. The aliquot of EDC powder should now be discarded. Use a fresh aliquot of EDC powder for each coupling episode.
7. The maximum count is 50,000 microspheres/μL. Counts above this number probably arise from an error in the calculation. Postcoupling stability of oligonucleotide-coupled microspheres (maintained discriminatory ability of the coupled microspheres) has been shown to extend more than 96 weeks by Luminex. The authors’ own experience corroborates this finding.
8. The critical recommendation for laboratory facilities is to maintain separate areas for pre-PCR, PCR, and post-PCR manipulations. These areas should ideally be in separate rooms or, at a minimum, the pre-PCR and post-PCR manipulations should be performed on separate lab benches within separate dedicated hoods.
9. For utensils, the critical recommendations are to use (1) fresh, sterilized filter tips for all pipetting throughout the assay; (2) separate sets of pipettes for pre-PCR and post-PCR reactions; (3) separate dedicated pipettes for addition of DNA controls and DNA extracted from samples; (4) preferably flat topped,

individually capped, thin-walled tubes when performing PCR. Strip caps may increase the possibility of micro-droplet contamination between tubes when the strip is removed.

10. There are several critical recommendations that should be adhered to during all procedures: (1) wear gloves when performing any PCR related work, e.g., preparing Master Mix, aliquoting of samples, etc. (2) change gloves before moving from one PCR area to another PCR area; (3) during the same day, do not return to pre-PCR or PCR areas after working in the post-PCR area; (4) add template DNA in a dedicated hood and/or in a separate area outside of the Master Mix hoods within the pre-PCR area; (5) centrifuge the capped tubes briefly to ensure any condensation on the cap or sides is brought down to the bottom of the tube after the PCR reaction is completed; (6) once thawed, keep all enzymes and nucleotide mixes on ice or in a cold block; (7) use dedicated distilled water for making buffers and controls; (8) discard transfer pipette after each use; (9) at the end of the day, thoroughly wipe the pipette tips with "DNA-away" (MβP Molecular Bioproducts) before storing; (10) clean all exposed surfaces with household bleach or "DNA-away." Follow with distilled water and wipe dry.
11. The present assay is an open system including unsealed procedures with a potential risk of contamination of the environment and samples. Handling unsealed PCR specimens while not being able to apply anticontamination procedures such as dUTP/Uracil glycosylase is a major concern regarding the Luminex assay, and common precautions against laboratory contamination should be taken. EPA's Office of Water and Office of Research and Development in USA developed and published a guidance document (11) in October 2004 on quality assurance procedures recommending that PCR facilities use at least three laboratories: (1) a reagent preparation room, (2) a sample preparation room, and (3) an amplification and product room. The post-PCR part of the Luminex assay presents two steps with a potential risk of contaminating the laboratory environment. Therefore, these two steps (the hybridization and the Luminex instrument analysis) should be separated from the amplification area.
12. As positive controls, DNA extracted from the following reference strains was used: *N. meningitidis* serogroup B (ATCC 13090), *S. pneumoniae* (ATCC 6301), *S. aureus* (ATCC 27217), *E. coli* (ATCC 25922), *S. agalactiae* (ATCC 12927), *L. monocytogenes* (VDL 148, National Veterinary Institute, Denmark), HSV1 (McIntyre-B strain), HSV2 (MS), and VZV (QCMD test strain). Concentrations of DNA used may be chosen to reflect a borderline positive (close to cutoff).

13. In a touchdown PCR, the annealing temperature is gradually decreased during the cycling process. The initial annealing temperature is set higher than the Tm of the primers. Higher temperatures favor only the most specific base pairing between the primers and the target DNA and therefore only specific products will be amplified. In the following cycles, the temperature is decreased stepwise until the annealing temperature is 2–5°C below the Tm. Specific products already being amplified and present in excess are then preferentially amplified at the lower, more permissive annealing temperatures.
14. Carry out bead hybridization in the post-PCR area.
15. Calibrate the instrument at least once a week. Use the sets of calibration and control microspheres provided by Luminex. Calibration is also needed if the delta calibration temperature exceeds ±3°C.
16. Except for *S. agalactiae*, where 200 microspheres were analyzed per specimen to improve the signal-to-noise ratio.

Acknowledgments

I am grateful to Birthe Kolmos and Lone Pødenphant for their excellent technical assistance in evaluating the Luminex technology for PCR detection of meningococcal meningitis. I thank Hongwei Zhang MD, PhD, Director, Research and Early Development, Luminex Molecular Diagnostics, Toronto, Canada for critically reviewing the manuscript.

References

1. Nigrovic LE, Malley R, Macias CG et al (2008) Effect of antibiotic pretreatment on cerebrospinal fluid profiles of children with bacterial meningitis. Pediatrics 122: 726–730
2. Kristiansen BE, Ask E, Jenkins, A et al (1991) Rapid diagnosis of meningococcal meningitis by polymerase chain reaction. Lancet 337: 1568–1569
3. Backman A, Lantz P, Radstrom P et al (1999) Evaluation of an extended diagnostic PCR assay for detection and verification of the common causes of bacterial meningitis in CSF and other biological samples. Mol Cell Probes 13: 49–60
4. Molling P, Jacobsson S, Backman A et al (2002) Direct and rapid identification and genogrouping of meningococci and porA amplification by LightCycler PCR. J Clin Microbiol 40: 4531–4535
5. Taha MK (2000) Simultaneous approach for nonculture PCR-based identification and serogroup prediction of Neisseria meningitidis. J Clin Microbiol 38: 855–857
6. Corless CE, Guiver M, Borrow R et al (2001) Simultaneous detection of Neisseria meningitidis, Haemophilus influenzae, and Streptococcus pneumoniae in suspected cases of meningitis and septicemia using real-time PCR. J Clin Microbiol 39: 1553–1558
7. Frosch M, Muller D, Bousset K et al (1992) Conserved outer membrane protein of Neisseria meningitidis involved in capsule expression. Infect Immun 60: 798–803

8. Boving MK, Pedersen LN, Moller JK (2009) Eight-plex PCR and liquid-array detection of bacterial and viral pathogens in cerebrospinal fluid from patients with suspected meningitis. J Clin Microbiol 47: 908–913
9. Dunbar SA (2006) Applications of Luminex xMAP technology for rapid, high-throughput multiplexed nucleic acid detection. Clin Chim Acta 363: 71–82
10. Mahony J, Chong S, Merante,F et al (2007) Development of a respiratory virus panel test for detection of twenty human respiratory viruses by use of multiplex PCR and a fluid microbead-based assay. J Clin Microbiol 45: 2965–2970
11. Sen K, Fout GS, Haugland R et al (2004) Quality Assurance/Quality Control Guidance for Laboratories Performing PCR Analyses on Environmental Samples (October 2004). EPA 815-B-04-001, p. 1-64, U.S. Environmental Protection Agency (EPA), Cincinnati, USA. http://www.epa.gov/nerlcwww/qa_qc_pcr10_04.pdf.

Chapter 4

Generating Knock-Out and Complementation Strains of *Neisseria meningitidis*

Vincent van Dam and Martine P. Bos

Abstract

The human-restricted pathogens *Neisseria meningitidis* and *Neisseria gonorrhoeae* are naturally competent for DNA uptake. This trait has been exploited extensively for genetic manipulation of these bacteria in the laboratory. Most transformation protocols were developed for *N. gonorrhoeae*, but appear to work also for *N. meningitidis*. In this chapter, we describe a number of protocols for genetic manipulation of *N. meningitidis*. Specifically, we describe how to (1) obtain knock-out mutants containing antibiotic-resistance markers, (2) generate markerless knock-out mutants, and (3) construct complementation strains. The generation of such mutants provides a valuable resource for studies of bacterial pathogenesis and vaccine development.

Key words: *Neisseria meningitidis*, Mutants, Genetic manipulation, Complementation

1. Introduction

Neisseria meningitidis, a causative agent of sepsis and meningitis, is a human-restricted Gram-negative bacterium which is naturally competent for transformation throughout its entire life cycle (1). This trait is illustrated by the extensive horizontal genetic exchange occurring in vivo and has also greatly facilitated the generation of mutants in the laboratory. Meningococci discriminate between self and nonself DNA, in that DNA containing a 10-bp sequence, called DUS for DNA Uptake Sequence, is taken up with much higher efficiency than DNA lacking this sequence. This DUS (5′ GCCGTCTGAA 3′) occurs around 2,000 times in the average 2.3 Mbp genomes of *N. meningitidis*, occupying as much as 1% of the chromosome. Addition of two semiconserved residues 5′ of the 10-mer DUS, yielding the 12-mer DUS 5′ATGCCGTCTGAA 3′, increases transformation efficiency in some strains (2).

Myron Christodoulides (ed.), *Neisseria meningitidis: Advanced Methods and Protocols*, Methods in Molecular Biology, vol. 799,
DOI 10.1007/978-1-61779-346-2_4,

Transformation involves multiple steps, including DNA uptake, processing, and chromosomal integration (3). The basis of transformation in the closely related species *N. gonorrhoeae* has been studied extensively (4, 5). Uptake of genus-specific DNA (6) is mediated by type IV pili at the bacterial surface. In *N. gonorrhoeae*, double-stranded plasmid DNA is linearized during DNA uptake and cleaved in a way which is probably not site specific (7). Next, incoming DNA is converted, in part, to a single-stranded intermediate (8), while the subsequent events are not well defined. Not as much is known about transformation in *N. meningitidis* (9), but it is generally thought to resemble the process observed in *N. gonorrhoeae*.

The most straightforward way to create mutants is to replace the relevant gene with an antibiotic resistance marker, allowing for easy selection. However, for obvious safety reasons, it is recommended to limit the number of different antibiotic resistance markers in a meningococcal strain. One way to avoid creating multiresistant strains is to construct markerless mutants. For *N. gonorrhoeae*, it was shown that incubation of a small number of bacteria with an excess amount of DNA results in such high transformation efficiencies that no selection is necessary as at least 20% of the resulting colonies is correctly transformed (10). Unfortunately, we did not obtain similar success with *N. meningitidis* using this approach. A different way to construct markerless mutants in *N. gonorrhoeae* was developed by Johnston and Cannon (11) by exploiting the counter-selective properties of the *rpsL* gene encoding ribosomal protein 12. This strategy involves the use of a two-gene cassette that contains both a selectable marker (*ermC*, conferring erythromycin resistance) and a counter-selectable marker (*rpsL*). Streptomycin resistance can be mediated by a particular allele of the *rpsL* gene, *rpsL*^{R}. The sensitive *rpsL* allele, *rpsL*^{S}, is dominant over the resistant one (12, 13). Thus, first *rpsL*^{S} is introduced together with *ermC* in a double gene cassette into the gene of interest in a streptomycin-resistant strain, causing the strain to become sensitive for streptomycin but resistant towards erythromycin. In a second step, the *ermC-rpsL*^{S} cassette is removed by transformation with DNA containing only flanking regions of the gene of interest and selecting for streptomycin resistance and erythromycin sensitivity (Fig. 1).

The pathogenic Neisseria species are notorious for their dynamic bacterial phenotype. This plasticity is in part due to high-frequency, reversible on/off switching of gene expression caused by the addition or deletion of nucleotide repeats caused by slippage of the DNA polymerase during replication. When these repeats are located in promoter regions or in open reading frames, this so-called slipped-strand mispairing results in variable or complete on/off switching of the expression of phase-variable genes (14). As a consequence, one can easily pick up a phase variant of the parent strain when analyzing single colonies after transformation. This may

Step 1: transformation of Sm^R strain with a linear stretch of DNA consisting of the flanks of the gene of interest with an *ermC-rpsL*^{S} cassette inserted

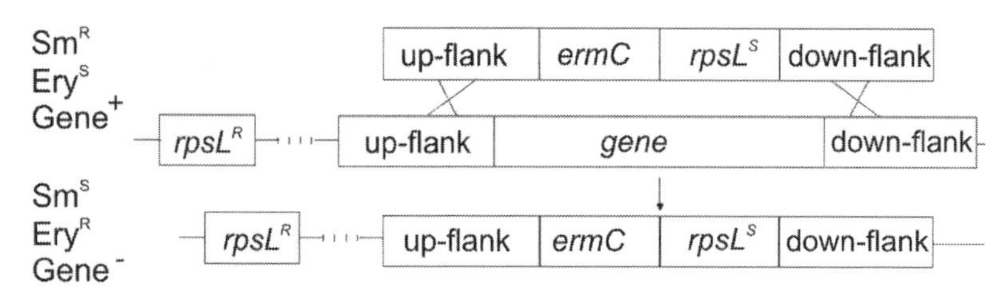

Step 2: transformation of one of the Ery^R, Sm^S recombinants with a linear piece of DNA consisting of the flanks of the gene of interest

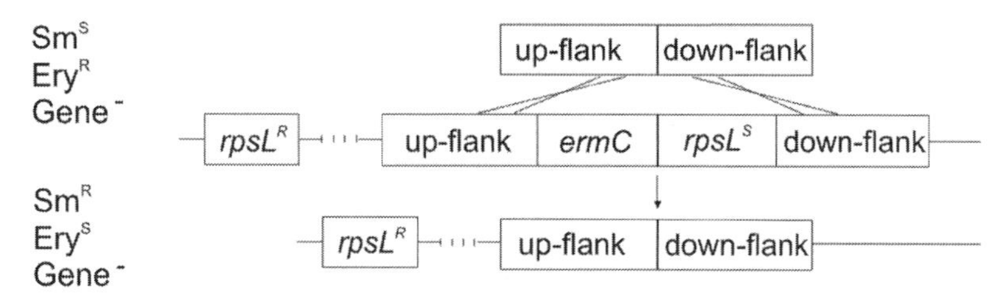

Fig. 1. Schematic representation of the strategy to obtain markerless knock-out strains using the counterselective properties of the *rpsL* gene. Streptomycin and erythromycin are referred to as Sm and Ery, respectively. Adapted from Johnston and Cannon (11).

obviously confound proper interpretation of mutant phenotypes. The solution to this problem would be to construct a complementation strain and test whether the wild-type phenotype is restored. But again, when complementation is done by reintroducing an intact copy of the defective gene in a mutant, one runs the risk of picking up a phase variant. Therefore, we prefer to use an inducible system for complementation analyses in which the effect of the absence and presence of a protein can be assessed in the same bacterium. We found the use of a tandem *lacUV5-tac* promoter-operator sequence (15) to result in tightly regulated gene expression in *N. meningitidis*. The complementing copy of the gene under study is introduced in the knock-out mutant under this tight promoter control, and phenotypes of bacteria grown with and without IPTG are analyzed, thereby ensuring a comparison of truly isogenic strains.

2. Materials

2.1. Bacterial Culture

1. GC-agar plates for culture of *N. meningitidis*. Stir 18 g of GC-agar base (Oxoid) in 500 mL of distilled H_2O (it will not dissolve). Autoclave for 15 min at 120°C and 2.7 kg/cm^2. After cooling to 60°C, add one bottle of Vitox supplement SR0090A (Oxoid) and antibiotics as required.

2. Luria Bertani (LB) medium for culture of *Escherichia coli*: stir 20 g of tryptone (Merck), 5 g of yeast extract (Merck), 5 g of NaCl, and 20 mg of thymine (Sigma-Aldrich) in 1 L of distilled H_2O. Adjust pH to 7.0 with NaOH and autoclave for 15 min at 120°C and 2.7 kg/cm^2. For solid medium, add 1.5% (w/v) agar (Oxoid).
3. Antibiotic stock solutions: 80 mg/mL kanamycin (Kan) (Sigma-Aldrich) in H_2O, filter sterilized through a 0.2-μM filter; 375 mg/mL streptomycin (Sm) (Sigma-Aldrich) in H_2O, filter sterilized; 100 mg/mL ampicillin (Amp) (Sigma-Aldrich) in H_2O, filter sterilized; 10 mg/mL chloramphenicol (Cam) (Sigma-Aldrich) in 96% (v/v) ethanol; 7 mg/mL erythromycin (Ery) (Sigma-Aldrich) in 96% (v/v) ethanol. Store all stocks in aliquots at −20°C.
4. Tryptic Soy Broth (TSB) for liquid culture of *N. meningitidis*: dissolve 15 g of TSB in 500 mL of distilled H_2O (this will become a clear solution) and autoclave for 15 min at 120°C and 2.7 kg/cm^2. To prepare −80°C stocks, add 17.5% (v/v) glycerol to TSB and filter sterilize.
5. Humidified CO_2 incubator or candle jar (see Note 1).

2.2. Agarose Gel Electrophoresis

1. Microwave.
2. Horizontal gel electrophoresis system (VWR).
3. Power supply, e.g., PowerPac 300 (Biorad).
4. Gel imaging system with an UV lamp (λ254 nm), e.g., Gel Doc (Biorad) or equivalent.
5. Solution of 0.5 M diaminoethane tetraacetic acid (EDTA): add 18.6 g EDTA (Sigma-Aldrich) to 100 mL of H_2O and adjust the pH to 8.0 using NaOH to dissolve the EDTA.
6. Tris-Borate-EDTA (TBE) solution, 5× stock: add 27.5 g boric acid and 54 g Tris base to 800 mL of H_2O and 20 mL of 0.5 M EDTA pH 8.0. After dissolving all the compounds, adjust the volume to 1 L with H_2O.
7. Agarose D1 LEEO (Hispanagar).
8. 6× loading buffer (Fermentas).
9. DNA Ladder (Fermentas).
10. Stock solution of ethidium bromide: dissolve 10 mg of ethidium bromide (Sigma-Aldrich) in 1 mL of H_2O. Dilute 1/10,000 in 0.5× TBE buffer to stain the gels. Caution: wear gloves and protective clothing as ethidium bromide is a known mutagen and toxic.

2.3. Isolation of DNA

1. Table top centrifuge, e.g., Eppendorf model 5424 or equivalent.
2. Closed glass Pasteur pipette: melt the pipette tip for a few seconds in the blue flame of a Bunsen burner.

3. Sterile cotton swabs.
4. Insulin syringe (Becton Dickinson).
5. Tris-EDTA (TE) solution: add 10 mL of 1 M Tris–HCl buffer, pH 7.5, and 2 mL of 0.5 M EDTA to 988 mL of H_2O.
6. 10% (w/v) sodium dodecyl sulfate (SDS) in H_2O.
7. 20 mg/mL of proteinase K (Sigma-Aldrich) in H_2O.
8. 10 mg/mL of RNAse A (Sigma-Aldrich) in H_2O.
9. Lysozyme (Sigma-Aldrich).
10. Phenol (100%).
11. Chloroform (100%).
12. Ethanol (100%).
13. Plasmid isolation kit, e.g., E.Z.N.A. plasmid mini kit (Omega Biotek).
14. Gel and PCR clean-up kit, e.g., Wizard SV gel and PCR clean-up kit (Promega).
15. Speed Vac concentrator (Savant Instruments Inc.).

2.4. Polymerase Chain Reaction (PCR) (see Note 2)

1. Thermocycler, e.g., Biometra T3000 (Westburg), or equivalent.
2. PCR tubes: 0.5 mL micro tubes (Sarstedt) or 0.2 mL PCR tubes (VWR).
3. Taq DNA polymerase with buffer (Fermentas).
4. Expand High Fidelity PCR system (Roche).
5. Deoxynucleotide (dNTPs) mixture, 2 mM stock solution (Fermentas).
6. Primers: 10 μM stock solutions.
7. Bacterial DNA template.

2.5. Creation of Knock-Out Constructs

1. Speed Vac concentrator (Savant Instruments Inc.).
2. Water bath at 42°C.
3. Sterile toothpicks: sterilize household toothpicks by autoclaving (15 min at 120°C and 2.7 kg/cm^2).
4. Restriction enzymes: SalI, EcoRI, HindIII with corresponding buffers (Fermentas).
5. T4 DNA ligase and 10× ligase buffer (Fermentas).
6. High Fidelity Expand proof-reading DNA polymerase supplied with the Expand High Fidelity PCR system (Roche).
7. Cloning vector pUC18 (New England Biolabs) or pCRII-TOPO (Invitrogen).
8. Chemically competent *E. coli* TOP10F' (Invitrogen).

9. Primers, ordered from a commercial oligonucleotide synthesis company (Biolegio). The M13For (−20) (GTAAAACGACGGCCAGT) and -M13Rev (−20) (GATAACAATTTCACACAGG) anneal on either side of the inserts in pUC18 and pCRII-TOPO.
10. Gel and PCR clean-up kit, e.g., Wizard SV gel and PCR clean-up kit (Promega).

2.6. Transformation of Neisseria meningitidis

1. TSB culture medium, GC-agar plates with and without the appropriate antibiotics.
2. Leica S6D Stereozoom binocular Microscope (Leica Microsystems GmbH).
3. Pharmacia Novaspec II Spectrophotometer (Amersham).
4. Sterile glass tubes, 0.85 × 15.5 cm (Beldico).
5. TSB supplemented with 10 mM $MgCl_2$.
6. Polystyrene flasks 25 cm^2 with canted neck (Corning).
7. Polypropylene Cellstar tubes 50 mL (Greiner).
8. Transformation solution (TS): supplement LB with 10% (w/v) polyethylene glycol (PEG) 8000 (Fluka), 5% (v/v) dimethyl sulfoxide (DMSO) (Sigma-Aldrich), 50 mM $MgCl_2$, and filter sterilize using a 0.2-μM filter. Store the sterile solution at 4°C (16).

2.7. Construction of Markerless Mutants

1. GC-agar plates, with and without Sm or Ery.
2. Knock-out constructs.

2.8. Verification of Mutants

1. Chromosomal DNA of *N. meningitidis.*
2. Primers (gene specific).

2.9. Complementation of Mutants

1. Cloning plasmid pCRII-TOPO (Invitrogen).
2. Complementation plasmid pEN11 (17).
3. Restriction enzymes NdeI and AatII with corresponding buffers (Fermentas).
4. Isopropyl β-D-1-thiogalactopyranoside solution (Sigma-Aldrich), 1 M in H_2O, filter sterilized through a 0.2-μM filter.

3. Methods

3.1. Bacterial Culture

1. Grow *N. meningitidis* on GC-agar plates at 37°C in a humidified incubator with 5% (v/v) CO_2 or in a candle jar. For mutant selection, add the following antibiotic concentrations to the plates: 80–100 μg/mL Kan, 7 μg/mL Ery, 5–10 μg/mL Cam, 750–1,500 μg/mL Sm. For growth of meningococci in liquid culture, inoculate TSB medium with bacteria at a starting optical density at λ550 nm (OD_{550}) of 0.1.

2. Preparation of –80°C *N. meningitidis* stocks: swab bacteria from an overnight plate into 1 mL of TSB containing 17.5% (v/v) glycerol and freeze at –80°C. To culture from this stock, scrap some bacteria from the frozen surface with a loop onto a GC-agar plate.
3. Grow *E. coli* on LB-agar plates at 37°C in a dry incubator. For *E. coli* mutant selection, plates should contain 50 μg/mL Kan, 200 μg/mL Ery, 25 μg/mL Cam, or 100 μg/mL Amp.
4. Preparation of –80°C *E. coli* stocks: swab bacteria from an overnight plate into 1 mL of LB containing 25% (v/v) glycerol and freeze at –80°C. To culture from this stock, scrap some bacteria from the frozen surface with a loop onto an LB-agar plate.

3.2. Agarose Gel Electrophoresis

1. Add 1.5 g agarose D1 LEEO to 100 mL of 0.5× TBE solution and boil the suspension in a microwave oven until the solution is completely clear.
2. Add 50 mL of 0.5× TBE solution and pour the gel into a horizontal gel electrophoresis system tray and let it solidify at room temperature, which usually takes 30 min. The gel should be 6–8 mm in thickness.
3. Immerse the agarose gel fully in 0.5× TBE solution in the horizontal gel electrophoresis system.
4. Prepare samples by adding an amount of 6× loading buffer that is one-fifth of the sample volume.
5. Load the samples into the gel and run it at 100 V for 1 h.
6. Transfer the gel to a solution of 1 μg/mL ethidium bromide in 0.5× TBE buffer.
7. Stain for 15–30 min and visualize the DNA with UV light (λ254 nm) using a gel imaging system.

3.3. Isolation of DNA

3.3.1. Isolation of Bacterial Chromosomal DNA (see Note 3)

1. Swab bacteria, using a sterile cotton swab, from half a plate of an overnight grown culture into 1.5 mL of TSB for meningococci or 1.5 mL of LB for *E coli*. Spin for 5 min at 6,000 ×*g* and suspend the pellet in 200 μL of TE buffer.
2. Add 22 μL of 10% (w/v) SDS and a few flakes of lysozyme and incubate for 1 h at 37°C in an incubator.
3. Next, add 33 μL of 20 mg/mL proteinase K and incubate for 1 h at 37°C.
4. Shear the DNA by pulling the suspension through an insulin syringe several times. The material should become less viscous.
5. Add 20 μL of 10 mg/mL RNase A and incubate for 1 h at 37°C.
6. Add 400 μL of phenol and mix by inverting the tube gently.
7. Centrifuge for 5 min at 20,000 ×*g*.

8. Take the upper phase into a new tube and add 400 μL of phenol and 400 μL of chloroform.
9. Mix again by gentle inversion and repeat step 7.
10. Take the upper phase and add 1 mL of ice-cold ethanol (100%). Mix by gentle inversion.
11. Take the cloud of DNA using a closed sterile Pasteur pipette into a new tube.
12. Add 1.5 mL of 70% (v/v) ethanol and repeat step 7. Remove the supernatant and let the pellet of DNA dry.
13. Dissolve the DNA in 100 μL of H_2O.

3.3.2. Plasmid Isolation from E. coli or N. meningitidis

1. Harvest bacteria, using a sterile cotton swab, from half a plate of an overnight grown culture into 1.5 mL of TSB for meningococci or 1.5 mL of LB for *E coli* in an Eppendorf tube (see Note 4).
2. Centrifuge the suspension for 5 min at 6,000 × *g*.
3. Use a commercially available mini-prep plasmid isolation kit to obtain plasmids in a volume of 50–100 μL, following the manufacturer's instructions.
4. If necessary, plasmids can be concentrated five times in approximately 30 min in a Speed Vac device.

3.3.3. Purification of DNA Fragments

1. Cut PCR products or restriction fragments from the agarose gels using a razor blade while illuminating the gel with UV light (see Note 5).
2. Use a commercially available PCR clean up/gel purification kit to obtain fragments in a volume of 30–50 μL, following the manufacturer's instructions.

3.4. Polymerase Chain Reaction (PCR)

1. Design primers (see Note 6). Download the sequenced genome of the *N. meningitidis* strain to be mutated into a bioinformatics software program, such as CloneManager. If this sequence is not available, use the genome of a closely related strain. Design primers using the primer design feature of the program, or design them manually using the following criteria adapted from Innis and Gelfand (18):
   (a) Preferred length: 17–28 bases.
   (b) G+C content should be between 50 and 60%.
   (c) The 3′ ends should be G or C, or preferably CG or GC. The more stable annealing of G and C bases increases the efficiency of priming.
   (d) Melting temperature (Tm) should be between 55 and 80°C. The Tm of the primers can be estimated by calculating 2°C for each A or T and 4°C for each G or C.

(e) The 3′ ends of a primer couple should not be complementary as this could cause primer dimers, which will be synthesized preferentially to any other product.

(f) The primer should not be able to form secondary structures as hairpins (this can be assessed using the bioinformatics software).

(g) Runs of three or more C or G residues at the 3′ ends should be avoided as these may promote mispriming at G or C-rich sequences.

2. Add the following into the PCR tubes in this order: 16.8 μL of H_2O; 2.5 μL of 10× PCR buffer supplied with the polymerase; 2.5 μL of a 2-mM solution of dNTPs; 1 μL of each primer (stock solution of 10 μM); 1 μL of template DNA and last of all, 0.2 μL of DNA polymerase.

3. Perform the following PCR steps in the thermocycler:

(a) DNA melting: 5 min at 95°C.

(b) DNA melting: 30 s at 95°C.

(c) Primer annealing: 30 s at a temperature 5°C below the Tm of the primer with the lowest Tm.

(d) Extension by the polymerase: the time depends on the size of the expected product. Use 1 min per 1,000 bp. Use the extension temperature recommended by the manufacturer of the polymerase, which is usually 72°C.

(e) Finalization of products: 5 min at 72°C.

(f) Repeat steps b–d for 29 cycles.

(g) Store the samples at 4°C until further analysis.

3.5. Creation of Knock-Out Constructs

Gene inactivation is accomplished by replacing all or most of the gene by an antibiotic resistance cassette (Fig. 2). To that end, two flanking regions of the gene to be inactivated that will direct homologous recombination are cloned. The size of these flanks should be at least 100 bp, but preferably longer flanks (up to 500 bp or longer) should be made to enable more efficient recombination. The flanking sequences may contain part of the targeted gene, but care should be taken that the interruption is at the 5′ end of the gene, as insertions at the 3′ end of the gene may yield truncated, but functional proteins.

1. Design primers, and take the following into account: for the specific purpose of cloning the PCR products into pUC18 vector and for the addition of an antibiotic-resistance cassette, the primers should contain at their 5′ends:

(a) Restriction sites for cloning into pUC18 for the Up-for and Down-rev primer (Fig. 2).

(b) SalI sites (CTGGAC) for the Up-rev and Down-for primers (Fig. 2) (see Note 7).

1. Amplification of up- and downstream flanks of gene B

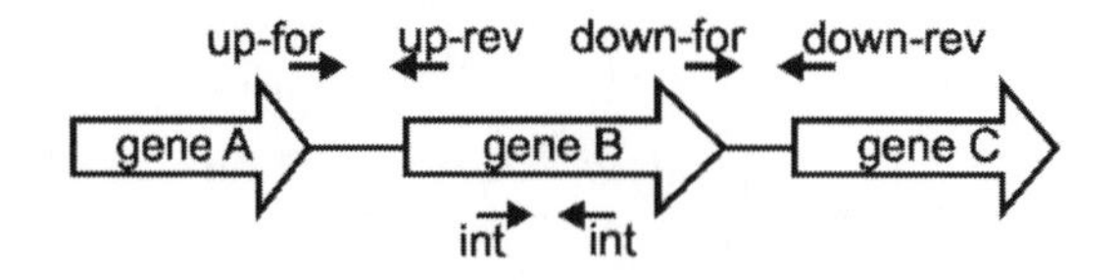

2. Flanks of gene B cloned into pUC vector, separated by a SalI restriction site

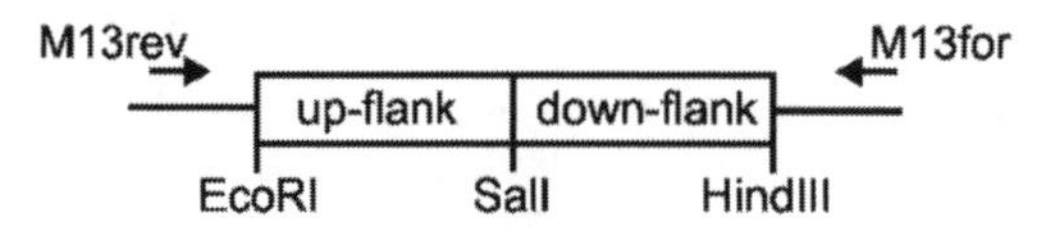

3. Antibiotic resistance cassette cloned in between flanks

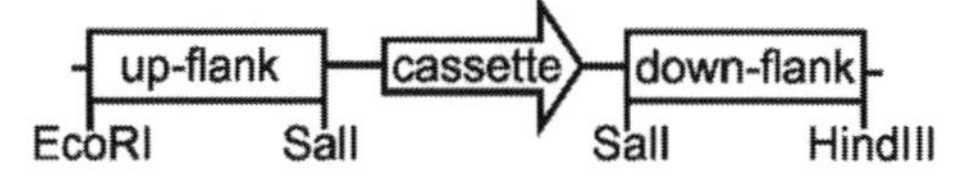

4. Removal of gene B after homologous recombination

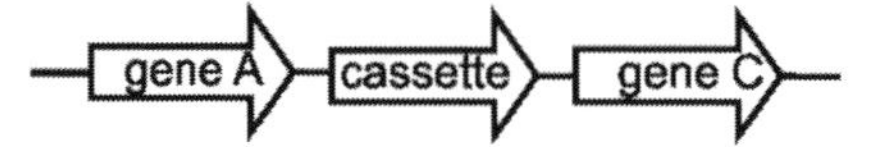

Fig. 2. Schematic representation of the described strategy to create knock-out mutants.

(c) 4–6 bases (e.g., ATCGCG) overhang to facilitate digestion of the PCR products.

2. Perform PCR using 1 μL of a 1/100 dilution of chromosomal DNA as template and a proof-reading DNA polymerase, such as the High Fidelity Expand.

3. Purify the PCR product using a commercial kit and agarose gel electrophoresis as described in Subheading 3.2 and cut 12 μL of the product with EcoRI+SalI enzyme mix or HindIII+SalI enzyme mix (0.2 U each) in a total volume of 15 μL in the buffer recommended by the supplier of the enzymes (see Note 8).

   Also, cut 8 μL of pUC18 vector with EcoRI and HindIII enzymes.

4. Cut the restricted pUC18 vector from the agarose gel, add the restricted PCR products to the gel slice, and purify all together. When using the Promega gel purification kit, the final volume of the purified mixture will be 50 μL. Bring the volume down to 17 μL using a Speed Vac device.

5. To the 17 μL of purified mixture add 2 μL of 10× ligation buffer plus 1 μL of T4 ligase. Incubate overnight at 14°C.

6. Transform the ligation mixture into *E. coli* TOP10F' according to the instructions of the manufacturer and plate the cells onto LB-plates containing 100 μg/mL Amp.

7. Test colonies by colony PCR using M13For and M13Rev primers. Set up a master PCR mix by multiplying the volumes of the PCR mix ingredients as described in step 2 of Subheading 3.4 by the number of colonies to be tested. Add H_2O instead of template and use the Taq polymerase. Deliver the master mix as 25 μL volumes into PCR tubes. Touch each colony with a toothpick, swirl the tip of the toothpick into the PCR tube, and subsequently onto a clean LB-agar plate containing 100 μg/mL Amp. Perform PCR with 58°C as the annealing temperature.
8. Check the size of the PCR products by agarose gel electrophoresis as described in Subheading 3.2.
9. Purify the plasmids from promising clones and determine the sequence of the insert using the services of a commercial company.
10. Cut 10 μL of a vector containing an appropriate antibiotic-resistance cassette (Fig. 3) (available from our lab upon request) and 6 μL of a verified pUC18 with flanks with SalI as described in step 3 of this section. Follow the purification, ligation, and transformation protocol as described in steps 4–6 of this section. Determine the orientation of the cassette by PCR using an internal primer in the cassette and a primer in one of the flanks (see Note 9).
11. Prepare approximately 1–3 μg of linear DNA containing the knock-out construct by restriction from pUC with the EcoRI and HindIII enzyme mix as described in step 3 of this section. The amount of DNA can be estimated by agarose gel electrophoresis by comparison to the DNA ladder marker, which

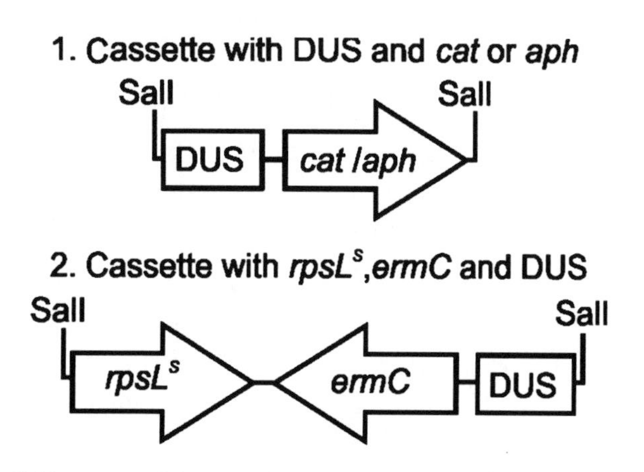

Fig. 3. Available cassettes for use in creating knock-out constructs. The cassettes contain DNA uptake sequences (DUS), a gene conferring Cam resistance: chloramphenicol acetyl transferase (*cat*); Kan resistance: aminoglucoside phosphotransferase (*aph*); or Ery resistance: rRNA adenine N-6-methyltransferase (*erm* C) and are flanked on the 5′ and 3′ ends by SalI restriction sites.

contains set amounts of DNA. Purify the DNA in H_2O, concentrate to 10 μL with a Speed Vac, and add 2 μL of a 30-mM $MgCl_2$ solution. This will be referred to as the knock-out DNA.

3.6. Transformation of *N. meningitidis*

3.6.1. Selection and Maintenance of Piliated (Pil^+) *N. meningitidis*

To select piliated (Pil^+) organisms, grow meningococci in TSB at 30°C without shaking for 16 h or more in sterile glass tubes. Collect piliated organisms from the air–water interface (see Note 10). Assess the piliation status by growing the bacteria on GC-agar plates at 30°C. Use a binocular microscope to judge their colony morphology: piliated colonies form small domed colonies with well-defined edges due to their auto-agglutinability while non-piliated organisms form flat, spreading colonies (19). *Neisseriae* are known to become nonpiliated upon repeated or prolonged (i.e., >48 h) subculture on agar plates (20). For highest transformation efficiencies, take bacteria from a −80°C stock that is known to be piliated and grow these only once on GC-agar for no longer than 18 h.

3.6.2. Transformation of Piliated (Pil^+) *N. meningitidis* on Agar Medium

1. Grow an overnight (~16 h) culture of meningococci on a GC-agar plate.
2. Draw a square of 1 × 1 cm on the back of a GC-agar plate using a marker and streak a small amount of the overnight grown bacteria onto this square: the streak should be barely visible.
3. Drop knock-out DNA onto the bacteria in this square and incubate the plate at 37°C with 5% (v/v) CO_2 for 6–8 h.
4. After this period, bacterial growth should be seen within the square. Plate all of this growth onto a fresh GC-agar plate containing the appropriate antibiotics.
5. Pick single colonies over the next 24–48 h and streak them onto fresh selection GC-agar plates (see Note 11).

3.6.3. Transformation of Piliated (Pil^+) *N. meningitidis* in Liquid Medium, Adapted from the Method of Gunn and Stein (10)

1. Suspend bacteria grown overnight on a GC-agar plate to an OD_{550} of 0.0007 (measured using a spectrophotometer) in TSB supplemented with 10 mM $MgCl_2$, prewarmed to 37°C. This corresponds to approximately 10^5 bacteria/mL.
2. Mix 50 μL of the suspension with 12 μL of knock-out DNA.
3. Incubate the mixture for 30 min at 37°C.
4. Plate out the mixture onto a nonselective GC-agar plate in a circle with a diameter of approximately 4 cm.
5. After overnight incubation, suspend the bacteria in TSB to an OD_{550} of 0.0007 and plate out 100 μL of the suspension onto GC-agar plates supplemented with an appropriate antibiotic.

3.6.4. Transformation of Nonpiliated (Pil⁻) N. meningitidis, Adapted from the Method of Bogdan et al. (16) (see Note 12)

1. Grow 5 mL of bacteria for a few hours in TSB in an upright standing Corning flask with shaking at 120 rpm at 37°C (start OD_{550} of 0.1). The end OD does not matter as long as the bacteria have grown.
2. Harvest the bacteria by centrifugation (10 min, 3,000 × *g*) and suspend them in 0.5 mL of ice-cold TS.
3. Transfer 190 μL of the suspension into a 50-mL plastic Greiner tube on ice.
4. Add 12 μL of knock-out DNA and leave for 15 min on ice.
5. Add 1.5 mL of TSB and shake at 175 rpm at 37°C for 2 h.
6. Plate the suspension onto GC-agar plates supplemented with the appropriate antibiotic.

3.7. Construction of Markerless Mutants

3.7.1. Selection of Streptomycin-Resistant Strains

1. Grow an overnight culture of *N. meningitidis* on a GC-agar plate.
2. Mark a square of 1 × 1 cm on a fresh GC-agar plate and streak a small and barely visible amount of the overnight growth onto this square.
3. After 5 h, streak all of the bacteria onto a GC-agar plate supplemented with 1.5 mg/mL Sm (GC-Sm) and grow overnight (see Note 13).
4. Grow the resultant colonies overnight on a fresh GC-Sm plate.
5. Analyze the sequence of the *rpsL* gene (see Note 14). Usually, the $rpsL^R$ allele encodes a lysine in codon 43, whereas $rpsL^S$ contains an arginine codon at this position.

3.7.2. Construction of Markerless Mutants

To construct markerless *N. meningitidis* mutants, carry out a first transformation step as described in Subheading 3.6.2 with knock-out DNA as shown in Fig. 1, Step 1. Select the transformants on GC-agar plates supplemented with Ery. For the second transformation step (Fig. 1, Step 2) with selection for Sm resistance, follow the protocol as described in Subheading 3.6.3, as we found this to work more efficiently for this step.

3.8. Verification of Mutants

When constructing knock-out mutants, it is very important to verify the absence of the inactivated gene and not just test for the presence of the mutated allele. *N. meningitidis* can exist as stable merodiploids, meaning they contain two different copies of one gene (21). We often obtained such merodiploids, next to real mutants, when we inactivated "semiessential" genes, for instance genes involved in the transport of lipopolysaccharide to the cell surface (17).

1. Prepare genomic DNA from the mutant and parent strains by boiling in water as described in Note 3. Set up a series of PCRs using this genomic DNA as template with different primer pairs using the guidelines described in steps 1–3 of Subheading 3.4:
   (a) Primer pair up-for and down-rev (Fig. 2). When the size of the antibiotic-resistance cassette is different from the size of the deleted sequence, differently sized PCR products should be detected. Preferably, the mutant product should be larger than the parent product to ensure that a wild-type PCR product does not go undetected, as the larger product is often amplified less efficiently than the smaller one. Merodiploids usually yield two PCR products in this test.
   (b) Internal primer pair (–int) that should no longer anneal in the mutant (Fig. 2) or a combination of one internal primer and the up-for or down-rev primer. A correct mutant should not yield any PCR product (see Note 15).

3.9. Complementation of Mutants

Complementation entails the reintroduction of the native gene into a mutant lacking this gene in order to demonstrate that the mutant phenotype is truly due to the absence of the gene of interest. This can be done by reintroducing the gene on a different location on the chromosome. The advantage of this approach is that only one copy is present, which is similar to the wild-type situation, but a disadvantage is that transcription of unrelated parts of the chromosome at the site of insertion may be affected. Alternatively, the gene may be reintroduced by inserting a copy on a plasmid that is able to replicate in the bacterium. Only a few plasmids have been described that replicate both in *E. coli* and in *Neisseria*. One example is the pFP10 plasmid (22) that contains the origin of replication of a naturally occurring gonococcal plasmid. We introduced a *lac*-derived promoter into this plasmid, which contains two promoter and operator sequences to provide tight promoter control, plus a LacI encoding gene. Next, we introduced the ribosomal binding site of the highly expressed *omp85* gene (23), and NdeI and AatII sites allowing for easy subcloning of any ORF into this construct. This plasmid is called pEN11 (17).

1. Amplify the gene of interest by PCR with a forward primer containing an NdeI site, such that the ATG present in the recognition site (CATATG) represents the start codon of the gene, and a reverse primer containing an AatII site after the stop codon. If a transcriptional terminator associated with this gene is present on the genome, one may choose to clone it along with the gene, as this may increase mRNA stability and thereby expression levels.

2. Cut and ligate the PCR product into pEN11 using NdeI/AatII restriction as described in step 3–5 of Subheading 3.5, and transform into *E. coli*. Verify the plasmids by commercial sequencing.
3. Introduce the plasmid preparation into *N. meningitidis* as described in Subheading 3.6.2 and select transformants on GC-agar plates containing 10 μg/mL Cam.
4. Correct transformants can be verified by testing for IPTG-inducible expression of the protein (see Note 16). The plasmid can be isolated from *N. meningitidis* and then sequenced commercially (see Note 17).

4. Notes

1. *N. meningitidis* grows better with elevated levels of CO_2. When standard CO_2 incubators are not available, candle jars can be used.
2. We only noted the suppliers we use for our molecular biology materials. Any other company will do, as well as other general molecular biology protocols, e.g., those described in (24).
3. A simple, quick, alternative is to suspend a few bacterial colonies in 50 μL of H_2O, boil these for 5 min, spin the suspension for 5 min at 20,000 × *g* in an Eppendorf centrifuge and use the supernatant as the source for DNA. A disadvantage of this method is that the DNA is not very stable, so it should not be stored for future use.
4. There is a limit to the amount of cells that can be used for mini-prep isolation. If a higher amount of plasmid material is desirable, use a midi or maxiprep kit.
5. Wear glasses and gloves, and work quickly, to avoid exposure of eye lenses and skin to UV light. Also, work quickly to avoid damage to the DNA as much as possible.
6. Neisserial genomes contain lots of repetitive DNA. Take this into account when designing primers: blast the 3′ ends of the primers on the whole genome to verify that they anneal only in the intended region.
7. Any other restriction site can be used, as long as it is not present in the sequence of the flanks. However, we found that the SalI site (GTCGAC) rarely occurs in meningococcal genome sequences and we have constructed a number of antibiotic cassettes flanked by SalI sites containing DUS sequences (Fig. 3).

8. Sometimes, restriction of PCR products does not work very efficiently. Alternatively, the PCR product can be cloned in a suitable vector first, such as pCRII-TOPO (Invitrogen) and then the fragment can be cut out of the resulting vector. This usually works better, and the restriction can be monitored using agarose gel electrophoresis.
9. It is best to use a knock-out construct with the antibiotic-resistance conferring gene having the same transcriptional direction as the gene to be replaced, to avoid potential polar effects on expression of downstream genes.
10. Piliated strains of *N. meningitidis* and other Gram-negative species form a pellicle when cultured in static broth media and such conditions can be used to selectively enrich for Pil^+ organisms.
11. It is important to streak the colonies from the transformation plate onto a new one before testing them by PCR. Colonies on the transformation plate may still contain nonincorporated DNA added for the transformation. Also, the growth on a new selection plate demonstrates that the resistance phenotype is stable.
12. This protocol was developed for Pil^- strains. It can be used when Pil^+ organisms cannot be obtained, or when a mutant, affected in piliation, needs to be transformed.
13. It is important to stick to the 5-h incubation period. We have tried various conditions to obtain these mutants such as plating an overnight GC-agar plate culture. This yielded almost no colonies and those that grew on GC-Sm were not mutated in the *rpsL* gene.
14. The *rpsL* gene can be amplified using primers annealing at the 5′ and 3′ ends of the gene and sequenced directly as a PCR product. Alternatively, the PCR product can be cloned first into pCRIITOPO followed by sequencing with M13For and M13Rev primers, as sequencing of plasmids often works better than sequencing of PCR products.
15. If an antibody or antiserum against the protein of interest is available, the absence of the protein can be verified by Western blotting.
16. Grow the complemented strain in TSB in the presence and absence of 1 mM IPTG for 4–6 h. Collect the cells and test for presence of the protein of interest by SDS-PAGE and/or Western blotting.
17. When parts of the plasmid-encoded gene are still present on the chromosome, recombination between plasmid and chromosome may occur. We noticed this phenomenon several times, but it appears to happen only during transformation. So once correct strains are obtained, they remain stable.

Acknowledgments

We would like to thank past and present members of the group of Jan Tommassen at the Department of Molecular Microbiology for help and stimulating discussions. This work was supported by the Research Council for Earth and Life Sciences with financial aid from the Netherlands Organization for Scientific Research (NWO) and by GlaxoSmithKline Biologicals, Rixensart, Belgium.

References

1. Jyssum K, Lie S (1965) Genetic factors determining competence in transformation of *Neisseria meningitidis*. 1. A permanent loss of competence. Acta Pathol Microbiol Scand 63:306–316
2. Ambur OH, Frye SA, Tønjum T (2007) New functional identity for the DNA uptake sequence in transformation and its presence in transcriptional terminators. J Bacteriol 189:2077–2085
3. Lorenz MG, Wackernagel W (1994) Bacterial gene transfer by natural genetic transformation in the environment. Microbiol Rev 58:563–602
4. Koomey M (1998) Competence for natural transformation in *Neisseria gonorrhoeae*: a model system for studies of horizontal gene transfer. APMIS Suppl 84:56–61
5. Kline KA, Sechman EV, Skaar EP et al (2003) Recombination, repair and replication in the pathogenic *Neisseriae*: the 3 R's of molecular genetics of two human-specific bacterial pathogens. Mol Microbiol 50:3–13
6. Elkins C, Thomas CE, Seifert HS et al (1991) Species-specific uptake of DNA by gonococci is mediated by a 10-base-pair sequence. J Bacteriol 173:3911–3913
7. Biswas GD, Burnstein KL, Sparling PF (1986) Linearization of donor DNA during plasmid transformation in *Neisseria gonorrhoeae*. J Bacteriol 168:756–761
8. Chaussee MS, Hill SA (1998) Formation of single-stranded DNA during DNA transformation of *Neisseria gonorrhoeae*. J Bacteriol 180:5117–5122
9. Alexander HL, Richardson AR, Stojiljkovic I (2004) Natural transformation and phase variation modulation in *Neisseria meningitidis*. Mol Microbiol 52:771–783
10. Gunn JS, Stein DC (1996) Use of a non-selective transformation technique to construct a multiply restriction/modification-deficient mutant of *Neisseria gonorrhoeae*. Mol Gen Genet 251:509–517
11. Johnston DM, Cannon JG (1999) Construction of mutant strains of *Neisseria gonorrhoeae* lacking new antibiotic resistance markers using a two gene cassette with positive and negative selection. Gene 236:179–184
12. Lederberg J (1951) Streptomycin resistance; a genetically recessive mutation. J Bacteriol 61:549–550
13. Breckenridge L, Gorini L (1970) Genetic analysis of streptomycin resistance in *Escherichia coli*. Genetics 65:9–25
14. Meyer TF, van Putten JP (1989) Genetic mechanisms and biological implications of phase variation in pathogenic *Neisseriae*. Clin Microbiol Rev Suppl2:S139–S145
15. Morales VM, Bäckman A, Bagdasarian M (1991) A series of wide-host-range low-copy-number vectors that allow direct screening for recombinants. Gene 97:39–47
16. Bogdan JA, Minetti CA, Blake MS (2002) A one-step method for genetic transformation of non-piliated *Neisseria meningitidis*. J Microbiol Methods 49:97–101
17. Bos MP, Tefsen B, Geurtsen J et al (2004) Identification of an outer membrane protein required for the transport of lipopolysaccharide to the bacterial cell surface. Proc Natl Acad Sci USA 101:9417–9422
18. Innis MA, Gelfand DH (1990) Optimisation of PCRs. In: PCR protocols: a guide to methods and applications. Ed: Innis MA, Gelfand DH, Sninsky JJ, White TJ. Academic, San Diego, Ca. p1–13

19. Blake MS, MacDonald CM, Klugman KP (1989) Colony morphology of piliated *Neisseria meningitidis*. J Exp Med 170:1727–1736
20. McGee ZA, Street CH, Chappell CL et al (1979) Pili of *Neisseria meningitidis*: effect of media on maintenance of piliation, characteristics of pili, and colonial morphology. Infect Immun 24:194–201
21. Tobiason DM, Seifert HS (2010) Genomic content of *Neisseria* species. J. Bacteriol. 192:2160–2168
22. Pagotto FJ, Salimnia H, Totten PA et al (2000) Stable shuttle vectors for *Neisseria gonorrhoeae*, *Haemophilus spp*. and other bacteria based on a single origin of replication. Gene 244:13–19
23. Voulhoux R, Bos MP, Geurtsen J et al (2003) Role of a highly conserved bacterial protein in outer membrane protein assembly. Science 299:262–265
24. Maniatis T, Fritsch, EF, Sambrook J (1982) *Molecular Cloning: A Laboratory Manual*. Cold Spring Harbor Laboratory, Cold Spring Harbor.

Chapter 5

Identification and Functional Characterization of sRNAs in *Neisseria meningitidis*

Yvonne Pannekoek and Arie van der Ende

Abstract

A riboregulated network, in which small RNAs (sRNAs) regulate the stability and thus translation of transcripts (mRNA), has only recently been discovered in prokaryotes. Yet, during the last 5 years, hundreds of sRNAs have been identified in various bacterial species by using a wide variety of both computational and experimental approaches. The majority of the sRNAs interact with the 5′-untranslated region (UTR) of target mRNAs, thereby influencing the stability of target mRNAs, or by either suppressing or upregulating the ribosome entry to the mRNAs influencing translation. Here, we describe experimental approaches successfully used in our laboratory to identify and functionally characterize sRNAs in vivo in our model micro-organism *Neisseria meningitidis*.

Key words: sRNA, *N. meningitidis*, Riboregulation, pEN11, Scan for matches, RNomics, IGR

1. Introduction

In the last decade, novel small RNAs (sRNAs) have been identified in a wide range of bacterial species and novel technologies as well as modification of standard techniques are being used to identify and functionally characterize sRNAs (1). Of the majority of the sRNAs identified to date, interacting mRNA targets and function are unknown. To identify and functionally characterize sRNAs it is important to realize that the synthesis of many (but not all!) of the sRNAs is controlled by the action of regulated promoters, which are often induced under stress conditions, thereby riboregulating the expression of sets of target genes in response to environmental signals, including iron homeostasis, quorum sensing, virulence, stationary stress and more (2, 3). Regulatory sRNAs, often encoded

Myron Christodoulides (ed.), *Neisseria meningitidis: Advanced Methods and Protocols*, Methods in Molecular Biology, vol. 799, DOI 10.1007/978-1-61779-346-2_5,

in intergenic regions (IGRs) of the bacterial genome are usually 50–300 nucleotides (nt) long followed by a Rho-independent terminator and a T stretch.

In this chapter we describe the use of a search pattern algorithm, Scan For Matches (3, 4), to identify sRNA candidates in the IGRs of *Neisseria meningitidis in silico*. This approach is directed by the universally conserved secondary structure of sRNAs and the fact that many sRNAs contain promoter signatures, for example, the recognition box of the ferric uptake regulation (Fur-box). Next, we describe how to verify expression of *in silico* predicted sRNAs in vivo by Northern blotting and, lastly, we describe how to create a meningococcal strain in which the sRNA can be overexpressed *in trans*. This construct can then be phenotypically analyzed in downstream experimental approaches to identify riboregulated mRNA targets.

2. Materials

Standard molecular biology equipment and molecular grade biochemicals and reagents are required. In addition, a bench top refrigerated microcentrifuge is needed. Unless stated otherwise, all solutions are prepared with distilled water (dH_2O).

2.1. In Silico Identification of Candidate sRNAs

1. ScanForMatches: http://blog.theseed.org/servers/2010/07/scan-for-matches.html
2. For the extraction of DNA sequences of IGRs of the strain of interest: http://cmr.jcvi.org/cgi-bin/CMR/shared/MakeFrontPages.cgi?page=intergenicregion&crumbs=genomes
3. Promoter prediction: http://www.fruitfly.org/seq_tools/promoter.html

2.2. Culture and Harvest of Meningococcal Cells

1. *N. meningitidis* strain H44/76 (see Note 1).
2. GC medium (Difco). Dissolve 26 g in 1 L of distilled water (dH_2O), autoclave to sterilize, cool to 55°C, supplement with Vitox (1% v/v, Oxoid). We will refer to this medium as GC broth. Prewarm the medium at 37°C before use.
3. Sterile Erlenmeyer flask (100 mL) stopped with a cotton plug.
4. Sterile cotton swabs.
5. Spectrophotometer (Pharmacia Biotech).

2.3. RNA Isolation

1. Nuclease-free filter tips for micropipettes.
2. Diethyl pyrocarbonate (DEPC)-treated H_2O (Fermentas).
3. RNase AWAY (Molecular Bioproducts).
4. Acidic phenol (pH 4) (Sigma), store at 4°C.

5. Acidic phenol:chloroform:isoamyl alcohol (50:49:1 v/v), store at 4°C.
6. Chloroform:isoamyl alcohol (49:1 v/v), store at 4°C.
7. Ice cold ethanol: absolute, 95 and 70%, store at −20°C.
8. 10% (w/v) SDS in DEPC H_2O, store at RT.
9. 5 M NaCl in DEPC H_2O, store at RT.
10. 0.5 M EDTA in DEPC H_2O pH 8.0.
11. 1 M Tris–HCl in DEPC H_2O pH 7.0.
12. 10 mM Tris–HCl-1 mM EDTA pH 7.0.
13. 15 mL PLG heavy tubes (5PRIME), prespun (2 min at 1,500 × *g*).
14. 0.2 μm filter (Millipore).
15. 10× stop solution: 95% ethanol, 5% acidic phenol.
16. Lysis buffer: 2% (w/v) SDS, 16 mM EDTA, 200 mM NaCl in DEPC H_2O pH 8.0. Filter sterilize and store at RT.
17. 50× TAE buffer pH 8.0: 242 g Tris–HCl in 500 mL dH_2O pH 8.0, add 57.1 mL glacial acid and 100 mL 0.5 M EDTA, adjust the solution to a final volume of 1 L with dH_2O, store at RT.
18. Ethidium bromide (Roche) 10 mg/mL in Milli-Q (see Note 2), store in dark at RT. (*Caution*: ethidium bromide is a mutagen; handle under appropriate guidelines).
19. 1% (w/v) agarose gel in 1× TAE containing 10 μg/mL ethidium bromide.
20. 10× DNA loading buffer (Fermentas), store at RT.
21. Quick-Load 100 bp DNA Ladder (New England Biolabs (NEB)).
22. Nanodrop spectrophotometer (Nanodrop).

2.4. Detection of sRNA Expression (see Note 3)

2.4.1. Radioactive Probe and Marker Labeling

1. Antisense oligonucleotide probe (30–50 nt, GC content ~50%) 5 pmol/μL in DEPC H_2O.
2. 100 bp Quick-Load 100 bp DNA ladder (NEB), 0.5 μg/μL.
3. (γ-^{32}P]ATP (Perkin Elmer) (see Note 4).
4. 10× Kinase buffer (Ambion).
5. T4 PNK (10 U/μL) (Ambion).
6. Illustra Microspin G-25 columns (GE Healthcare).

2.4.2. RNA PAGE/Urea Gel Electrophoresis

1. 0.5 M EDTA pH 8.0 in DEPC H_2O.
2. 10× Tris Boric Acid EDTA (TBE) buffer, pH 8.0: 108 g Tris base, 55 g boric acid, dissolve in 900 mL DEPC H_2O, add 40 mL of 0.5 M EDTA and adjust the solution to a final volume of 1 L DEPC H_2O, store at RT (see Note 5).

3. 40% (w/v) acrylamide (acryl:bisacryl 19:1; Bio Rad) stored in the dark at 4°C. (*Caution*: acrylamide is a neurotoxin when unpolymerized and care should be taken to avoid exposure).
4. N,N,N′,N′-Tetramethyl-ethylenediamine (TEMED) (Merck).
5. Ammonium persulfate (APS): 10% (w/v) in DEPC H_2O, immediately freeze in aliquots (200 μL) for single use and store at −20°C.
6. RNA loading dye without ethidium bromide (Sigma).
7. RNA samples, [γ-^{32}P]ATP end-labeled molecular weight 100 bp marker (see above).
8. Bucket filled with ice.

2.4.3. Northern Blotting

1. Biodyne B Nylon membrane.
2. 3 MM paper (Bio Rad).
3. 0.5× TBE in DEPC H_2O, store at 4°C.
4. Saran Wrap (Saran).
5. Hybridization curtain (Biometra).
6. ULTRAhyb-Oligo hybridization solution (Ambion), preheated at 68°C.
7. [γ-^{32}P]ATP end-labeled antisense oligonucleotide probe.
8. 10% (w/v) SDS in dH_2O.
9. 20× SSC buffer, pH 7.0: dissolve 70 g NaCl and 35 g sodium citrate in 2 L of dH_2O, store at RT.
10. Low stringency buffer: 2×SSC, 0.1% (w/v) SDS in dH_2O (use at RT).
11. 50 mL low stringency buffer prewarmed at 37°C.

2.5. Construction of a Shuttle Plasmid Expressing sRNA Gene Under Control of an Inducible Promoter

2.5.1. Isolation of Chromosomal DNA and Plasmid pEN11-pldA

1. PUREGENE DNA Isolation Kit (Gentra systems).
2. Luria Bertani (LB) medium: dissolve 10 g Bacto-Tryptone (Difco), 5 g Bacto-yeast extract (Difco) 10 g NaCl in 1 L of dH_2O. Autoclave to sterilize, cool to 55°C, add antibiotic(s) if desired.
3. Chloramphenicol (Cm) (stock solution 5 mg/mL, dissolved in absolute ethanol, stored at −20°C).
4. *E. coli* TOP10F’ transformed with pEN11-pldA.
5. Wizard SV 96 Plasmid DNA Purification System (Promega).

2.5.2. PCR Amplification of pEN11-pldA and the sRNA Gene

1. Shuttle vector pEN11-*pldA* (5).
2. Chromosomal DNA.
3. dNTPs: 10 mM stock solutions (Promega).
4. 10 μM stock solution of primers (high-performance liquid chromatography purified).

5. *Pfu* proofreading DNA polymerase (2 U/μL) (Promega).
6. 10× *Pfu* DNA polymerase buffer (Promega).
7. Thermal cycler (Biometra).
8. Thin-walled PCR tubes (Thermoscientific).
9. 50× TAE buffer, pH 8.0.
10. Ethidium bromide (Roche) 10 mg/mL.
11. 1% (w/v) agarose gel in 1× TAE containing 10 μg/mL ethidium bromide.
12. 10× DNA loading buffer (Fermentas), store at RT.
13. 1 kb DNA ladder (Fermentas).
14. Sterile razor blades or scalpels (see Note 6).
15. QIAEX II Gel Extraction Kit (Qiagen).

2.5.3. Ligation of pEN11 Fragment to sRNA Amplicon and Selective Amplification

1. T4 ligase (1 U/μL) (Roche).
2. 10× T4 ligase buffer (Roche).
3. All reagents for selective amplification are the same as specified in items 3–13 in Subheading 2.5.2 but using 10 μM stock solutions of gene-specific primers.

2.5.4. Digestion of pEN11-pldA and the pEN11-sRNA Fragment with MauBI and BspHI

1. Restriction enzymes MauBI (Fermentas) and BspHI (NEB).
2. 10× Digestion buffer, delivered with enzyme.
3. All other reagents needed are the same as specified in items 7–13 in Subheading 2.5.2.

2.5.5. Ligation of the Digested sRNA-pEN11 Fragment in the Digested pEN11 Fragment and Transformation of pEN11 + sRNA into E. Coli TOP10F'

1. T4 ligase (1 U/μL) (Roche).
2. 10× T4 ligase buffer (Roche).
3. Chemically competent *E. coli* TOP10F' cells (Invitrogen), stored in aliquots at −70°C.
4. SOC medium (Invitrogen).
5. LB medium.
6. Chloramphenicol (Cm) stock solution (5 mg/mL).
7. LB agar plates containing Cm (5 μg/mL).

2.5.6. Colony Screening of Transformants by PCR Using Primer 5 and Primer 6

1. dNTPs: 10 mM stock solutions (Promega).
2. 25 mM $MgCl_2$ (Promega).
3. 10 μM stock solution of gene-specific primers.
4. Taq polymerase (5 U/μL) (Promega).
5. 5× Go-Taq buffer (Promega).
6. LB agar plates containing Cm (5 μg/mL).
7. LB broth containing Cm (5 μg/mL).

2.5.7. Plasmid Isolation and Sequence Analysis

1. Wizard SV 96 Plasmid DNA Purification System (Promega).
2. BigDye Terminator Cycle Sequencing Ready Reaction kit v1.1 (Applied Biosystems).

2.5.8. Transformation of pen11-sRNA into Meningococci and Assessment of sRNA Gene Expression in Trans

1. 1 M $MgCl_2$ solution.
2. GC broth.
3. Cm (stock solution 5 mg/mL).
4. GC agar plates containing Cm (5 μg/mL).
5. 25 mL sterile cell culture flask (Falcon).
6. Isopropyl β-D-1-thiogalactopyranoside (IPTG) is dissolved at 100 mM in Milli-Q and stored in aliquots at −20°C.
7. For Northern blotting, use all the reagents specified in Subheading 2.4.3.

3. Methods

3.1. In Silico Identification of Candidate sRNAs

For the *in silico* identification of sRNAs located in IGRs, the algorithm Scan For Matches is used. A fasta file of the IGRs of meningococcal strain MC58 is scanned for the presence of the following DNA pattern: ANTGATAATNATTATCANT[2,0,0] 50…400 p1 = 6…12 3…9 ~ p1[1,0,0] TTTTT[1,0,0]. This pattern scans for a FUR consensus with a maximum of two mismatches, followed by a spacer of 50–400 nt, followed by a stem loop and a T stretch (Fig. 1a). An example of the output generated by Scan for Matches is produced in Fig. 1b. Promoter and transcriptional start can be predicted using a promoter prediction tool (e.g., http://www.fruitfly.org/seq_tools/promoter.html).

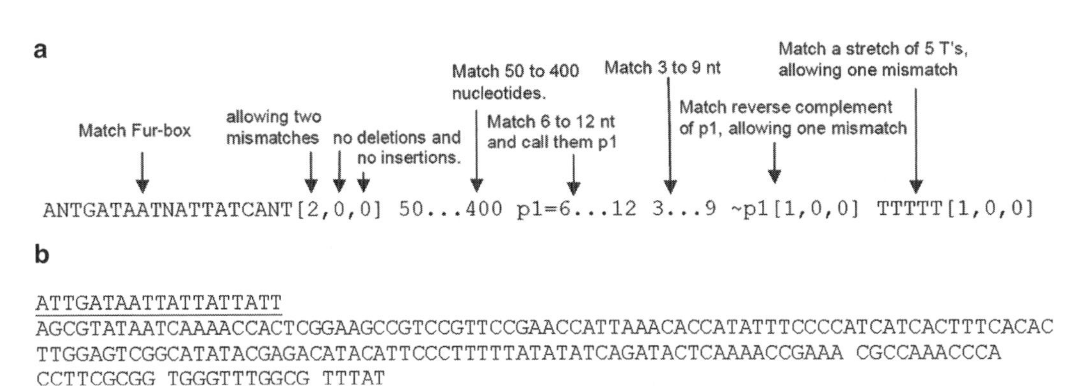

Fig. 1. Scanning intergenic regions (IGRs) for sRNA. (**a**) Example of a search pattern to scan IGRs for sRNA using the pattern search algorithm scan for matches (4) and (**b**) example of the output. The Fur-box is *underlined.*

3.2. Culture and Harvest of Meningococcal Cells

1. Colonies of a fresh overnight culture plate of meningococci are collected by a sterile cotton swab and resuspended in 5 mL of GC medium (see Note 7). This suspension is used to inoculate prewarmed GC medium (25–50 mL) in a 100-mL Erlenmeyer flask (see Note 8). The start optical density at λ530 nm (OD_{530}) should be about 0.05.
2. Broth cultures are incubated on a gyratory shaker (180 rpm) at 37°C. Growth is monitored by measuring the OD_{530} of 1 mL culture samples and when the mid logarithm phase is reached (usually after approximately 2–3 h) (OD_{530} = 0.4–0.6), transfer 18 mL of culture to a sterile 50 mL Falcon tube and add 2 mL of 10× stop solution. Mix by inversion, centrifuge for 3 min at 5,000 × *g* at 4°C, remove the supernatant and immediately continue with the RNA isolation procedure or snap freeze pellets in liquid nitrogen and store at –70°C.

3.3. RNA Isolation (see Note 9 and 10)

Total meningococcal RNA can be isolated using a hot phenol-based method for RNA extraction (see Note 11). *Caution*: Phenol and its vapors are corrosive to the eyes, the skin, and the respiratory tract. Since hot phenol is used in this protocol, be very careful and perform the isolation procedure in a chemical fume hood.

1. Prespin 15 mL PLG heavy tubes at 1,500 × *g* for 2 min.
2. Resuspend the cells in 1.3 mL DEPC H_2O, add 700 μL of boiling lysis buffer, mix well by inversion and heat for 3 min at 100°C.
3. Add 2.0 mL of 65°C acidic phenol, mix well by inversion, incubate at 65°C for 5 min: mix by inversion every 2 min and then put on ice.
4. Add 600 μL of chloroform:isoamyl alcohol (49:1 v/v), mix by inversion and add to a prespun 15 mL PLG heavy tube.
5. Keep on ice for 10 min and then spin at 1,500 × *g* for 5 min (4°C).
6. Add 2–2.5 mL (1:1 with aqueous volume) of acidic phenol:chloroform:isoamyl alcohol (50:49:1 v/v) to the same tube, mix thoroughly by repeated gentle inversion. Do not vortex.
7. Again, keep on ice for 10 min, spin at 1,500 × *g* for 5 min at 4°C and move the aqueous layer to four separate 2 mL Eppendorf tubes.
8. Add 2.5 volumes of ice-cold 100% ethanol, vortex and let stand for 30 min at –20°C. (The addition of sodium acetate is not necessary because of the high NaCl concentration in the lysis buffer).
9. Spin at 13,000 × *g* (4°C) for 30 min and remove the supernatant carefully.
10. Wash the pellet with 2.5 volumes of ice-cold 95% ethanol and then spin for 20 min at 14,000 × *g* (4°C) (see Note 12). Make sure that the pellet loosens when washing. Repeat step 10.

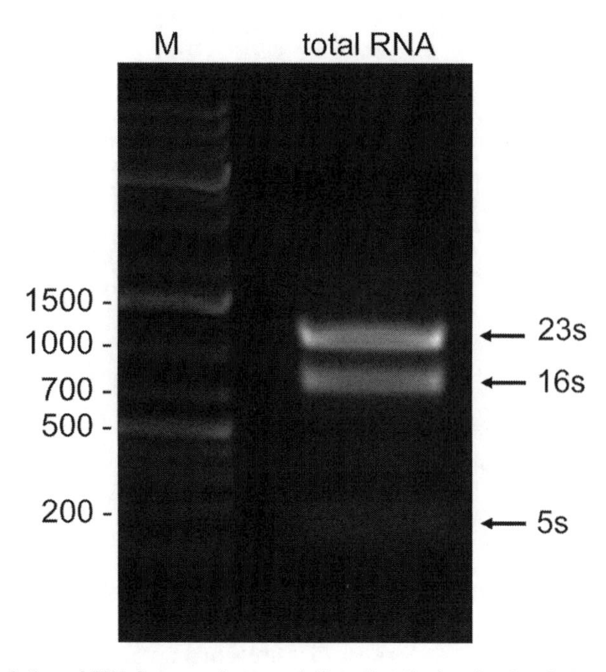

Fig. 2. Isolation of RNA from meningococci. Note that the bands of 16S rRNA and 23S rRNA are very sharp, indicating high-quality RNA. In addition, the 5S rRNA (115 nt) band is clearly visible. The latter indicates that the RNA isolation method used here is very efficient in the extraction of small-sized RNA molecules and thus well suited when working with sRNAs. M; DNA marker, sizes indicated in base pairs.

11. Carefully remove the supernatant, short spin and carefully pipette off the remaining supernatant.
12. Dry the pellets in a flow cabinet for ~5 min (RT).
13. Resuspend the RNA in 100 μL of 10 mM Tris–HCl–1 mM EDTA pH 7.0 buffer. Quantify and determine the λ260:280 nm ratio using a NanoDrop spectrophotometer (NanoDrop) according to the manufacturer's instructions. The total yield should be ~400 μg and the λ260:280 ratio should be ~2.0.
14. Analyze 1 μL of sample on a 1% (w/v) agarose gel. A representative result is shown in Fig. 2.

 Aliquot and store the RNA at −70°C (see Note 13).

3.4. Detection of sRNA Expression

3.4.1. Radioactive Probe and Marker Labeling (see Note 14)

1. For probe labeling, add the following reagents to an Eppendorf tube: 11 μL of DEPC H_2O, 5 μL of antisense oligonucleotide (5 pmol/μL), 1 μL of [γ-^{32}P]ATP, 2 μL of 10× kinase buffer and 1 μL of T4 PNK. For marker labeling, add in a second tube: 14 μL of DEPC, 2 μL of 100 bp marker (1 μg), 1 μL of [γ-^{32}P]ATP, 2 μL of 10× kinase buffer and 1 μL of T4 PNK. Mix carefully by pipetting, spin down, and incubate both tubes for 1 h at 37°C.
2. Microspin G-25 columns are used to remove unincorporated nucleotides. Resuspend the resin in the columns by vortexing. Loosen the cap by one-quarter turn and snap off the bottom

closure. Place the columns in the supplied collection tubes and spin for 1 min at 735×*g*. Place the columns into two new 1.5 mL Eppendorf tubes, add 20 μL of DEPC H_2O to both the probe sample and the marker sample, carefully mix and then pipette the complete sample(s) onto the top of the resin. Spin the columns for 2 min at 735×*g*, retain the eluates and store them at −20°C.

3.4.2. RNA PAGE/Urea Gel Electrophoresis (see Note 15)

1. Prepare a 1.5-mm thick 8% PAGE/Urea gel by mixing 8.4 g of urea, 2 mL of 10× TBE, 4 mL of 40% (w/v) acrylamide (19:1), 14 mL of DEPC H_2O, 13.3 μL of TEMED and 100 μL of APS (the amount prepared here is sufficient for three gels). Pour the gel and insert the comb. The gel should polymerize within 30 min but let it stand for at least 45 min or more.
2. Carefully remove the comb and use a micropipette to wash the wells with running buffer. Assemble the gels in the running system, add 1× TBE running buffer to the upper and lower chambers of the gel unit and prerun for a minimum of 30 min at 40–50 V.
3. During the prerun phase prepare the samples and the marker. Take 10 μg of RNA sample and adjust the volume to 5 μL with DEPC H_2O. Add 6 μL (5+1 to counter evaporation) of RNA loading dye to a total of 11 μL. Take 1 μL of the radiolabeled 100 bp marker, add 4 μL of nuclease-free H_2O and 6 μL of RNA loading dye to a total of 11 μL. Heat the samples and marker for 5 min at 95°C and then chill on ice immediately for ≥2 min (max. 30 min).
4. Disconnect the system from the power supply. Wash the wells using a micropipette (see Note 16). Load the wells with the samples and marker and run the gel at 40–50 V (~2 h).

3.4.3. Northern Blotting

1. Samples that are separated by gel running are transferred to Biodyne B Nylon Membrane electrophoretically. The membrane sheet as well as the blotting papers should be cut just larger than the gel.
2. Disconnect the system from the power supply and disassemble the gel unit carefully.
3. Equilibrate the gel, the membrane and the two extra thick blotting papers in cold (4°C) 0.5× TBE for 15 min.
4. Build the blotting sandwich. On the bottom of the transfer machine add one piece of 0.5× TBE soaked 3 MM paper, membrane and gel. Mark the edges of the gel on the membrane with a wooden pencil and add one piece of 0.5× TBE-soaked 3 MM paper on top. Make sure all ingredients stay wet and that no air bubbles are trapped in the resulting sandwich. Gently squeeze out any bubbles by rolling with a plastic 5 mL pipette. Put the lid on and close the transfer machine and blot for 1 h at 0.4 A (maximum 25 V).

5. After transfer, disassemble the sandwich, wrap the membrane in Saran Wrap and UV crosslink on a Stratalinker 1800 (Stratagene), autocrosslink (1,200 μJ × 100). Crosslinking will be complete in ~25–50 s.
6. Clean hybridization tubes with dH_2O.
7. After cross linking, remove the Saran Wrap and wrap the membrane in a hybridization curtain (RNA facing outwards) and put everything rolled up in a hybridization tube.
8. Add 10 mL of preheated (68°C) ULTRAhyb-Oligo and prehybridize for ≥1 h at 37°C in a hybridization oven. Make sure that the "curtain" folds open so that it attaches completely to the inner wall of the glass tube and that there is no leakage.
9. During prehybridization prepare the hybridization buffer: mix the [γ-^{32}P]ATP-labeled oligo-probe with 1 mL of preheated (68°C) ULTRAhyb-Oligo (Ambion).
10. Add the probes to the hybridization buffer (it is not necessary to replace the buffer) and hybridize overnight at 37°C. Again, do not forget to check for leakage.
11. Next day, prewarm 50 mL of Low Stringency Buffer for 20 min at 37°C (see Note 17).
12. Decant the hybridization buffer with probe into a radioactive waste vessel and wash the membrane with 50 mL of Low Stringency Buffer for 5 min at RT (low stringency wash).
13. After 5 min, decant buffer into the radioactive waste container and wash the membrane for 2 min (not longer!) with prewarmed Low Stringency Buffer at 37°C.
14. Check the buffer for residual radioactivity and wash at RT with Low Stringency Buffer until radioactivity is no longer detected.
15. Wrap the membrane in Saran Wrap.
16. Erase a phosphor image screen (extended, more than 20 min), place the membrane in the cassette and cover with the screen.
17. Develop for 1–3 days. Then, remove the phosphor screen (face down) and place it on the glass plate of the Typhoon Imager in the lower left corner and use the acquisition mode "Storage Phosphor." Then select the scan area, choose "best sensitivity": (NOT "best resolution") and set the pixel size to 100 μm. An example of the expected results is shown in Fig. 3.

3.5. Construction of a Shuttle Plasmid Expressing sRNA Gene Under Control of an Inducible Promoter

To overexpress sRNA genes in *N. meningitidis*, a shuttle plasmid is constructed in which the sRNA gene is under control of an inducible promoter. This allows pulse overexpression, as well as long-term expression of the sRNA gene. The protocol described here is based on the use of shuttle vector pEN11-*pldA* (5) and an outline of the procedure is depicted in Fig. 4 and the primers are listed in Table 1. Of importance, the

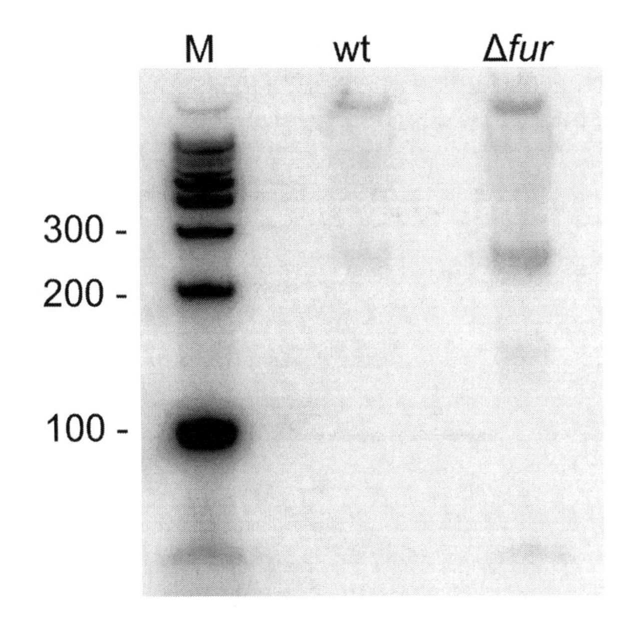

Fig. 3. Northern blot demonstrating Fur-dependent expression of a sRNA. Deletion of Fur in meningococci (Δ*fur*) results in increased sRNA expression compared to wild type cells (wt). M: radiolabeled DNA marker sizes indicated in bp.

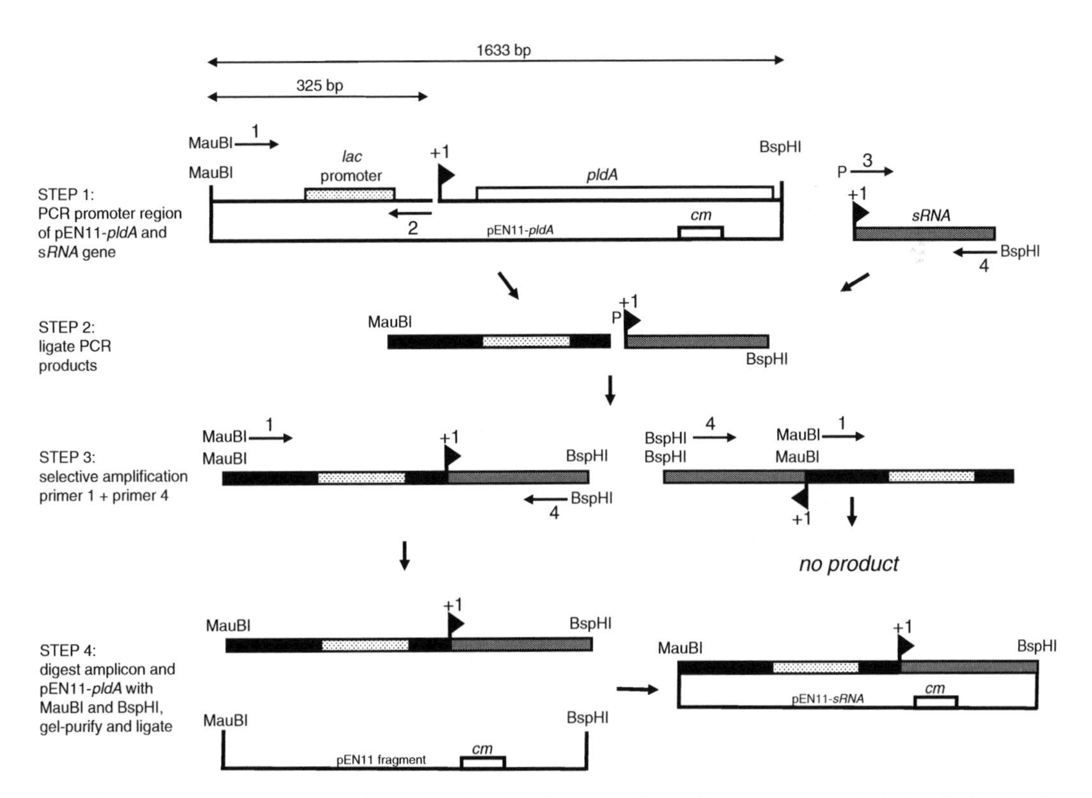

Fig. 4. Strategy for the construction of a shuttle plasmid expressing the sRNA gene under control of an IPTG inducible promoter. Step 1: PCR the promoter region of pEN11-*pldA* and the sRNA gene. Step 2: ligation of the amplicons. Step 3: selective amplification of the ligated product. Step 4: restriction digestion of the final PCR product and ligation into pEN11.

Table 1
Primers used for the construction of a shuttle plasmid expressing a sRNA gene under control of an IPTG inducible promoter

Name	Target	Modification	Orientation	Sequence 5′–3′
Primer 1	pEN11 promoter region	–	Forward	GTTTTTTCGCGCGCGAATTG CAAGCTGATCCGG[a]
Primer 2	pEN11 promoter region	–	Reversed	CCACACAACATACG AGCCGGAAGC
Primer 3	sRNA	Phosphorylated	Forward	Gene-specific, primer starts on +1
Primer 4	sRNA	–	Reversed	Gene-specific, primer ends at terminator of sRNA and contains the BspH1 recognition sequence and a GTTTTTT at the 5′ end
Primer 5	pEN11	–	Forward	GACAATTAATCATCGGCTCGT
Primer 6	pEN11	–	Reversed	AGCAAAAACAGGAAGGCAAA

The recognition site of MauBI of primer 1 is *underlined*
[a]To ensure optimal restriction enzyme activity, a short nt stretch (GTTTTTT) is added to the primer before (forward primers) or after (reverse primer) the recognition sequence. See also Fig. 4

use of a proof-reading DNA polymerase for all PCR steps described in Subheadings 3.5.2 and 3.5.3 is essential (see Note 18).

3.5.1. Isolation of Chromosomal DNA and Plasmid pEN11-pldA

1. Isolate chromosomal DNA of strain H44/76 using the PUREGENE DNA Isolation Kit (Gentra systems) according to the manufacturer's instructions.
2. Isolate pEN11-*pldA* from *E. coli* TOP10F' using the Wizard SV 96 Plasmid DNA Purification System (Promega) according to the manufacturer's instructions.

3.5.2. PCR Amplification of pEN11-pldA and the sRNA Gene

1. Set up two PCR reactions: for the first PCR use pEN11-*pldA* DNA as the template in combination with primer 1 and primer 2 (see Fig. 4). Mix 1 μL of dNTPs (final concentration 200 μM), 5 μL of 10× *Pfu* DNA polymerase buffer, 1.25 μL of primer 1 and 1.25 μL of primer 2 (final concentration of each primer 250 nM), 1 μL of pEN11-*pldA* template, 0.5 μL *Pfu* DNA Taq polymerase (1 U) (see Note 18) and adjust the final reaction volume to 50 μL with 40 μL of MilliQ. For the second PCR, use the amounts described above but now use 1 μL of genomic DNA as the template and primer 3 and primer 4. (Ensure that primer 3 is 5′ phosphorylated.) As a negative control, omit adding DNA template and add 41 μL of MilliQ instead. Mix, quick spin, and amplify using thermal conditions as follows: 2 min of denaturation at 95°C followed by 30 cycles

of 0.5 min at 95°C, 0.5 min at 50°C and 2 min at 72°C, ending with 5 min at 72°C and pausing at 10°C.

2. Analyze the amplicons by 1% (w/v) agarose gel electrophoresis: an amplicon of 325 bp should be obtained with primer pair 1–2, and a product of the calculated size should be obtained with primer pair 3–4. There should be no product present in the negative control reaction.
3. Purify the amplicons using QIAEX II Gel Extraction Kit and quantify the amount of DNA using a NanoDrop spectrophotometer (NanoDrop) according to the manufacturer's instructions.

3.5.3. Ligation of pEN11 Fragment to sRNA Amplicon and Selective Amplification

1. For ligation, mix 1 μL (usually 10–40 ng) of amplicon generated with primer pair 1 and 2 with 1 μL of amplicon generated with primer pair 3 and 4, 1 μL of 10× ligation buffer, 1 μL of T4 ligase (1 U) and adjust the volume to 10 μL with 7 μL of MilliQ. Incubate for 1–3 h at RT or overnight at 16°C.
2. Selectively amplify the ligation product by PCR, using primer pair 1 and 4. Use 1 and 1 μL of a five-time diluted ligation reaction as DNA template. Prepare the mixture as outlined in step 1 in Subheading 3.5.2. Analyze the PCR reaction by 1% (w/v) agarose gel electrophoresis, gel-purify and quantify the DNA as described in steps 2 and 3 in Subheading 3.5.2. The DNA yield is usually ~0.1–0.3 μg in total.

3.5.4. Digestion of pEN11-pldA and the pEN11-sRNA Fragment with MauBI and BspHI

1. Digest ~5 μg of pEN11-pldA and ~250–500 ng of the pEN11-sRNA fragment with MauBI and BspHI enzymes according the manufacturer's instructions.
2. Analyze the digestion by 1% (w/v) agarose gel electrophoresis. Restriction of pEN11-pldA should result in two linear fragments, one representing the *pldA* gene (~1,600 bp) and a second representing the core plasmid (~8,600 bp). The size of the restricted pEN11-sRNA fragment should be 325 + length sRNA bp. Gel purify the ~8,600 bp pEN11 fragment and the pEN11-sRNA fragment and quantify as described in steps 2 and 3 in Subheading 3.5.2.

3.5.5. Ligation of the Digested sRNA-pEN11 Fragment in the Digested pEN11 Fragment and Transformation of pEN11 + sRNA to E. coli TOP10F'

1. Ligate the digested and purified sRNA and pEN11. Mix ~100–200 ng of the sRNA-pEN11 fragment with 200–400 ng of pEN11 fragment and set up a ligation reaction as described in step 1 in Subheading 3.5.3.
2. Add 5 μL of the pEN11 + sRNA ligation reaction to *E. coli* TOP10F' competent cells following the manufacturer's instructions. Plate onto LB agar plates containing Cm and incubate overnight at 37°C.

3.5.6. Colony Screening of Transformants by PCR Using Primer 5 and Primer 6

1. On the next day, isolated colonies on the plates are numbered on the plastic bottom of the Petri dish, picked with a toothpick or 200 μL pipette point, resuspended in PCR tubes containing 50 μL MilliQ, vortexed and heated for 5 min at 100°C.
2. Pellet cell debris by centrifugation (13,000 × *g*, 2 min) and use 2.5 μL of the supernatant as template for PCR in a total volume of 25 μL. For colony screening it is not necessary to use a proof-reading DNA polymerase. For one PCR mix, prepare 0.5 μL of dNTPs (final concentration 200 μM), 5 μL of 10× Go-Taq buffer, 0.625 μL of primer 5 and 0.625 μL of primer 6 (final concentration of each primer 250 nM), 2.5 μL of DNA template, 2.5 μL of $MgCl_2$ (final concentration 2 mM), 0.2 μL of Taq DNA polymerase (1 U) and adjust the final reaction volume to 25 μL with 13.55 μL MilliQ. Multiply the reagents' volumes listed by the number of PCR reactions required +1 up to a single master mix which can then be dispended into 22.5 μL amounts into PCR tubes combined with 2.5 μL of template. Testing ten colonies per transformation is usually sufficient to obtain one or more clones with inserts of the expected calculated size.
3. After amplification, analyze the PCR products by 1% (w/v) agarose gel electrophoresis. Clones containing the appropriate-sized inserts are selected and inoculated onto a fresh LB plate containing Cm and grown overnight. A single isolated colony from this plate is selected and used to inoculate 5 mL LB broth containing Cm. Bacteria are cultured overnight at 37°C on a gyratory shaker (200 rpm).

3.5.7. Plasmid Isolation and Sequence Analysis

1. Isolate plasmid DNA of selected clones using the 5 mL overnight cultures and the Wizard SV 96 Plasmid DNA Purification System (Promega) following the manufacturer's instructions.
2. Plasmid DNA of selected clones is sequenced according to the manufacturer's instructions using Big Dye terminator chemistry. Only plasmid DNA of clones containing inserts without point mutations or deletions (wild type sequences) is used for natural transformation of meningococci. We will refer to this plasmid as pEN11-sRNA.

3.5.8. Transformation of pEN11-sRNA into Meningococci and Assessment of sRNA Gene Expression in Trans

1. Add 100 μL of 1 M $MgCl_2$ solution (final conc. 10 mM) to 10 mL of GC broth.
2. Add 5 mL of GC broth + $MgCl_2$ to a 25-mL sterile cell culture flask and prewarm in a CO_2 incubator with the lid loosened.
3. From an overnight culture of meningococci on a GC plate, resuspend cells (use isolated single colonies as much as possible as inoculum) in 2 mL of GC broth + $MgCl_2$ to a density of OD_{530} = 0.3–0.5.

4. Add 35 μL of this suspension to 1 mL of GC broth + $MgCl_2$ in a 2-mL Eppendorf tube and add ~3–5 μg of plasmid. Incubate for 1 h at 37°C in an end-over-end rotator.
5. Add 500 μL of the transformation mix to the prewarmed cell culture flask and incubate until the medium is cloudy (usually 3–4 h).
6. Plate 10 μL onto a GC plate containing no antibiotics (control for contamination and viability of cells).
7. Plate 10 and 100 μL onto GC plates containing 5 μg/mL Cm.
8. Centrifuge the remainder of the 5 mL volume for 1 min at 13,000 × *g*, resuspend the pellet in 300 μL of GC broth and plate 3 × 100 μL onto three separate GC plates containing Cm.
9. Let the plates dry at RT and then incubate overnight at 37°C in a humidified atmosphere of 5% (v/v) CO_2.
10. The transformants are usually visible 1–2 days after transformation (but up to 4 days is possible!).
11. Expression of the sRNA gene *in trans* in meningococci is assessed by Northern blotting. Inoculate two GC broth cultures with H44/76 + psRNA containing Cm as described in step 1 in Subheading 3.2. Add IPTG to a final concentration of 1 mM to one of the culture flasks. Culture and harvest cells as described in step 2 in Subheading 3.2 and isolate RNA as described in Subheading 3.3. Assess sRNA levels using Northern blotting as described in Subheading 3.4.3.

4. Notes

1. The only restriction for meningococcal strain selection is that the strain(s) used can be transformed.
2. Milli-Q refers to water that has been purified and deionized to a high degree by a water purification system manufactured by Millipore Corporation. It is also a registered trademark of Millipore.
3. RNA detection takes several days and can be carried out according to this time schedule: Day 1, label the probe and marker; Day 2, run the gel, blot, prehybridize and hybridize overnight; Day 3, wash and expose the film (for 1–3 days) and then develop.
4. Since the half-life of ^{32}P is 14 days, use the stock ^{32}P within 5 days after arrival.
5. 10× TBE: this solution can be stored at room temperature but a precipitate will form in older solutions. Store the buffer in glass bottles and discard if a precipitate has formed.

6. Use a new sterile razor blade or scalpel to cut each amplicon from the agarose gel. This is to avoid cross-contamination of samples.
7. For the inoculation of GC broth cultures, use isolated colonies as much as possible, since, in general, these contain the most viable bacteria.
8. Optimal growth conditions of meningococci are achieved with sufficient aeration by using relative large volume Erlenmeyer flasks and using a relative small volume of medium. We routinely use a 100-mL Erlenmeyer flask with a maximum volume of 30 mL of broth.
9. RNA is generally very sensitive to the action of ribonucleases. For successful isolation of high-quality RNA, it is essential that all solutions and equipment used should be free of RNAses. Hands are the major source of nuclease contamination and powder-free gloves should be worn at all times. Bench surfaces should be cleaned thoroughly with detergent and after that cleaned with RNase AWAY. Pipettes should also be treated with RNAseOUT and it is essential that the gel system used for PAGE/urea electrophoresis of RNA is not used for other applications. Change gloves several times during the course of the procedure, because the outside of gloves themselves can become contaminated through contact with items in the lab. Use sterile disposable plastic ware as much as possible and if needed, rinse sterilized glass ware or plastic tanks with DEPC water before use. Use nuclease-free filter tips for micropipettes.
10. Northern analysis is still the gold standard for the detection and quantitation of mRNA levels. This is because Northern blot analysis allows a direct comparison of the mRNA or sRNA abundance between samples on a single membrane. In addition, Northern blots are very useful in determining the number of different transcripts produced by a single gene and their relative size, even if both are unknown.
11. It is important to realize that most commercially available RNA isolation procedures, in which columns are used, are not sufficient generally for the isolation of small-sized RNAs (<200 nt). Thus, when working with sRNAs, we highly recommend the procedure based on hot phenol. This method ensures efficient sRNA isolation.
12. 95% (v/v) ethanol is used to improve recovery of RNA smaller than 200 nt.
13. In general, RNA isolated in this way will be stable for 3–6 months. It should be noted that precipitated RNA stored under ethanol at –70°C remains stable indefinitely.
14. Working with radioactivity requires a separate laboratory containment area. Avoid contact with radioactive material and

radiation. All radioactive waste should be collected and disposed according to national laws.

15. These instructions assume the use of the mini-gel system of Bio Rad and the Bio-Rad semi-dry transblotter but are easily adapted to other formats. It is critical that the glass plates are scrubbed with a rinsable detergent and are extensively rinsed with dH_2O and finishing with DEPC H_2O. Make sure that all buffers and reagents used are made with DEPC H_2O or autoclaved MilliQ up until the UV cross-linking of the semi-dry transblotted membrane.
16. This is to remove excess urea. Urea can prevent proper filling of the samples into the slots.
17. Only preheat the amount of buffer needed for washing and do not heat the stock repeatedly in order to prevent the SDS from hydrolyzing and thereby decreasing the pH.
18. A proofreading DNA polymerase, like *Pfu* DNA polymerase, is used to ensure blunt ends and to avoid second-site mutations.

Acknowledgements

We thank MSc. R.A.G. Huis in't Veld for critical reading of the chapter. This research was partly funded by the Sixth Framework Programme of the European Commission, Proposal/Contract no.: 512061 (Network of Excellence "European Virtual Institute for Functional Genomics of Bacterial Pathogens," http://www.noe-epg.uni-wuerzburg.de).

References

1. Sharma CM, Vogel J (2009) Experimental approaches for the discovery and characterization of regulatory small RNA. Curr Opin Microbiol 12: 536–546.
2. Romby P, Vandenesch F, Wagner EG (2006) The role of RNAs in the regulation of virulence-gene expression. Curr Opin Microbiol 9: 229–236.
3. Waters LS, Storz G (2009) Regulatory RNAs in bacteria. Cell 136: 615–628.
4. Dsouza M, Larsen N, Overbeek R (1997) Searching for patterns in genomic data. Trends Genet 13: 497–498.
5. Bos MP, Tefsen B, Voet P et al (2005) Function of neisserial outer membrane phospholipase A in autolysis and assessment of its vaccine potential. Infect Immun 73: 2222–2231.

Chapter 6

Expression, Purification, and Crystallization of Neisserial Outer Membrane Proteins

Muhammad Saleem, Jeremy Moore, and Jeremy P. Derrick

Abstract

Integral outer membrane proteins (OMPs) play key roles in solute transport, adhesion, and other processes. In *Neisseria*, they can also function as major protective antigens. Structural, biophysical, and immunological studies of Neisserial OMPs require their isolation in milligram quantities. Purification of any OMP directly from *Neisseria* would require the growth of large quantities of cell mass, with attendant concerns about safety and convenience. As a result, many investigators have developed methods for expression of OMPs into inclusion bodies in *E. coli*, followed by refolding of the resolubilized protein. Here we describe such a method, as optimized for the PorA porin but which can be applied, with suitable adaptation, to other OMPs. We also describe an approach to the crystallization of PorA.

Key words: *Neisseria meningitidis*, Outer membrane protein, PorA porin, Refolding, Crystallization

1. Introduction

Structural studies on porins and other integral outer membrane proteins (OMPs) have established that they generally adopt a transmembrane β-barrel structure, with an even but varying number of β-strands. This creates a highly stable and versatile structure, which can be adapted for use as a pore-forming protein, an energized transporter or an adhesin for recognition of host cell surface receptors (1). In comparison with their transmembrane helix-containing counterparts, OMPs are generally more stable in the detergent-solubilized state and their relative abundance in the Protein Data Bank suggests that they are more tractable to structure determination, traditionally by X-ray crystallography and, more recently, by solution state NMR. Investigators may also wish to isolate specific OMPs

Myron Christodoulides (ed.), *Neisseria meningitidis: Advanced Methods and Protocols*, Methods in Molecular Biology, vol. 799, DOI 10.1007/978-1-61779-346-2_6, © Springer Science+Business Media, LLC 2012

for immunological purposes, particularly as vaccine components. The application of all these methods requires that the protein is isolated in high yield, and in its native state. A method which has been widely adopted for OMP isolation is to direct expression into inclusion bodies (IBs), resolubilize (e.g. using urea or guanidine HCl) and refold, by addition of detergent and simultaneous removal of chaotrope. This general method, with some modifications, has been reported for many of the main integral OMPs from *Neisseria*, including PorA (2), PorB (3), FetA (4), NspA (5), OpcA (6), and the opacity proteins (7). In some cases, crystal structures have been reported (3, 5, 6), providing strong evidence that the correct native state was obtained from refolding. Our experience has been that particular attention needs to be paid to the conditions for refolding, but that there may be several ways to achieve optimally refolded protein in high yield. Thus, although there are a variety of methods reported in the literature, it is possible to give some general guidance in devising a refolding protocol. Here, we have taken the PorA porin as an example; its importance as a meningococcal vaccine component is well established (8), and the methods employed can be generalized to other OMPs relatively easily. We also show how the purified OMP can then be fed into crystallization trials, and give an example of crystals obtained by this approach.

2. Materials

2.1. Expression of the Neisseria Outer Membrane Protein, PorA, as Inclusion Bodies (IBs)

1. PorA plasmid expression vector (see Notes 1 and 2).
2. T7 Express Competent *E. coli* (New England Biolabs).
3. Luria Bertani (LB) agar and LB liquid medium broth (Fisher).
4. 2xYT liquid medium broth: add 16 g of tryptone, 10 g of yeast extract, and 5 g of NaCl to 1 L of deionized H_2O. Alternatively, purchase 2xYT broth powder (Fisher).
5. Kanamycin (30 mg/mL stock concentration, stored at –20°C).
6. Isopropyl β-D-1-thiogalactopyranoside, IPTG (1 M stock).
7. Baffled Erlenmeyer flasks (2 L volume).
8. Spectrophotometer.
9. Centrifuge.
10. Orbital shaker.

2.2. Preparation and Refolding of Inclusion Bodies (IBs)

1. Lysozyme from chicken egg white (Fluka).
2. Deoxyribonuclease (DNAse) type I from bovine pancreas, stock solution of 7 mg/mL (Sigma-Aldrich).
3. 50 mM Tris–HCl buffer, pH 7.9.

4. Lauryldimethylamine-oxide (LDAO): stock solution of 30% (v/v) (Sigma-Aldrich).
5. Lysis buffer: 50 mM Tris–HCl buffer, pH 7.9 containing 5 mg of lysozyme/gram of cell paste and 5 μL of DNAse I type IV stock/gram of cell paste.
6. Denaturing buffer: 10 mM Tris–HCl buffer, pH 7.5 containing 1 mM ethylenediamine tetraacetic acid (EDTA) and 8 M urea.
7. Refolding buffer: 20 mM Tris–HCl buffer, pH 7.9 containing 1 M NaCl and 5% (v/v) LDAO.
8. Dialysis buffer: 20 mM Tris–HCl buffer, pH 7.9 containing 0.5 M NaCl and 0.1% (v/v) LDAO.
9. Soniprep 150 probe sonicator (MSE).
10. Centrifuge (Eppendorf).

2.3. Purification and Folding of Recombinant PorA Protein

1. Imidazole: prepare a stock solution of 4 M in deionized water and store at room temperature.
2. HISTrap HP High-performance Column, bed volume 5 mL, height 1.6 cm, width 2.5 cm (GE Healthcare).
3. HiPrep™ 26/10 Desalting column, bed volume 53 mL, height 10 cm, width 2.6 cm (GE Healthcare).
4. Superdex 200 10/300 GL, bed volume 24 mL, height 30 cm, width 1 cm (GE Healthcare).
5. P1 peristaltic pump (GE Healthcare).
6. AKTA protein purification system for desalting and size exclusion chromatography (SEC) (GE Healthcare).
7. Vivaspin 100 concentrators (Sartorius).
8. Centricon 100 kDa concentrator (Generon).
9. Buffer A: 20 mM Tris–HCl buffer, pH 7.9 containing 0.5 M NaCl, 0.1% (v/v) LDAO and 10 mM imidazole.
10. Buffer B: 20 mM Tris–HCl buffer, pH 7.9 containing 0.5 M NaCl, 0.1% (v/v) LDAO and 40 mM imidazole.
11. Buffer C: 20 mM Tris–HCl buffer, pH 7.4 containing 0.5 M NaCl, 0.1% (v/v) LDAO and 500 mM imidazole.
12. Buffer D: 20 mM Tris–HCl buffer, pH 7.4 containing 150 mM NaCl and 0.1% (v/v) LDAO.
13. Buffer E: 50 mM Tris–HCl buffer, pH 7.9 containing 150 mM NaCl, 0.5% (w/v) *n*-octyl-hydroxypenta(oxyethylene) (C_8E_5: Bachem) and 0.5 M imidazole.
14. Buffer F: 50 mM Tris–HCl buffer, pH 7.9 containing 150 mM NaCl and 0.5% (w/v) C_8E_5.
15. Deoxycholate, sodium (DOC).

16. n-Octylpentaoxyethylene (C8E5; Bachem).
17. Trichloroacetic acid (TCA).
18. Total protein assay kit, Micro Lowry (Sigma-Aldrich).
19. NanoDrop ND1000 Spectrophotometer (Thermo Scientific).

2.4. Analysis of Protein by Sodium Dodecyl Sulphate (SDS) Polyacrylamide Gel Electrophoresis (PAGE) (NuPAGE)

1. Bis–tris gel NuPAGE Novex 15 well format high performance for SDS-PAGE 12% and 1 mm thickness (Invitrogen).
2. X-cell Sure-lock gel running apparatus (Invitrogen).
3. SDS-PAGE sample loading buffer: 62.5 mM Tris–HCl buffer, pH 6.2 containing 10% (v/v) glycerol, 2% (w/v) SDS and 2% (v/v) β-mercaptoethanol. Prepare a 4× stock solution and store in aliquots at −20°C. Add β-mercaptoethanol after thawing. Caution: β-mercaptoethanol is highly toxic and should be stored and used with extreme care.
4. NuPAGE MOPS SDS Running Buffer: 50 mM 3-(N-morpholino) propanesulfonic acid, (MOPS), 50 mM Tris, 0.1% (w/v) SDS, 1 mM EDTA, pH 7.7. Prepare a 20× stock solution by dissolving 104.6 g of MOPS, 60.6 g of Tris Base, 10 g of SDS and 3.0 g of EDTA in 400 mL of ultrapure water, making the final volume to 500 mL after adjusting the pH.
5. SDS-PAGE molecular standards, low range (Bio-Rad).
6. InstantBlue stain (Expedeon).
7. Laboratory flat-bed scanner.

2.5. Verification of the Folded State of PorA

1. Superdex 200 10/300 GL bed volume 24 mL (GE Healthcare) for SEC.
2. Detergents: LDAO, n-Dodecyl-β-maltoside (DDM) (Anatrace), Octaethylene glycol mono-n-dodecyl ether ($C_{12}E_8$) (Anatrace), Zwittergent 3–14 (Anatrace), Tween 20(Fisher).
3. SDS-PAGE sample loading buffer: 62.5 mM Tris–HCl buffer, pH 6.2 containing 10% (v/v) glycerol, 0.1% (w/v) SDS and 2% (v/v) β-mercaptoethanol. Prepare a 4× stock solution and store in aliquots at −20°C (for semi-native NuPAGE).
4. NuPAGE MOPS SDS Running Buffer: 50 mM MOPS, 50 mM Tris, 0.1% (w/v) SDS, 1 mM EDTA, pH 7.7.
5. SDS-PAGE molecular standards, low range (Bio-Rad).
6. InstantBlue stain (Expedeon).
7. Laboratory flat-bed scanner.

2.6. Crystallization of PorA

1. Crystallization screens: Morpheus™ HT-96, MemStart+MemSys HT-96, MemGold™ HT-96, PGA Screen.
2. The MRC Crystallisation Plate™.
3. Morpheus™ MPD_P1K_P3350 Mix.

4. Morpheus™ single reagent.
5. Morpheus™ Buffer System 3.
6. Morpheus™ Ethylene Glycols Mix.
7. Morpheus™ Monosaccharides Mix. All Morpheus reagents are from Molecular Dimensions.
8. Detergents: hexyl glucoside, hexyl maltoside, heptyl glucoside. All detergents are from Anatrace.
9. Mosquito automated solution dispenser (TTP LabTech).

3. Methods

The following methods are a description of a protocol which we have optimized for *Neisseria meningitidis* PorA variants, which can be readily applied to other (bacterial) OMPs. In the following protocol, we expect that the gene encoding the OMP of interest has already been cloned into an expression plasmid vector, using any of the expression systems available commercially. For our OMPs, we have used the T7 expression system (see Note 1).

The preparation of OMP crystals requires the following four sequential stages:

1. The isolation of recombinant OMP as IBs: if the method is to be applied to other OMPs, we have found that that little deviation from this procedure is necessary.
2. Refolding of the isolated recombinant OMP, by solubilization in urea and rapid dilution into a detergent-containing buffer. We have found this to be a useful general approach, but conditions would need to be optimized for other OMPs. Variable conditions to consider include the detergent used, the final protein concentration after dilution, the type of chaotrope employed, and the dilution ratio (see Note 2).
3. Purification of the recombinant OMP, in this case using metal chelate affinity chromatography, followed by SEC. Again, variations from this approach are possible. We have used ion exchange chromatography in the past, for example, for purification of Opa proteins (7). Following purification, it is important to confirm that the protein has folded to a native conformation. Various approaches have been reported in the literature, including gel shift assays and circular dichroism (4). Due to the nature of the OMP under consideration, convenient biological assays, e.g. enzyme assays, are often unavailable, so most approaches have used a biophysical measurement of some kind or another. This is probably the most difficult part of the procedure for which to give general advice; in this chapter,

we have illustrated the approach by showing examples of the measurement of trimer formation through the use of native gels and SEC. Both these methods allow a measurement of the relative amounts of different oligomers in the PorA OMP preparation.

4. Process of optimization of diffraction of crystals of a PorA OMP variant. Clearly, this precise process is particular to the individual OMP; indeed, we find that different antigenic variants of PorA vary in their crystallization behaviour. We have therefore opted to show the process of optimization of diffraction for a single crystal form, as a way of illustrating how such a procedure would work.

3.1. Expression of the Neisseria Outer Membrane Protein, PorA, as Inclusion Bodies (IBs)

1. To initiate the transformation, add 10 ng of PorA plasmid vector to a 50 μL aliquot of competent *E. coli* T7 express cells in an Eppendorf tube and leave on ice for 30 min.
2. Heat-shock the cells by transfer of the Eppendorf tube for 20 s to 42°C and then immediately replace the tube on ice for 2 min. Then, add 950 μL of LB medium to the cells and incubate at 37°C for 1 h with shaking at 200 rpm in an orbital shaker. Spread 100–200 μL of the transformed cells onto an LB agar plate containing 50 μg/mL of kanamycin and incubate the plate(s) overnight at 37°C.
3. Inoculate two to three individual bacterial colonies from the agar plates into 250 mL baffled Erlenmeyer flasks containing 50 mL of 2xYT broth medium and kanamycin (50 μg/mL) and grow the bacterial "seed" culture overnight in an orbital shaker at 37°C and 180 rpm.
4. Divide 2 L of 2xYT media equally between four 2 L Erlenmeyer flasks and sterilize in an autoclave for 15 min at 121°C and 2.68kg/cm^2. Add kanamycin to a final concentration of 50 μg/mL. Inoculate each flask with 10 mL of the overnight bacterial "seed" culture and grow the cells at 37°C in an orbital shaker at 180 rpm to an optical density (OD) at λ_{600}nm of 0.8–1.0, measured using a spectrophotometer. Add IPTG to a final concentration of 1 mM and grow the cells for a further 3 h. Harvest the cells by centrifugation at 7,000 × *g* for 20 min at 4°C.

3.2. Preparation and Refolding of Inclusion Bodies (IBs)

1. Suspend the bacterial cells in Lysis buffer, using 8 mL of buffer per gram wet weight of cell paste. Disrupt the cells by sonication using a probe sonicator for 5–6 min (5 s on and 10 s off), maintaining the cells on ice.
2. Centrifuge the sonicated preparation at 14,000 × *g* for 20 min at 4°C to sediment the IBs.
3. Wash the IBs by suspending them in 50 mL of 50 mM Tris–HCl buffer, pH 7.9 containing 1.5% (v/v) LDAO for each 1–1.5 g

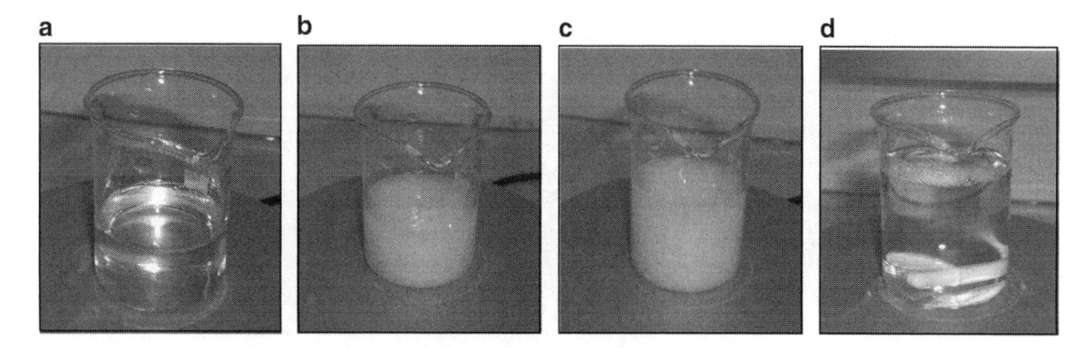

Fig. 1. Refolding of PorA OMP. The figure records the change in appearance during refolding of PorA: (**a**) before addition of refolding buffer; (**b**) after addition of 12.5% of the refolding buffer volume; (**c**) after addition of 25% of the refolding buffer volume; (**d**) after addition of 100% of the refolding buffer volume.

wet weight of IBs (see Note 3). Incubate the suspended IBs at 20°C on a shaking platform for 1 h. Centrifuge at 16,000×*g* and 4°C for 30 min to sediment the IBs.

4. Wash the IBs twice by suspending in 50 mM Tris–HCl buffer, pH 7.9 (20 mL for each 1–1.5 g of IBs) and centrifuge at 16,000×*g* at 4°C for 30 min. Discard the supernatant, flash-freeze the IBs in a suitable cryo-vial in liquid nitrogen and store at −20°C.
5. The first step to refolding the recombinant OMP is to thaw the IBs to room temperature and dissolve 500 mg in 40 mL of Denaturing buffer (see Note 3).
6. Next, centrifuge the IB solution at 14,000×*g* for 20 min to remove any undissolved material and retain the supernatant.
7. Add 40 mL of Refolding buffer drop-wise to the dissolved IB solution with rapid stirring to produce a final 1:1 volume ratio. Maintain continuous stirring for 1 h at 20°C (Fig. 1; see Note 4).
8. Dialyse the solution at 4°C against two changes of 4 L of Dialysis buffer, with each change of buffer at 6–8 h.

3.3. Purification and Folding of Recombinant PorA Protein

3.3.1. Metal Chelate Affinity Chromatography

1. Equilibrate the HISTrap HP High-performance Column (5 mL) with ten column volumes of Buffer A.
2. Load the dialyzed IB solution, prepared as described in step 8 of Subheading 3.2, onto the column using a peristaltic pump at a flow rate of 1 mL/min.
3. After application, wash the column with ten column volumes of Buffer B and collect the flow-through volume in two 25 mL fractions.
4. Elute the bound PorA protein by addition of eight column volumes of Buffer C to the column and collect the protein in

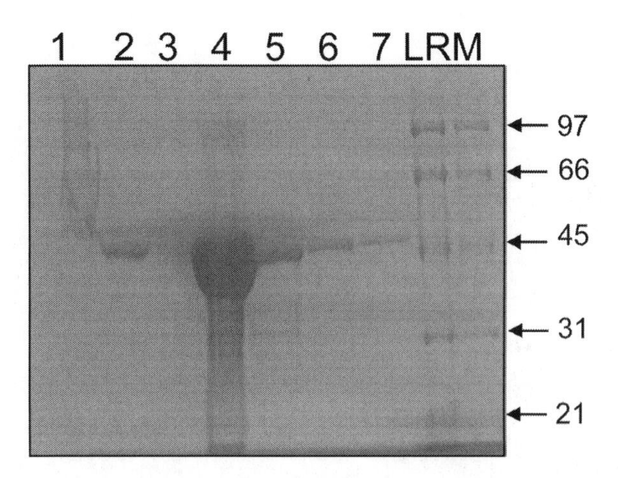

Fig. 2. Purification of PorA by metal chelate affinity chromatography. SDS-PAGE of eluted fractions from the metal chelate affinity column during PorA purification. Lane 1: flow through from the column. Lanes 2 and 3: wash fractions. Lanes 4–7: elution fractions LRM, low-range mass markers (kDa).

four 10 mL fractions (Fig. 2; see Note 5). Analyze all the samples by SDS-PAGE (NuPAGE) as described in Subheading 3.4.

3.3.2. Desalting of PorA Protein on HiPrep™ 26/10 Desalting Column

1. After SDS-PAGE (NuPAGE) analysis of the elution fractions from the HISTrap column, pool all fractions containing PorA and apply to a HiPrep™ 26/10 Desalting Column on the automated AKTA protein purification system, previously equilibrated in Buffer D.
2. Analyze the elution fractions by SDS-PAGE (NuPAGE) and measure the protein concentration using the Micro-Lowry protein assay as described in Subheading 3.3.4.
3. Store the PorA samples in 5 mL aliquots at a concentration of 1 mg/mL at –20°C. PorA can also be desalted by dialysis (see Note 6).

3.3.3. Purification of PorA by Size Exclusion Chromatography (SEC)

1. After desalting, concentrate the PorA samples to about 2 mg/mL using a Centricon 100 kDa concentrator (Generon), following the manufacturer's instructions.
2. Estimate the protein concentration by SDS-PAGE (NuPAGE) as described in Subheading 3.4.
3. Pre-equilibrate the Superdex 200 column on the automated AKTA protein purification system with Buffer D, inject 0.5 mL of PorA (about 1 mg) for each run and collect 0.5 mL fractions. Pool the peak fractions – shown in a typical elution profile in Fig. 3.

3.3.4. Detergent Exchange and Determination of Protein Concentration

1. Apply the pooled fractions from step 3 of Subheading 3.3.3 onto a 1 mL HISTrap column. Wash the column with ten-column volumes of Buffer D and elute the PorA protein in

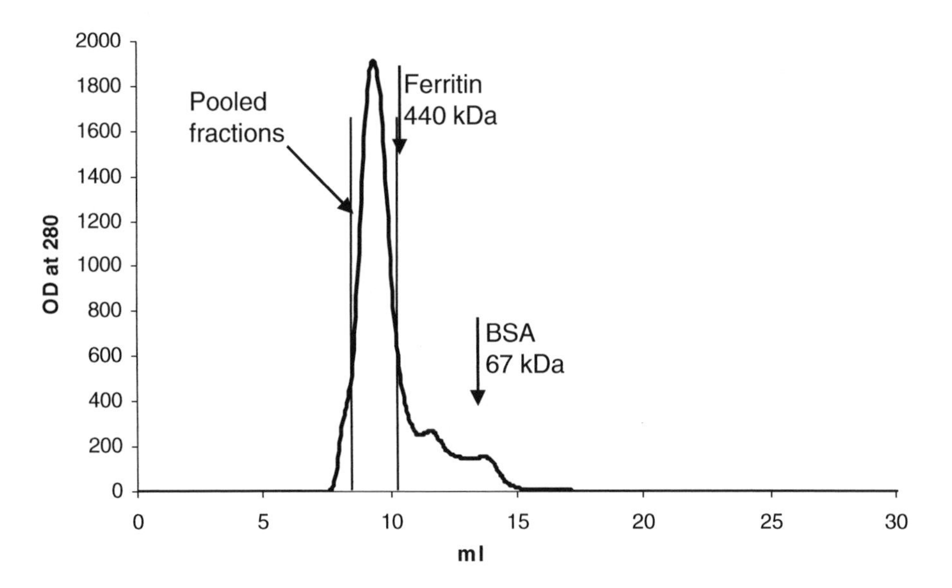

Fig. 3. Size exclusion chromatography of PorA. An elution absorption profile at 280 nm is shown. The Superdex 200 SEC column (GE Healthcare) is run in buffer D (50 mM Tris–HCl, pH 7.9, 150 mM NaCl, 0.1% LDAO) at a flow rate of 0.5 mL/min. Fractions are pooled as indicated.

four 1 mL fractions with Buffer E. Pool the first two fractions and dilute them ten times with Buffer F.

2. Concentrate the diluted sample using a Vivaspin 100 concentrator to ~10 mg/mL for use in crystallization trials.
3. Measure the protein concentration by using the Micro-Lowry protein assay with Peterson's modification (9) (see Note 7) as follows:
   (a) Prepare a standard concentration range of bovine serum albumin at 0.1, 0.2, 0.3, 0.4, 0.5, 0.8, and 1.0 mg/mL.
   (b) Prepare PorA samples up to 20 μL and add 2 μL of DOC (0.15% (w/v) stock solution) to each sample; mix well and allow to stand for 10 min at 20°C.
   (c) Add 2 μL of TCA (72% (v/v) stock solution) to each tube and mix.
   (d) Centrifuge the solutions for 10 min at 13,000 × *g* to sediment the precipitate. Decant and blot away the supernatant.
   (e) Dissolve the pellets in 20 μL of Lowry Reagent, mix well and allow to stand for 20 min.
   (f) Next, add 10 μL of Folin & Ciocalteu's Phenol Reagent and allow the colour to develop for 30 min at room temperature.
   (g) Measure the absorbance with a NanoDrop ND 1000 spectrophotometer. Alternatively, the assay can be scaled up

proportionately to give a final volume of 1 mL, and the absorption at 750 nm recorded using a standard spectrophotometer.

3.4. Analysis of Protein by Sodium Dodecyl Sulphate (SDS) Polyacrylamide Gel Electrophoresis (PAGE) (NuPAGE)

For denaturing SDS-PAGE analysis of protein samples during and after purification, we use the Nu-PAGE ready-made gel system.

1. Suspend the PorA protein samples in 4× SDS-PAGE loading buffer at a volume ratio of 3:1. For each gel, add a sample of SDS-PAGE low molecular mass standard.
2. Transfer the tube to a heating block at 100°C for 5 min to denature the proteins before loading onto a NuPAGE gel (see Note 8).
3. Run the gel in NuPAGE MOPS SDS Running Buffer using the X-cell Sure-lock gel running apparatus, at 180 V for 45 min, following the manufacturer's instructions.
4. After electrophoresis, remove the gel from the tank and transfer it directly into 20–30 mL of InstantBlue stain to visualize the protein bands. Although protein bands will start to develop immediately, incubate the gel for 15 min at room temperature with gentle shaking for appropriate intensity.
5. Wash the gel with distilled H_2O and scan it with a laboratory flat-bed scanner.
6. The protein concentration of the applied protein samples can also be estimated by SDS-PAGE by reference to the molecular mass standards of known concentration (0.5, 1, and 2 μg) (see Note 9).

3.5. Verification of the Folded State of PorA

Following purification of the recombinant OMP, it is essential to confirm folding of the protein to a native conformation. This can be done in two ways:

3.5.1. Size Exclusion Chromatography (SEC)

SEC is a convenient and reliable method for studying the oligomerization of proteins. Its utility for membrane proteins, such as PorA, lies in its independence from the nature of the solvent used; it is therefore possible to analyze the effect of different detergents, for example, on the oligomeric state of PorA. The choice of detergent can be crucial for maintaining a particular membrane protein in a stable state, and avoiding aggregation which would be refractory to crystallization.

Figure 4 shows a comparison of the SEC elution profiles of PorA in different detergents. SEC was carried out as described in step 3 of Subheading 3.3.3. We routinely observe that PorA elutes at a much higher molecular mass than would be expected from a trimer (*circa* 120 kDa), even once allowance has been made for additional mass from the detergent micelle. The main elution peak of PorA for all

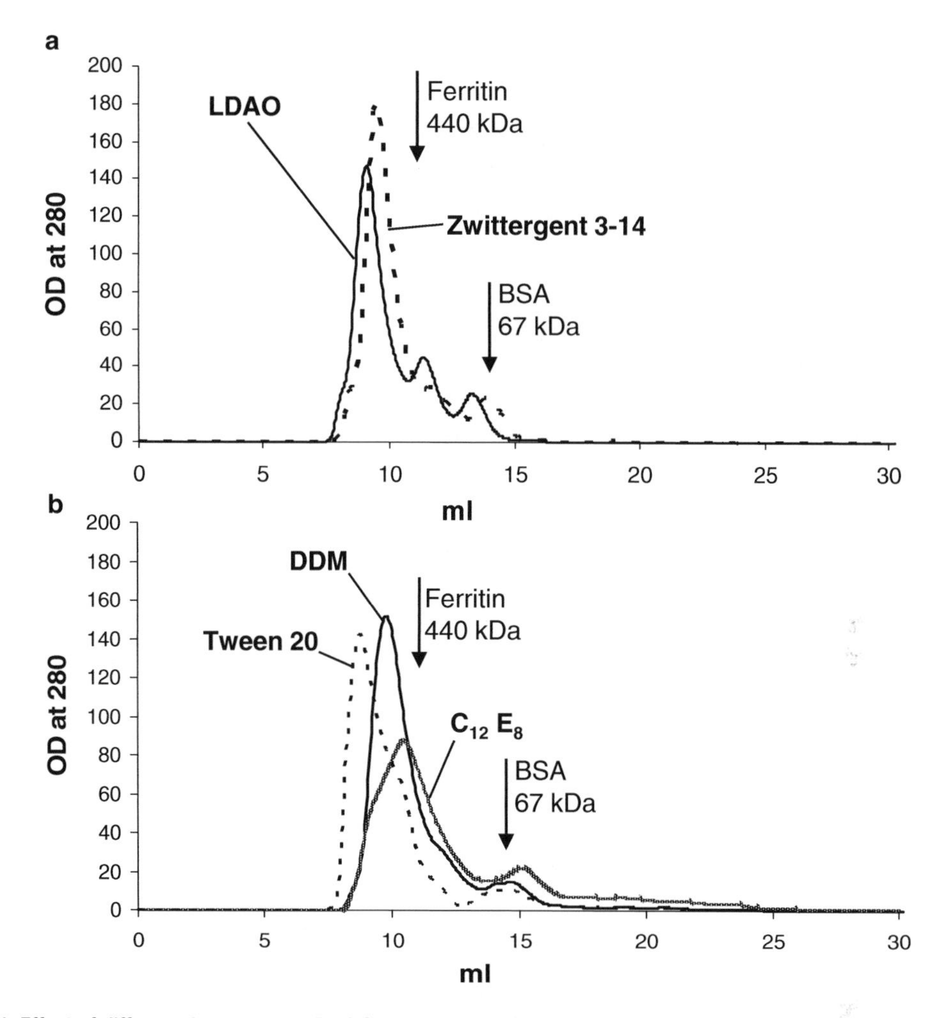

Fig. 4. Effect of different detergents on PorA fractionation by SEC. Elution absorption profiles at 280 nm are shown. A Superdex 200 SEC column (GE healthcare) is run in 50 mM Tris–HCl buffer, pH 7.9, 150 mM NaCl plus detergent, as indicated, at the following concentrations: LDAO 0.1%, zwittergent 3–14 0.05%, Tween 20 0.1%, DDM 0.04%, $C_{12}E_8$ 0.05%. For each run, 0.5 mg of PorA was loaded onto the column and the column flow rate was 0.5 mL/min. (**a**) LDAO and Zwittergent 3-14; (**b**) Tween 20, DDM and C12E8.

five detergents is consistently higher than the ferritin marker at 440 kDa, for example (Fig. 4). The reason for this is unclear; nevertheless, we find SEC a useful method for comparison of the effects between different detergents. PorA in Tween20, for example, elutes at a significantly lower elution volume, indicating a tendency of the protein to aggregate under these conditions. The other four detergents give peaks at approximately the same elution volume, although some differences in long-term stability were noted: PorA is stable in LDAO for 2–3 months at 20°C compared to DDM, which causes it to precipitate after few days (see Note 10).

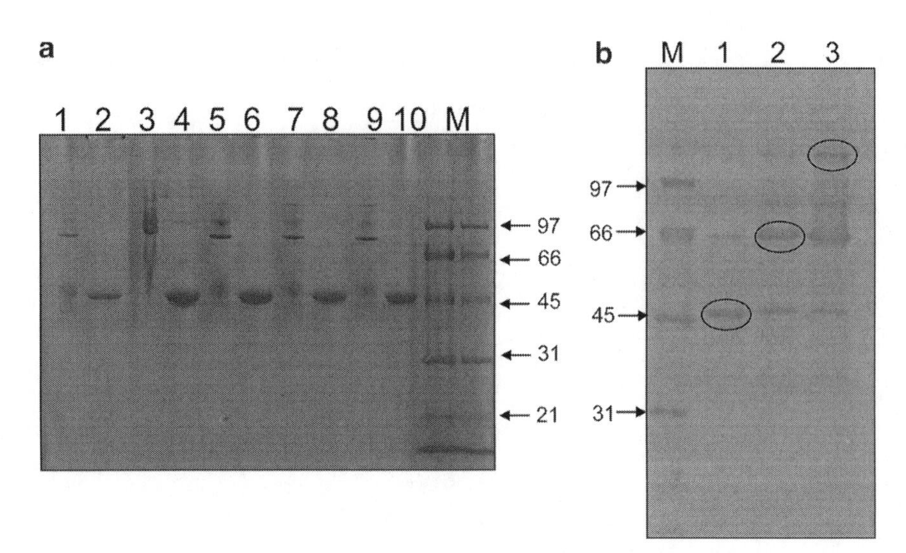

Fig. 5. Analysis of PorA oligomers by native PAGE. (**a**) Samples in odd numbered lanes were not pre-treated at 100°C and included 0.1% (w/v) SDS in the gel sample buffer; samples in even numbered lanes were treated at 100°C for 5 min before loading and included 2% (w/v) SDS in the gel sample buffer. Detergents used were lanes 1 and 2, $C_{12}E_8$; 3 and 4, LDAO; 5 and 6, zwittergent 3–14; 7 and 8, Tween 20; 9 and 10, DDM. (**b**) Effect of varying ratios of SDS:β-octyl glucoside on migration: lane 1 1:0.25%, lane 2 1:0.5%, lane 3 1:1%. Samples were not pre-treated at 100°C. Bands equivalent to monomer, dimer, and trimer are circled.

3.5.2. Semi-Native SDS-PAGE (NuPAGE)

In addition to SEC, the collected fractions can by analyzed with semi-native SDS-PAGE, which is often used to examine the assembly of oligomeric proteins.

1. For semi-native PAGE, suspend the protein samples in the loading buffer that contains 0.1% (w/v) SDS, but do not heat the samples at 100°C before loading.
2. Run the gel at 150 V for 60 min using the same running buffer as described for denaturing gels and then follow the procedure as outlined in steps 4–6 of Subheading 3.4.
3. Examine the migration of PorA in the presence of the same five detergents used for SEC (Fig. 5a). In each case there is a clear difference in migration between the boiled and un-boiled samples, with the latter exhibiting three bands at lower migration than the single band from the boiled sample. Note also that LDAO tends to disrupt the migration of PorA on these gels (lane 3 of Fig. 5a). This kind of assay can therefore be readily used to follow refolding efficiency and examine stability in different detergents. Through use of the non-charged detergent β-octyl glucoside and modification of the ratio of SDS:β-octyl glucoside, it is also possible to modify the relative quantities of PorA in each of the three major bands (Fig. 5b). As we increase the β-octyl glucoside concentration, more PorA partitions into species that migrate a shorter distance (circled in Fig. 5b). The obvious interpretation is that the three bands correspond

to monomer, dimer and trimer, and higher concentrations of β-octyl glucoside promote trimer stability (see Note 11).

3.6. Crystallization of PorA

The availability of crystallization robots, which require 100 nL or less per trial, in combination with the availability of custom-made screens, has greatly accelerated the initial screening process. For OMPs, we would recommend carrying out initial trials using the MemStart/MemSys, MemGold (10) and Morpheus (11) crystallization screens, available from Molecular Dimensions, as follows:

1. Start the initial sitting drop trials using 100 nL of protein (10 mg/mL concentration) plus 100 nL from the reservoir solution in a standard MRC Crystallisation Plate™ using the Mosquito automated solution dispenser.
2. Maintain the plates in an incubator at 20°C and inspect daily during the first week and weekly thereafter.
3. Repeat the initial hits and confirm with solutions prepared in-house.
4. For the purposes of illustration, the detergent trials for crystals of the PorA variant P1.7-2,4. are described. The first trials were conducted using PorA in the detergent LDAO; crystals were found under some conditions in all the screens, but exhaustive testing failed to show any diffraction from them. The next step was to change the detergent used, in this case to the polyoxyethylene detergent C_8E_5, using the detergent exchange method outlined in Subheading 3.3.4, and to repeat the screen. In this second round, fewer "hits" were observed but, in the case of the Morpheus screen, they could be extracted and frozen directly (all conditions in the Morpheus screen are cryoprotecting). Crystals were tested for diffraction quality at a synchrotron radiation source. Diffraction was now observed, but weak, to between 15 and 7 Å resolution. The best crystals obtained were from condition E12 of the Morpheus screen which consistently diffracted to about 7 Å (see Note 12). An example of those crystals is shown in Fig. 6a.
5. Following successful detergent trials, diffraction can be optimized in two ways:
   (a) Adjusting the concentrations of the existing components of the crystallization conditions. Adjustment of pH was found to have relatively little effect on crystal size or diffraction quality. Increasing the concentration of ethylene glycol inhibited crystallization but increasing the MPD/PEG1000/PEG3350 concentration from 37.5 to 42% improved the size of the crystals.
   (b) Investigating the effect of addition of other components. Two additives were found to increase crystal size and diffraction quality: the addition of imidazole to 50 mM and

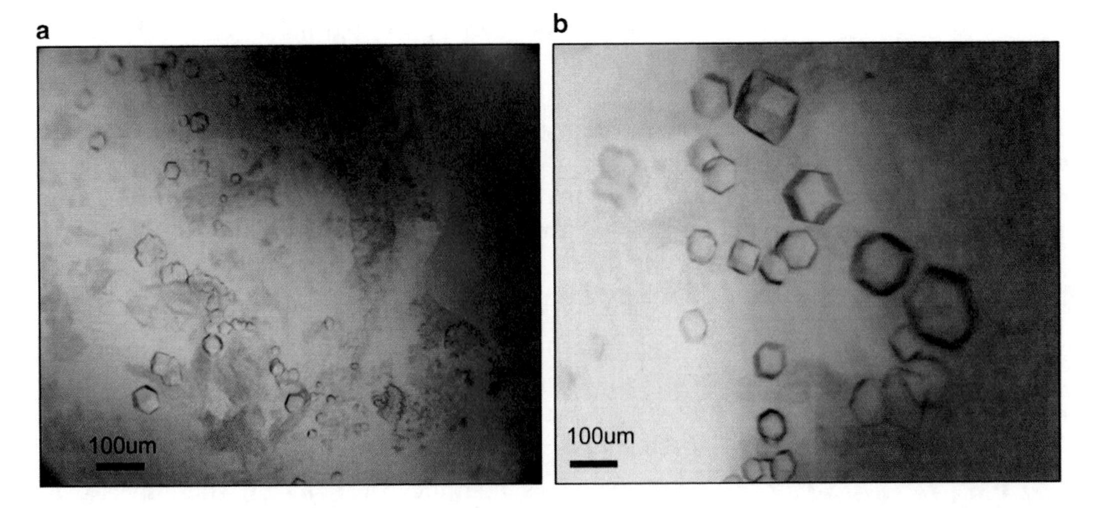

Fig. 6. Examples of the optimization of crystals of the PorA variant P1.7–2, 4. (**a**) Crystals of PorA at an initial stage from the Morpheus screen (see Subheading 2.5 and Note 12). (**b**) PorA crystals after optimization of initial conditions.

the inclusion of 0.5% (w/v) hexyl-glucoside. The latter is an example of a small amphiphilic molecule, which are often used to modulate crystallization conditions, particularly membrane proteins. The size of the largest crystals was increased to about 150–200 μm across (Fig. 6b). These improvements resulted in a substantial improvement in the diffraction limit, from 7 to 3.2 Å resolution.

6. Final structure determination will require (1) collection of a native dataset, (2) acquisition of phases, (3) model building, and (4) refinement. A variety of strategies will be required for experimental phase determination; traditional multiple isomorphous replacement methods can be used, which require soaking of heavy atom derivatives, collection of derivative data, and identification of the location of heavy atoms within the unit cell. Frequently, such approaches are now combined with anomalous diffraction methods, which can be readily implemented at synchrotron radiation sources, and from which high-quality phase information can be extracted.

4. Notes

1. There are a variety of ways in which the expression vector can be designed; the minimum requirements are a strong, inducible promoter, and the absence of the native signal sequence. Signal sequences can be reliably predicted using the SignalP server (http://www.cbs.dtu.dk/services/SignalP/). We frequently use the pET series of expression vectors from Novagen.

2. We have found that the methods reported here are suitable for all naturally occurring PorA variants which we have encountered. We also successfully applied them to other *Neisseria* OMPs, such as PorB, FetA, and OpcA. However, for OpcA, the dilution ratio (IBs to refolding buffer) is higher (1:20).
3. For optimal suspension, use a pestle and mortar.
4. Figure 1 shows the visual changes in the PorA solution during refolding. Initially, upon drop-wise addition of Refolding buffer to the IB solution, the turbidity and density of the solution increases and reaches a maximum (Fig. 1c) before declining upon further addition of refolding buffer to become transparent (Fig. 1d).
5. The concentration of imidazole can be increased from 250 to 500 mM in buffer C for better elution. Generally, we use 250 mM imidazole for the first two elution fractions and 500 mM for the next two fractions.
6. Dialyze against two changes of 2 L of buffer D.
7. We find that the Micro-Lowry protein assay to be the most convenient and reliable method for estimation of protein concentration in our OMP preparations.
8. It is better to centrifuge the samples for 30 s in a micro-centrifuge after boiling the samples, in order to avoid any sample loss by evaporation before loading them onto SDS-PAGE gels.
9. Although this is not the most accurate way to determine protein concentration, we find it a quick and valuable method to employ during purification.
10. We find that the elution volume of PorA in LDAO is independent of sequence variant type.
11. Native gel electrophoresis results can vary with different conditions, including running apparatus, voltage and percentage of gel.
12. The E12 reservoir solution uses a mixture of organic precipitants: 12.5% (w/v) polyethylene glycol (PEG) 1000(P1K), 12.5% (w/v) PEG 3350 (P3350), and 12.5% (v/v) 2-methyl2,4-pentanediol (MPD). Other components are 0.1 M TrisBicine buffer pH 8.5 and 0.03 M each of diethylene glycol, triethylene glycol, tetraethylene glycol, and pentaethylene glycol.

Acknowledgments

We thank Dr Hannah Chan and Prof. Ian Feavers (National Institute for Biological Standards and Control, South Mimms) for supply of PorA expression vectors.

References

1. Wimley WC (2003) The versatile beta-barrel membrane protein. Curr Opin Struct Biol 13: 404–411.
2. Jansen C, Wiese A, Reubsaet L et al (2000) Biochemical and biophysical characterization of *in vitro* folded outer membrane porin PorA of *Neisseria meningitidis*. Biochim Biophys Acta-Biomem 1464: 284–298.
3. Tanabe M, Nimigean CM, Iverson TM (2010) Structural basis for solute transport, nucleotide regulation, and immunological recognition of *Neisseria meningitidis* PorB. Proc Natl Acad Sci USA 107: 6811–6816.
4. Kortekaas J, Muller SA, Ringler P et al (2006) Immunogenicity and structural characterisation of an *in vitro* folded meningococcal siderophore receptor (FrpB, FetA). Microb Infect 8: 2145–2153.
5. Vandeputte-Rutten L, Bos MP, Tommassen J et al (2003) Crystal structure of Neisserial surface protein A (NspA), a conserved outer membrane protein with vaccine potential. J Biol Chem 278: 24825–24830.
6. Prince SM, Achtman M, Derrick JP (2002) Crystal structure of the OpcA integral membrane adhesin from *Neisseria meningitidis* Proc Natl Acad Sci USA 99: 3417–3421.
7. Moore J, Bailey SES, Benmechernene Z et al (2005) Recognition of saccharides by the OpcA, OpaD, and OpaB outer membrane proteins from *Neisseria meningitidis* J Biol Chem 280: 31489–31497.
8. Feavers IM, Pizza M (2009) Meningococcal protein antigens and vaccines. Vaccine 27: B42–B50.
9. Peterson GL (1977) A simplification of the protein assay method of Lowry *et al.* which is more generally applicable. Anal Biochem 83: 346–356.
10. Newstead S, Ferrandon S, Iwata S (2008) Rationalizing alpha-helical membrane protein crystallization. Prot Sci 17: 466–472.
11. Gorrec F (2009) The MORPHEUS protein crystallization screen. J Applied Crystallog 42: 1035–1042.

Chapter 7

Genome-Scale Metabolic Models: Reconstruction and Analysis

Gino J.E. Baart and Dirk E. Martens

Abstract

Metabolism can be defined as the complete set of chemical reactions that occur in living organisms in order to maintain life. Enzymes are the main players in this process as they are responsible for catalyzing the chemical reactions. The enzyme–reaction relationships can be used for the reconstruction of a network of reactions, which leads to a metabolic model of metabolism. A genome-scale metabolic network of chemical reactions that take place inside a living organism is primarily reconstructed from the information that is present in its genome and the literature and involves steps such as functional annotation of the genome, identification of the associated reactions and determination of their stoichiometry, assignment of localization, determination of the biomass composition, estimation of energy requirements, and definition of model constraints. This information can be integrated into a stoichiometric model of metabolism that can be used for detailed analysis of the metabolic potential of the organism using constraint-based modeling approaches and hence is valuable in understanding its metabolic capabilities.

Key words: Genome-scale metabolic network reconstruction, Metabolic networks, Metabolic flux analysis, Flux balance analysis, Constraint-based modeling

1. Introduction

The vast amount of available genome sequence databases for organisms and the development of high throughput bioinformatics and experimental techniques have led to intense efforts to functionally characterize these genomes and build genome-scale datasets for a variety of organisms like *Neisseria meningitidis* (1). Genome-scale metabolic network models are an example of this. Whereas enzyme kinetics are often unknown, the stoichiometry of the reaction the enzyme catalyzes is in most cases well established. The reaction stoichiometry of the biochemical reactions that are present forms

Myron Christodoulides (ed.), *Neisseria meningitidis: Advanced Methods and Protocols*, Methods in Molecular Biology, vol. 799, DOI 10.1007/978-1-61779-346-2_7, © Springer Science+Business Media, LLC 2012

the basis of a genome-scale network in which the different metabolic reactions are connected through their common substrates and products. Whether a metabolic reaction can take place inside an organism can primarily be reconstructed using the information provided in the annotated genome combined with information from curated databases and literature. Such a genome-scale network thus represents the metabolic potential of the organism and hence is valuable in understanding its metabolic capabilities. Genome-scale metabolic models have been built for several organisms (2, 3) and have been used to analyze cultivation data, get a better understanding of cellular metabolism (1, 4), develop metabolic engineering strategies (5–8), design media and processes (1, 7, 9, 10), assess theoretical capabilities (2), and even for online control of the process (11), which illustrates its usefulness for process development and optimization. In this chapter, an introduction to genome-scale metabolic network reconstruction and the constraint-based modeling approach is presented.

2. Materials

For the genome-scale metabolic network reconstruction and constraint-based modeling methods described below, the following software materials can be used.

1. Major online bioinformatics databases (see Table 1).
2. Software for metabolic network analysis (see Table 2).
3. Tools for assigning putative gene function, e.g., BLAST or InParanoid.
4. Automated tools for metabolic network reconstruction, e.g., Pathologic (Pathway Tools) and AUTOGRAPH.

3. Methods

3.1. Metabolic Network Reconstruction

Before a metabolic model for a given microorganism (or specific phenotype) can be used to gain new insights into its metabolic capabilities or evolutionary history, it must first be built from the available (scattered) genomic, biochemical, and physiological information. This process is known as model reconstruction and has been extensively reviewed before (12, 13). The process of model reconstruction is schematically shown in Fig. 1.

It starts with the functional annotation of a sequenced genome, meaning that for each open reading frame (ORF) a function is sought using literature and pathway databases. Part of the functionally annotated genes will code for enzymes involved in (primary) metabolism.

Table 1
Major online databases that can be used as resource for model reconstruction

Database	URL	Description
GOLD	http://www.genomesonline.org	Monitoring of genome sequencing projects around the world
TIGR	http://www.tigr.org	Information about genomes of sequence organisms
NCBI	http://www.ncbi.nlm.nih.gov/Genbank	Contains sequence data of microbial and higher organisms
EBI	http://www.ebi.ac.uk/embl	Similar as GenBank
KEGG	http://www.genome.ad.jp/kegg	Database that includes all microorganisms with publically available genome sequence. Holds both genomic and metabolic information
BioCyc	http://biocyc.org	Similar as KEGG, also includes MetaCyc
Expasy	www.expasy.org	Protein sequence database that provides organism-specific annotation
EMP	http://www.empproject.com	Generic information on enzymes and metabolism
BRENDA	http://www.brenda.uni-koeln.de	Information about enzymes (organism related)
UniProt	http://www.uniprot.org	Universal Protein Resource database
MIPS	http://mips.gsf.de	Information about genomic structure and integration of data for several organisms
TCDB	http://tcdb.ucsd.edu	Classification system for membrane transport proteins
TransportDB	http://www.membranetransport.org	Predictions of membrane transport proteins for fully sequenced genomes
Reactome	http://www.reactome.org	Curated database of biological pathways
BiGG	http://bigg.ucsd.edu	Repository of reconstructed genome-scale metabolic models
TargetP	http://www.cbs.dtu.dk/services/TargetP/	Predicts subcellular location of eukaryotic proteins
SignalP	http://www.cbs.dtu.dk/services/SignalP/	Predicts the presence and location of signal peptide cleavage sites in amino acid sequences from different organisms
PSort	http://psort.hgc.jp	Similar as TargetP

Table 2
Software for metabolic network analysis

Package	URL	Reference
Flux balance analysis/metabolic flux analysis		
CellNetAnalyzer	http://www.mpi-magdeburg.mpg.de/projects/cna/cna.html	(71, 72)
MetaFluxNet	http://mbel.kaist.ac.kr/lab/mfn/index.html	(73, 74)
FBA, Cobra Toolbox	http://gcrg.ucsd.edu/Downloads/Flux_Balance_Analysis http://gcrg.ucsd.edu/Downloads/Cobra_Toolbox	(75)
CellDesigner	http://celldesigner.org/index.html	(76, 77)
Fluxor	http://sourceforge.net/projects/fluxor/	
Simpheny (commercial)	http://www.gtlifesciences.com/technology/Simpheny.html	(78)
Isotope-based metabolic flux analysis		
FiatFlux	http://www.imsb.ethz.ch/researchgroup/nzamboni/Software/fiatflux	(79)
13C flux	http://www.13cflux.net/omix/	(80)
OpenFLUX	http://web.aibn.uq.edu.au/cssb/Resources.html	(81)
Elementary flux modes/extreme pathways		
Metatool	http://pinguin.biologie.uni-jena.de/bioinformatik/networks/	(82)
Yana	http://yana.bioapps.biozentrum.uni-wuerzburg.de	(83, 84)
ExpA (extreme pathways)	http://gcrg.ucsd.edu/Downloads/Extreme_Pathway_Analysis	(85)

For these enzymes the reaction stoichiometry is obtained as well as information on the reversibility of the reaction, which can next be included in the model. In case of different cellular compartments also the localization of the enzyme can be determined. A special reaction is the reaction for biomass formation, which is a lumped reaction, not connected to a single enzyme but the sum of many biochemical reactions. This results in a basic reaction set, which usually still contains errors like a wrong reaction stoichiometry and pathway gaps, which must be manually corrected. Next, the model must be experimentally validated, which includes estimation of the unknown energy parameters, being the efficiency of ATP generation and the amount of ATP required for maintenance and biomass formation. This process is iterative meaning that in case the model does not agree with experimental data, the model is critically evaluated again starting at the functional annotation. The different aspects of the model reconstruction are discussed in more detail below. The process of reconstruction of the (genome-scale) metabolic network of a microorganism requires a significant input

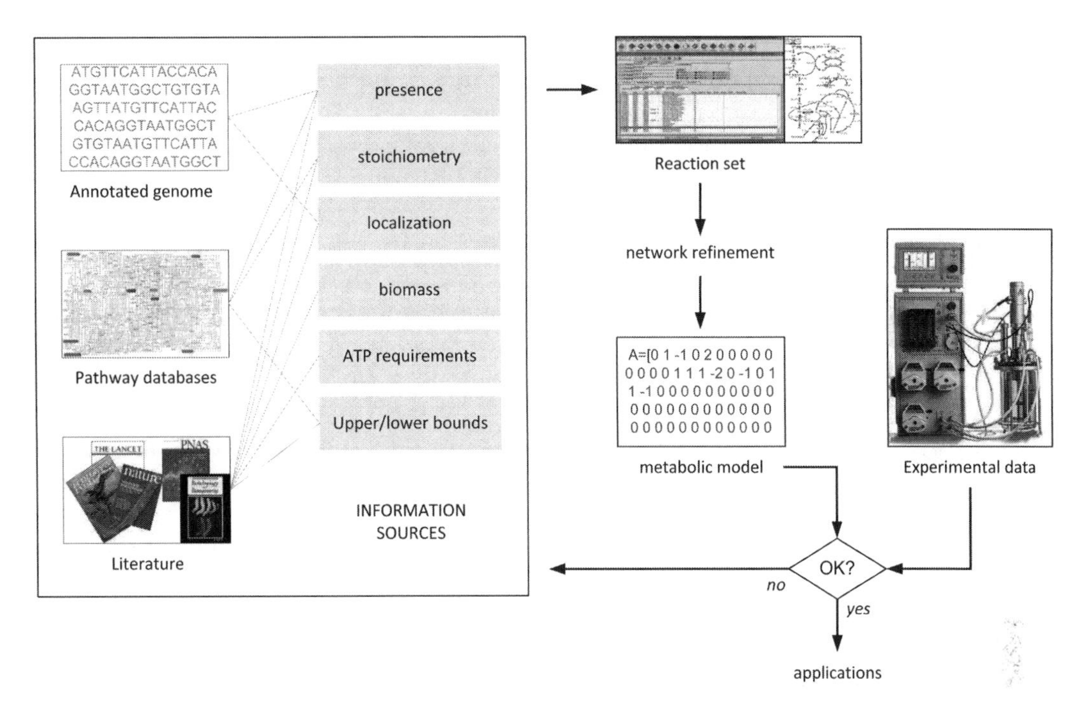

Fig. 1. The iterative process of metabolic network reconstruction. The process starts with gathering the current knowledge about the microorganism from multiple sources of information. Next the construction and debugging of the reaction set, the building of a stoichiometric matrix, and the comparison of the *in silico* simulation results using experimental data are carried out. Once the experimental data (i.e., measured rates) are in agreement with the *in silico* predictions, the model can be used for further applications like, e.g., development of metabolic engineering strategies. Adapted from Rocha et al. (69).

from online bioinformatics databases, where information regarding genome sequencing and metabolic reactions can be found. Some important online resources are listed in Table 1.

3.2. Identification of Reactions and Their Stoichiometry

An important requisite in genome-scale model reconstruction is the availability of a well-annotated genome. Meaning that the process of genome assembly, identification of ORFs, using a good gene prediction training set and their localization on the genome (structural annotation), and the assignment of possible biological roles (functional annotation) should have been done carefully. In all cases, detailed inspection of the genome (functional) annotation is required prior to or concurrently with reconstruction of the network. For this, reliable databases, such as GOLD, TIGR, NCBI and EBI (Table 1) can be consulted. Important data to be extracted from the various databases include gene or ORF, assigned function(s), (translated) sequence and for protein coding genes the Enzyme Commission (EC) number that corresponds with the reaction(s) encoded by the gene. In the case where an ORF has no function assigned, the sequence can be analyzed aiming at assigning a putative function.

Using orthology-based methods like BLAST or Inparanoid (14), putative functions can be found and based on the similarity score between orthologous genes one can decide to include the associated reaction in the model, hereby assigning the putative function. However, caution is required when including putative functions, especially if shortened genes are identified, since some gene regions coding for specific protein domains that are related with the annotated function can be lost. In some cases, genes encoding proteins with multiple functions may lose the domain(s) responsible for some of these activities. These circumstances might lead to the assignment of a gene to an incorrect functional category, which could lead to an erroneous determination of the set of functions an organism is able to perform. Next to the inclusion of genes or ORFs encoding enzymes (with an EC number), also membrane transport reactions (that lack an EC number) should be included. Hence, annotated genomes should also be analyzed for transporter functions. Genes involved in regulatory processes of metabolic reactions or signal transduction cascades are currently excluded from model reconstruction. Nevertheless, it is useful to store information related to these processes as they might be interesting for future exploration.

The result of the above mining of the genome is a set of genes with an identified function. Notably, in *Neisseria meningitidis* strain MC58 the percentage of protein coding genes with a known function varies from 54 to 60% (the latter value includes putative functions), meaning that many of the genes still have an unknown function. Metabolic network reconstruction is time-consuming and automated methods to accelerate the process have been proposed (15–18). Two examples are Pathologic, which is part of Pathway Tools (17) and AUTOGRAPH (15), both of which can be used to predict as much metabolic information as possible for a given species. The Pathologic algorithm takes annotated genomes as input and predicts gene–reaction associations based on name-matching and EC-codes with the idea to curate the predictions afterwards (17). The EC-code matching approach to link metabolic information to genes is similar to other methods like metaSHARK, IdentiCS, and Pandora (19–21). Notably, Pathologic is the first step in the construction of a so-called PGDB (pathway-genome database) which consists of gene–reaction associations.

Several organism-specific PGDBs have been constructed and manually curated (9, 22–26). Despite the significant time saving using the automated methods, manual curation and inspection is still required to ensure elimination of possible errors and inconsistencies. Once the genome has been adequately mined, databases that describe gene-reactions (e.g., KEGG, BioCyc) and relevant biochemical literature should be consulted for determination of the exact reaction stoichiometry including cofactor requirements, reaction reversibility/irreversibility and consistency by inspection of the elemental balances of each reaction. When it is unclear whether NADH or NADPH is used as cofactor in a particular reaction, most times both reactions are included. In this case, the name of the gene stays the same but the names of the reactions differ. Notably, in some cases the cofactor specificity can be determined from the sequence and phylogenetic analysis (27). Similarly, when an enzyme accepts more than one substrate, the gene is associated with several reactions. For reactions that are catalyzed by enzyme complexes, several genes are associated with one (or more) reactions. In such cases, the overall or net reaction is often included. Isoenzymes are encoded by different genes but catalyze the same reaction(s) and should be considered separately. In order to determine whether a reaction is reversible or irreversible, the standard Gibb's free energy change associated with the reaction can be used (28). This reversibility/irreversibility information leads to definition of appropriate thermodynamic model constraints on the direction of reactions. Finally, like for the overall modeling approach the reconstruction process (Fig. 1) can also be seen as an iterative process where information obtained from several sources is combined to construct a preliminary set of reactions and constraints that are then analyzed to detect potential errors, for example, erroneous reactions in terms of stoichiometry, errors related to duplication of metabolite names or gaps in biochemical pathways.

3.3. Compartments and Localization

Another important aspect for metabolic network reconstruction is the definition of compartments and the assignment of the localization of enzymes inside these compartments. Software tools like SignalP, TargetP, and PSort (Table 1) deduce such information from the amino acid sequence of the enzymes. If a particular metabolite can be present in different compartments (e.g., by translocation or biochemical reactions), it is important to distinguish between these identical metabolites by using compartment-specific naming (e.g., M_mit for metabolite M in the mitochondrion and M_cyt for the same metabolite M in the cytosol). The translocation of such metabolites between compartments should be included as separate reactions in the model.

3.4. Biomass Formation

The next step in genome-scale model reconstruction is to add the reactions that describe the formation of biomass. The biomass formation reaction can be described by the assembly of macromolecules that constitute the biomass, e.g., protein, RNA, DNA, lipids and carbohydrates, which in turn are formed from precursor metabolites, e.g., amino acids, nucleotides, fatty acids, and sugars. It is also possible to define a reaction where biomass is directly formed from the precursor metabolites. The reaction stoichiometry is determined by the composition of the biomass and the biopolymers. For example, the reaction equation for protein can be directly derived from the measured amino acid composition. This illustrates the importance of accurate determination of the biomass composition. The biomass composition of several well-studied organisms like *Escherichia coli* (29, 30), *Neisseria meningitidis* (1, 4), *Lactococcus lactis* (31), *Saccharomyces cerevisiae* (32–38), *Penicillium chrysogenum* (39), and *Pichia pastoris* (40) is available. An important component in the biomass synthesis reaction is the number of ATP molecules that need to be included per unit of biomass, i.e., ATP stoichiometry, as described below.

3.5. Network Refinement

Once the complete set of reactions has been extracted and the stoichiometry has been checked the network may still contain gaps due to annotation errors, incomplete or incorrect EC-codes or annotation errors in protein databases. In addition, some reactions proceed spontaneously without being catalyzed by an enzyme. It is important to correct these errors by manual investigation of the network using biochemical literature, comparative genomics approaches (41), transcriptomics data (42) or by direct measurements. In addition, automated approaches like Pathologic can be used.

3.6. Energy Requirements

Although the ATP stoichiometry coefficients of many reactions that include ATP are known, with respect to ATP generation in mitochondria or the bacterial membrane and ATP usage for growth and maintenance the stoichiometry is not fixed and must be estimated from experiments. Part of the energy required for growth is accounted for in the model, like for example the ATP required to polymerize amino acids into protein. However, for another part related to, for example, the assembly of the biopolymers into functional biomass, and growth-related turnover of macromolecules and maintenance of gradients over the membrane this is not known. This part of the required energy, also called the growth-associated maintenance requirements represents all energy consumption related to growth not accounted for in the model, which must be determined experimentally. The same holds for the nongrowth-associated maintenance requirements for ATP, which represents the ATP required to keep the organism alive. ATP consumption for nongrowth-associated maintenance is usually included in the model by a separate irreversible reaction that converts ATP into ADP and orthophosphate. Growth and nongrowth-associated maintenance values of *Neisseria meningitidis* (4) and other organisms can be found in the literature (30, 43) or determined experimentally (4, 44).

In addition to the growth- and nongrowth-associated energy requirements also the ATP stoichiometry of oxidative phosphorylation is not fixed and is unknown. As mentioned earlier, the plasma membrane of bacteria contains ATP synthases that catalyze the formation of ATP in response to a transmembrane proton gradient. This gradient is formed when electrons are transferred from mostly NADH or $FADH_2$ to oxygen, where in the case of NADH more protons are translocated than for $FADH_2$. The amount of ATP formed per oxygen atom (P/O ratio) thus depends on the ratio of NADH and $FADH_2$, and the translocation stoichiometry (H^+/ATP). Thus to obtain the right flux of ATP formation in oxidative phosphorylation determination of the P/O ratio (45, 46) is important. The unknown parameters being the P/O ratio, and the growth- and nongrowth-associated parameters can be estimated from experiments with different carbon substrates or ratios of mixed substrates for pseudo steady-state growth at different growth rates as described extensively elsewhere (44).

3.7. Metabolic Network Analysis

When the pathway gaps and other possible errors have been fixed and the stoichiometry of all reactions is defined, the set of reactions can be used as a metabolic model to describe the function of the metabolic network in a quantitative manner. There are three types of metabolic models:

1. Steady-state models, which only take into account the flow of metabolites through the system in (pseudo) steady state.
2. Steady-state kinetic models, which take into account the flow of metabolites through the system in (pseudo) steady state and contain at least one kinetic equation that relates the concentration of a metabolite to a reaction rate.
3. Dynamic kinetic models, which take into account the kinetics of all different enzymes involved in the reaction network that are valid under dynamic conditions.

In an ideal situation, the model contains different levels of information, including reaction stoichiometry, reaction kinetics and regulatory information. Despite several studies in this direction (47–52), the required kinetic and regulatory information is currently lacking. For this reason a (pseudo) steady-state approximation, using the reaction stoichiometry in combination with mass balancing of the fluxes (1, 4, 30, 45, 53–57) is generally applied. The metabolic capabilities of the constructed network may be calculated using constraint-based computer simulation methods like metabolic flux analysis (MFA), flux balance analysis (FBA), elementary flux modes (58), or extreme pathways (59). An overview of available flux analysis software packages is provided in Table 2 and a brief summary of modeling techniques (i.e., MFA and FBA) is provided below (see Note 1).

From the set of reactions, including the transport reactions over the membrane, a stoichiometric matrix, S ($m \times n$), is constructed where m is the number of metabolites and n the number of reactions. Each column of S specifies the stoichiometry of the metabolite in a given reaction from the metabolic network. The construction of a network is relatively easy for small networks. However, if the networks become more complex mistakes are easily made. The consistency of a network may be checked by looking at the elemental balances for each reaction. The elemental composition is written in an elemental composition matrix *E*, with columns containing the different compounds and rows containing the different elements. Since each reaction must obey the law of elemental conservation, the following equation should always hold:

$$E \cdot S = 0 \tag{1}$$

Next, the mass balance for each intracellular metabolite can be constructed resulting in:

$$\frac{\mathrm{d}C}{\mathrm{d}t} = S \cdot v \tag{2}$$

where C is the vector of concentrations of intracellular metabolites (mole.(unit biomass)$^{-1}$), v is the vector of conversion rates (mole. unit biomass^{-1}.h^{-1}), including the transport rates over the cell membrane. Since metabolism operates on a much faster timescale than for instance cell division, it is reasonable to assume that metabolic dynamics have reached a so-called pseudo steady state, where metabolite concentrations are constant in time or, in other words, that there is no accumulation of the intracellular metabolites (law of conservation of mass). Hence, their net accumulation rate is zero giving:

$$S \cdot v = 0 \tag{3}$$

From the above (3) the solution space, i.e., all possible combinations of flux values that satisfy the equation, can be derived. If a flux does not contribute to this solution space meaning its value is always zero, the reaction is called a dead end. Dead ends can never carry a flux and are usually caused by a gap in a pathway or a wrong annotation. Usually part of the transport rates over the membrane, being the specific consumption and production rates of substrates and products are estimated from measurements resulting in the next equation.

$$S_c \cdot v_c = S_m \cdot v_m \tag{4}$$

where v_c contains the unknown and v_m contains the measured fluxes. S_c and S_m are the corresponding parts of the matrix *S*. If (4) has one unique solution, the system is called determined. The unknown fluxes can then be calculated from the measured rates using the next equation in case S_c is a square matrix:

$$v_c = -S_c^{-1} \cdot S_m \cdot v_m \tag{5}$$

where S_c^{-1} is the inverse of the matrix. In case the matrix S_c is not a square matrix, the inverse cannot be calculated and the following equation can be used:

$$v_c = -(S_c^{T} \cdot S_c)^{-1} \cdot S_c^{T} \cdot S_m \cdot v_m \tag{6}$$

where S_c^T is the transpose of matrix S_c. Calculation of fluxes based on measurements is called MFA. However, often the system of equations is not determined and a unique solution of (6) does not exist. This can be due to two reasons. First of all insufficient rates are measured (see Note 1). Second, even if all rates that can be measured are measured it is still possible that a unique solution does not exist due to the presence of different pathways with the same overall stoichiometry, i.e., parallel and cyclic pathways (see Note 2). For both cases S_c^{-1} and $(S_c^T \cdot S_c)^{-1}$ cannot be calculated. For both situations usually part of the fluxes can be calculated from the measurements, while in principle at least two of the fluxes cannot be calculated from the measurements. In case of insufficient measurements, the problem can of course be simply solved by measuring the additional fluxes. For the second situation this is not possible. In some cases ^{13}C MFA, which uses ^{13}C labeling distribution of intracellular metabolites, can provide additional constraints and can yield a determined model system for this situation for which a unique solution can be calculated (60). Whereas this is the preferred method to estimate the unknown fluxes, it is also experimentally and mathematically complex and time-consuming. Another option is using the so-called Moore–Penrose pseudo inverse the solution given by:

$$v_c = -S_c^{\#} \cdot S_m \cdot v_m \tag{7}$$

where $S_c^{\#}$ is the (left) pseudo inverse of matrix S_c. For the determined part of the network, this equation calculates the flux values. For the under-determined parts that do not have a unique solution, this equation calculates the minimum norm solution. The minimum norm solution is only one of the infinite possible solutions and minimizes the sum of the fluxes squared (see Note 1). However, the biological relevance of this solution remains questionable. To get a biologically more meaningful solution, linear optimization can be used for both mentioned situations (see Notes 1 and 2). Within the solution space defined by the connectivity (stoichiometry) and capacity constraints (upper and lower boundary values of fluxes and transport rates lb and ub, respectively), a solution is calculated based on a defined objective function. This approach is called FBA. For this, constraints are defined for the different flux values, including the transport rates over the cell membrane and an objective function is defined resulting in the next definition of the problem (see Note 1):

$$\begin{aligned} &\text{Objective function}: && \text{maximize}(c \cdot v_i) \\ &\text{Constraints}: && S \cdot v = 0 \\ & && \text{lb}_i \le v_i \le \text{ub}_i, i = 1, \ldots N \end{aligned} \tag{8}$$

where $c.v_i$ contains the combination of fluxes to be optimized (usually a single flux).

The main objection against the technique of linear optimization (i.e., linear programming) to find a solution for an under-determined network is that the objective function may not be valid for the biological system and one should realize that a mathematical solution is calculated that may not represent the actual situation. Using FBA, the original solution space of the under-determined parts of the model system is reduced to a single solution. Notably, this single solution is not necessarily unique and there can be more than one flux distribution that reaches the optimal value of the objective function.

3.8. Objective Functions and Constraints

Some examples of objective functions described in the literature are maximization of biomass formation, maximization of ATP production, minimization of glucose uptake, and minimization of the production of NADH and NADPH (56, 61, 62). Successful prediction of growth rate and metabolite secretion rate has been done using the FBA approach combined with maximization of biomass formation as an objective function (63–66). Nevertheless, as stated, the calculated intracellular flux distribution in these cases may not represent the true values. An extensive comparison of FBA-based *in silico* flux predictions, using different objective/constraint combinations, with in vivo fluxes from ^{13}C-experiments demonstrated that prediction of flux distributions is, within limits, possible. The essential element in this case is to identify constraints and the most relevant objective for a specific condition, since no single objective predicted the experimental data for wild-type *E. coli* under all conditions (62). Constraints can be set on certain enzymatic reactions on the basis of information found in the literature or determined experimentally. For example, measured production and consumption rates can be included as constraints on the transport fluxes over the cell membrane. The main constraints to be added to the set of reactions are the so-called thermodynamic constraints, which are related to the reversibility/irreversibility of the reactions. Such thermodynamic information can be found in online databases (e.g., Brenda, UniProt, Kegg, and BioCyc) but should be used with caution since the information in the databases is not organism-specific. It is worth noting that a software tool like anNET (28) can be used to test quantitative data sets for thermodynamic consistency.

4. Notes

1. Flux analysis in a nutshell
   Consider the simplified metabolic network in Fig. 2 in which a substrate (S) can be converted into biomass (B) and a by-product (P) through several chemical conversions with a stoichiometry of one, and which involves five intracellular metabolites (Mi).

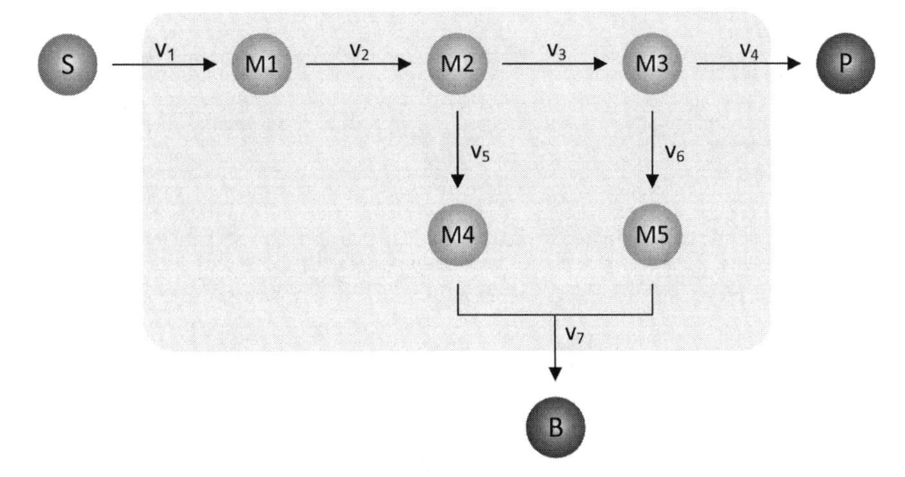

Fig. 2. Simplified metabolic network. Substrate (S) can be converted into biomass (B) and product (P) through several chemical conversions involving intracellular (*light grey* area) metabolites (Mi).

In the network in Fig. 2, there are five intracellular metabolites and seven reactions (five equations and seven unknowns). Therefore, the nullspace (i.e., solution space) is two-dimensional (Fig. 3a). If values for the biomass and product formation rate (v_7, v_4) are known from measurement, the degree of freedom of the network is zero, which means that the system is determined and a single unique solution for all fluxes and the exchange rates can be calculated (Fig. 3b) using the linear-algebra techniques described in the text. A special situation occurs when v_1 is also measured. In this case, more rates are measured than is needed for solving the system. In this case, the measurements are called redundant and a procedure called data reconciliation can be used to check the consistency of the measurements. This is described in detail elsewhere (67, 68). Notably, solving determined and redundant metabolic networks is often referred to as MFA. In the case when no measurements are available, the system is under-determined. A solution can be calculated (i.e., estimated) by assuming that the system fulfills an optimality condition. By using a particular objective function, for example maximizing the product formation rate (v_4), the optimal solution (i.e., optimal flux distribution) that optimizes this object function can be found using linear programming. To find a particular maximum, one or more constraints are required, otherwise the solution to the optimization will be unbounded (Fig. 3c). The technique of solving an under-determined metabolic model is commonly referred to as FBA and is often used in genome-scale models, in which the number of unknown fluxes is too large to calculate from measurements (57). Measurements of uptake and production rates are then added as constraints on the transport fluxes over the membrane.

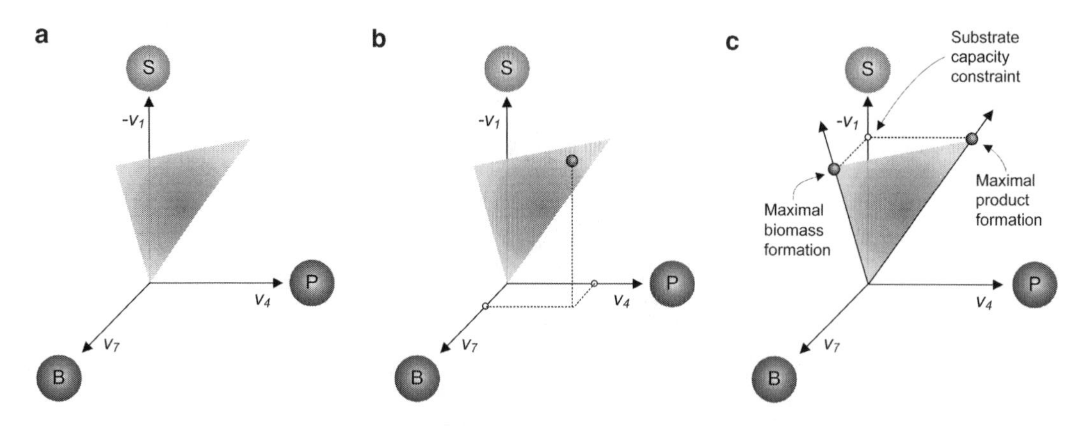

Fig. 3. Metabolic flux and flux balance analysis (FBA). (**a**) The null space of the example network is two-dimensional. (**b**) Metabolic flux analysis (MFA): if consumption and production rates are measured a unique solution can be calculated. (**c**) FBA: if only the substrate conversion rate (S) is known from measurement, a solution that optimizes a particular objective, like maximize biomass formation or maximize product formation can be calculated using linear programming. Adapted from Teusink et al. (57).

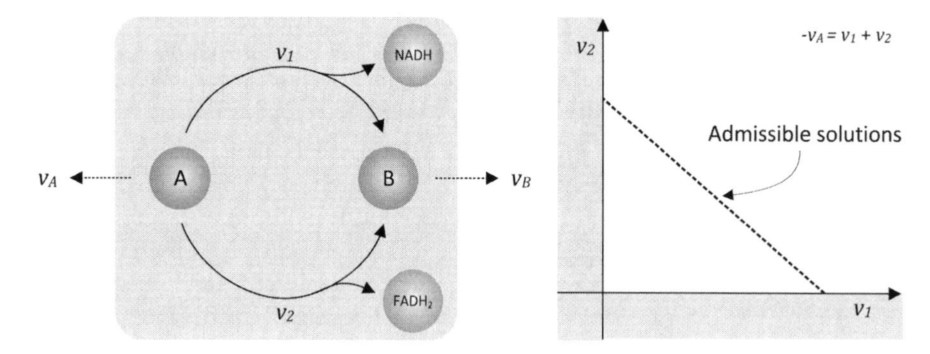

Fig. 4. Example of an under-determined network. Adapted from Bonarius et al. (70).

2. Inherently under-determined networks

In the simple under-determined metabolic network shown in Fig. 4, compound A is consumed by a cell and is metabolized to compound B via two different irreversible routes and then secreted. One route produces redox potential in the form of NADH, the other in the form of $FADH_2$.

The only flux balance in this system is that the sum of the two internal fluxes must equal the exchange flux. Once v_A is known from measurement, the solution space forms a straight line in the v_1 v_2 plane. Even if all rates are measured, a unique flux solution does not exist. Since the fluxes are constrained to be positive, we can only be in the positive quadrant of the plane, and thus the solution space is the dashed line shown in Fig. 4. A solution obtained using the Moore–Penrose pseudo-inverse,

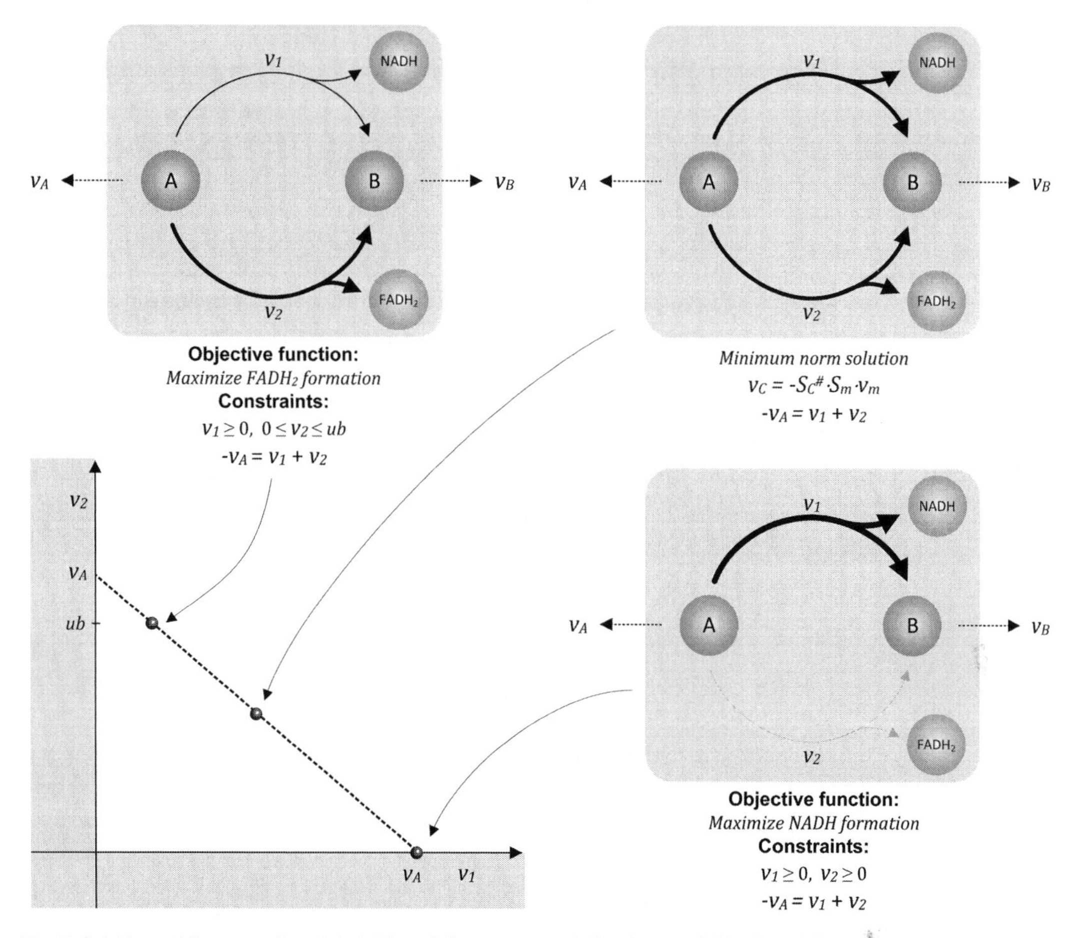

Fig. 5. Solutions of the example network. The minimum norm solution (*upper right*) adapted from Bonarius et al. (70), the solution obtained using linear programming and the objective function maximize NADH formation and constraints $v_1 \geq 0$ and $v_2 \geq 0$ (*lower right*) and the solution obtained using linear programming and the objective function maximize $FADH_2$ formation and constraints $v_1 \geq 0$ and $0 \leq v_2 \leq$ ub (*upper left*).

also called the minimum norm solution (or minimum norm constraint), minimizes the length of the solution vector without any further restrictions, i.e., it minimizes the sum of the squares of the fluxes. The Moore–Penrose pseudo-inverse gives the one solution vector that has the smallest Euclidian norm (e.g., the smallest square of its length) and is shown in Fig. 5. In linear programming, one uses an objective function (and constraints) to calculate an optimal result. If one uses the objective function that maximizes NADH production, it is clear that v_2 should go to zero and v_1 to the maximum value equal to the uptake rate. That solution lies at the extreme point of the solution space to the right (Fig. 5).

Conversely, if one maximizes the redox production from this metabolite in the form of $FADH_2$, and the upper boundary value for v_2 equals ub (e.g., obtained from measured maximal enzyme activity), then the optimal solution is at the opposite end of the solution space (Fig. 5). These simple examples illustrate how optimal solutions in linear spaces are found at the extremities of the allowable solution space.

To determine which elements of v_c have a single solution and to determine the relationship between the infinity of solutions for the other elements of v_c, the nullspace of S_c can be calculated. The nullspace or solution space or kernel matrix is defined as the set of linear independent vectors v_n that fulfill the equation:

$$S_c \cdot v_n = 0. \tag{9}$$

The number of independent nullspace vectors is equal to the number of columns in S_c minus the rank of S_c. Each nullspace vector can be added an arbitrary number of times to the base solution given in the above equation. Thus the complete solution is:

$$v_c = -S_c^{\#} \cdot S_m \cdot v_m + \text{nullspace}(S_c) \cdot \lambda \tag{10}$$

where λ is a vector with as many elements as there are vectors (columns) in the nullspace of S_c. For all possible values of λ, the solution remains valid. In other words, the nullspace vectors contain combinations of reactions of which the sum is always zero.

Parallel pathways have a common net reaction and cause under-determinancies in the metabolic network. To mathematically detect parallel pathways, they are considered to run in the opposite direction so that the net production rate is zero. These are called nullcycles and the best way to identify these nullcycles is to determine the nullspace of *S*. The number of nullspace vectors (i.e., columns) equals the number of nullcycles. If a unique solution is wanted, one of the parallel pathways must be removed or at least one reaction rate should be measured (e.g., by using ^{13}C labeled substrate) or depending on the conditions, fluxes should be set to zero. Identification of nullcycles is often called a network sensitivity check and the number of parallel pathways in a metabolic network is a measure of its robustness. It is worth noting that futile cycles are not the same as parallel pathways as they have ATP consumption as net effect.

Acknowledgments

We are grateful to The Netherlands Vaccine Institute (NVI) for supporting this work.

References

1. Baart GJ, Zomer B, de Haan A et al (2007) Modeling *Neisseria meningitidis* metabolism: from genome to metabolic fluxes. Genome Biol 8: R136.
2. Price ND, Papin JA, Schilling CH et al (2003) Genome-scale microbial in silico models: the constraints-based approach. Trends Biotechnol 21: 162–169.
3. Kim HU, Kim TY, Lee, SY (2007) Metabolic flux analysis and metabolic engineering of microorganisms. Mol BioSyst 4: 113–120.
4. Baart GJE, Willemsen M, Khatami E et al (2008) Modeling *Neisseria meningitidis* B metabolism at different specific growth rates. Biotechnol Bioeng 101: 1022–1035.
5. Hua Q, Joyce AR, Fong SS et al (2006) Metabolic analysis of adaptive evolution for in silico-designed lactate-producing strains. Biotechnol Bioeng 95: 992–1002.
6. Fong SS, Burgard AP, Herring CD et al (2005) In silico design and adaptive evolution of *Escherichia coli* for production of lactic acid. Biotechnol Bioeng 91: 643–648.
7. Smid EJ, Molenaar D, Hugenholtz J et al (2005) Functional ingredient production: application of global metabolic models. Curr Opin Biotechnol 16: 190–197.
8. Baart GJ, Langenhof M, van de Waterbeemd B et al (2010) Expression of phosphofructokinase in *Neisseria meningitidis*. Microbiology 156: 530–542.
9. Teusink B, van Enckevort FH, Francke C et al (2005) In silico reconstruction of the metabolic pathways of *Lactobacillus plantarum*: comparing predictions of nutrient requirements with those from growth experiments. Appl Environ Microbiol 71: 7253–7262.
10. Xie L, Wang DIC (1994) Stoichiometric analysis of animal cell growth and its application in medium design. Biotechnol Bioeng 43: 1164–1174.
11. Provost A, Bastin G (2004) Dynamic metabolic modelling under the balanced growth condition. J Proc Control 14: 717–728.
12. Covert MW, Schilling CH, Famili I et al (2001) Metabolic modeling of microbial strains in silico. Trends Biochem Sci 26: 179–186.
13. Francke C, Siezen RJ, Teusink B (2005) Reconstructing the metabolic network of a bacterium from its genome. Trends Microbiol 13: 550–558.
14. Ostlund G, Schmitt T, Forslund K et al (2010) InParanoid 7: new algorithms and tools for eukaryotic orthology analysis. Nucleic Acids Res 38: D196–203.
15. Notebaart RA, van Enckevort FH, Francke C et al (2006) Accelerating the reconstruction of genome-scale metabolic networks. BMC Bioinformatics 7: 296.
16. Herrgard MJ, Fong SS, Palsson BO (2006) Identification of genome-scale metabolic network models using experimentally measured flux profiles. PLoS Comput Biol 2: e72.
17. Karp PD, Paley S, Romero P (2002) The Pathway Tools software. Bioinformatics 18 Suppl 1: S225–232.
18. Moriya Y, Itoh M, Okuda S et al (2007) KAAS: an automatic genome annotation and pathway reconstruction server. Nucleic Acids Res 35: W182–185.
19. Pinney JW, Shirley MW, McConkey GA et al (2005) metaSHARK: software for automated metabolic network prediction from DNA sequence and its application to the genomes of *Plasmodium falciparum* and *Eimeria tenella*. Nucleic Acids Res 33: 1399–1409.
20. Sun J, Zeng AP (2004) IdentiCS – identification of coding sequence and in silico reconstruction of the metabolic network directly from unannotated low-coverage bacterial genome sequence. BMC Bioinformatics 5: 112.
21. Zhang KX, Ouellette BF (2010) Pandora, a pathway and network discovery approach based on common biological evidence. Bioinformatics 26: 529–535.
22. Caspi R, Altman T, Dale JM et al (2010) The MetaCyc database of metabolic pathways and enzymes and the BioCyc collection of pathway/genome databases. Nucleic Acids Res 38: D473–479.
23. Keseler IM, Collado-Vides J, Gama-Castro S et al (2005) EcoCyc: a comprehensive database resource for *Escherichia coli*. Nucleic Acids Res 33: D334–337.
24. Mueller LA, Zhang P, Rhee SY (2003) AraCyc: a biochemical pathway database for Arabidopsis. Plant Physiol 132: 453–460.
25. Romero P, Karp P (2003) PseudoCyc, a pathway-genome database for Pseudomonas aeruginosa. J Mol Microbiol Biotechnol 5: 230–239.
26. Romero P, Wagg J, Green ML et al (2005) Computational prediction of human metabolic pathways from the complete human genome. Genome Biol 6: R2.
27. Zhu G, Golding GB, Dean AM (2005) The selective cause of an ancient adaptation. Science 307: 1279–1282.
28. Zamboni N, Kummel A, Heinemann M (2008) anNET: a tool for network-embedded

thermodynamic analysis of quantitative metabolome data. BMC Bioinformatics 9: 199.

29. Neidhardt FC, Umbarger HE (1996) Chemical composition of *Escherichia coli*, In *Escherichia coli and Salmonella typhimurium: Cellular and Molecular Biology* (Neidhardt FC, Curtiss R, Ingraham JL, Brooks Low K, Magasanik B, Reznikoff WS, Riley M, Schaechter M, Umbarger HE, Eds.) 2 ed., pp 13–16, American Society for Microbiology, Washington.
30. Taymaz-Nikerel H, Borujeni AE, Verheijen PJ et al (2010) Genome-derived minimal metabolic models for *Escherichia coli* MG1655 with estimated in vivo respiratory ATP stoichiometry. Biotechnol Bioeng 107: 369–381.
31. Novak L, Loubiere P (2000) The metabolic network of *Lactococcus lactis*: distribution of (14) C-labeled substrates between catabolic and anabolic pathways. J Bacteriol 182: 1136–1143.
32. Albers E, Larsson C, Andlid T et al (2007) Effect of nutrient starvation on the cellular composition and metabolic capacity of *Saccharomyces cerevisiae*. Appl Environ Microbiol 73: 4839–4848.
33. Cortassa S, Aon JC, Aon MA (1995) Fluxes of carbon, phosphorylation, and redox intermediates during growth of *Saccharomyces cerevisiae* on different carbon sources. Biotechnol Bioeng 47: 193–208.
34. Herwig C, Von Stockar U (2003) Quantitative comparison of transient growth of *Saccharomyces cerevisiae, Saccharomyces kluyveri, and Kluyveromyces lactis*. Biotechnol Bioeng 81: 837–847.
35. Hjersted JL, Henson MA (2009) Steady-state and dynamic flux balance analysis of ethanol production by *Saccharomyces cerevisiae*. IET Syst Biol 3: 167–179.
36. Nissen TL, Schulze U, Nielsen J et al (1997) Flux distributions in anaerobic, glucose-limited continuous cultures of *Saccharomyces cerevisiae*. Microbiology 143 (Pt 1): 203–218.
37. Verduyn C, Postma E, Scheffers WA et al (1990) Energetics of *Saccharomyces cerevisiae* in anaerobic glucose-limited chemostat cultures. J Gen Microbiol 136: 405–412.
38. Wisselink HW, Cipollina C, Oud B et al (2010) Metabolome, transcriptome and metabolic flux analysis of arabinose fermentation by engineered *Saccharomyces cerevisiae*. Metab Eng 12(6):537–5178.
39. Nasution U, van Gulik WM, Ras C et al (2008) A metabolome study of the steady-state relation between central metabolism, amino acid biosynthesis and penicillin production in *Penicillium chrysogenum*. Metab Eng 10: 10–23.
40. Carnicer M, Baumann K, Toplitz I et al (2009) Macromolecular and elemental composition analysis and extracellular metabolite balances of *Pichia pastoris* growing at different oxygen levels. Microb Cell Fact 8: 65.
41. Osterman A, Overbeek R (2003) Missing genes in metabolic pathways: a comparative genomics approach. Curr Opin Chem Biol 7: 238–251.
42. Kharchenko P, Vitkup D, Church GM (2004) Filling gaps in a metabolic network using expression information. Bioinformatics 20 Suppl 1: i178–185.
43. Forster J, Famili I, Fu P et al (2003) Genome-scale reconstruction of the *Saccharomyces cerevisiae* metabolic network. Genome Res 13: 244–253.
44. van Gulik WM (2010) Metabolic models for growth and product formation, In *The Metabolic Pathway Engineering Handbook* (Smolke CD, Ed.), CRC press, Boca Raton.
45. van Gulik WM, Antoniewicz MR, deLaat WT, et al (2001) Energetics of growth and penicillin production in a high-producing strain of *Penicillium chrysogenum*. Biotechnol Bioeng 72: 185–193.
46. Vanrolleghem PA, Heijnen JJ (1998) A structured approach for selection among candidate metabolic network models and estimation of unknown stoichiometric coefficients. Biotechnol Bioeng 58: 133–138.
47. Covert MW, Knight EM, Reed JL et al (2004) Integrating high-throughput and computational data elucidates bacterial networks. Nature 429: 92–96.
48. Bruggeman FJ, Snoep JL, Westerhoff HV (2008) Control, responses and modularity of cellular regulatory networks: a control analysis perspective. IET Syst Biol 2: 397–410.
49. Bruggeman FJ, Westerhoff HV (2006) Approaches to biosimulation of cellular processes. J Biol Phys 32: 273–288.
50. De Mey M, Taymaz-Nikerel H, Baart G et al (2010) Catching prompt metabolite dynamics in *Escherichia coli* with the BioScope at oxygen rich conditions. Metab Eng 12: 477–487.
51. Tomita M (2001) Whole-cell simulation: a grand challenge of the 21st century. Trends Biotechnol 19: 205–210.
52. Young JD, Henne KL, Morgan JA et al (2008) Integrating cybernetic modeling with pathway analysis provides a dynamic, systems-level description of metabolic control. Biotechnol Bioeng 100: 542–559.
53. Edwards JS, Covert M, Palsson B (2002) Metabolic modelling of microbes: the flux-balance approach. Environ Microbiol 4: 133–140.
54. Henriksen CM, Christensen LH, Nielsen J et al (1996) Growth energetics and metabolic fluxes in continuous cultures of *Penicillium chrysogenum*. J Biotechnol 45: 149–164.

55. Kayser A, Weber J, Hecht V et al (2005) Metabolic flux analysis of *Escherichia coli* in glucose-limited continuous culture. I. Growth-rate-dependent metabolic efficiency at steady state. Microbiology 151: 693–706.
56. Oliveira AP, Nielsen J, Forster J (2005) Modeling *Lactococcus lactis* using a genome-scale flux model. BMC Microbiol 5: 39.
57. Teusink B, Smid EJ (2006) Modelling strategies for the industrial exploitation of lactic acid bacteria. Nat Rev Microbiol 4: 46–56.
58. Schuster S, Dandekar T, Fell DA (1999) Detection of elementary flux modes in biochemical networks: a promising tool for pathway analysis and metabolic engineering. Trends Biotechnol 17: 53–60.
59. Schilling CH, Schuster S, Palsson BO et al (1999) Metabolic pathway analysis: basic concepts and scientific applications in the post-genomic era. Biotechnol Prog 15: 296–303.
60. Christensen B, Gombert AK, Nielsen J (2002) Analysis of flux estimates based on ^{13}C-labelling experiments. Eur J Biochem 269: 2795–2800.
61. Edwards JS, Ibarra RU, Palsson BO (2001) In silico predictions of *Escherichia coli* metabolic capabilities are consistent with experimental data. Nat Biotechnol 19: 125–130.
62. Schuetz R, Kuepfer L, Sauer U (2007) Systematic evaluation of objective functions for predicting intracellular fluxes in *Escherichia coli*. Mol Syst Biol 3 (article 119): 1–15.
63. Edwards JS, Palsson BO (2000) Metabolic flux balance analysis and the in silico analysis of *Escherichia coli* K-12 gene deletions. BMC Bioinformatics 1: 1.
64. Edwards JS, Ramakrishna R, Palsson BO (2002) Characterizing the metabolic phenotype: a phenotype phase plane analysis. Biotechnol Bioeng 77: 27–36.
65. Famili I, Forster J, Nielsen J et al (2003) *Saccharomyces cerevisiae* phenotypes can be predicted by using constraint-based analysis of a genome-scale reconstructed metabolic network. Proc Natl Acad Sci U S A 100: 13134–13139.
66. Forster J, Famili I, Palsson BO et al (2003) Large-scale evaluation of in silico gene deletions in *Saccharomyces cerevisiae*. Omics 7: 193–202.
67. van der Heijden RTJM, Romein B, Heijnen JJ et al (1994) Linear constraint relations in biochemical reaction systems: I. Classification of the calculability and the balanceability of conversion rates. Biotech Bioeng 43: 3–10.
68. van der Heijden RTJM, Romein B, Heijnen JJ et al (1994) Linear constraint relations in biochemical reaction systems: II. Diagnosis and estimation of gross errors. Biotech Bioeng 43: 11–20.
69. Rocha I, Forster J, Nielsen J (2008) Design and application of genome-scale reconstructed metabolic networks, In *Microbial Gene Essentiality: Protocals and Bioinformatics* (Osterman AL, Gerdes SY, Eds.), pp 409–431, Humana Press, Totowa.
70. Bonarius HPJ, Schmid G, Tramper J (1997) Flux analysis of underdetermined metabolic networks: the quest for the missing constraints. Trends Biotechnol 15: 308–314.
71. Klamt S, Stelling J, Ginkel M et al (2003) FluxAnalyzer: exploring structure, pathways, and flux distributions in metabolic networks on interactive flux maps. Bioinformatics 19: 261–269.
72. Klamt S, Saez-Rodriguez J, Gilles ED (2007) Structural and functional analysis of cellular networks with CellNetAnalyzer. BMC Syst Biol 1: 2.
73. Lee DY, Yun H, Park S et al (2003) MetaFluxNet: the management of metabolic reaction information and quantitative metabolic flux analysis. Bioinformatics 19: 2144–2146.
74. Lee SY, Lee DY, Hong SH, et al (2003) MetaFluxNet, a program package for metabolic pathway construction and analysis, and its use in large-scale metabolic flux analysis of *Escherichia coli*. Genome Inform 14: 23–33.
75. Becker SA, Feist AM, Mo ML et al (2007) Quantitative prediction of cellular metabolism with constraint-based models: the COBRA Toolbox. Nat Protoc 2: 727–738.
76. Kitano H, Funahashi A, Matsuoka Y et al (2005) Using process diagrams for the graphical representation of biological networks. Nat Biotechnol 23: 961–966.
77. Funahashi A, Tanimura N, Morohashi M et al (2003) CellDesigner: a process diagram editor for gene-regulatory and biochemical networks. BIOSILICO 1: 159–162.
78. Schilling C, Thakar R, Travnik E et al (2008) SimPheny™: A Computational Infrastructure for Systems Biology, In Genomics: GTL Contractor—Grantee Workshop III, pp 67–68, U.S. Department of Energy, Washington.
79. Zamboni N, Fischer E, Sauer U (2005) FiatFlux – a software for metabolic flux analysis from 13C-glucose experiments. BMC Bioinformatics 6: 209.
80. Wiechert W, Mollney M, Petersen S et al (2001) A universal framework for 13C metabolic flux analysis. Metab Eng 3: 265–283.
81. Quek LE, Wittmann C, Nielsen LK et al (2009) OpenFLUX: efficient modelling software for 13C-based metabolic flux analysis. Microb Cell Fact 8: 25.

82. Pfeiffer T, Sanchez-Valdenebro I, Nuno JC et al (1999) METATOOL: for studying metabolic networks. Bioinformatics 15: 251–257.
83. Schwarz R, Liang C, Kaleta C et al (2007) Integrated network reconstruction, visualization and analysis using YANAsquare. BMC Bioinformatics 8: 313.
84. Schwarz R, Musch P, von Kamp A et al (2005) YANA - a software tool for analyzing flux modes, gene-expression and enzyme activities. BMC Bioinformatics 6: 135.
85. Bell SL, Palsson BO (2005) Expa: a program for calculating extreme pathways in biochemical reaction networks. Bioinformatics 21: 1739–1740.

Chapter 8

TMT Labelling for the Quantitative Analysis of Adaptive Responses in the Meningococcal Proteome

Karsten Kuhn, Christian Baumann, Jan Tommassen, and Thorsten Prinz

Abstract

In addition to standard gel-based proteomic approaches, gel-free approaches using isobaric label reagents, such as Tandem Mass Tags (TMT), provide a straightforward method for studying adaptations in microbial proteomes to changing environmental conditions. This approach does not have the known difficulties of 2-D gel electrophoresis with proteins of extreme biochemical properties. The workflow described here was designed to study adaptive responses in bacteria and has been applied to study the response of meningococci to iron limitation. The supplemental use of western blotting allows the confirmation of certain changes in protein abundance identified within the TMT study.

Key words: Proteomics, Tandem mass tags, LC-MS/MS, Protein quantification, Iron regulation

1. Introduction

Both proteomic and genomic approaches are used to study the adaptive responses of microorganisms to changing environmental conditions. A proteomic technology often used is based on the two-dimensional separation of proteins by gel electrophoresis (2-DE) and delivers valuable results. However, the 2-DE technology suffers from the incapability of analysing proteins with extreme biochemical properties such as hydrophobic membrane proteins or very basic proteins. Alternatively, quantitative gel-free proteomic approaches with isobaric label reagents, such as Tandem Mass Tags (TMT®; Thermo Scientific) and isobaric tags for relative and absolute quantification (iTRAQ®; Applied Biosystems), can be applied which do not have these limitations (1–3). Of the TMT label reagents two different sets of tags are available, which share the

Myron Christodoulides (ed.), *Neisseria meningitidis: Advanced Methods and Protocols*, Methods in Molecular Biology, vol. 799, DOI 10.1007/978-1-61779-346-2_8, © Springer Science+Business Media, LLC 2012

chemical structure, but differ in the number of incorporated heavy isotopes, i.e. TMTduplex, which allows for the routine investigation of two different samples in parallel, and TMTsixplex, which allows for the simultaneous analysis of six samples. Routinely, raw protein samples are denatured, reduced, and alkylated, and digested with a protease such as trypsin prior to labelling with TMT reagents (Fig. 1a). The TMT label attaches to amino groups (both N-terminal α-amino groups and ε-amino groups of lysine residues) of peptides that are generated by the tryptic digestion. After labelling, the samples are mixed, which conserves the ratios of the

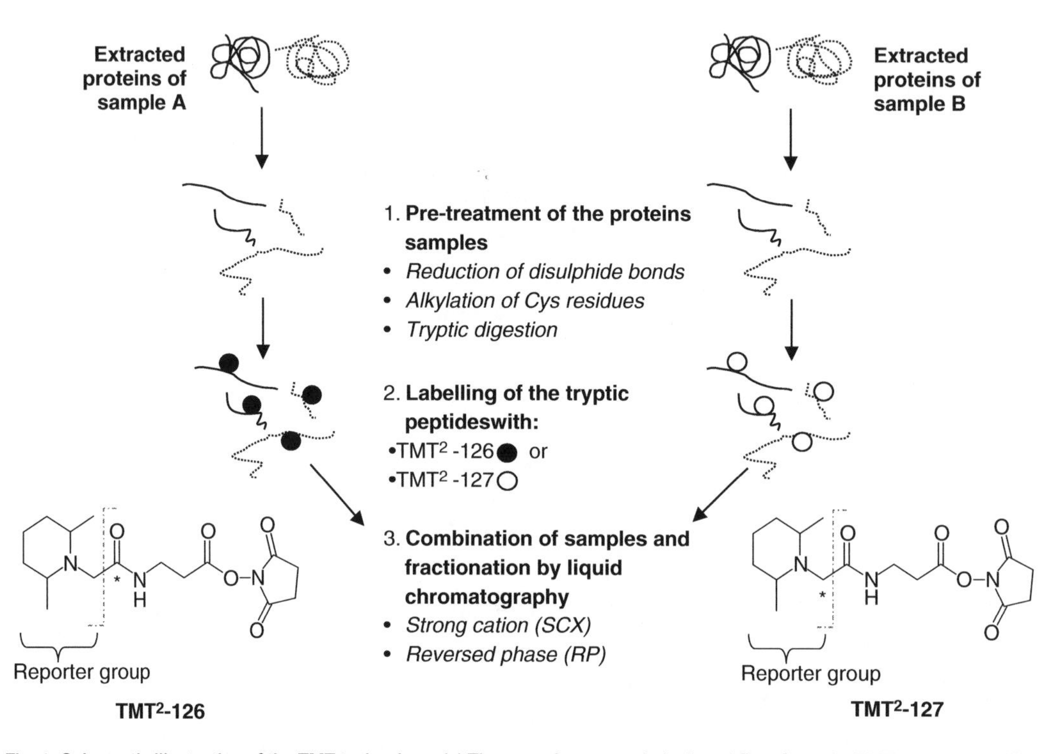

Fig. 1. Schematic illustration of the TMT technology. (**a**) The generic pre-analytical workflow from the TMT technology. After solubilisation, the bacterial samples are denatured, the disulphide bridges are reduced, and cysteine residues are alkylated. Then, the pre-treated proteins are digested with trypsin and the resulting peptides are labelled either with the TMT^2-126 or TMT^2-127 label. The structures of the two isobaric TMTduplex reagents are shown. The *asterisks* indicate the positions of the heavy isotope ^{13}C. The *dotted line* indicates the fragmentation site; the reporter group is released to the *left*-hand side of the cleavage site upon collision-induced dissociation (CID) (see (**b**)). Prior to MS analysis, the peptide mixtures are further purified and fractionated by strong cation (SCX) and reversed phase (RP) chromatography. (**b**) The principle of isobaric labelling during the mass spectrometric analysis. On the *left*-hand side, this is shown by schematic figures; whereas on the *right*-hand side MS and MS/MS spectra are given that show experimental examples. The analysed peptides derived from bacterial sample A and B (see (**a**)) coelute during chromatographic separations and combine into one peak in the MS-mode due to the identical overall mass of isobaric labels. Upon CID of a selected mass signal, several fragments are generated in the MS/MS-mode. In the same MS/MS spectrum, both structural fragments and different reporter fragments are obtained, due to the different positioning of the heavy C and/or N isotopes within the label. The structural fragments allow for the identification of the peptide and, thus, of the represented protein, whereas the reporter fragments allow for a relative quantification of that peptide. In the shown example, the higher peak of the reporter ion with m/z 127.1 originating from the TMT^2-127-labelled peptide indicates that the protein concerned is more highly expressed in the original bacterial sample B than in sample A.

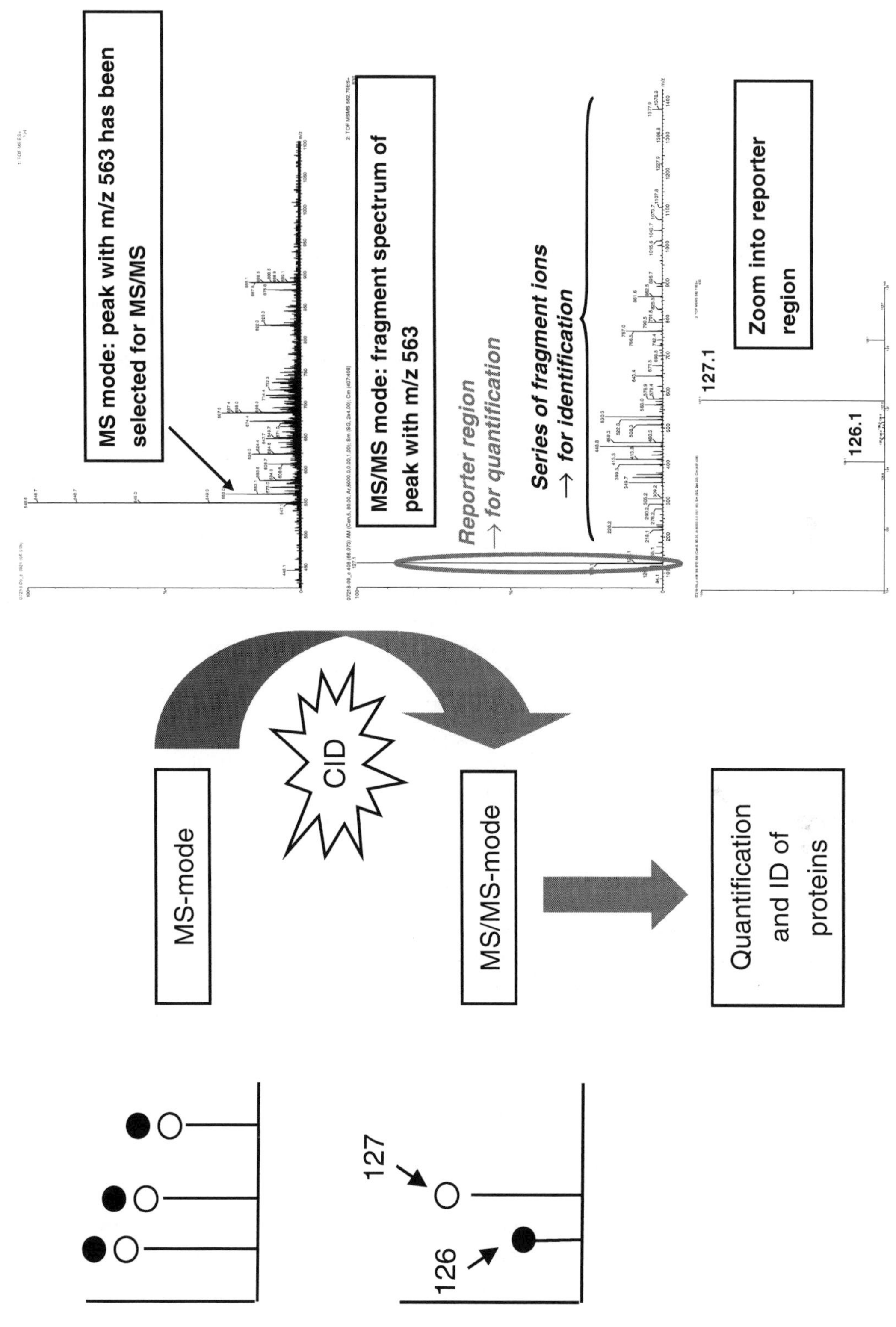

Fig. 1. (continued)

protein abundances for liquid chromatography (LC) methods like strong cation exchange (SCX) and reversed phase (RP). Mixing of samples early in a given workflow allows for an increased reproducibility of experiments. The subsequent LC-tandem mass spectrometry (LC-MS/MS) and data analysis allows for simultaneous peptide identification and quantitation (Fig. 1b).

The TMT workflow described here was applied to study the response of *Neisseria meningitidis* to iron deprivation. However, apart from comparing the proteome of a single bacterial strain grown under two different conditions, the approach can also be applied to compare the proteomes of two bacterial strains, e.g. a wild-type and a mutant strain after growth under the same condition.

N. meningitidis colonises the mucosal surfaces of its human host thereby encountering a restricted access to iron, which is essential for microbial growth. To overcome iron limitation, iron-acquisition systems are expressed, which enable the bacterium to gain access to different iron sources such as the human transport and storage proteins lactoferrin and transferrin (4). The TMT analysis of meningococcal samples resulted in 609 identified proteins of which 35 proteins were found to be modulated in their expression under iron-limiting conditions, including, e.g. components of the transferrin and lactoferrin receptors and of an Fe^{2+} transport system, the haemoglobin receptor HmbR and an unknown receptor, designated TdfK (5).

In addition to the usage of isobaric TMT reagents in discovery studies, a selected reaction monitoring (SRM)-based approach can be used to perform a targeted quantitation of selected proteins. Therefore, isotopic versions of the TMT reagents are applied that are structurally identical to the isobaric ones but have different numbers of heavy isotopes incorporated, referred to as light TMT and heavy TMT. We used this SRM-based approach in combination with light and heavy TMT to analyse crude cell lysate samples without prior or subsequent fractionation (e.g. SCX chromatography) to validate the regulations for 12 proteins represented by 33 peptides found in the discovery study (6). For all selected proteins, including, e.g. components of the multidrug efflux pump MtrCDE, their regulation found during discovery could be confirmed.

2. Materials

Unless otherwise specified all buffers are stored at room temperature. Routinely, HPLC-grade acetonitrile and ultrapure water (type 1) are used unless otherwise specified.

2.1. Cell Culture and Lysis

1. GC agar supplemented with Vitox (Oxoid).
2. Tryptic soy broth (TSB) (Becton-Dickinson).

3. Ethylenediamine di-*o*-hydroxyphenylacetic acid (EDDHA) (Sigma-Aldrich).
4. Ultracentrifuge Optima LE-80K (Beckman Coulter).
5. Tabletop centrifuge 5804 (Eppendorf).
6. UV spectrometer one (Unicam).
7. Hank's Balanced Salt Solution (plus Ca/Mg) (Invitrogen).
8. 100 m*M* triethylammonium bicarbonate, pH 8.4–8.6 (TEAB) (Sigma-Aldrich), freshly prepared prior to use.
9. 20 m*M* *Tris*[2-carboxyethyl]phosphine*HCl (TCEP) (Perbio Science).
10. Cell lysis buffer: 100 m*M* TEAB, 1 m*M* TCEP, 1% (w/v) sodium dodecyl sulphate (SDS).
11. Protein assay dye reagent (Bio-Rad).
12. Ultrasonic cell disruptor Sonifier 450 with Microtip (Branson Ultrasonics Corp.).
13. 1.5 mL polypropylene Safe-Lock tubes (Eppendorf).

2.2. SDS-Polyacrylamide Gel Electrophoresis (SDS-PAGE) and Western Blotting

1. Mini-Protean precast gels with 7.5% acrylamide in the resolving part (Bio-Rad) (see Note 1).
2. Mini-Protean 3 gel system with precast gels (Bio-Rad) for SDS-PAGE and TE62 tank blot system (Hoefer) for western blotting, or equivalent.
3. Phosphate-buffered saline (PBS), pH 7.4.
4. Laemmli buffer (2×): 125 m*M* Tris–HCl, pH 6.8, 20% (v/v) glycerine, 4% (w/v) SDS, 0.2 M dithiothreitol (DTT), 0.02% (w/v) bromophenol blue. Store in aliquots at –20°C.
5. Thermo mixer comfort (Eppendorf).
6. Running buffer (do not adjust pH): 25 m*M* Tris, 192 m*M* glycine, 0.1% (w/v) SDS.
7. Pre-stained molecular weight marker: RPN800 (GE Healthcare).
8. Transfer buffer: 20 m*M* Tris, 150 m*M* glycine, 0.025% (w/v) SDS, 20% (v/v) methanol, freshly prepared prior to use.
9. Protran nitrocellulose (NC) transfer membrane, ∅ 0.2 μm (Whatman).
10. Gel blotting paper (Whatman).
11. Blot staining solution: 0.1% (w/v) Ponceau S (Sigma-Aldrich) in water with 5% (v/v) acetic acid.
12. Tris-buffered saline (TBS): 100 m*M* Tris–HCl, pH 7.5, 300 m*M* NaCl.
13. Blocking buffer: TBS, 0.1% (v/v) Tween-20 (Sigma-Aldrich), 5% (w/v) skimmed milk powder. Store the buffer at 4°C.

14. Specific antisera against target proteins.
15. Secondary antibodies, e.g. horseradish peroxidase-conjugated goat anti-rabbit IgG.
16. Substrate for chemiluminescence reaction: SuperSignal West Femto (Perbio Science).
17. Polypropene bags, 200×300 mm (Roth).
18. Versadoc 5000 CCD camera and Quantity One 1-D analysis software (Biorad), or equivalent.

2.3. Reduction, Alkylation, Digestion, and TMTduplex Labelling

1. 100 m*M* TEAB, pH 8.4–8.6, freshly prepared prior to use.
2. 20 m*M* TCEP in ultrapure water.
3. 150 m*M* iodoacetamide (Sigma-Aldrich) in 100% acetonitrile, freshly prepared prior to use.
4. Trypsin solution: 0.4 μg/μL trypsin (sequencing grade, modified, from porcine) in 100 m*M* TEAB buffer, freshly prepared prior to use.
5. TMT solution: 60 m*M* TMTduplex reagents TMT^2-126 or TMT^2-127 (TMTduplex Isobaric Label Reagent Set; Perbio Science) in 100% acetonitrile, freshly prepared prior to use.
6. 5% (w/v) aqueous hydroxylamine solution (Sigma-Aldrich).
7. 2 mL polypropylene Safe-Lock tubes (Eppendorf) for all reactions.
8. Thermo mixer: MKR23 (HLC).

2.4. Strong Cation Exchange (SCX) Fractionation

1. Materials for prior desalting of samples: HLB Oasis cartridges (1 cc, 30 mg) (Waters); RP buffer 1 (conditioning): 95% (v/v) acetonitrile+0.1% (v/v) trifluoroacetic acid; RP buffer 2 (loading and washing): 5% (v/v) acetonitrile+0.1% (v/v) trifluoroacetic acid; RP buffer 3 (elution): 50% (v/v) acetonitrile+0.1% (v/v) trifluoroacetic acid.
2. SCX eluant A: 5 m*M* KH_2PO_4 in 25% (v/v) acetonitrile, pH 3 (adjusted with H_3PO_4).
3. SCX eluant B: 5 m*M* KH_2PO_4 in 25% (v/v) acetonitrile, 500 m*M* KCl, pH 3 (adjusted with H_3PO_4).
4. Column for SCX chromatography: Polysulfoethyl A™ column (100 mm length×4.6 mm inner diameter, 5 μm particle size, 200 Å pore size) (PolyLC).
5. 5 mL polypropylene tubes (Greiner) for collection of samples and fractions.
6. Vacuum concentrator: Univapo 150H (Uniequip).

2.5. LC-MS/MS

1. LC-MS/MS instrument: Q-Tof 2 coupled to a CapLC system (Waters).
2. Trapping column, e.g. C18 Pepmap100, 0.3×5 mm (Dionex).

3. Analytical column 75 μm inner diameter × 15 cm length packed with standard C18 material, e.g. ReproSil-Pur C18-AQ, 3-μm particles (Dr. Maisch HPLC) (see Note 2).
4. PCR plates Thermo-Fast 96, skirted (Perbio Science).
5. Solvent A: 99.9% (v/v) acetonitrile, 0.1% (v/v) formic acid.
6. Solvent B: 99.9% (v/v) water, 0.1% (v/v) formic acid.
7. Solvent C: 98.9% (v/v) water, 1% (v/v) acetonitrile, 0.1% (v/v) formic acid. All reagents for Solvents A, B, and C are ULC/MS grade.

2.6. Data Analysis

1. SEQUEST (TurboSEQUEST, Version 27, ThermoFisher Scientific).
2. PeptideProphet and ProteinProphet (7, 8), Version 2.0, (download from http://tools.proteomecenter.org/software.php).
3. Own-developed scripts based on the program "R" (see http://www.r-project.org).

3. Methods

Proteomics provides us with good methods for studying changes within bacterial cells under certain stress conditions (e.g. iron-limitation). However, a high technical reproducibility during sample preparation and analysis is the most critical issue in proteomic research. In respect of bacterial samples it is essential that different cultures serving as single samples are grown to the same growth phase. It is also recommended that at least three biological replicates are analysed in parallel. Applying the TMTduplex reagents for the analysis of proteomic changes in meningococci upon iron-limitation, we studied also the technical reproducibility of this method (5). Indeed, the expression values of individual proteins were highly consistent. This was not only due to the above mentioned sample standardisation, but also to the high reproducibility of the labelling procedure and the SCX chromatography. Overall, the reproducibility aspects need special care and attention. Results of the TMT study concerning the changed expression of certain proteins can be confirmed by western blotting. In addition, western blotting can be used prior to the TMT analysis to verify whether the bacteria responded to the applied conditions, in our example by detecting known iron-regulated proteins. Since our original study (5), we have further optimised the protocols and these slight modifications are included here.

3.1. Cell Culture and Lysis

1. Grow the pre-culture of *N. meningitidis* strain MC58 (see Note 3) overnight on GC agar plates supplemented with Vitox in an humidified incubator at 37°C, with 5% (v/v) CO_2. For the proteomics study, scrape the cells from the plates and resuspend them in either TSB or, to achieve iron limitation, in TSB supplemented with 20 μg/mL EDDHA to an optical density at λ600 nm (OD_{600}) of 0.1. Grow the cells at 37°C with shaking at 110 rpm to an OD_{600} of 2.5, independent of iron availability. For each condition, grow three cultures as biological replicates.
2. Harvest the cells that correspond to 30 OD units in a tabletop centrifuge for 20 min at ~3,200 × *g*, wash them once with Hank's Balanced Salt Solution, heat-kill for 1 h at 56°C in a water bath, and collect finally by ultracentrifugation for 1 h at 184,000 × *g* at 4°C (see Note 4).
3. To each cell pellet add 1 mL of lysis buffer and resuspend by vortexing. For complete cell lysis, incubate the suspension under continuous shaking for 10 min at 95°C and subsequently expose the samples to sonication with a ultrasonic cell disruptor for 20 s on ice (see Note 5).
4. Centrifuge the samples for 15 min at 20,600 × *g* at 4°C in order to remove unbroken cells and then transfer the supernatant to a fresh reaction tube.
5. Determine the total protein concentration of the cell lysates using the protein assay dye reagent according to the manufacturer's instructions (see Note 6).

3.2. SDS-PAGE, Western Blotting, and Immunodetection

1. These instructions assume the use of a Mini-Protean three gel system with precast gels (Bio-Rad) for SDS-PAGE and the use of a TE62 tank blot system (Hoefer) for western blotting.
2. Assemble the gel system with the precast gel and add the running buffer to the buffer chambers of the gel unit.
3. From each meningococcal cell pellet, collected as described in step 2 of Subheading 3.1, use a spatula tip to transfer to a reaction tube, add 1 mL of PBS and measure the $OD_{600.}$ If necessary, adjust the OD to ten with PBS. Mix 10 μL of this cell suspension with 10 μL of 2× Laemmli buffer, heat the samples for 10 min at 95°C in a thermomixer, and then cool on ice prior to electrophoresis. Load 20 μL of each sample to the wells of the gel and to a single well add the pre-stained molecular weight marker.
4. Connect the gel unit to a power supply and carry out the electrophoresis at 200 V and at 8–10°C until the dye front (blue) reaches the end of the gel.

5. During the run of the gel, prepare the NC membrane (see Note 7) and the gel blotting paper and cool the transfer buffer in the tank to a temperature of 10°C.
6. Soak the gel, the NC membrane, and the gel blotting paper for 5 min in transfer buffer.
7. Assemble the transfer cassette with sponges, gel, NC membrane, and blotting paper as described in the manufacturer's application guide.
8. Move the assembled transfer cassette to the tank that is filled with transfer buffer.
9. After the assembly of the transfer unit is completed, connect it to a power supply and carry out the transfer at constant 1 A for 1.5 h.
10. After the transfer is complete, remove the NC membrane carefully and then discard the gel and blotting papers.
11. Mark with care, using a soft pencil, the positions on the NC membrane of the pre-stained bands of the molecular weight marker.
12. Subsequently, stain the NC membrane reversibly with a Ponceau S solution for 5 min at room temperature, in order to confirm the transfer of the proteins from the gel.
13. Incubate the membrane with blocking buffer for 1 h at room temperature under continuous shaking, in order to prevent non-specific binding of the antibodies to the membrane.
14. After blocking, incubate the membrane with the diluted primary antiserum in blocking buffer for 1 h at room temperature under continuous shaking. Since antibody preparations vary in their levels of purity and specific binding properties, the working dilution has to be titrated for each antiserum.
15. Wash the membrane four times for 5 min each at room temperature with TBS containing 0.1% (v/v) Tween-20.
16. Incubate the membrane with diluted secondary antibody in blocking buffer for 1 h at room temperature under continuous shaking. Host species-specific secondary antibodies coupled to horseradish peroxidase are available from several manufacturers; we recommend antibody conjugates from Jackson Immunoresearch, which should be diluted to 1/50,000 for use.
17. Repeat the washing steps with TBS containing 0.1% (v/v) Tween-20 with a final wash with TBS only.
18. Apply the substrate solution for chemiluminescence reaction to the blot, incubate for 5 min at room temperature, and finally discard the solution before placing the wet membrane between the leaves of an appropriate polypropene bag.
19. View the chemiluminescence signals under a Versadoc 5000 CCD camera and quantify the signals using the QuantityOne 1-D analysis software, following the manufacturer's instructions.

3.3. Reduction, Alkylation, Digestion, and TMTduplex Labelling

The labelling protocol given here has been successfully applied to several sample types. Do not vary the ratios of the stock solutions to be added relative to the start volume of the cell lysate samples (here 300 μL), since the actual concentrations of the reactants are essential for guaranteeing a homogeneous digestion and a complete labelling (see also Notes 8 and 9). All incubations have been performed in a thermo mixer at the given temperatures and at a speed of 500 rpm.

1. Dilute all cell lysate solutions obtained from step 4 in Subheading 3.1 with 100 m*M* TEAB buffer, in order to reduce the SDS concentration to 0.1% (v/v) (i.e. at least tenfold since the lysis buffer contains 1% (v/v) SDS) and to ensure the same protein concentration in each sample (to be maximally 1 μg/μL).
2. Use 300 μL of each cell lysate sample that contains 100 μg of total protein for further processing. Add 16 μL of the TCEP stock solution to each sample, and incubate at 55°C for 1 h.
3. Allow the samples to cool to room temperature, add 17 μL of the iodoacetamide stock solution to each sample and then incubate at 23°C for 1 h.
4. Add 30 μL of trypsin solution to each sample and incubate overnight at 37°C.
5. For each sample (300 μL start volume), 121 μL of the respective TMT solution has to be added to ensure complete labelling (see Note 9). The samples of cells grown under different conditions are labelled with different TMT reagents. In our example, lysates of cells grown in the presence of iron (TSB) are labelled with the reagent TMT^2-126 and those grown under iron-limitation (TSB supplemented with EDDHA) are labelled with TMT^2-127. Incubate all samples at 23°C for 1 h.
6. Next, add 26 μL of the hydroxylamine stock solution to each sample and incubate at 23°C for 15 min, in order to inactivate residual labelling reagents and to reverse the occasional side labelling of Tyr, Ser, and Thr residues. Mix one sample labelled with TMT^2-126 with one sample labelled with TMT^2-127 and incubate for a further 15 min. The combined samples are then stored at –20°C until further processing is carried out.

3.4. Strong Cation Exchange (SCX) Fractionation

Since every laboratory has its own experience and equipment for SCX chromatography, we provide here a general procedure for an HPLC-assisted fractionation based on our workflow (see Note 10).

1. In order to ensure a proper binding of the peptides to the stationary phase, the combined peptide samples have to be desalted before application to SCX chromatography. This can be done in a simple and straightforward manner by using RP cartridges for single use, e.g. HLB Oasis cartridges.

2. Activate the HLB Oasis cartridge with 1 mL of RP buffer 1 and subsequently equilibrate with 2 mL of RP buffer 2. Dilute the mixed peptide sample with 3 mL of RP buffer 2 to decrease the acetonitrile content. Load the diluted peptide sample onto the prepared HLB Oasis cartridge and then wash the cartridge with 4 mL of RP buffer 2. Then, elute the peptides from the cartridge with 2 mL of RP buffer 3 into a new tube.
3. Equilibrate the SCX column by purging with at least five column volumes of SCX eluant A. Load the eluate from the HLB Oasis cartridge and apply the following gradient at a flow rate of 2 mL/min: 0–7 min at 100% A (to wash the bound peptides), 7–40 min to 50% SCX eluant A, and 50% SCX eluant B to elute the peptides. Collect fractions every minute during the 7–40 min phase and dry the fractions with a vacuum concentrator. For LC-MS/MS analysis, dissolve the fractions in an appropriate volume (here: 200 μL for 200 μg processed protein that has been loaded onto the SCX column) of 5% acetonitrile (v/v) + 0.1% (v/v) trifluoroacetic acid.

3.5. LC-MS/MS

The equipment used for LC-MS/MS analyses differ between laboratories (see Note 11). The following instructions describe our measurement procedure on a Q-Tof 2 (Waters) coupled to a CapLC system (Waters). For the analysis of TMT-labelled peptides, standard LC and MS/MS settings for tryptic peptides can be used.

1. Transfer the digested, labelled, and fractionated samples into a PCR plate and introduce this plate into the autosampler.
2. Load 5 μL of each SCX fraction onto a trapping column for desalting and peptide concentration, using a flow rate of 10 μL/min of solvent C for 10 min.
3. After trapping and desalting of the sample, a standard HPLC gradient is started to separate the peptides on the analytical column at a flow rate of about 300 nL/min, e.g. in 85 min from 5% (v/v) solvent A/95% (v/v) solvent B to 30% (v/v) solvent A/70% (v/v) solvent B.
4. Acquire the mass spectrometric data by electrospray ionisation in positive mode using standard settings for MS and MS/MS analysis.

3.6. Data Analysis

The qualitative and quantitative analysis of MS/MS data derived from peptides being labelled with isobaric reagents is feasible by different approaches and software packages, respectively (see Note 12). In our laboratories, SEQUEST (TurboSEQUEST, Version 27) in combination with the algorithms PeptideProphet and ProteinProphet (7, 8) are usually used for qualitative analysis, and own-developed scripts based on the program "R" (http://www.r-project.org) are used for quantitative analysis and statistical calculations (see Notes 12 and 13).

1. Process the acquired MS/MS spectra as SEQUEST-compatible *.dta files after smoothing and centroiding, to ensure for optimal mass accuracy and peak areas for quantitation of the TMT reporter ions.
2. Submit the resulting processed *.dta files to a database search using SEQUEST, with the following parameters:
   - Database: *N. meningitidis* strain MC58 (Genbank accession no AE002098.2).
   - Enzyme: Trypsin, three miscleavages allowed.
   - Mass tolerance: 150 ppm.
   - Cysteine: +57.0215 (Carbamidomethylation) static.
   - Lysine: +225.1558 (TMTduplex) static.
   - N-terminus peptides: +225.1558 (TMTduplex) differential.

4. Notes

1. Alternatively, Tris-Glycine PAGEr® precast gels (Lonza) can be recommended. For the detection of meningococcal proteins with a broad range of molecular masses, a 4–12% gradient gel would be optimal.
2. Although standard RP18 capillary columns are available commercially for the separation of tryptic peptides, in our laboratory we prefer to prepare much cheaper self-packed capillary analytical columns, as follows:
   (a) Cut a fused silica capillary (e.g. TSP075375 from polymicro) into pieces of about 10 cm longer than the desired column length (e.g. 25 cm for a 15-cm column).
   (b) Mix a potassium silicate solution (Kasil 1624; PQ Europe) and formamide in a 75:25% (v/v) ratio and dip the capillaries into the mixture for about a second. Due to the capillary action, a small portion of the mixture gets into the capillary. Use a tissue to wipe off the solution on the outside and check against a light source to confirm that the mixture has indeed entered the capillary.
   (c) Place the prepared capillaries in an oven at 100°C for about 2–3 h. After the sintering step, the frit appears white and can easily been seen.
   (d) Use a ceramic cutter to cut the typically 1–2 cm-long frit down to 1–2 mm. An empty capillary with a frit of 2 mm has nearly no additional backpressure.

(e) Next, prepare a slurry of the desired packing material in 100% methanol or acetonitrile. Introduce this slurry into an empty HPLC column (150 × 1 mm), with the metal frit removed at the exit side and use a HPLC pump to transfer the slurry (using acetonitrile at a flow rate of about 10 μL/min) into the capillary with the frit at the end.

(f) Cut the packed capillary column to the desired length and equilibrate with typical HPLC starting conditions prior to use.

3. The strain MC58 was chosen for two reasons. First of all, the complete genome sequence of this strain is known (9), which is a precondition for any proteomic analysis. Secondly, this strain is the ideal sample for a proteomic investigation concerning its response to iron limitation, because microarray data are available as well (10).

4. Meningococci are known to shed off membranous protrusions called blebs. Therefore, it is important that meningococci are collected by ultracentrifugation. Otherwise, the study results could be biased in an unfavourable way.

5. For an enhanced analysis of membrane proteins, it is recommended to isolate bacterial membranes first and then purify them: in our laboratory we use a sodium carbonate extraction procedure (11, 12), but other extraction methods are also applicable. Briefly, the collected cell pellet is taken up in 100 m*M* Na_2CO_3 and sonicated on ice until the cell solution appears clear. The cell membranes are collected by ultracentrifugation at 100,000 × *g* at 4°C for 2 h and washed once again with 100 m*M* Na_2CO_3. The re-collected cell pellet can be taken up in cell lysis buffer and further be used for TMT labelling as described in our protocol.

6. In our hands, the best-performing protein assay for these samples is the Coomassie dye-based Bradford assay. By contrast, the commonly used BCA (bicinchoninic acid) protein assay delivered inaccurate protein concentrations, which is probably due to interfering substances in the sample.

7. In general, polyvinylidene fluoride (PVDF) is regarded as the superior material for transfer membranes because of its mechanical strength. However, both NC and PVDF have to be tested together with each antibody in order to find the optimal combination regarding signal intensity and background.

8. The given workflow (cell lysis, pre-analytical sample processing) has been developed and optimised to be simple and straightforward to avoid any advanced purification steps during the pre-analytical processing, since these are prone to introduce artificial deviations if done prior to the sample mixing. Thus,

the cell lysis has been done with buffers and detergents that are compatible only with the chemical sample processing, i.e. no additives that interfere with the digestion (e.g. guanidinium hydrochloride) and no usage of buffers that contain primary or secondary amino groups, which would interfere with the labelling reaction (e.g. Tris).

9. To achieve complete labelling of the peptides, the given amounts and volumes of label reagent stock solution relative to the volume of the protein sample must not be changed, independently of the actual protein amount. In fact, during the labelling reaction the acetonitrile content must be 25% (v/v) and the actual concentration of the TMT label reagents must be 15 m*M*. Both parameters are fulfilled when preparing the stock solutions of the label reagents and addition to the protein samples, as described.

10. SCX chromatography performs optimally when carried out in an HPLC-assisted manner with an appropriate SCX column and applying a linear salt gradient. The gradient and fraction collection scheme given here have been developed to concentrate most of the peptides in one fraction. In fact, as tested with several synthetic peptides, the average peak width at half height is about 25% of a fraction size. Such a performance is decreased when using SCX cartridges only and when applying a step-wise salt gradient.

11. Quantitation of peptides labelled with TMT requires a mass spectrometer capable of MS/MS fragmentation, such as an ion trap FT (like the LTQ-Orbitrap), quadrupole time of flight (Q-TOF), time of flight-time of flight (TOF-TOF), or triple quadrupole (QQQ) instrument. Please note that TMT or iTRAQ reagent reporter ions are not visible in ion traps (like the LTQ) in collision-induced dissociation (CID) fragmentation mode due to the known low-mass cut-off.

12. Several software packages support the modifications by TMT reagents and the relative quantitation of reporter ions released from labelled peptides, including Mascot 2.2 (Matrix Science), Proteome Discoverer 1.1 (Thermo Scientific), and Scaffold Q+(Proteome Software). Additionally, any custom-made software packages may be used to consider specific user-defined solutions.

13. Prior to the actual analysis, a representative set of the acquired LC-MS/MS data is analysed with differential modifications for both Lys residues and the N-terminus of peptides. Of the identified MS/MS spectra, ≥95% needs to be assigned to completely labelled peptides to be accepted for final analysis.

Acknowledgments

Work in the lab of J.T. is supported by the Netherlands Organization for Health Research and Development (ZonMw).

References

1. Thompson A, Schäfer J, Kuhn K et al (2003) Tandem mass tags: A novel quantification strategy for comparative analysis of complex protein mixtures by MS/MS. Anal Chem 75:1895–1904.
2. Dayon L, Hainard A, Licker V et al (2008) Relative quantification of proteins in human cerebrospinal fluids by MS/MS using 6-plex isobaric tags. Anal Chem 80:2921–2931.
3. Ross PL, Huang YN, Marchese JN et al (2004) Multiplexed protein quantitation in *Saccharomyces cerevisiae* using amine-reactive isobaric tagging reagents. Mol Cell Proteomics 3:1154–1169.
4. Pettersson A, Poolman JT, van der Ley P et al (1997) Response of *Neisseria meningitidis* to iron limitation. Antonie van Leeuwenhoek 71:129–136.
5. van Ulsen P, Kuhn K, Prinz T et al (2009) Identification of proteins of *Neisseria meningitidis* induced under iron-limiting conditions using the isobaric tandem mass tag (TMT) labeling approach. Proteomics 9:1771–1781.
6. Byers HL, Campbell J, van Ulsen P et al (2009) Candidate verification of iron-regulated *Neisseria meningitidis* proteins using isotopic versions of tandem mass tags (TMT) and single reaction monitoring. J Proteomics 73:231–239.
7. Keller A, Nesvizhskii AI, Kolker E et al (2002) Empirical statistical model to estimate the accuracy of peptide identifications made by MS/MS and database search. Anal Chem 74:5383–5392.
8. Nesvizhskii AI, Keller A, Kolker E et al (2003) A statistical model for identifying proteins by tandem mass spectrometry. Anal Chem 75:4646–4658.
9. Tettelin H, Saunders N, Heidelberg J et al (2000) Complete genome sequence of *Neisseria meningitidis* serogroup B strain MC58. Science 287:1809–1815.
10. Grifantini R, Sebastian S, Frigimelica E et al (2003) Identification of iron-activated and -repressed Fur-dependent genes by transcriptome analysis of *Neisseria meningitidis* group B. Proc Natl Acad Sci USA 100:9542–9547.
11. Fujiki Y, Hubbard AL, Fowler S et al (1982) Isolation of intracellular membranes by means of sodium carbonate treatment: application to endoplasmic reticulum. J Cell Biol 93:97–102.
12. Fujiki Y, Fowler S, Shio H et al (1982) Polypeptide and phospholipid composition of the membrane of rat liver peroxisomes: comparison with endoplasmic reticulum and mitochondrial membranes. J Cell Biol 93:103–110.

Chapter 9

Meningococcal Ligands and Molecular Targets of the Host

Darryl J. Hill and Mumtaz Virji

Abstract

Meningococcal mechanisms of adhesion are complex, involving multiple adhesins and their respective target receptors on host cells. Three major surface structures – pili, Opa, and Opc – have been known for some time to mediate meningococcal adhesion to target human cells. More recently, several other relatively minor adhesins have also come to light. The literature on bacterial adhesion mechanisms provides numerous examples of various adhesins acting cooperatively in an apparently hierarchical and sequential manner; in other instances, adhesins may act in concert leading to high avidity interactions, often a prelude to cellular invasion and tissue penetration. Such examples are also present in the case of meningococci, although our knowledge of adhesin cooperation and synergy is far from complete. Meningococcal mechanisms used to target the host, which are often specific for the host or a tissue within the host, include both lectin-like interactions and protein–protein interactions; the latter tend to determine specificity in general. Understanding (a) what determines specificity (i.e. molecular features of adhesins and receptors), (b) encourages cellular penetration (i.e. adhesin pairs, which act in concert or synergistically to deliver effective signals for invasion and induce other cellular responses), (c) level of redundancy (more than one mechanisms of targeting host receptors), (d) host situations that encourage tissue penetration (inflammatory situations during which circulating cytokines upregulate target cell receptors, effectively encouraging greater adhesion/invasion), and (e) down-stream effects on host functions in general are all clearly important in our future strategies of controlling meningococcal pathogenesis.

Key words: *Neisseria meningitidis*, Host cell receptors, Polymeric adhesins, β-barrel proteins, Autotransporters

1. Introduction

The last two decades have seen a considerable advancement in our understanding of the complexities of *Neisseria meningitidis* (meningococcus) pathogenic mechanisms. The research has been fuelled by developments in the areas of genomics, transcriptomics, and proteomics. Such advances have improved our understanding of how meningococci alter their metabolism in response to different

Myron Christodoulides (ed.), *Neisseria meningitidis: Advanced Methods and Protocols*, Methods in Molecular Biology, vol. 799, DOI 10.1007/978-1-61779-346-2_9, © Springer Science+Business Media, LLC 2012

environments, acquire factors such as iron, as well as alter the array of surface molecules, which can be frequent and complex. Our understanding of how such molecules interact with host components promoting adhesion and influencing the immune response has also been enhanced by such studies. However, while many new surface molecules of the pathogen have been revealed by genome sequencing and mining, their functions remain to be fully defined. Indeed, the molecular mechanisms of even the traditional and long-known major adhesins are still unclear and some of their host cell targets remain undefined. This compendium provides background on meningococcal adhesins and their (putative) receptors as a prelude to the following chapters that describe some of the approaches used in studying pathogen–host cell interactions.

2. Meningococcal Adhesins and Cell Surface Located Human Receptors

Meningococcal adhesins can be divided into three broad structural classes, the polymeric hair-like pili, the integral outer membrane proteins (OMP), including the opacity proteins Opa and Opc, which are usually beta barrel structures, and the autotransporters (including meningococcal serine protease (Msp)A), meningococcal surface fibril (Msf, or Neisseria hia homolog (Nhh)A), and Neisseria adhesin (Nad)A, which have only recently emerged as significant contenders as adhesins. In addition, surface carbohydrates can be classified as adhesins as well as anti-adhesins as they can interfere with integral OMP function. Further, enzymes and envelope-associated proteins have also been postulated to be involved in cellular interaction of meningococci (Table 1). This compendium focuses principally on bacterial structures supporting adhesion to host cells. It should be noted that the following chapters also provide details of host receptors and cells involved in host immune function such as pattern recognition receptors and dendritic cells. Whilst several meningococcal adhesins have now been described, information on receptors utilised by them is often lacking. However, key methods such as receptor overlay of western blots containing bacterial proteins or *vice versa* (Far-Western blotting) and coprecipitation in combination with techniques to visualise and/or quantify bacterial adhesion (immunofluorescence microscopy, viable count assays) has led to the identification of cognate receptors for some of the neisserial adhesins. Examples of methods used in the identification of receptors for adhesive ligands of meningococci are provided below.

2.1. Pili and Their Putative Receptors

Pili are primarily composed of repeating pilin (PilE) subunits assembled in a helical array. They are dynamic structures able to extend and retract, giving bacteria a twitching motility. Some 15 genes are

Table 1
Meningococcal components identified with a role in adhesion to and or invasion of human cells

Surface structure/ properties	Structural features	Function	Target	Receptor(s)	Reference(s)
Pili (type 4)/associated proteins	Polymeric protein with associated protein (PilC), extended via outermemprane pore (PilQ)	Adhesion/invasion	Human epithelial and endothelial cells, RBC	CD46? Laminin receptor (PilQ)	(1, 7, 26)
Opa	8 stranded β-barrel	Adhesion/invasion	Human epithelial cells, fibronectin	CEACAMs HSPG ECM/integrins Saccharides	(16, 20, 51, 52)
Opc	10 stranded β-barrel	Adhesion/invasion	Human epithelial and endothelial cells Activated vitronectin Fibronectin	ECM/integrins HSPG α-actinin	(18, 19, 22–24)
PorA	16 stranded β-barrel	Porin/adhesin	Endothelial cells	Laminin receptor	(26)
Msf (meningococcal surface fibril, NhhA)	Trimeric autotransporter	Adhesion	Human epithelial cells	Laminin HSPG Vitronectin	(27, 53)
App (adhesion and penetration protein)	Autotransporter	Adhesion/spread	Human epithelial cells		(54)
HrpA	Haemagglutinin/haemolysin related protein	Adhesion	Some human epithelial cell lines		(55)
NadA (Neisserial adhesin A)	Trimeric autotransporter	Adhesion/invasion	Human epithelial cells Monocytes Macrophages		(56, 57)

(continued)

Table 1 (continued)

Surface structure/ properties	Structural features	Function	Target	Receptor(s)	Reference(s)
MspA (meningococcal serine protease)	Autotransporter	Adhesion	Human epithelial and endothelial cells (when expressed in *E. coli*)		(58)
TspA	Unknown, Fim-V like	Adhesion, T-cell stimulation	Epithelial and primary meningothelial cells		(59)
Fructose 1,6-bisphosphate aldolase	Enzyme	Adhesion	Epithelial and endothelial cells		(60)
GapA-1	Enzyme	Adhesion	Epithelial and endothelial cells		(61)
LPS (sialylated or PE modified)	Lipopolysaccharide	Adhesion	Endothelial cells epithelial cells, macrophages	Siglecs (macrophages via sialic acid)	(33, 62)

involved in meningococcal pilus biogenesis (reviewed in ref. (1)). Meningococcal pili were first recognised as adhesins in the 1980s using human nasopharyngeal organ cultures (2). Many studies have ensued since that time, describing pilus adhesion to a variety of cell types (1, 3, 4). It should be noted that studies have focused on pili from *Neisseria. gonorrhoeae* as well as *N. meningitidis*, with the inference that similar adhesive events and receptor utilisation may be extrapolated between species due to similarities in pilus molecular composition. However, differences are observed between PilE type and PilC types expressed by different neisserial species and the roles of each in adhesion remain to be fully appreciated (1). The efficiency of pilus-mediated adhesion of capsulate meningococci appears to be dependent on the target cell type as well as post translational modifications of the pilus structure (1, 3, 5, 6).

The greatest enigma within the field of neisserial pilus research relates to the pilus receptor. By overlaying western blots of human epithelial cell proteins with purified pili, CD46 was identified as a possible pilus receptor for both *N. meningitidis* and *N. gonorrhoeae* (7, 8). In addition, piliated Neisseria were shown to coaggregate with CD46-coated *Staphylococcus aureus* (7). However, not all studies support the involvement of CD46 in pilus-mediated adhesion; for example, studies examining the adhesion of gonococcal PilC2 to CD46 expressing cells did not find the involvement of CD46 (9). In addition, observations using FACS and viable count-based adhesion assays showed an inverse correlation with CD46 expression levels and gonococcal pilus-mediated adhesion to epithelial cells (10). Fluorescence microscopy was used to show that another receptor was internalised and subsequently recycled to the surface following gonococcal PilC2 interaction (11). In this study, biochemical analyses revealed the receptor to be proteinaceous in nature and that glycostructures were not involved. Recently, a CD46 transgenic mouse model has been used to study the potential CD46-pilus interplay in neisserial disease by whole animal infection studies and bioimaging (12).

2.2. β-Barrel Surface Proteins

The Opa family of adhesins are beta barrel proteins and are characterised by their extensive structural variations in three of the four surface loops (13). In addition, three to four different Opa proteins can be produced by a single meningococcal strain. In contrast, Opc (OpcA), also a beta barrel protein but with five surface loops, is encoded by a single gene and is largely invariant (14, 15).

CEACAMs (formerly CD66) were identified as receptors for meningococcal Opa proteins by the use of non-human cell lines transformed with cloned cDNA encoding human cellular receptors. The identity was confirmed using purified receptor overlays in combination with blocking antibodies (16, 17). The use of specific integrin inhibitors and RGD peptides in conjunction with purified matrix proteins revealed a bridging mechanism involving Opc interaction primarily with vitronectin and to a lesser extent fibronectin, in binding to integrins on human endothelial cells (18). Fibronectin-mediated binding to human brain microvascular cells was also shown by similar methods (19). Techniques using radiolabelled heparin for direct binding assessment and the use of vitronectin-derived sulphated peptides has elucidated the roles for heparan sulphate proteoglycans as receptors for Opa and Opc and defined the mechanisms of vitronectin targeting by Opc leading to the facilitation of epithelial and endothelial cell adhesion and invasion (20–23). Ligand overlay and immunoprecipitation also identified α-actinin as an intracellular target for Opc (24).

Of the two major outer membrane β-barrel porins, PorA and PorB that have been described (25), PorA has been shown to bind to laminin receptor on microvascular endothelial cells using contact-dependent cross-linking (retagging) amongst other techniques (26). These studies also observed a similar role of PilQ in binding to laminin receptor.

2.3. Autotransporters

The trimeric autotransporter, Msf (NhhA, also known as Hsf) has been shown to bind to HSPG and laminin by direct ELISA as well as to epithelial cells by immunofluorescence microscopy (27). Binding of Msf to activated vitronectin has been demonstrated recently by vitronectin capture from normal human serum and immunoblot analysis (28). Other meningococcal autotransporter adhesins have also been identified (Table 1), but their respective binding partners are not known.

Whilst the molecular partners for a number of meningococcal adhesins remain to be identified, some studies have eluded to the cooperative effects of adhesins in infection of human cells and tissues including synergism between the opacity proteins and pili (21, 29–31). In addition to the receptors for surface proteins described above, additional receptors have been identified (e.g. siglecs and asialoglycoprotein receptor) that bind to non-protein surface structures of *N. meningitidis* and *N. gonorrhoeae* (32, 33). In addition, gonococcal adhesion via PorB may involve the human heat shock glycoprotein Gp96 (34) and L12 (a ribosomal protein that can be surface expressed) may bind to lutropin receptor (35). Whether similar interactions occur for meningococci remain to be determined.

2.4. Toll-Like Receptors (TLRs) in Meningococcus–Host Interactions

TLRs form an important arm of the host innate immune system against multiple pathogens and accordingly, innate recognition of meningococci also involves specific interaction between meningococcal ligands and host TLRs. For example, meningococcal LPS (specifically lipid A) is recognised by a MD2-TLR-4 complex (36, 37). In addition, PorB of *N. meningitidis* binds to TLR-2 (38, 39). Both TLR-2 and TLR-4 have been recently shown to recognise group B meningococcal capsular polysaccharide (40). Besides

TLR-2 and TLR-4, the only other TLR demonstrated to recognise meningococci is TLR-9, which recognises repetitive extragenic regions of DNA (41). Related to the innate immune response receptors, the intracellular NOD-like receptors (NLRs) may also be involved in meningococcal antigen recognition (42).

3. Conclusion

The use of the methods mentioned above and those described in more detail in subsequent chapters, has increased our current understanding of meningococcal adhesion to, and invasion of, human cells and tissues. Such studies on the roles of surface expressed and associated proteins in host colonisation and disease are important for two principal reasons. Firstly, adhesins and invasins involved in the pathogenic processes could be useful as components of cross-protective vaccines; and while such surface proteins can be highly variable, several studies point to a limited repertoire of protein/adhesin variants in hypervirulent lineages of meningococci, raising the possibility of their utility as vaccine components (43–45). Secondly, besides mediating adhesion and/or invasion, meningococcal surface proteins also serve to subvert host signalling pathways to the advantage of the bacterium. Such subversion may include the manipulation of novel pathways to cross endothelial barriers (46) and blocking innate immune mechanisms such as epithelial shedding (47). Further, some variants in envelope structures such as LPS have been reported to induce low cytokine production and coagulopathy during disease (48). Understanding which surface structures are involved in subversion of host processes could also inform vaccine design and highlight signalling pathways which could be of therapeutic importance for disease control. Further advances in animal models and transcriptomics coupled with improvements in proteomics (49, 50), open up the prospect in the coming years of some exciting discoveries relating to the molecular interplay of *N. meningitidis* with its unique host during colonisation and disease.

References

1. Carbonnelle E, Hill DJ, Morand P et al (2009) Meningococcal interactions with the host. Vaccine 27: B78–B89.
2. McGee ZA, Stephens DS (1984) Common pathways of invasion of mucosal barriers by *Neisseria gonorrhoeae* and *Neisseria meningitidis*. Surv Synth Path Res 3: 1–10.
3. Virji M, Alexandrescu C, Ferguson DJP et al (1992) Variations in the expression of pili: the effect on adherence of *Neisseria meningitidis* to human epithelial and endothelial cells. Mol Microbiol 6: 1271–1279.
4. Nassif X, Lowy J, Stenberg P et al (1993) Antigenic variation of pilin regulates adhesion of *Neisseria meningitidis* to human epithelial cells. Mol Microbiol 8: 719–725.

5. Virji M, Saunders JR, Sims G et al (1993) Pilus-facilitated adherence of *Neisseria meningitidis* to human epithelial and endothelial cells: modulation of adherence phenotype occurs concurrently with changes in primary amino acid sequence and the glycosylation status of pilin. Mol Microbiol 10: 1013–1028.
6. Marceau M, Forest K, Beretti J et al (1998) Consequences of the loss of o-linked glycosylation of meningococcal type iv pilin on piliation and pilus-mediated adhesion. Mol Microbiol 27: 705–715.
7. Kallstrom H, Liszewski MK, Atkinson JP et al (1997) Membrane cofactor protein (MCP or CD46) is a cellular pilus receptor for pathogenic *Neisseria*. Mol Microbiol 25: 639–647.
8. Rytkonen A, Johansson L, Asp V et al (2001) Soluble pilin of *Neisseria gonorrhoeae* interacts with human target cells and tissue. Infect Immun 69: 6419–6426.
9. Kirchner M, Heuer D, Meyer TF (2005) CD46-independent binding of neisserial type IV pili and the major pilus adhesin, PilC, to human epithelial cells. Infect Immun 73: 3072–3082.
10. Tobiason DM, Seifert HS (2001) Inverse relationship between pilus-mediated gonococcal adherence and surface expression of the pilus receptor, CD46. Microbiology-Sgm 147: 2333–2340.
11. Kirchner M, Meyer TF (2005) The PilC adhesin of the Neisseria type IV pilus - binding specificities and new insights into the nature of the host cell receptor. Mol Microbiol 56: 945–957.
12. Sjolinder H, Jonsson AB (2007) Imaging of disease dynamics during meningococcal sepsis. Plos One 2:e241.
13. Aho EL, Dempsey JA, Hobbs MM et al (1991) Characterization of the opa (class-5) gene family of *Neisseria meningitidis*. Mol Micro 5: 1429–1437.
14. Achtman M (1995) Epidemic spread and antigenic variability of *Neisseria meningitidis*. Trend Microbiol 3: 186–192.
15. Zhu PX, Morelli G, Achtman M (1999) The opcA and psi opcB regions in Neisseria: genes, pseudogenes, deletions, insertion elements and DNA islands. Mol Micro 33: 635–650.
16. Virji M, Makepeace K, Ferguson DJP et al (1996) Carcinoembryonic antigens (CD66) on epithelial cells and neutrophils are receptors for Opa proteins of pathogenic neisseriae. Mol Microbiol 22: 941–950.
17. Virji M, Watt SM, Barker S et al (1996) The N-domain of the human CD66a adhesion molecule is a target for Opa proteins of *Neisseria meningitidis* and *Neisseria gonorrhoeae*. Mol Microbiol 22: 929–939.
18. Virji M, Makepeace K, Moxon R (1994) Distinct mechanisms of interactions of Opc-expressing meningococci at apical and basolateral surfaces of human endothelial cells; the role of integrins in apical interactions. Mol Microbiol 14: 173–184.
19. Unkmeir A, Latsch K, Dietrich G et al (2002) Fibronectin mediates Opc-dependent internalization of *Neisseria meningitidis* in human brain microvascular endothelial cells. Mol Micro 46: 933–946.
20. Chen T, Belland RJ, Wilson J et al (1995) Adherence of pilus- Opa+ gonococci to epithelial cells *in vitro* involves heparan sulfate. J Exp Med 182: 511–517.
21. Virji M, Makepeace K, Peak IRA et al (1995) Opc- and pilus-dependent interactions of meningococci with human endothelial cells: molecular mechanisms and modulation by surface polysaccharides. Mol Microbiol 18: 741–754.
22. deVries FP, Cole R, Dankert J et al (1998) *Neisseria meningitidis* producing the Opc adhesin binds epithelial cell proteoglycan receptors. Mol Microbiol 27: 1203–1212.
23. Cunha CSE, Griffiths NJ, Virji M (2010) *Neisseria meningitidis* Opc invasin binds to the sulphated tyrosines of activated vitronectin to attach to and invade human brain endothelial cells. Plos Pathogens 6: e1000911.
24. Cunha CSE, Griffiths NJ, Murillo I et al (2009) *Neisseria meningitidis* Opc invasin binds to the cytoskeletal protein alpha-actinin. Cell Micro 11: 389–405.
25. Frasch CE, Zollinger WD, Poolman JT (1985) Serotype antigens of *Neisseria meningitidis* and a proposed scheme for designation of serotypes. Rev Infect Dis 7: 504–510.
26. Orihuela CJ, Mahdavi J, Thornton J et al (2009) Laminin receptor initiates bacterial contact with the blood brain barrier in experimental meningitis models. J Clin Invest 119: 1638–1646.
27. Scarselli M, Serruto D, Montanari P et al (2006) *Neisseria meningitidis* NhhA is a multifunctional trimeric autotransporter adhesin. Mol Micro 61: 631–644.
28. Virji M, Griffiths NJ, Hill DJ et al, *Neisseria meninigitidis* Msf (NhhA) interacts directly with human vitronectin: the interplay between meningococcal Hsf and Opc in host cell adhesion and serum resistance, in: 17th International Pathogenic Neisseria Conference Canada., 2010
29. Griffiths NJ, Bradley CJ, Heyderman RS et al (2007) IFN-gamma amplifies NF kappa B-dependent *Neisseria meningitidis* invasion of epithelial cells via specific upregulation of

CEA-related cell adhesion molecule 1. Cell Micro 9: 2968–2983.

30. Rowe HA, Griffiths NJ, Hill DJ et al (2007) Co-ordinate action of bacterial adhesins and human carcinoembryonic antigen receptors in enhanced cellular invasion by capsulate serum resistant *Neisseria meningitidis*. Cell Micro 9: 154–168.
31. Schielke S, Frosch M, Kurzai O (2010) Virulence determinants involved in differential host niche adaptation of *Neisseria meningitidis* and *Neisseria gonorrhoeae*. Med Micro Immunol 199: 185–196.
32. Jones C, Virji M, Crocker PR (2003) Recognition of sialylated meningococcal lipopolysaccharide by siglecs expressed on myeloid cells leads to enhanced bacterial uptake. Mol Micro 49: 1213–1225.
33. Harvey HA, Porat N, Campbell CA et al (2000) Gonococcal lipooligosaccharide is a ligand for the asialoglycoprotein receptor on human sperm. Mol Micro 36: 1059–1070.
34. Rechner C, Kuhlewein C, Muller A et al (2007) Host glycoprotein Gp96 and scavenger receptor SREC interact with PorB of disseminating *Neisseria gonorrhoeae* in an epithelial invasion pathway. Cell Host & Microbe 2: 393–403.
35. Spence JM, Tyler RE, Domaoal RA et al (2002) L12 enhances gonococcal transcytosis of polarized Hec1B cells via the lutropin receptor. Micro Path 32: 117–125.
36. Zughaier SM, Tzeng YL, Zimmer SM et al (2004) *Neisseria meningitidis* lipooligosaccharide structure-dependent activation of the macrophage CD14/toll-like receptor 4 pathway. Infect Immun 72: 371–380.
37. Zimmer SM, Zughaier SM, Tzeng YL et al (2007) Human MD-2 discrimination of meningococcal lipid A structures and activation of TLR4. Glycobiol 17: 847–856.
38. Massari P, Henneke P, Ho Y et al (2002) Cutting edge: Immune stimulation by neisserial porins is toll-like receptor 2 and MyD88 dependent. J Immunol 168: 1533–1537.
39. Wetzler LM (2010) Innate immune function of the neisserial porins and the relationship to vaccine adjuvant activity. Future Microbiol 5: 749–758.
40. Zughaier SM (2010) *Neisseria meningitidis* capsular polysaccharides induce inflammatory responses via TLR2 and TLR4-MD-2. J Leuk Biol Epub ahead of print December 29.
41. Magnusson M, Tobes R, Sancho J et al (2007) Cutting edge: Natural DNA repetitive extragenic sequences from Gram-negative pathogens strongly stimulate TLR9. J Immunol 179: 31–35.
42. Chauhan VS, Sterka DG, Furr SR et al (2009) NOD2 plays an important role in the inflammatory responses of microglia and astrocytes to bacterial CNS pathogens. Glia 57: 414–423.
43. Urwin R, Russell JE, Thompson EAL et al (2004) Distribution of surface protein variants among hyperinvasive meningococci: Implications for vaccine design. Infect Immun 72: 5955–5962.
44. Callaghan MJ, Jolley KA, Maiden MCJ (2006) Opacity-associated adhesin repertoire in hyperinvasive *Neisseria meningitidis*. Infect Immun 74: 5085–5094.
45. Feavers IM, Pizza M (2009) Meningococcal protein antigens and vaccines. Vaccine 27: B42-B50.
46. Coureuil M, Lecuyer H, Scott MG et al (2010) Meningococcus hijacks a β2-adrenoceptor/β-Arrestin pathway to cross brain microvasculature endothelium. Cell 143: 1149–1160.
47. Muenzner P, Bachmann V, Zimmermann W et al (2010) Human-restricted bacterial pathogens block shedding of epithelial cells by stimulating integrin activation. Science 329: 1197–1201.
48. Fransen F, Heckenberg SGB, Hamstra HJ et al (2009) Naturally occurring lipid A mutants in *Neisseria meningitidis* from patients with invasive meningococcal disease are associated with reduced coagulopathy. PLoS Pathog 5: e1000396.
49. Byers HL, Campbell J, van Ulsen P et al (2009) Candidate verification of iron-regulated *Neisseria meningitidis* proteins using isotopic versions of tandem mass tags (TMT) and single reaction monitoring. J Proteomics 73: 231–239.
50. Bumann D (2010) Pathogen proteomes during infection: A basis for infection research and novel control strategies. J Proteomics 73: 2267–2276.
51. Virji M, Evans D, Hadfield A et al (1999) Critical determinants of host receptor targeting by *Neisseria meningitidis* and *Neisseria gonorrhoeae*: identification of Opa adhesiotopes on the N-domain of CD66 molecules. Mol Microbiol 34: 538–551.
52. Moore J, Bailey SES, Benmechernene Z et al (2005) Recognition of saccharides by the OpcA, OpaD, and OpaB outer membrane proteins from *Neisseria meningitidis*. J Biol Chem 280: 31489–31497.
53. Griffiths NJ, Virji M, Meningococcal vitronectin binding phenotypes: Sialylation, serum resistance and cellular interactions, in: 17th International Pathogenic Neisseria Conference, 2010.
54. Serruto D, du-Bobie J, Scarselli M et al (2003) *Neisseria meningitidis* App, a new adhesin with

autocatalytic serine protease activity. Mol Micro 48: 323–334.

55. Schmitt C, Turner D, Boesl M et al (2007) A functional two-partner secretion system contributes to adhesion of *Neisseria meningitidis* to epithelial cells. J Bacteriol 189: 7968–7976.
56. Comanducci M, Bambini S, Brunelli B et al (2002) NadA, a novel vaccine candidate of *Neisseria meningitidis.* J Exp Med 195: 1445–1454.
57. Franzoso S, Mazzon C, Sztukowska M et al (2008) Human monocytes/macrophages are a target of *Neisseria meningitidis* Adhesin A (NadA). J Leuk Biol 83: 1100–1110.
58. Turner DPJ, Marietou AG, Johnston L et al (2006) Characterization of MspA, an immunogenic autotransporter protein that mediates adhesion to epithelial and endothelial cells in *Neisseria meningitidis.* Infect Immun 74: 2957–2964.
59. Oldfield NJ, Bland SJ, Taraktsoglou M et al (2007) T-cell stimulating protein A (TspA) of *Neisseria meningitidis* is required for optimal adhesion to human cells. Cell Micro 9: 463–478.
60. Tunio SA, Oldfield NJ, Berry A et al (2010) The moonlighting protein fructose-1, 6-bisphosphate aldolase of *Neisseria meningitidis*: surface localization and role in host cell adhesion. Mol Micro 76: 605–615.
61. Tunio SA, Oldfield NJ, Ala'Aldeen DAA et al (2010) The role of glyceraldehyde 3-phosphate dehydrogenase (GapA-1) in *Neisseria meningitidis* adherence to human cells. BMC Microbiol 10:280.
62. Takahashi H, Carlson RW, Muszynski A et al (2008) Modification of lipooligosaccharide with phosphoethanolamine by LptA in *Neisseria meningitidis* enhances meningococcal adhesion to human endothelial and epithelial cells. Infect Immun 76: 5777–5789.

Chapter 10

In Vivo Imaging of Meningococcal Disease Dynamics

Hong Sjölinder and Ann-Beth Jonsson

Abstract

Neisseria meningitidis is a human specific organism that causes severe sepsis and/or meningitis with high mortality. The disease scenario is rapid and much remains unknown about the disease process and host–pathogen interaction. In this chapter, we describe a protocol for generating a bioluminescently labeled *N. meningitidis* strain in order to advance our understanding of meningococcal disease progression. We also describe how in vivo bioluminescence imaging (BLI) can be used to observe novel features of the disease dynamics during meningococcal infection.

Key words: *Neisseria meningitidis*, In vivo bioluminescence imaging, Animal model, Disease dynamics, Sepsis, Meningitis

1. Introduction

Luciferase enzymes, present in certain bacteria, marine crustaceans, fish, and insects, can generate light through the oxidation of an enzyme-specific substrate (commonly known as luciferin) in the presence of oxygen and energy (1). Photon emission can be detected by a light sensitive apparatus, such as a charge-coupled device (CCD) camera. A significant advantage of luciferases as optical indicators is the sensitivity and high signal intensity in live mammalian cells and tissues. Bioluminescent reporters have been applied intensively to monitor biological and disease processes in cell lines and live animal models (2–5).

The *luxCDABE* operon, originating from the soil bacterium *Photorhabdus luminescens*, carries all the genes essential for luminescence production. *luxAB* encodes the alpha and beta subunits of the heterodimeric luciferase and *luxCDE* encodes the fatty acid reductase complex responsible for synthesizing fatty aldehydes for the luminescence reaction (6). The *luxCDABE* operon can be

Myron Christodoulides (ed.), *Neisseria meningitidis: Advanced Methods and Protocols*, Methods in Molecular Biology, vol. 799, DOI 10.1007/978-1-61779-346-2_10, © Springer Science+Business Media, LLC 2012

efficiently expressed in Gram-negative bacteria and the encoded luciferase is functional at temperatures as high as 45°C. It is therefore ideally suited for real-time monitoring of microbial physiology and infectious disease in vivo (7). In this chapter, we describe a protocol for generating a bioluminescently labeled *Neisseria meningitidis* strain based on the *luxCDABE* operon. We also show how in vivo BLI revealed intriguing disease dynamics during meningococcal infection (8). Bioluminescent *N. meningitidis* strains combined with suitable animal disease models provide a potent tool for in vivo investigations of the pathogenesis of meningococcal infection, the role of pathogenic factors, as well as the efficacy of vaccine candidates.

2. Materials

2.1. Bacteria and Growth Conditions

1. *N. meningitidis* strain FAM20 is stored in GC medium containing 30% (v/v) glycerol at –80°C (see Note 1).
2. *E. coli* strain DH5α is stored in Luria Bertani (LB) medium with 30% (v/v) glycerol at –80°C.
3. Kellogg's supplement (9): 40% (w/v) D-glucose, 0.5% (w/v) L-glutamine, 0.05% (w/v) ferric nitrate, and 0.01% (w/v) cocarboxylase. Sterilize by passage through a 0.2-μm filter and store at 4°C.
4. GC agar plates: dissolve 36 g of GC agar powder (Acumedia) in 1 L of distilled water. Autoclave for 15 min at 121°C and 2.68 kg/cm^2 and then cool to around 50°C, add 10 mL of Kellogg's supplement before pouring into sterile plastic Petri dishes.
5. GC medium: dissolve 15 g of protease peptone (Oxoid), 1 g of soluble starch, 4 g of K_2HPO_4, 1 g of KH_2PO_4, and 5 g of NaCl in 1 L of distilled water. Autoclave and store at room temperature.
6. Luria Bertani (LB) medium: dissolve 25 g of LB broth powder (Acumedia) in 1 L of distilled water, autoclave, and store at room temperature.
7. LB agar plates: dissolve 37 g of LB agar powder (Acumedia) in 1 L of distilled water. Autoclave, cool to around 50°C, and pour a thin layer of agar into Petri dishes.
8. Kanamycin (Sigma-Aldrich) is dissolved at 50 mg/mL in distilled water, sterilized by passage through a 0.2-μm filter, and stored in aliquots at –20°C. The working concentration in agar plates or medium is 50 μg/mL (see Note 2).
9. Ampicillin (Sigma-Aldrich) is dissolved at 100 mg/mL in distilled water, sterilized by passage through a 0.2-μm filter, and stored

in aliquots at –20°C. The working concentration in LB agar plates or medium is 100 μg/mL.

10. CO_2 incubator.

2.2. Preparation of Template DNA

1. QIAGEN Plasmid Miniprep kit (QIAGEN, see Note 3).
2. Wizard Genomic DNA purification kit (Promega, see Note 3).
3. Phosphate buffered saline (PBS), pH 7.4: dilute one part of 10× PBS (VWR International, see Note 3) with nine parts of distilled water.
4. Spectrophotometer.

2.3. Generation of Linear Inserts for Cloning by Polymerase Chain Reaction

1. Sterile distilled water.
2. dNTP mix, 25 mM each (Fermentas, see Note 3).
3. Platinum® *Pfx* DNA polymerase with 50 mM magnesium sulfate, 10× *Pfx* amplification buffer, and 10× enhancer solution (Invitrogen, see Note 4).
4. Forward and reverse primers: the sequence of the genes of interest and the backbone plasmid are analyzed by CLC sequence viewer software, and restriction sites are designed into each primer for downstream molecular cloning. The primers used in this instruction are listed in Table 1 and are synthesized commercially. Primers are dissolved at 10 pmol/μL in sterile distilled water and stored in aliquots at –20°C.
5. Polymerase chain reaction (PCR) thermal cycler.
6. Thin walled PCR reaction tubes (0.2 mL).
7. Genomic DNA of *N. meningitidis* strain FAM20.
8. Plasmid pZERO-2.1 (Invitrogen) containing the kanamycin resistance gene is carried by *E. coli* DH5α. Bacterial growth requirements are LB agar or LB medium containing 50 μg/mL kanamycin with incubation at 37°C.

2.4. Agarose Gel Electrophoresis

1. UltraPure™ Agarose (Invitrogen, see Note 3).
2. 1× TBE buffer: dilute one part of 10× TBE buffer (Bio-Rad, see Note 5) with nine parts of distilled water.
3. Illustra™ GFX PCR DNA and Gel Band purification kit (GE Healthcare, see Note 3).
4. GeneRuler™ 100 bp and 1 kb DNA ladder (Fermentas, see Note 3).
5. 6× DNA loading dye (Fermentas, see Note 3).
6. Electrophoresis apparatus and power supply.
7. Transilluminator.
8. Ethidium bromide. Caution: this reagent is a mutagen and must be handled under appropriate safety guidelines recommended by the supplier.

Table 1
Primer pairs used for construction of the plasmid. (Reproduced from (8))

Nr.	Gene	Primer sequence	Restriction site[a]	Size
1	*kanR*[b]	Forward 5′-gatgaat**gtcgac**tactgggccgtctgaa caagggaaaacg-3 Reverse 5′-ctggaacaacac**ggatcc**ctatcgcggt ctattcttttg-3′	*Sal* I *Bam*HI	1,140 bp
2	*porA*	Forward 5′-aatagtac**gg atcc**gattca cttggtgctt cagcacc-3′ Reverse 5′-cgagggcggtaagttttttcgc**agatct** gcttcc-3′	*Bam*HI *Bgl* II	379 bp
3	*gapdh*	Forward 5′-ttcgcaggccggccg**taa**gg cttgaacaaa cctgtgg-3′ Reverse 5′-tttgttcgcca**gcggccgc**aatatcaag ttatagcgg-3′	*Eae* I *Not* I	190 bp
4	UHS	Forward 5′-aacaatagagc**agtact**tcccggcagg tcaaattg-3′ Reverse 5′-gctaacagaaaa**ctcgag** tccactattgttagggg-3′	*Sca* I *Xho* I	890 bp
5	DHS	Forward 5′-cttggttaca**gcggccgc**atgattg caaataatgag-3′ Reverse 5′-cgtaaatggtttca**gagctc**ataa tttttctctttc-3′	*Not* I *Sac* I	835 bp

[a]Restriction enzyme cleavage sites are marked in *bold*

[b]The DNA uptake sequence is underlined in the forward primer

2.5. Construction of N. meningitidis Transformation Plasmid pLKMp

1. Plasmid pXen-13 (Caliper Life Sciences) containing the *luxCDABE* operon for engineering bioluminescent Gram-negative bacteria is carried by *E. coli* DH5α. Bacterial growth requirements are LB agar or medium containing 100 μg/mL ampicillin with incubation at 37°C.
2. Illustra™ GFX PCR DNA and Gel Band purification kit (GE Healthcare, see Note 3).
3. Restriction endonuclease (Fermentas, see Note 3 and Table 1).
4. T4 DNA ligase with 10× ligation buffer (Fermentas, see Note 3).
5. Shrimp alkaline phosphatase (SAP) with 10× reaction buffer (Fermentas, see Note 3).
6. *E. coli* DH5α competent cells (see Notes 6 and 7).
7. QIAGEN Plasmid Miniprep kit (QIAGEN).
8. Electroporator and 1 mm gap electroporation cuvettes.
9. Heating-block and low temperature water bath.
10. Sterile inoculating loops (1 and 10 μL).

2.6. Generation of a Bioluminescent *N. meningitidis* Strain

1. GC agar plates and GC medium with Kellogg's supplement. For the selection of transformants add 50 μg/mL of Kanamycin.
2. Wizard Genomic DNA purification kit (Promega, see Note 3).
3. Sterile inoculating loops (1 μL).
4. Restriction enzyme *EcoR*I.
5. DNA Depurination solution: 0.25 M HCl in distilled H_2O.
6. DNA Denaturing Solution: 0.5 M NaOH/1.5 M NaCl in distilled H_2O.
7. DNA Neutralizing Solution: 1 M Tris–HCl buffer, pH 7.5 containing 1.5 M NaCl, prepared in distilled H_2O.
8. Plastic tray.
9. Biotin-High Prime (Roche, see Note 8).
10. Biotin Luminescent Detection Kit (Roche, see Note 8).
11. Biodyne® Plus Membrane, 0.45 μm (Pall Gelman Laboratory, see Note 3).
12. UV cross-linker (Stratagene, see Note 3).
13. Vacu-Blot System (Biometra GmbH, see Note 3).
14. Optical In Vivo Imaging System (IVIS® Spectrum, Caliper Life Sciences). The system principally contains an imaging chamber mounted with a cooled CCD camera, a gas anesthesia system, an acquisition computer, and a monitor. The bioluminescent photons emitted by the luciferase reactions are detected by the CCD camera and the data acquired can then be displayed and analyzed with the Living Image® software supplied with the system (7).

2.7. In Vivo Imaging of Meningococcal Disease Dynamics

1. The hCD46Ge transgenic mouse line harbors the complete human CD46 gene and expresses CD46 in a human-like pattern (10, 11). CD46 is a human cell-surface protein involved in regulation of complement activation and also interacts with a number of important pathogenic bacteria and viruses, including *N. meningitidis*. It is believed that CD46 might facilitate meningococcal growth in the blood and pilus-dependent interactions of meningococci with the epithelial mucosa. Mice of 6–10 weeks old, matched by age and gender, are used for experiments. The animal experiments should be approved by appropriate ethical committees and all animal procedures should be performed in accordance with institutional guidelines.
2. Sterile 1-mL syringe and 27-gauge 1/2 in. needle.
3. Sterile 0.5-mL syringe with 29-gauge 1/2 in. needle (BD Micro-Fine™, see Note 3).
4. Isoflurane anesthetic.

5. Scalpel blade.
6. Ethanol (100%).
7. Optical In Vivo Imaging System (IVIS® Spectrum, Caliper Life Sciences) with Living Image® software.
8. GC agar plates.

3. Methods

The plasmid pXen-13 containing the *luxCDABE* operon can be modified to a *N. meningitidis* transformation plasmid. A promoter sequence of the *porA* gene (P) is inserted upstream of the *luxCDABE* operon in order to increase the expression level of bioluminescent signals in *N. meningitidis* and a terminator sequence (T) of the glyceraldehyde-3-phosphate dehydrogenase (*gapdh*) gene is added to eliminate possible read through of the *luxCDABE* operon. A kanamycin resistance gene (*kanR*) is cloned into the vector for transformant selection and a 10-bp DNA uptake sequence required for efficient DNA transformation of *Neisseria* (12) is introduced into the forward primer (see Table 1). Two stretches of meningococcal DNA, called UHS and DHS, are cloned into 5′ and 3′ flanking region of the *luxCDABE* expression cassette for plasmid integration by homologous recombination. The schematic plasmid maps are shown in Fig. 1a, b.

3.1. Bacteria and Growth Conditions

1. Grow *N. meningitidis* on GC agar plates in a humidified incubator at 37°C for 16–20 h in an atmosphere of 5% (v/v) CO_2. For the selection of transformants add 50 μg/mL of kanamycin.
2. Grow *E. coli* strains on LB agar medium plates and in LB liquid medium in a dry 37°C incubator. For the selection of transformants add 50 μg/mL kanamycin.

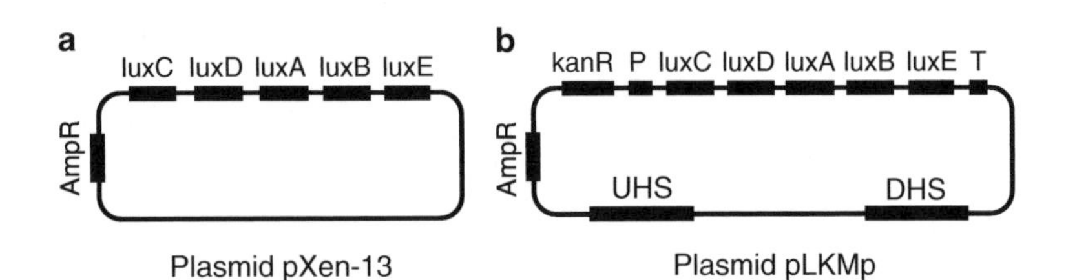

Fig. 1. Generation and characterization of a bioluminescent *N. meningitidis* strain. (**a**) Schematic diagram of the plasmid pXen-13. (**b**) Schematic diagram of the *N. meningitidis* transformation plasmid pLKMp. *Neisseria* specific *porA* promoter (P) and *gapdh* terminator (T) sequences flank the *luxCDABE* operon. UHS and DHS represent two sequences homologous to noncoding sequences of the FAM20 genome. The *luxCDABE* expression cassette was integrated into the FAM20 genome by homologous recombination and the $FAM20^{LU}$ transformants were selected on kanamycin containing GC agar plates.

3.2. Preparation of Template DNA

1. To prepare *E. coli* plasmid DNA, grow the bacteria overnight at 37°C with shaking in LB medium. Collect 1 mL of the bacterial solution into Eppendorf tubes and isolate plasmid DNA using a commercially available plasmid miniprep kit, according to manufacturer's guidelines.
2. To prepare *Neisseria* genomic DNA, inoculate the bacteria onto a GC agar plate and incubate the plate overnight in a humidified incubator at 37°C with 5% (v/v) CO_2. Collect one full loop of the bacteria from the agar plate into an Eppendorf tube and wash once with 1× PBS. Isolate meningococcal genomic DNA using a commercial kit, according to the manufacturer's guidelines.
3. Determine the concentration of DNA by measuring the absorbance of light (λ260 nm) in a spectrophotometer. An absorbance of 1.0 corresponds to 50 μg/mL of pure double stranded DNA.

3.3. Generation of Linear Inserts for Cloning by Polymerase Chain Reaction

1. Amplify the kanamycin resistance gene by PCR from the plasmid pZERO-2.1 using primer pair nr. 1 (see Table 1). Set up the PCR reaction in a 0.2 mL thin wall PCR reaction tube containing the following components: 2.5× *Pfx* amplification buffer, 0.3 mM dNTP mix, 1 mM $MgSO_4$, 0.3 μM of each primer, 50 ng of plasmid DNA, and 1 unit of Platinum® *Pfx* DNA polymerase. Add sterile distilled water to a final volume of 50 μL. Mix the components well by pipetting. The amplification conditions are one cycle of 94°C for 4 min and 30 cycles of 94°C for 15 s, 65°C for 40 s, and 68°C for 2 min.
2. Amplify the meningococcal *por*A promoter sequence, *gapdh* terminator sequence, and UHS and DHS integration sequences from the strain FAM20 using primer pair nr. 2, 3, 4, and 5 (see Table 1), respectively. Each PCR reaction contains 1× *Pfx* enhancer solution, 2.5× *Pfx* amplification buffer, 0.3 mM dNTP, 1 mM $MgSO_4$, 0.3 μM of each primer, 500 ng of genomic DNA, and 1 unit of Platinum® *Pfx* DNA polymerase. Add sterile distilled water to a final volume of 50 μL and mix the components well by pipetting.
   (a) The amplification conditions for the *porA* promoter are one cycle of 94°C for 4 min and 30 cycles of 94°C for 15 s, 65°C for 40 s, and 68°C for 45 s.
   (b) The amplification conditions for the *gapdh* terminator are one cycle of 94°C for 4 min and 30 cycles of 94°C for 15 s, 60°C for 40 s, and 68°C for 30 s.
   (c) The amplification conditions for HUS and HDS are one cycle of 94°C for 4 min and 30 cycles of 94°C for 15 s, 65°C for 40 s, and 68°C for 1 min.

3.4. Agarose Gel Electrophoresis

1. Prepare a 1% (w/v) agarose gel by dissolving 1 g of agarose powder in 100 mL of 1× TBE buffer and heating in a microwave oven until completely melted. After cooling the solution to about 50°C, add ethidium bromide to the gel to a final concentration of 0.5 μg/mL, pour the solution into a casting tray containing a sample comb and allow the gel to solidify at room temperature.
2. Mix 10 μL of the PCR product(s) generated in Subheading 3.3 with 2 μL of 6× loading buffer and load the DNA sample together with an appropriate molecular weight DNA ladder onto the agarose gel. Run the gel at 100 V for 1–2 h and observe the migration of DNA with a transilluminator to make sure the reaction is successful.
3. Cut out the PCR band(s) by using a clean scalpel blade.
4. Purify the PCR product(s) from the agarose gel using the Illustra™ GFX PCR DNA and Gel Band purification kit, according to manufacturer's guidelines.
5. Determine the DNA concentration by measuring the absorbance of light (λ260 nm) in a spectrophotometer.

3.5. Construction of *N. meningitidis* Transformation Plasmid pLKMp

We use restriction endonuclease enzymes to digest both insert and vector DNA. After dephosphorylation of the linearized vector, we use a T4 DNA ligase to anneal the insert and the vector. We start with the kanamycin resistance gene that is cloned into the *Sal*I/*Bam*HI site of the plasmid pXen-13.

1. DNA Digestion: mix 1 μg of vector or insert DNA with 1 unit of selected restriction endonuclease enzyme, 1× reaction buffer, and sterile distilled water to a total reaction volume of 50 μL. Incubate at 37°C for 2 h, purify the digested DNA excised from the agarose gel using the Illustra™ GFX PCR DNA and Gel Band purification kit according to manufacturer's guidelines and then determine the concentration of DNA as described in Subheading 3.4, step 5.
2. Dephosphorylate the 5′-ends of the linearized vector DNA with Shrimp Alkaline Phosphatase prior to ligation of the DNA insert. To do this, mix purified vector DNA with 1 unit of SAP, 1× SAP reaction buffer, and sterile distilled water to a final volume of 15 μL. Incubate at 37°C for 1 h followed by incubation at 65°C for 20 min to inactivate the SAP.
3. Perform DNA ligation by incubating the dephosphorylated vector with DNA insert (at a molar ratio of 1:3, see Note 9) in the presence of 1 unit of T4 DNA ligase, 1× T4 DNA ligase buffer, and sterile distilled water to a final volume of 15 μL. Incubate the reaction mixture overnight at 15°C. For every reaction, one ligation reaction plus one negative control with no insert DNA are prepared.

4. Transform 2–3 μL of the ligate into *E. coli* DH5α competent cells by electroporation (1.8 KV/25 μF/200 Ω) and screen for transformants by spreading onto LB agar plates containing 50 μg/mL of kanamycin (see Note 2). After overnight incubation, pick up ten single colonies, respectively, from the plates and inoculate into 4 mL of LB medium containing 50 μg/mL of kanamycin for plasmid isolation.
5. Isolate the plasmid by using the QIAGEN Plasmid Miniprep kit. In order to identify the correct recombinant plasmid, digest the plasmid by mixing 500 ng of the plasmid DNA, 0.5 μL of *SalI* restriction enzyme, 0.5 μL of *Bam*HI restriction enzyme, 1 μL of 10× Buffer, and sterile distilled water to a total reaction volume of 10 μL in an Eppendorf tube and incubate the reaction for 1 h at 37°C. Separate the digested DNA on a 1% (w/v) agarose gel as described in Subheading 3.4 and two fragments should be observed: the inserted *KanR* fragment which is around 1.2 Kbp in length and the linearized plasmid vector pXen-13 which is around 8.7 Kbp in length. Sequence the plasmid DNA using the services of a commercial company to confirm the presence of the correct *kanR* gene sequence.
6. Using the same strategy as described in steps 1–5, we then clone fragment UHS and DHS, *porA* promoter, and *gapdh* terminator into the *Sca*I/*Xho*I, *Not*I/*Sac*I, *Bam*HI, and *Not*I sites of the plasmid pXen-13, respectively. The final plasmid, designated pLKMp, will be used for transformation of meningococci, to generate a bioluminescent strain.

3.6. Generation of a Bioluminescent N. meningitidis Strain

1. Grow *N. meningitidis* strain FAM20 on GC agar plates for 16–20 h and suspend the bacteria in GC medium to a concentration of 1×10^8 bacteria/mL. Mix 20 μL of this bacterial suspension with 100 μL of GC medium containing 100 ng to 1 μg of plasmid pLKMp DNA (see Note 10). Incubate the reaction for 30 min in a humidified incubator at 37°C with 5% (v/v) CO_2, then add 0.9 mL of GC medium containing 1% (v/v) of Kellogg's supplement to the reaction and continue the incubation for a further 5 h. Then, streak 50–200 μL of the bacterial solution onto GC agar plates containing 50 μg/mL kanamycin (GC-KanR) and incubate the plates overnight in a humidified incubator at 37°C with 5% (v/v) CO_2 (see Note 11).
2. Pick 5–10 single colonies using inoculating loops and streak the colonies onto GC-KanR plates for 2–3 times to obtain pure clones.
3. Isolate genomic DNA of the transformed strain ($FAM20^{LU}$) using the Wizard genomic DNA purificiation kit, following the manufacturer's instructions. Isolates containing the correct

luciferase expression cassette can be identified by Southern blot, which is carried out as follows:

(a) Digest 0.5–1 μg of genomic DNA with *EcoRI* enzyme at 37°C overnight.

(b) Separate the digested DNA samples on a 1% (w/v) agarose gel as described in Subheading 3.4.

(c) Place the gel in a plastic tray and depurinate the DNA by rocking in 100 mL of DNA Depurination solution at room temperature for 15 min. Remove the Depurination solution and rinse the gel with distilled water. Then, add 100 mL of DNA Denaturating solution and rock for 30 min. Remove the Denaturating solution, add 100 mL of DNA Neutralizing solution and rock the gel for further 30 min. Rinse the gel again with distilled water.

(d) Blot the DNA onto a Biodyne® Plus Membrane by using the Vacu-Blot system (Biometra) according to manufacturer's instructions.

(e) After UV-linking with a UV cross-linker, hybridize the membrane with a probe that is biotin-labeled single stranded DNA synthesized from the *kanR* gene. The probe is generated using the Biotin-High Prime kit following the manufacturer's guidelines. In brief, 1 μg of PCR-amplified *KanR* gene in a total volume of 16 μL is denatured by heating at 100°C for 10 min and chilling quickly in an ice/water bath. Then, add 4 μL of Biotin-High Prime to the denatured DNA and incubate for 18 h at 37°C. Stop the reaction by heating to 65°C for 10 min.

(f) Use the Biotin Luminescent Detection Kit to detect the signals, following the manufacturer's guidelines. The correct single copy integration of the plasmid will generate one band, which is around 3 Kbp in length. An example of the Southern blot result is shown in Fig. 2. (see Note 12).

4. Detect the luciferase bioluminescence from the FAM20^{LU} strain using an Optical In Vivo Imaging System, by direct observation of luminescence emitted from the bacterial colonies grown on GC agar plate (see Note 13). Analyze the data with the Living Image® software following the manufacturer's instructions. An example of the result is shown in Fig. 3.

5. Following direct observation of bioluminescence, it is important to demonstrate that the bioluminescent FAM20^{LU} strain behaves like the wild-type strain. (see Note 14). This is necessary since insertion of external genes into the genome could affect the transcription of downstream genes. Furthermore, spontaneous phase variation of certain virulent proteins could happen during genomic modification.

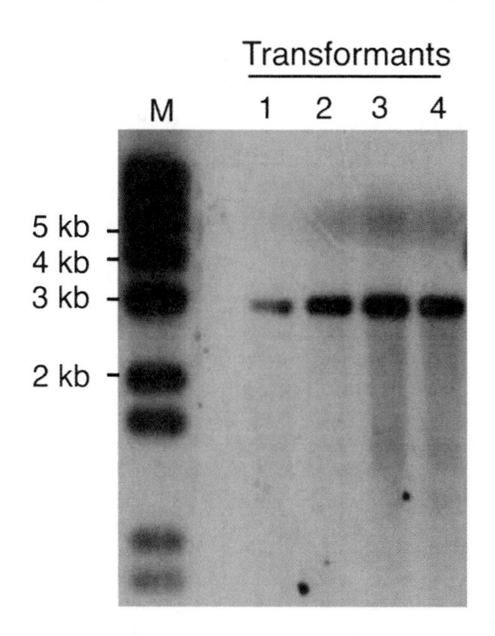

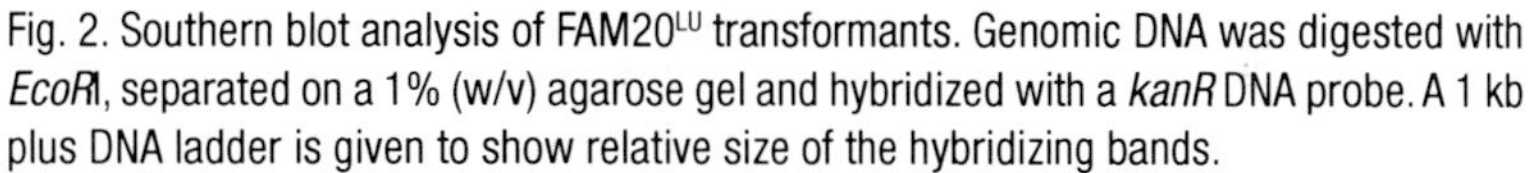

Fig. 2. Southern blot analysis of FAM20^{LU} transformants. Genomic DNA was digested with *EcoRI*, separated on a 1% (w/v) agarose gel and hybridized with a *kanR* DNA probe. A 1 kb plus DNA ladder is given to show relative size of the hybridizing bands.

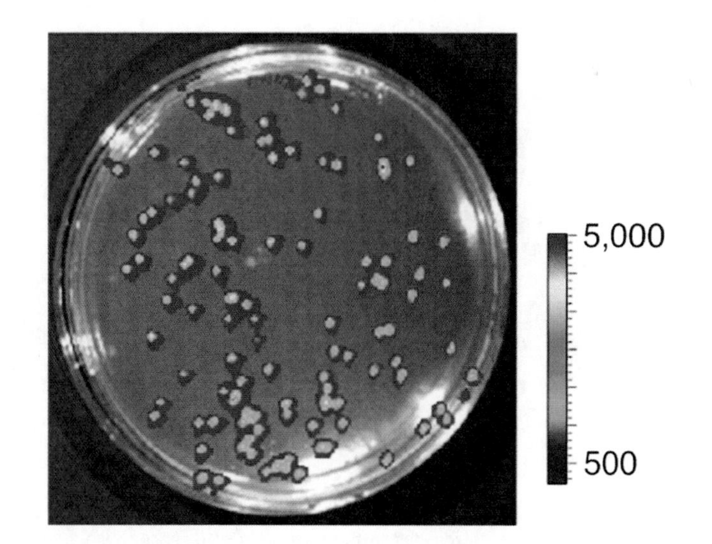

Fig. 3. Bioluminescence of *N. meningitidis* strain FAM20^{LU} on GC agar plate. The color scale bar shown next to the image represents the photons/second emitted from the bacteria. *Red* represents the highest photons/second emission, whereas *blue* represents the lowest photons/second emission. (Reproduced from (8), open access).

3.7. In Vivo Imaging of Meningococcal Disease Dynamics

1. Grow the FAM20^{LU} strain on GC-KanR agar plates (see Note 15) for 16–20 h and suspend the bacteria into GC medium to a final concentration of 10^{10} bacteria/mL.
2. Infect the hCD46Ge transgenic mice intravenously (i.v.) with 10^{9} bacteria/mouse through the tail vein (see Note 16). At various time points after challenge, anaethetise the mice with isoflurane, put them into the imaging chamber, and image the animals for a maximum of 2 min using the IVIS® Spectrum Optical In Vivo Imaging System, following the manufacturer's guidelines (see Note 17). Quantify the total bioluminescence emission from a whole mouse or defined areas of each mouse using the Living Image® software, according to the manufacturer's instructions (see Note 18). Images of the bioluminescence signal from infected mice showing different disease patterns and disease dynamics are shown in Figs. 4 and 5.
3. Determine the numbers of bacteria in blood by collecting 5 μL of blood from the tail vein using a pipette tip and diluting this volume immediately in GC liquid medium (245 μL). Plate out 100 μL volumes of serial dilutions onto GC agar plates and incubate overnight in a humidified incubator at 37°C with 5% (v/v) CO_2. Count the number of colony forming units (CFU) on the next day.
4. Determine bacteria in cerebrospinal fluid (CSF) by sacrificing each mouse by prolonged inhalation of isoflurane. Then, disinfect

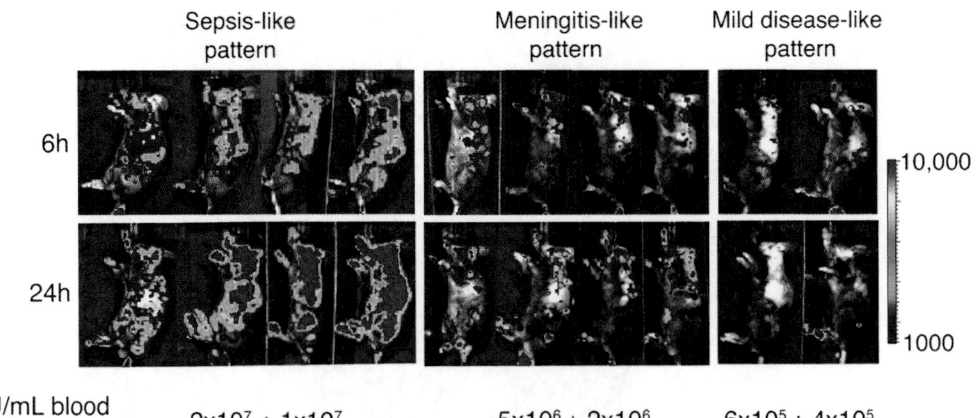

Fig. 4. Meningococcal disease patterns in CD46 transgenic mice. Intravenous challenge of CD46 transgenic mice with a dose of 1×10^{9} FAM20^{LU}/mouse resulted in three patterns of disease. Mice with a "sepsis-like" pattern showed strong bioluminescence signals in the whole mouse body. In mice with a "meningitis-like" pattern, strong signals were mainly detected in the brain and spinal region. In mice with a "mild disease-like" pattern, no obvious signal was detected. Bacterial numbers in the blood at 24 h postinfection are shown as mean CFU/mL ± standard deviation. The color scale bar shown next to the images represents the photons/second emitted from the mice and was applied to all images. *Red* represents the highest photons/second emission, whereas *blue* represents the lowest photons/second emission. (Reproduced from (8), open access).

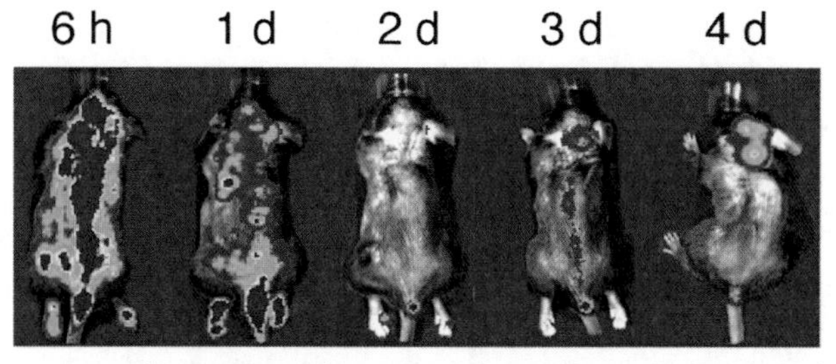

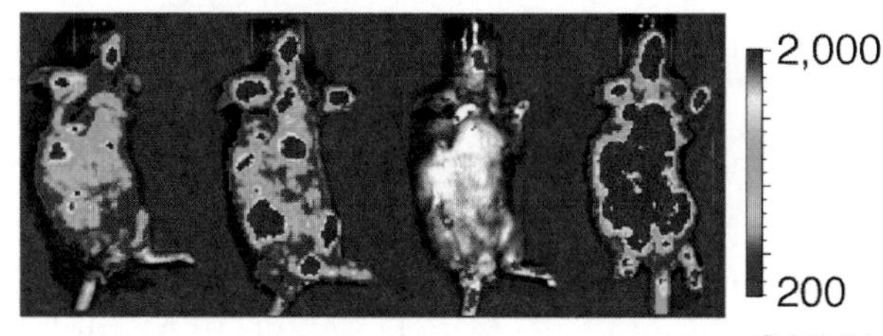

Fig. 5. Dynamic meningococcal disease progression in CD46 transgenic mice. CD46 transgenic mice, challenged i.v with FAM20[LU], were imaged at indicated time points postinfection. Bioluminescence signals from the mice were strong on day 1, but were undetectable by 2 days postinfection. On day 3, the signals reappeared either in the mouse brain area (upper panel) or in the whole mouse body (lower panel). Each panel follows one mouse over time. Bacterial numbers in blood and CSF at indicated time points post-challenge are shown. The color scale bar shown next to the images represents the photons/second emitted from the mouse and was applied to all images. represents the highest photons/second emission, whereas represents the lowest photons/second emission. (Reproduced from (8), open access).

the skin of the back of the neck with ethanol, expose the skull through a midline incision with a scalpel blade and puncture the *cisterna magna* with a syringe fitted with a 29 gauge, 1/2 in. needle. Collect approximately 1–2 μL of clear CSF and immediately suspend in 100 μL of GC liquid. Check this suspension for the absence of red blood cells under a microscope before spreading the samples onto GC agar plates. Determine the number of bacterial CFU after an overnight incubation at 37°C in a 5% (v/v) CO_2 atmosphere.

4. Notes

1. This protocol can be adapted for other meningococcal strains or *Neisseria* species. The DNA sequence mentioned in this instruction for plasmid integration by homologous recombination may need modification to be strain-specific.
2. We have found that certain *N. gonorrhoeae* strains are more sensitive to kanamycin selection. It is important to test a suitable

kanamycin concentration prior to the transformation experiment.

3. Reagents, kits, or instruments from other commercial sources are acceptable.
4. Other DNA polymerases with proof-reading ability are acceptable.
5. Other types of electrophoresis buffer, such as Tris-acetate-EDTA (TAE) buffer and sodium borate (SB) buffer can also be used.
6. Other types of competent cells can be used for cloning. We recommend using competent cells from commercial sources if the cloning efficiency is very low.
7. Both electro competent and chemically competent bacterial cells can be used for cloning.
8. Other labeling and detection methods are acceptable.
9. Formula for double stranded DNA (dsDNA) Molar conversion: pmol of dsDNA = μg (of dsDNA) × 10^6 pg/1 μg × 1 pmo l/660 pg × 1/Number of base pairs (of dsDNA).
10. We noticed that using higher amounts of the plasmid DNA does not increase transformation efficiency.
11. As an alternative, we also perform spot transformation, which might enhance transformation frequency. Spot 2–3 μL (containing 0.1–1 μg) of plasmid DNA onto a GC agar plate and leave the plate at room temperature for a short time until the spot becomes dry. Streak a thin layer of fresh FAM20 strain onto the GC agar plate and incubate at 37°C in a 5% (v/v) CO_2 atmosphere overnight. On the next day, pick up bacteria from the spot area, spread them onto GC-KanR plates and incubate at 37°C in a 5% (v/v) CO_2 atmosphere overnight. Pick up single colonies to identify transformants.
12. Correct integration of the plasmid can also be identified by PCR using primers flanking the UHS and DHS regions.
13. Genes encoding luciferase enzymes and their substrates are located on the *luxCDABE* operon. The labeled bacteria do not require exogenous addition of substrate for light production.
14. Examples of strain characterization for *N. meningitidis* are presented in this instruction. The protocols may need to be adapted when other strains or species are studied. The expression of major surface-associated virulence factors in both strains is carefully examined (8). For example, the expression levels of the pilus-associated adhesin PilC and the opacity-associated (Opa) proteins are detected by western blot; the expression level of capsule is detected by enzyme-linked immunosorbent assay (ELISA); lipooligosaccharide (LOS) is isolated and

detected using tricine SDS-polyacrylamide gel electrophoresis and visualized by silver staining and the pilE gene is amplified by PCR and sequenced by using primers 5′-GATGCCGCAAATTTCCAATC-3′ and 5′-TCACGACCGGGTCAAACC-3′. To compare the bacterial growth rates of both strains, grow each bacterial strain on GC agar plates for 16–20 h and then suspend the bacteria in 50 mL of GC medium containing Kellogg's supplement to an optical density at λ600 nm ($OD_{600\ nm}$) of 0.2. Divide each sample into three small tissue culture flasks and incubate with gentle shaking at 37°C in a humidified incubator with 5% (v/v) CO_2. Remove 1 mL of bacterial culture every hour and measure the $OD_{600\ nm}$. Record the growth for at least 10 h and measure a sample on the next day at the 24 h time point. Plot $OD_{600\ nm}$ readings against time. In addition, bacterial adherence and invasion capabilities of both strains can be determined by using the saponin lysis/gentamicin assay as described previously (8).

15. Labeled bacteria can grow on GC agar plates without kanamycin selection at this stage, since the kanamycin resistance gene is integrated into the genome. However, we notice that bacteria grow quicker and form bigger colonies on GC agar plates alone.
16. Other infection routes can also be used, but the choice is determined by the study aims. The most suitable infection dose needs to be titrated in advance when other strains or species are used.
17. The dark fur of the hCD46Ge transgenic mice is carefully shaved away in order to increase the sensitivity of photon detection. Bioluminescence light is transmitted more efficiently through the light-colored or nude mice.
18. To account for background luminescence, one uninfected mouse is imaged alongside the infected mice.

Acknowledgments

We thank Caliper Life Sciences for kindly providing the plasmid pXen-13, which was used to construct the plasmids described in this protocol. This work was funded by The Swedish Research Council, Swedish Cancer Society, Magnus Bergvalls Foundation, Ake Wibergs Foundation, Swedish Society for Medicine, Torsten och Ragnar Söderderbergs Foundation, Knut och Alice Wallenbergs Foundation, Laerdal Foundation for Acute Medicine, and Uppsala University.

References

1. Contag CH, Bachmann MH (2002) Advances in *in vivo* bioluminescence imaging of gene expression. Annu Rev Biomed Eng 4: 235–260.
2. Hardy J, Francis KP, DeBoer M et al (2004) Extracellular replication of *Listeria monocytogenes* in the murine gall bladder. Science 303: 851–853.
3. Monack DM, Bouley DM, Falkow S (2004) *Salmonella typhimurium* persists within macrophages in the mesenteric lymph nodes of chronically infected Nramp1$^{+/+}$ mice and can be reactivated by IFNγ. J Exp Med 199: 231–241.
4. Wiles S, Clare S, Harker J et al (2004) Organ specificity, colonization and clearance dynamics *in vivo* following oral challenges with the murine pathogen *Citrobacter rodentium*. Cell Microbiol 6: 963–972.
5. Yoshimitsu M, Sato T, Tao K et al (2004) Bioluminescent imaging of a marking transgene and correction of Fabry mice by neonatal injection of recombinant lentiviral vectors. Proc Natl Acad Sci USA 101: 16909–16914.
6. Meighen EA (1991) Molecular biology of bacterial bioluminescence. Microbiol Rev 55: 123–142.
7. Doyle TC, Burns SM, Contag CH (2004) *In vivo* bioluminescence imaging for integrated studies of infection. Cell Microbiol 6: 303–317.
8. Sjolinder H, Jonsson AB (2007) Imaging of disease dynamics during meningococcal sepsis. PLoS One 2: e241.
9. Kellogg DS Jr., Cohen IR, Norins LC et al (1968) *Neisseria gonorrhoeae*. II. Colonial variation and pathogenicity during 35 months *in vitro*. J Bacteriol 96: 596–605.
10. Johansson L, Rytkonen A, Bergman P et al (2003) CD46 in meningococcal disease. Science 301: 373–375.
11. Mrkic B, Pavlovic J, Rulicke T et al (1998) Measles virus spread and pathogenesis in genetically modified mice. J Virol 72: 7420–7427.
12. Elkins C, Thomas CE, Seifert HS et al (1991) Species-specific uptake of DNA by gonococci is mediated by a 10-base-pair sequence. J Bacteriol 173: 3911–3913.

Chapter 11

Methods for Studying *Neisseria meningitidis* Biofilms

Michael A. Apicella, Jianqiang Shao, and R. Brock Neil

Abstract

Neisseria meningitidis is an organism whose environmental niche is limited to the human host. It can frequently colonize the human nasopharynx and has the ability to cause severe systemic infections. These infections can be sporadic, endemic or occur in outbreaks associated with more virulent meningococcal strains. Studies have demonstrated that the meningococcus can form biofilms both in vivo and ex vivo. In this chapter, we discuss methods to establish biofilms in the laboratory for in-depth biochemical, genetic, or microscopic studies.

Key words: *Neisseria meningitidis*, Biofilm, Bronchial epithelial cells, Stable isotope labeling

1. Introduction

Biofilm research has gained considerable attention in the last 10–15 years. A biofilm is a community of microorganisms attached to a surface and encased in an exopolymeric matrix. Biofilms are formed under conditions of environmental stress and are often associated with shear forces. The members of the biofilm are afforded more protection from environmental factors than, if they were, in an individual planktonic state. These include protection from antimicrobials, osmolarity fluctuation, predation, UV light, reactive oxygen species, and dehydration (1). A model organism for biofilm development is *Pseudomonas aeruginosa*, which forms biofilms in the lungs of cystic fibrosis patients. Biofilm formation occurs in five stages for this organism and these stages are generally applicable to other biofilm-forming organisms: (1) Loose association of the bacterium with a surface; (2) tight bacterial adherence and transcriptional changes to adapt to the new environment; (3) bacterial aggregation and microcolony formation as defined by a small number

Myron Christodoulides (ed.), *Neisseria meningitidis: Advanced Methods and Protocols*, Methods in Molecular Biology, vol. 799, DOI 10.1007/978-1-61779-346-2_11,

of bacteria encased in a matrix; (4) mature biofilm development, often seen as mats or tall structures that have channels to disperse nutrients to all the bacteria; (5) detachment of bacteria from the biofilm to colonize other locations (1).

Neisseria spp. are obligate human pathogens and commensal organisms. The nasopharyngeal flora members, *N. meningitidis* and *Neisseria lactamica*, are able to colonize human hosts for long periods of time without causing obvious symptomatic disease (2, 3). This requires evasion of both innate and adaptive immune defense mechanisms.

Neisseria gonorrhoeae is the etiologic cause of the sexually transmitted disease, gonorrhoeae. In men, gonococcal infection results in acute urethritis, 2–5 days postinfection, whereas in women, infection is often asymptomatic. The undetected infection can ascend the female genital tract and cause more serious infections such as pelvic inflammatory disease, which can lead to sterility. The ascending infection can also escape the female reproductive tract to cause disseminated gonococcal infection (4). The ability of the gonococcus to remain undetected in women led to speculation that *N. gonorrhoeae* was persisting as a biofilm. Gonococcal biofilms can form on cervical tissue ex vivo and persist up to 8 days without obvious damage to the cervical tissue (5). These biofilms can also be maintained in a flow cell on a glass surface (5). Gonococcal biofilms have been demonstrated in vivo, including on in-dwelling intrauterine devices of women with reproductive tract infections (6).

Electron micrographs of gonococcal or meningococcal biofilms show the presence of long membrane-like structures throughout biofilms grown in vitro on glass or on cervical tissue culture cells in flow cells, as well as on archival cervical biopsies of gonococcal culture positive patients. These membrane-like structures label with antibody 2C3, which reacts to the outer membrane protein H.8 present in both the gonococcus and meningococcus (7). The pathogenic *Neisseria* are known to bleb the outer membrane and release it into the extracellular environment. A mutation in the *msbB* gene results in a penta-acyl lipid A and the resulting mutant organism has a reduced ability to both bleb and form biofilms in flow cells (7). The experimental evidence obtained using the *msbB* mutant demonstrates that the blebbing process and possibly membrane contributions from deceased gonococci aid in matrix formation that stabilizes the gonococcal biofilm structure.

N. meningitidis is predominantly a commensal organism present in the nasopharynx of 5–10% of the population (8), but in rare circumstances the organism can cause invasive disease characterized by sepsis and meningitis. Carriage of *N. meningitidis* can last an average of 9.6 months and during this time, person-to-person spread can occur (9, 10). Long-term carriage of the organism suggests the possibility that biofilm formation may be occurring.

Clinical evidence of the presence of meningococcal biofilms during carriage is limited and a comprehensive study of tissue excised during tonsillectomy, adenoidectomy, or in surgery to repair sinus passages would be beneficial. A small-scale study was previously performed that examined nasopharyngeal swabs in order to identify carriers of *N. meningitidis* from 32 patients attending for tonsillectomy. This study reported that 10% of the patients were carriers. However, immuno-labeling for the PorA protein of the meningococcus on 5 μm frozen sections of the tonsils revealed that 45% of the patients carried the meningococcus (11). Interestingly, the labeled bacteria appeared to be in microcolonies just below the epithelial surface, which could explain why as few as 10% of the patients were identified as carriers by nasopharyngeal swab. Another study analyzed laryngo-tracheal stents from 21 patients for biofilm formation. The average age of the patients in this study was 37 months and all the stents were positive for 2–5 different bacterial organisms and 52% of the patients carried *Neisseria spp*. However, no effort was made to establish if these species were *N. meningitidis*, *N. lactamica*, or one of the oral *Neisseriae* (12).

Purpura fulminans is a disease state associated with meningococcal sepsis in which cutaneous hemorrhagic lesions occur. In one study of patients with sepsis, skin biopsies from these lesions were immuno-labeled to determine if capsule, pili, and PorA protein were expressed during infection. All three virulence factors were expressed and interestingly the bacteria were mostly grouped in microcolonies in these lesions (13). This indicated the possibility of biofilm formation during symptomatic disease and not just during chronic nasopharyngeal infection. In another study of symptomatic meningococcal sepsis, histological sections from septic infection demonstrated the presence of microcolonies of meningococci in brain capillaries likely to have low blood flow (14). Further in vitro studies showed that meningococci were unable to attach to surfaces at high flow rates, whereas attachment was achieved at low flow rates. Once attached, the bacteria could withstand higher shear forces. *In situ* imaging of rat brain capillaries demonstrated that meningococci could attach when blood flow temporarily decreased and that the adherent bacteria could then multiply into microcolony-like structures in these capillaries (14). These studies with brain tissue and *purpura fulminans* necrotic lesions demonstrated the ability of meningococci to form small biofilms in vivo during septic infection (13, 14). However, none of the above studies provided high resolution images to determine if a matrix was present in these microcolonies and if the matrix was composed of membranes, as our laboratory has demonstrated for gonococcal biofilms in vivo (7).

Our laboratory has developed a modified flow cell that accommodates a coverslip with a monolayer of transformed airway epithelial cells to study if encapsulated meningococci can form biofilms and to identify bacterial factors involved in biofilm formation. Encapsulated meningococci can form mature biofilms on SV-40 transformed human bronchial epithelial (HBE) cells in the flow cell by 48 h. These biofilms have significantly more biomass and a taller average height than biofilms grown on collagen-coated coverslips or on glass alone in the same chambers. In situ immunolabeling of the biofilms demonstrated that organisms within the biofilm were encapsulated. Furthermore, the average biofilm height of wild-type and isogenic unencapsulated mutants was statistically indistinguishable at 48 h on HBE cells (15, 16). These studies demonstrate the need for a proper substrate for the meningococci to adhere to and to form biofilms. This finding is significant because meningococcal infection as a biofilm of encapsulated organisms would provide a population of bacteria more capable than unencapsulated organisms, in evading host immune defenses as well as in crossing epithelial or endothelial cell barriers (1, 17). In this chapter, we describe methods for the assembly of continuous growth biofilm chambers and for the growth of bacterial biofilms on inert and human host cell substrates.

2. Materials

2.1. Assembly of Continuous Growth Biofilm Chambers

1. 205U/CA multichannel peristaltic pump (Watson Marlow) or equivalent.
2. Manifold tubing 3 mm outer diameter × 1.8 mm inner diameter (Watson Marlow).
3. Silastic tubing 1.98 mm inner diameter × 3.18 mm outer diameter (Dow Corning).
4. Biofilm chambers: manufactured in-house (Figs. 1 and 2). Alternatively, chambers can be purchased from Biosurfaces Technologies (http://cu.imt.net/~mitbst/) (see Note 1).
5. Glass coverslips.
6. Connectors (Cole Palmer Instrument Company).
7. Nalgene polypropylene sterilization pans for constructing a protected environment for biofilm chambers (US Plastics Corporation).
8. Containment box made of polycarbonate or purchased commercially (Cole Palmer Instrument Company).
9. Sheet of polypropylene, 53 × 43 × 1 cm as a cover for the sterilization pan.
10. Environmental Chamber (Thermo Electron Corporation).
11. Solution of 2% (v/v) sodium dodecyl sulphate (SDS) in water for spillage containment.

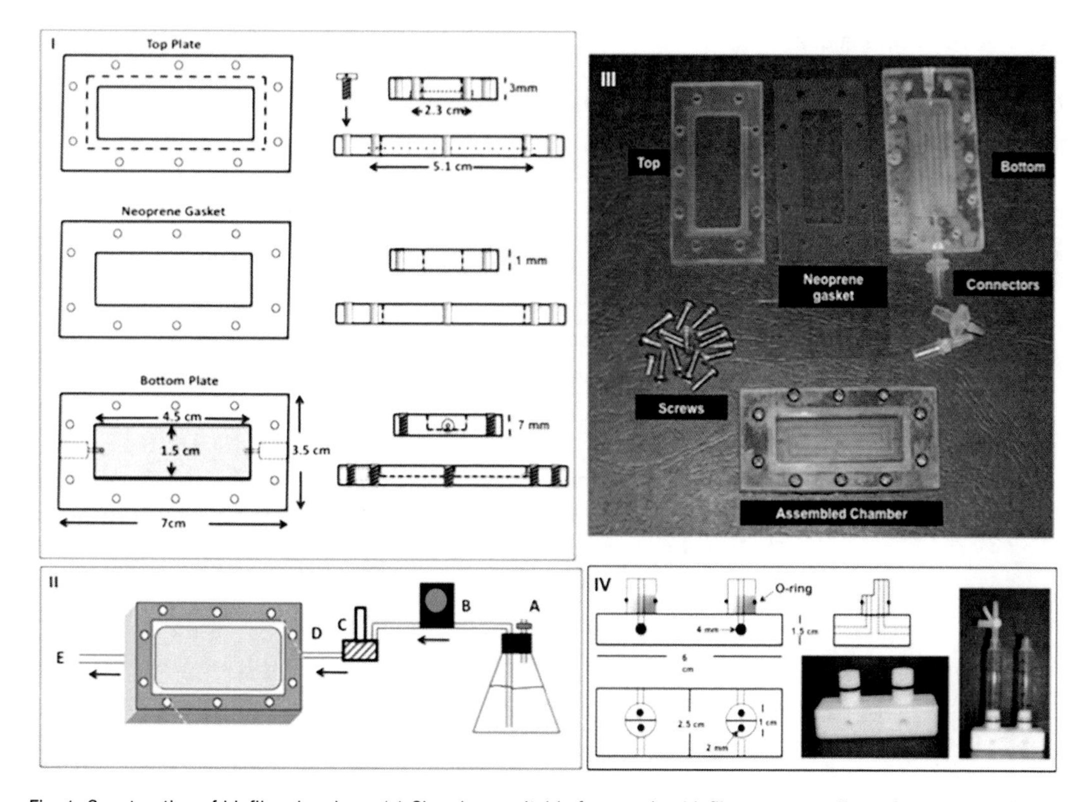

Fig. 1. Construction of biofilm chambers. (**a**) Chambers suitable for growing biofilms on coverslip surfaces or over human cells on coverslips. (**b**) The arrangement of the components of the biofilm apparatus, (A) medium reservoir, (B) peristaltic pump, (C) bubble trap, (D) the chamber and (E) waste collection vessel. (**c**) The components of the chamber and an assembled chamber. (**d**) The dimensions of the bubble trap necessary to remove air bubbles from the system.

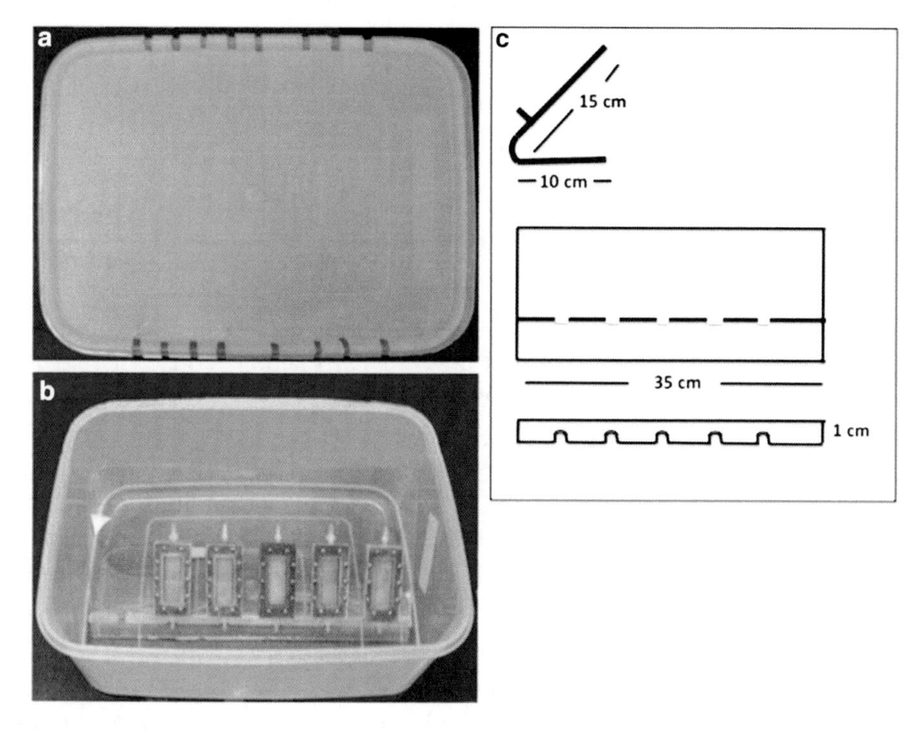

Fig. 2. The containment box for growth of meningococcal biofilms. The box is used to contain any accidental aerosols or leaks. (**a**) and (**b**) show the lid and box respectively and (**c**) the rack on which the biofilm chamber is placed, sloped at 45° to permit movement of bubbles.

2.2. Growth of Meningococcal Biofilms in Continuous-Flow Chambers Over Glass Coverslips

1. *N. meningitidis* spp. stored at −80°C in brain heart infusion (BHI) with 25% (v/v) glycerol.
2. GC agar plates for growing *Neisseria* spp.
3. Bacto Brain Heart Infusion (BHI) nutrient broth (Becton Dickinson) containing 1% (v/v) IsoVitaleX enrichment (Becton Dickinson) and 100 μM sodium nitrite.
4. BHI agar plates: prepared by adding agar (1.5% w/v) to BHI broth and autoclaving at 121°C for 15 min with 2.68 kg/cm^2 pressure.
5. Roswell Park Memorial Institute (RPMI)-1640 growth medium with glucose, supplemented with 1% (v/v) IsoVitaleX enrichment, 100 μM sodium pyruvate, 370 μM hypoxanthine, 450 μM uracil, and 100 μM sodium nitrite. The medium is filtered sterilized through a Nalgene 0.2 μm127 Filter Unit.
6. Sterile phosphate buffered saline (PBS), pH 7.4.
7. Humidified CO_2 incubator (Forma Scientific or equivalent).
8. Continuous growth biofilm chamber assembly, including glass coverslips.
9. Zeiss 720 confocal microscope, or equivalent.
10. Spectrophotometer for measuring optical density of bacterial growth.
11. 4′,6-diamidino-2-phenylindole (DAPI) (Fisher Scientific).
12. 4% (v/v) paraformaldehyde or glutaraldehyde in PBS for fixing cells.

2.3. Growth of Meningococcal Biofilms in Continuous-Flow Chambers Over Human Cells in Culture

1. SV-40 transformed HBE cells (ATCC).
2. Complete medium for HBE cells: serum-free keratinocyte growth media (K-SFM, Invitrogen) supplemented with 12.5 mg bovine pituitary extract (BPE), 0.08 μg of epidermal growth factor (EGF) per 500 mL bottle, and a final concentration of 1% (v/v) penicillin/streptomycin.
3. RPMI 1640 growth medium.
4. 0.25% trypsin/1 mM ethylenediamine tetraacetic acid (EDTA) for cell detachment.
5. 1:5 JEM medium: prepared by mixing two parts of serum-free hybridoma medium, one part of McCoy's 5A medium, and one part defined K-SFM, which is then diluted 1/5 with sterile PBS containing 1% (v/v) IsoVitaleX, 100 μM sodium nitrite, and 0.5 g/L of sodium bicarbonate to buffer the medium.
6. Continuous-flow chamber assembly adapted for tissue culture.
7. Collagen-coated glass coverslips (22 × 50 mm) for growing cells. Prepare the coverslips by autoclaving for 15 min at 121°C and 2.68 kg/cm^2 pressure in bovine tendon collagen type I

(Worthington Biochemical Corp). Then, remove the coverslips from the solution and allow to dry for 30 min at room temperature in 100 mm sterile tissue culture plates.

8. Zeiss 720 confocal microscope, or equivalent.
9. Humidified CO_2 incubator.
10. Cell Tracker Orange™ (Molecular Probes, Invitrogen).
11. DAPI.

2.4. Processing of Biofilms for Sectioning and Microscopy

1. Biofilm chamber assemblies for growth over either glass coverslips or human cells.
2. 4% (v/v) paraformaldehyde in PBS, pH 7.4.
3. Sucrose: 20% (w/v) in PBS, pH 7.4.
4. Low-melt agarose (Fischer Scientific).
5. Optimal Cutting Temperature (OCT) Compound (Ted Pella).
6. Cell Tracker Orange™ and DAPI for cell staining.
7. LR White embedment (Ted Pella).
8. Leica EM-UC6 Microtome with a Leica EM FC6 cryo-attachment.

2.5. Quantitative Proteomics of Meningococcal Biofilms Using Stable Isotope Labeling with Amino Acids in Cell Culture

1. C^{13} labeled amino acids (Cambridge Isotopes Laboratories).
2. Biofilm chamber assemblies for growth over either glass coverslips or human cells.
3. Minimal defined medium (18) (MDM) for *Neisseria spp.*, prepared as a 4× stock (see Note 2).
4. MDM agar plates: prepare 100 mL of 2× MDM medium (omitting the single labeled amino acid, which will be added just before the plate is used) and 100 mL of 2× agar (100 mL water containing 3 g of agar). Autoclave the agar at 121°C for 15 min with 2.68 kg/cm^2 pressure and filter the 2×MDM solution. When the agar has cooled to 56°C in a water bath, mix with the filtered MDM solution and pipette 10 mL of the mixture into 60×15 mm Petri plates. Once the agar has solidified, wrap the plate in aluminum foil and store at 4°C until ready for use.
5. Nicotinamide.
6. Sterile Erlenmeyer flasks for bacterial culture.
7. Spectrophotometer for measuring bacterial growth.
8. PBS, pH 7.4.
9. 10% (w/v) sodium azide, as preservative.
10. Balance, accurate to within ±10–100 μg, e.g., Mettler XP105DR10 or equivalent.
11. Trypsin.
12. ProGest automatic digester (Genomic Solutions).

3. Methods

3.1. Assembly of Continuous Growth Biofilm Chambers

1. Figure 1a shows the construction of chambers suitable for growing biofilms on coverslip surfaces or over human cells on coverslips. The chamber is constructed in two sections of polycarbonate plastic separated by a 1 mm neoprene gasket so it can be autoclaved. The top plate is grooved ~1 mm on its lower surface to accommodate a 50×22 mm coverslip on which the biofilm is formed. The lower plate is solid plastic with the biofilm chamber which is 4.5×1.5×2 mm routed out of its upper surface. At each end of the chamber, ports are drilled to accommodate a syringe-type fitting that narrows to a 1 mm opening prior to entering the chamber.
2. Figure 1b shows the arrangement of the components of the biofilm apparatus consisting of a medium reservoir (A), a peristaltic pump (B), a bubble trap (C), the chamber (D), and a waste collection vessel (E). Screws are used to secure the three components together.
3. Figure 1c shows the components of the chamber and an assembled chamber.
4. Figure 1d shows the dimensions of the bubble trap necessary to remove air bubbles from the system. The bubble trap is constructed so that a 5 mL syringe can be used with a three-way value to purge the system. When using this system, approximately 1–1.5 mL of medium is allowed to enter the syringe at the start of the experiment when the pump is first started. The valve is closed for the remainder of the experiment.
5. Figure 2 shows the containment box for growth of meningococcal biofilms to contain any accidental aerosols or leaks. Slits are cut into the top of the box to allow passage of the tubing into and out of the chambers. The biofilm chamber is placed on a rack (Fig. 2c), which is sloped at 45° to permit movement of any bubbles, which might pass through the bubble trap out of the chamber. A solution of 2% (w/v) SDS is placed in the box covering the base of the rack holding the chambers (see Note 3).
6. Alternatively, place the chambers in an environmental chamber that does not require the use of a containment box. Figure 3 shows this set up consisting of (A) perfusion media, (B) the peristaltic pump, (C) the bubble traps, (D) the biofilm chambers on the slanted rack, and (E) the run-off bottles.

3.2. Growth of Meningococcal Biofilms in Continuous-Flow Chambers Over Glass Coverslips

1. Grow *N. meningitidis* from frozen culture onto a plate of GC agar overnight in a humidified incubator at 37°C with 5% (v/v) CO_2. Culture the bacterium in BHI nutrient broth to a density of 10^8 colony forming units (CFU)/mL or an absorbance of 0.1 measured using a spectrophotometer to measure optical density (OD) at $\lambda_{600\ nm}$.

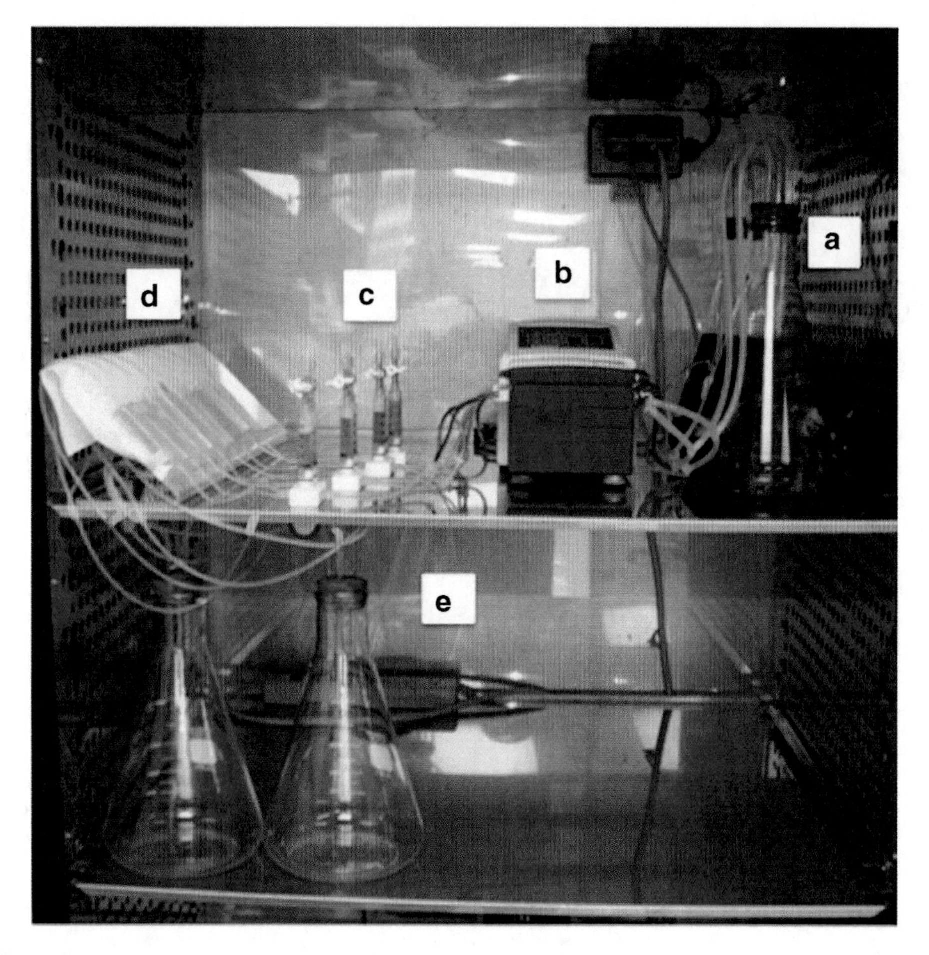

Fig. 3. Setup of the chambers without the protective box in the environmental chamber. (**a**) Perfusion medium, (**b**) the peristaltic pump, (**c**) bubble traps, (**d**) the chambers on the slanted rack and (**e**) the run-off bottles.

2. Meningococcal biofilms are grown in continuous-flow chambers assembled as shown in Figs. 1 and 2 in a 37°C environmental incubator. Use either BHI growth medium or RPMI 1640 growth medium in the chamber assembly, ensuring that these solutions are diluted 1/10 with sterile PBS for use during the continuous-flow experiments.
3. Inoculate the chamber with 1 mL of *N. meningitidis* culture (containing 10^8 CFU) and leave the culture in the chamber for 1 h. Then, start the flow of medium at a rate of 150 μL/min.
4. The meningococcal biofilm will form on the coverslip over the next 48 h and perfusion of medium continues for the duration of the experiment.
5. Visualize the biofilm using the differential interference contrast (DIC) function of the confocal microscope at any time during the experiment (see Note 4).

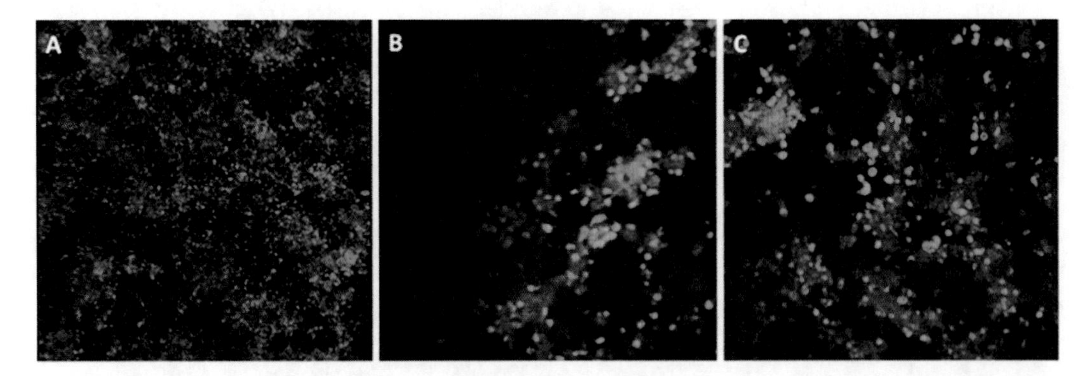

Fig. 4. Confocal microscopy images of a *Neisseria meningitidis* biofilm. (**a**) Intact 48 h biofilm of meningococcal strain NMBgfp (expressing green fluorescent protein) grown over human airway epithelial cells. The epithelial cells are stained with Cell Tracker Orange. (**b**, **c**) Confocal images of 20 μm thick sections of a 48 h meningococcal strain NMBgfp biofilm grown on a glass coverslip stained with monoclonal antibody 2C3 and counter-stained with DAPI. The matrix DNA stained by the DAPI can be clearly seen surrounding the organisms.

6. At the end of the experimental period, stain the biofilm with an antibody specific for the meningococcal surface (e.g., monoclonal antibody 2C3, which reacts with the H.8 protein) and DAPI (see Notes 5 and 6). If fixation is indicated, add either 4% (v/v) paraformaldehyde or glutaraldehyde to the chamber for 30 min. Next, wash the chamber gently with PBS prior to further steps. If electron microscopy is indicated and there is a need to maintain the biofilm in situ, the process of embedment in low-melt agar can be used, as described in Subheading 3.4. Figure 4 shows representative confocal microscopy images of a green fluorescent protein-expressing meningococcal biofilm grown over human airway epithelial cells, with the bacteria also labeled with specific antibodies and the human cells counter-stained with cell tracker dyes.
7. During the experiment, the dissolved oxygen content present in the biofilm medium collected from the medium reservoir and biofilm outflow can be measured (see Note 7). At the end of the experiment, sample the effluent and plate out onto BHI agar plates to assure that purity of the meningococcal culture has been maintained.

3.3. Growth of Meningococcal Biofilms in Continuous-Flow Chambers Over Human Cells in Culture

1. Culture SV-40 transformed HBE cells on coverslips in complete medium in 100 mm tissue culture plates in a humidified incubator at 37°C with 5% (v/v) CO_2.
2. Once confluent, split the cells to collagen-coated coverslips as follows:
   (a) Add 2 mL of 0.25% (v/v) tryspin/1 mM EDTA solution to a confluent 100 mm tissue culture plate for 4 min at room temperature, aspirate the solution, and then incubate the plate for an additional 5 min in an incubator at 37°C with 5% (v/v) CO_2.

(b) Collect the cells in 5 mL of K-SFM supplemented with 5% (v/v) fetal bovine serum to neutralize the trypsin/EDTA, and then centrifuge the cell suspension at 2,000 × *g* for 3 min. Suspend the cell pellet in a final volume of K-SFM to give a 1/8 dilution per glass coverslip.

(c) Add 0.5 mL of this cell suspension to each collagen-coated coverslip, covering the entire surface.

(d) Incubate the coverslips for 3–4 h in an incubator at 37°C with 5% (v/v) CO_2 to allow the cells to adhere, and then flood the plate(s) with 10 mL of K-SFM medium.

(e) Grow the cells until confluent (usually for 2 days) and then stain with Cell Tracker Orange®, according to the manufacturer's instructions, just prior to infection (see Note 8).

3. Assemble the culture biofilm chambers (5 × 22 × 5 mm tissue, approximately 3 mL volume) with the confluent monolayer of transformed HBE cells on coverslips, which are placed between the top and bottom portions of the chamber, and then seal with a rubber gasket and screws that fasten the top and bottom portions together (Fig. 1). The number of HBE cells in an average confluent monolayer on a coverslip is ~5×10^5 cells.
4. Grow *N. meningitidis* as described in steps 1 and 2 of Subheading 3.2.
5. Inoculate the biofilm chamber with *N. meningitidis* at a multiplicity of infection (MOI) of 100:1 using bacterial cell suspensions prepared in 1:5 JEM biofilm medium, which is subsequently used for the duration of the experiment(s).
6. Incubate the flow chambers under static conditions at 37°C with 5% (v/v) CO_2 for 1 h after bacterial infection, in order to allow meningococci to adhere to the human cells.
7. Then, incubate the flow chambers for 48 h at 37°C with a medium flow rate of 180 μL/min of medium. After 48 h, culture the biofilm effluent to assure culture purity as described in step 7 of Subheading 3.2.
8. Assess biofilm formation by confocal microscopy as described in steps 5 and 6 of Subheading 3.2 and by electron microscopy as described in Subheading 3.4.

3.4. Processing of Biofilms for Sectioning and Microscopy

1. Grow the meningococcal biofilms in the continuous-flow chambers for 48 h on either glass coverslips alone or adherent to human cells as described in Subheadings 3.2 and 3.3, respectively.
2. Carefully displace the fluid in the chambers with a solution of 4% (v/v) paraformaldehyde in PBS and leave for 30 min in order to fix the biofilms.

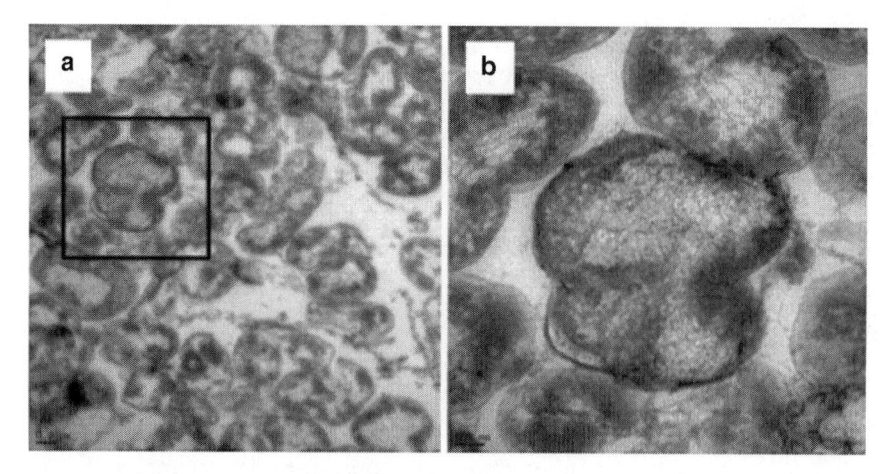

Fig. 5. An electron micrograph of an in situ embedded 48 h biofilm cryopreserved with 20% (w/v) sucrose. (**a**) 5,000× magnification (**b**) 15,000× magnification of the *boxed* area (**a**).

3. Remove the coverslips with attached biofilms carefully from the chambers and overlay them with 20% (w/v) sucrose in PBS in situ, followed by 0.5% (w/v) low-melt agarose in PBS.
4. After the agarose solidifies at 4°C, which usually takes 5–10 min, carefully remove the coverslip and embed the agarose containing the biofilm in situ in OCT medium and freeze by placing it in a –70°C freezer. Cut the OCT frozen block into 20 μm sections with a cryostat.
5. Stain the biofilm as described in step 6 of Subheading 3.2.
6. Alternatively, the samples can be processed for transmission electron microscopy (TEM). Place the agarose containing biofilm in LR White embedment and cut the block into 80 nm sections on an ultramicrotome for viewing on a TEM, following the instrument manufacturers' instructions. Figure 5 shows a sample of such a section.

3.5. Quantitative Proteomics of Meningococcal Biofilms Using Stable Isotope Labeling with Amino Acids in Cell Culture

If you plan to perform proteomics it is important to determine the auxotroph pattern of the strain used in the analysis. Based on this pattern, the amino acid requirements of the strain for growth can be determined. For transcriptional studies, this step can be omitted and RMPI 1640 can be used as the medium.

1. Make up components for MDM for *Neisseria* spp. (suitable for both the gonococcus and meningococcus) (18) (see Notes 2 and 9). Prepare two amino acid stocks: in one stock, omit the particular naturally occurring amino acid for which the strain is an auxotroph and add the amino acid with the universal C^{13} label. The second amino acid stock solution will contain the non-isotopic naturally occurring amino acid. Use the non-isotopic version of the amino acid stock for making the medium

to be used for growing the biofilm sample and add the isotopic version of the amino acid stock for the medium used for growing the planktonic sample.

2. Make up MDM agar plates. These will be used on "Day 2" to subculture the bacteria from a GC agar plate. Prior to culture of bacteria to the MDM plate, warm the plate to 37°C in an incubator and then add the appropriate amount of the single amino acid (making sure that the plate is labeled correctly, in order to identify whether the isotopic or non-isotopic version of the amino acid has been added) and NAD to the plate and spread.
3. On Day 1, streak out bacteria from a frozen stock onto a GC agar plate or a BHI agar plate and grow the organism overnight in a humidified incubator at 37°C with 5% (v/v) CO_2.
4. On Day 2, subculture the bacterial growth from the GC or BHI agar plate onto two MDM plates: one plate has the non-isotopic amino acid and the other contains the C^{13} isotopic amino acid added to the medium. Grow the organisms overnight in an incubator at 37°C with 5% (v/v) CO_2.
5. Prepare a starter culture by inoculating a 15–20 mL volume of 1× MDM in sterile 100 mL Erlenmeyer flasks with enough bacteria from an overnight MDM agar plate to give an OD $\lambda_{600\ nm}$ of 0.1. It is important to add bacteria from the isotopic MDM agar plate to the corresponding isotopic 1× MDM starter culture and similarly the bacteria from the non-isotopic MDM agar plate to the non-isotopic 1× MDM starter culture. Inoculate the starter cultures with a loopful of organisms to an OD$\lambda_{600\ nm}$ of 0.05 and grow with shaking (200 rpm) at 37°C.
6. When bacterial growth has reached the required optical density (OD$\lambda_{600\ nm}$ of approximately 0.25), inoculate the biofilm chambers with this suspension. Incubate the chambers for 1 h at 37°C with the coverslip side facing downwards and then attach the chambers to the appropriate medium flasks.
7. Grow the biofilm from which the planktonic cells (in isotopic MDM) will be collected for 24 h. Then, on Day 4, divert the run-off tubes through the glass columns containing glass wool and into a flask containing 100 μL of 10% (w/v) sodium azide. Place 2–4 run-off tubes (from 2 to 4 biofilm chambers) through each glass wool column. If the run-off from two chambers is collected, add 100 μL of 10% (w/v) sodium azide to the flask and correspondingly, 200 μL of preservative if collecting from four chambers. Collect these planktonic cells for another 24 h. The run-off rate is 150 μL/min or 9 mL/h and is not diluted.
8. On Day 5, collect the non-isotopic biofilm samples into 5 mL of PBS made with HPLC grade water. Centrifuge the samples at 5,000 × *g* for 15 min at room temperature and suspend the

pellet in 2 mL of HPLC grade water. Repeat the centrifugation step, remove the supernatant, and freeze the pellet at –80°C. Collect the isotopic planktonic samples from the run-off flask and centrifuge at 5,000 × *g* for 15 min at room temperature. Suspend the pellet in 1/4 of the volume of PBS made with HPLC grade water and repeat the centrifugation spin. Suspend the pellet in 2 mL of HPLC grade water, repeat the centrifugation spin, and finally freeze the pellet at –80°C. Finally, freeze-dry the samples.

9. Weigh the portions of each pellet and combine equal amounts, e.g., 2 mg of biofilm and 2 mg of planktonic material.
10. For proteomic analysis, run the combined samples on a one-dimensional polyacrylamide gel and stain with Coomassie Blue dye (see Note 10).
11. Place the gels on a clean glass plate on a light box and cut the gels by rastering the stained regions of selected lanes. Depending on the banding pattern, 38–48 spots can be cut per gel lane.
12. Digest the excised gel pieces with trypsin using a ProGest automatic digester, according to the manufacturer's instructions.
13. As the resulting tryptic peptide solutions can vary slightly in volume, adjust the volumes to approximately 10–15 μL each with 0.1% (v/v) formic acid.
14. The samples are now ready for proteomic analysis by mass spectrometry.

4. Notes

1. If you plan to make your own chambers for growth over cells, you will need polycarbonate sheets for milling of chambers as well as teflon blocks for milling of bubble traps. Most local machine shops can easily construct these chambers and bubble traps.
2. MDM was originally developed for the growth of gonococci and is also suitable for meningococci. The recipe is extensive and the reader is referred to the original article of Morse and Bartenstein (18).
3. This will limit any spread of the organism should leakage occur and will result in lysis of bacteria that could be released from the chamber. If such gross contamination occurs, the entire chamber can be detached from the pump and can be autoclaved.
4. Care must be taken when disconnecting the tubing, in order to maintain sterility during transfer to the microscope stage.

5. In addition to the dyes suggested in the protocol, the biofilm can be stained with a variety of other dyes, e.g., toluidine blue, Gram's stain, and a plethora of specific dyes available from Molecular Probes including CellTrace™ calcein red-orange AM, CellTrace™ calcein violet AM, BODIPY® 630/650-X SE, and FM® 1-43.
6. For visualizing DNA in biofilms, the use of Vectorshield containing DAPI (Vector Laboratories) is recommended.
7. A GEM Premier™ 3000 blood gas meter (Instrumentation Laboratory Company) can be used to measure the dissolved oxygen content present in the biofilm medium collected from the medium reservoir and biofilm outflow during growth. The medium is collected under mineral oil to prevent gas exchange during collection. Samples are transferred to the blood gas meter in sealed syringes for analysis and approximately 0.2 mL of each sample is analyzed immediately following collection.
8. CellTracker dyes are vital stains and must be used prior to fixation of the eukaryotic cells. This is done by injecting the stain into the biofilm chamber at the conclusion of the experiment allowing it to set for 15 min, washing the chamber out very gently with 4 mL of PBS, and then fixing the sample with 4% (v/v) paraformaldehyde in PBS.
9. Prepare a sufficient volume of MDM medium required for the planned experiments. When making the "4× stock" add the omitted amino acid to the correct concentration. Add non-isotopic amino acid to one medium preparation and the isotopic amino acid to the other preparation. In order to avoid contamination as much as possible, since MDM cannot be autoclaved, we recommend autoclaving the water used for making the final 1× MDM medium, ahead of time as an extra precaution. In addition, filter sterilize the 1× MDM before use.
10. The methods for proteomics analysis, i.e., 1D and 2D polyacrylamide gel electrophoresis and sample preparation for mass spectrometry and outside the scope of the current protocol and readers are referred to the chapter from Williams and colleagues in this volume.

Acknowledgements

These methods were developed with the support of US Public Health Service NIAID grant AI045728.

References

1. Hall-Stoodley L, Costerton JW, Stoodley P (2004) Bacterial biofilms: from the natural environment to infectious diseases. Nat Rev Microbiol 2:95–108.
2. Ala'Aldeen DA, Neal KR, Ait-Tahar K et al (2000) Dynamics of meningococcal long-term carriage among university students and their implications for mass vaccination. J Clin Microbiol 38:2311–2316.
3. Bennett JS, Griffiths DT, McCarthy ND et al (2005) Genetic diversity and carriage dynamics of *Neisseria lactamica* in infants. Infect Immun 73:2424–2432.
4. Handsfield HH, Sparling PF (2005) *Neisseria gonorrhoeae*. In: Mandell GL, Bennett JE, Dolin R (Eds.) Principles and Practices of Infectious Diseases. 4th ed. New York: Churchill Livingstone; p2514–2529.
5. Greiner LL, Edwards JL, Shao J et al (2005) Biofilm formation by *Neisseria gonorrhoeae*. Infect Immun 73:1964–1970.
6. Pruthi V, Al-Janabi A, Pereira BM (2003) Characterization of biofilm formed on intrauterine devices. Indian J Med Microbiol 21:161–165.
7. Steichen CT, Shao JQ, Ketterer MR et al (2008) Gonococcal cervicitis: a role for biofilm in pathogenesis. J Infect Dis 198:1856–1861.
8. Broome CV (1986) The carrier state: *Neisseria meningitidis*. J Antimicrob Chemother 18:25–34.
9. Wilder-Smith A, Barkham TM, Ravindran S et al (2003) Persistence of W135 *Neisseria meningitidis* carriage in returning Hajj pilgrims: risk for early and late transmission to household contacts. Emerg Infect Dis 9:123–126.
10. Greenfield S, Sheehe PR, Feldman HA (1971) Meningococcal carriage in a population of "normal" families. J Infect Dis 123:67–73.
11. Sim RJ, Harrison MM, Moxon ER et al (2000) Underestimation of meningococci in tonsillar tissue by nasopharyngeal swabbing. Lancet 356: 1653–1654.
12. Simoni P, Wiatrak BJ (2004) Microbiology of stents in laryngotracheal reconstruction. Laryngoscope 114:364–367.
13. Harrison OB, Robertson BD, Faust SN et al (2002) Analysis of pathogen-host cell interactions in purpura fulminans: expression of capsule, type IV pili, and PorA by *Neisseria meningitidis in vivo*. Infect Immun 70: 5193–5201.
14. Mairey E, Genovesio A, Donnadieu E et al (2006) Cerebral microcirculation shear stress levels determine *Neisseria meningitidis* attachment sites along the blood-brain barrier. J Exp Med 203:1939–1950.
15. Neil RB, Apicella MA (2009) Clinical and laboratory evidence for *Neisseria meningitidis* biofilms. Future Microbiol 4:555–563.
16. Neil RB, Apicella MA (2009) Role of HrpA in biofilm formation of *Neisseria meningitidis* and regulation of the hrpBAS transcripts. Infect Immun 77:2285–2293.
17. Spinosa MR, Progida C, Tala A et al (2007) The *Neisseria meningitidis* capsule is important for intracellular survival in human cells. Infect Immun 75:3594–3603.
18. Morse SA, Bartenstein L (1980) Purine metabolism in *Neisseria gonorrhoeae*: the requirement for hypoxanthine. Can J Microbiol 26:13–20.

Chapter 12

A Laminar-Flow Chamber Assay for Measuring Bacterial Adhesion Under Shear Stress

Magali Soyer and Guillaume Duménil

Abstract

Shear stress levels generated by circulating blood have a strong impact on biological processes taking place in the vasculature. It is therefore important to take them into account when studying infectious agents targeting the endothelium. Here we describe a protocol using disposable laminar-flow chambers and video microcopy to study bacterial infections in an environment that mimics the bloodstream. We initially focused on the interaction of *Neisseria meningitidis* with human endothelial cells and determined that shear stress is an important factor for the pathogen's initial adhesion and for the formation of micro-colonies. The experimental set-up can be used to investigate other pathogens that interact with the endothelium as well as with other sites where shear stress is present.

Key words: *Neisseria meningitidis*, Microcirculation, Shear stress, Septicemia, Bacterial adhesion, Bacterial proliferation

1. Introduction

The importance of blood flow generated drag forces can be clearly illustrated by the physiological interaction of lymphocytes with endothelial cells. Lymphocytes and platelets have developed efficient systems allowing adhesion to the vascular wall despite high forces, through the interaction of selectins with its ligands (1). In addition, the presence of shear stress for extended periods of time alters endothelial cell physiology, and studies in vascular biology extensively take advantage of experimental designs that allow the study of cell behavior under shear stress conditions (2). Blood flow in the vessels generates hydrodynamic forces on vessel walls and on circulating objects. The shear stress reports the tangential force exerted per unit area by a fluid moving near a stationary wall.

Myron Christodoulides (ed.), *Neisseria meningitidis: Advanced Methods and Protocols*, Methods in Molecular Biology, vol. 799, DOI 10.1007/978-1-61779-346-2_12, © Springer Science+Business Media, LLC 2012

This is a quantity similar to a pressure and may be expressed in dynes/cm^2. It depends on blood velocity, viscosity, and vessel diameter. An analogous situation is a river bank that is submitted to a shear stress that depends on how fast the river flows (3). All blood components are submitted to forces induced by blood flow that tend to prevent their attachment to the vessel wall. Such forces are called hydrodynamic forces and depend on shear stress and cell geometry. Since most objects do not display regular shapes, hydrodynamic forces are not easy to determine and shear stress is more commonly used as a reference during laminar-flow assays.

The importance of mechanical forces is not only found for physiological processes, but also in the case of pathological processes such as septicemia and meningitis where infectious agents interact with endothelial cells in the bloodstream. In particular, these forces have an impact on contact establishment between the pathogen and the endothelium. Although the general interest in studying the connection between microcirculation and infectious diseases (4) is rising, there are still only a few appropriate experimental models for these specific infection processes.

To address this issue we have designed a simple in vitro experimental approach using disposable flow channels to study the bloodstream phase of bacterial infections (Fig. 1). Our main study was focused on *Neisseria meningitidis* (meningococcus) infection and we have shown that initial adhesion of individual bacteria only occurs in small brain capillaries, where transient blood flow arrests take place, even in physiological conditions (5). The mechanical properties of microcirculation are therefore a determining factor for pathogenesis. We also found that this adhesion is “irreversible,” as adhering bacteria resist blood flow returning to normal values. Shear stress does not interfere with bacterial growth, as large micro-colonies grow out of individual adhering bacteria over extended periods of time (5). In another study, we highlighted the importance of bacteria-induced plasma membrane reorganization in the resistance to shear stress of micro-colonies growing on cell surfaces (6).

The importance of shear stress has been shown also for other pathogens and other infection sites. Bacterial adhesion conditioned by shear stress has been described by the study of the FimH adhesin found on uropathogenic *Escherichia coli* (UPEC). Similar to selectins, the interaction between FimH and its host cell receptor was shown to be strengthened by shear-induced mechanical forces (7) and the CfaE adhesin of enterotoxigenic *E. coli* has been reported also to mediate adhesion to intestinal epithelial cells via a shear-dependant mechanism (8). We reported, using the laminar-flow chamber assay protocol described in this chapter that *Streptococcus agalactiae* pili were essential for adherence of this pathogen to epithelial cells under flow conditions (9). Such studies confirm that our flow chamber assay is a useful tool for investigating host cell–pathogen interactions under conditions of shear stress.

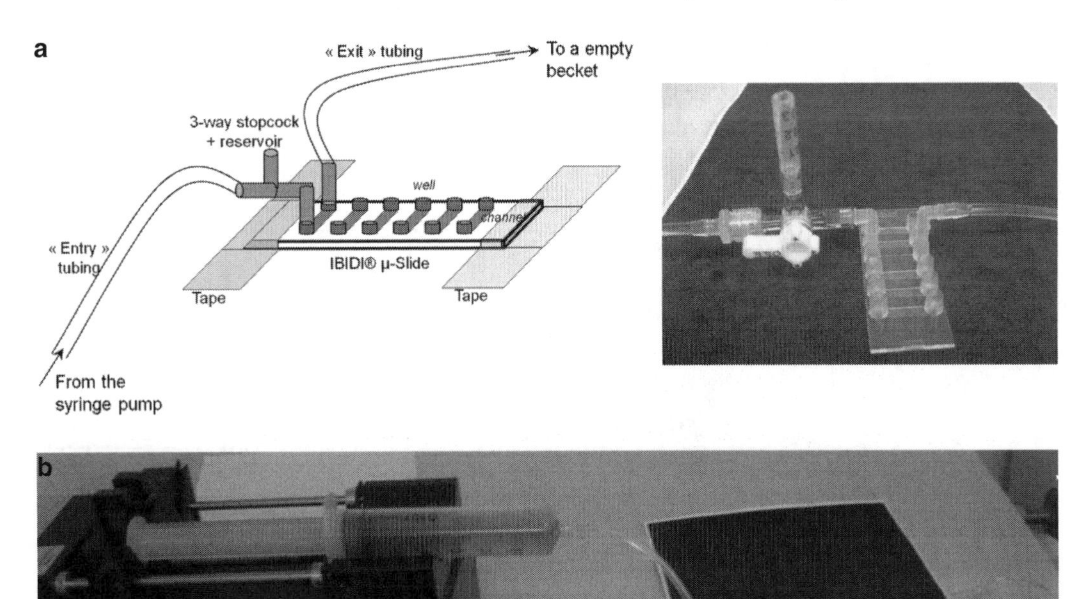

Fig. 1. Description and illustration of the disposable flow chamber assay. (**a**) Schematic and illustration of the slides and tubing. (**b**) Illustration of the entire set-up which is placed on the heated platform of the microscope and connected to the syringe.

2. Materials

2.1. Human Host Cell and Bacterial Culture

1. Human umbilical vein endothelial cells (HUVECs; Promocell).
2. Human endothelial serum free medium (Endo-SFM; Gibco) supplemented with 10% (v/v) heat-inactivated fetal bovine serum (FBS), 2 mM L-Glutamine, 0.5 IU/mL of heparin and 40 µL/mL of endothelial cell growth supplement (ECGS, Gibco). Dilute ECGS to a concentration of 10 mg/mL with Hank's balanced salt solution (HBSS) to prepare a stock solution, which is stable for a month at 4°C. Heat-inactivate FBS at 56°C for 1 h to decomplement.
3. Cell culture flasks, 75 cm^2 (Corning).
4. Solution of trypsin–ethylenediamine tetraacetic acid (EDTA), 1/250 dilution.
5. µ-Slides VI^{0,4} ibiTreat, tissue culture treated (IBIDI® GmbH) (see Notes 1 and 2).

6. *N. meningitidis* strain 8013 expressing the green fluorescent protein (GFP) (see Note 3).
7. GC Base Medium (GCB Medium, Difco) supplemented with 0.12% (w/v) agar (Sigma-Aldrich).
8. Kellogg's supplements (10):

   Supplement 1: 88.9% (v/v) of glucose 45% (w/v) (Sigma-Aldrich) and 1% (v/v) thiamine (2 mg/mL sterile solution, Sigma-Aldrich) in sterile MilliQ water. Sterilize the thiamine stock solution through a 0.22-μm filter and prepare the 100× stock solution under sterile conditions.

   Supplement 2: 0.5% (w/v) $Fe(NO_3)_3 \cdot 9H_2O$ (Sigma-Aldrich) in sterile MilliQ water. Sterilize the 1,000× stock solution through a 0.22-μm filter.

   Supplement 3: 1% (w/v) L-Glutamine (Sigma-Aldrich) in sterile MilliQ water. Sterilize the 100× stock solution through a 0.22-μm filter.

   All stock solutions are stored at −20°.
9. Chloramphenicol (Sigma-Aldrich).
10. Cell culture incubators with 5% (v/v) CO_2 for overnight cell culture and for on-plate bacterial culture.
11. Malassez cell counting chamber, or equivalent.
12. Heraeus Sepatech Omnifuge 2.0 RS, or equivalent centrifuge.

2.2. Shear Stress Experiments

1. Cell culture incubator for bacterial culture with 5% (v/v) CO_2 equipped with an orbital shaker; or alternatively an incubation Shaker Minitron (Infors-HT).
2. Isopropyl-beta-D-thiogalactopyranoside (IPTG).
3. Spectrophotometer for measuring optical density (ODλ = 600 nm) of bacterial suspensions.
4. Phosphate buffered saline (PBS), pH 7.4.
5. μ-Slide VI$^{0.4}$ flow kit (IBIDI®).
6. Syringe Plastipak 50 mL Luer-Lock (Becton Dickinson).
7. Tygon® Tubing R3603 3.2×4.8 mm (Fisher-Scientific).
8. 3-way stopcock, two female luer to male luer (Bio-Rad).
9. Syringe pump, e.g., PHD 2000 (Harvard Apparatus).
10. Inverted microscope, e.g., Olympus CKX41 or Nikon Eclipse T*i*.
11. CCD camera, e.g., ORCA 285 CCD or ORCA 3-CCD (Hamamatsu) and software, e.g., NIS-Elements (Nikon), to control the system.
12. ThermoPlates MATS heated platform (TokaiHit) or LIS Microscope Temperature Control System incubation chamber, compatible with microscope.
13. ImageJ software for image analysis (http://rsbweb.nih.gov/ij/).

3. Methods

The protocol described in this chapter is reliant on video-microscopy. Therefore, it is essential that the microscope set-up is suitable for automatic acquisition of images with a controlled camera and shutters. We recommend the use of an inverted microscope equipped with a 20× fluorescence objective and filters adapted to the fluorescent protein used for the study (in our case it was the GFP), a CCD camera and software controlling the system (see Note 4). As some of the experiments can last for several hours, it is important to keep the cells at a controlled temperature using either a heated platform or better still an incubation chamber. All the equipment has to be placed in an appropriate containment room, which is chosen on the basis of the bio-safety level (BSL) of the pathogen studied. For experiments with *N. meningitidis*, the microscope set-up has to be located in a BSL-2 containment laboratory. A video demonstration of the protocol can be found in reference (11).

3.1. Human Host Cell and Bacterial Culture

1. Culture HUVECs between passages 1 and 9 (see Note 5) at 37° in a humidified incubator with 5% (v/v) CO_2. Passage the cells when they approach 80% confluence with trypsin–EDTA to provide maintenance cultures on 75 cm^2 culture flasks and experimental cultures in disposable flow chambers (μ-Slides $VI^{0,4}$).
2. Withdraw medium from the flasks to be passaged, wash the cells with 10 mL of PBS, withdraw and replace it with 1.5 mL of trypsin–EDTA. Allow the cells to detach in a cell culture incubator for 5 min at 37°C.
3. Collect the cells with 10 mL of cell culture medium into a 15 mL collection tube. Pellet the cells with a 5 min centrifugation (at 200*g*), at room temperature. Suspend the cells in 4 mL of Endo-SFM supplemented with 10% (v/v) FBS and count them on a Malassez chamber, according to the manufacturer's instructions.
4. Introduce 30 μL of a 1×10^6 HUVECs per mL suspension into the channel (see Note 6) and allow the cells to adhere for 3 h at 37° in a humidified incubator with an atmosphere of 5% (v/v) CO_2. Add an additional volume of 120 μL of Endo-SFM to fill the wells.
5. Grow *N. meningitidis* strain 8013, expressing GFP under the control of an IPTG-inducible promoter (see Note 7), on GCB agar plates containing Kellogg's supplements and 5 μg/mL of chloramphenicol at 37°C in a moist atmosphere containing 5% (v/v) CO_2 for 16 h.

3.2. Shear Stress Experiments

A series of four protocols are described for measuring the effects of shear stress on meningococcal interactions with host cells.

3.2.1. Measuring the Initial Adhesion of Individual Bacteria to Host Cells

1. Adjust the concentration of bacteria grown on GCB agar plates to an OD_{600} of 0.05 with pre-warmed Endo-SFM containing 10% (v/v) FBS and incubate for 120 min at 37°C in a moist atmosphere containing 5% (v/v) CO_2, under gentle shaking (130 rpm). Induce GFP expression by adding 1 mM IPTG into the culture medium for the whole incubation period.
2. Place the disposable μ-Slide on the stage of an inverted microscope equipped with a heated platform to maintain the sample temperature at 37°C (see Note 8).
3. Pour pre-warmed Endo-SFM supplemented with 2% (v/v) FBS (see Note 9) into a sterile glass beaker and fill a sterile 50 mL syringe with this medium. Attach the "entry" tubing to the syringe and fill by introducing medium. A schematic and a picture of the set-up are shown in Fig. 1.
4. Connect the "entry" tubing to the μ-Slide, with care, in order to avoid introducing air into the chamber (see Note 10). Then, affix the "exit" tube to the other end and carefully fill the channel with medium to approximately 1 cm distance from the chamber "exit" (see Note 11).
5. Measure the bacterial OD_{600} and adjust to 0.15 in Endo-SFM containing 2% (v/v) FBS. In order to look at individual bacteria, any aggregates must be disrupted by vigorous vortexing of the bacterial sample.
6. Place 100 μL of Endo-SFM supplemented with 2% (v/v) FBS medium into the reservoir (see Note 12) and add 100 μL of the bacterial solution taken from the top of the vortexed solution to avoid sampling any remaining bacterial aggregates.
7. Carefully introduce the 200 μL volume into the μ-Slide by turning the stopcock to inject the bacteria into the chamber.
8. Introduce Endo-SFM containing 2% (v/v) FBS, maintained at 37°C with the heated platform, into the chamber using a syringe pump with a shear stress compatible with adhesion of 0.044 dyn/cm^2 (see Note 13) for 15 min (see Note 14).
9. At the end of those 15 min, acquire images of ten fields randomly (see Note 15) while the flow is still circulating (see Note 16). An example of the acquired images is shown in Fig. 2a.
10. Next, analyze the acquired images using the ImageJ software (12) in order to access the mean number of adherent bacteria per field. This is done as follows: first of all, each image is thresholded using the "Threshold" window found in the "Image" menu to highlight individual adhering fluorescent bacteria, while minimizing the background from the cells. Then, every single bacterium is counted, by using the Plug-In "Cell Counter" (http://rsb.info.nih.gov/ij/plugins/cell-counter.html). The data for each image are reported in an Excel spreadsheet and are averaged to quantify the mean number of adherent bacteria per field.

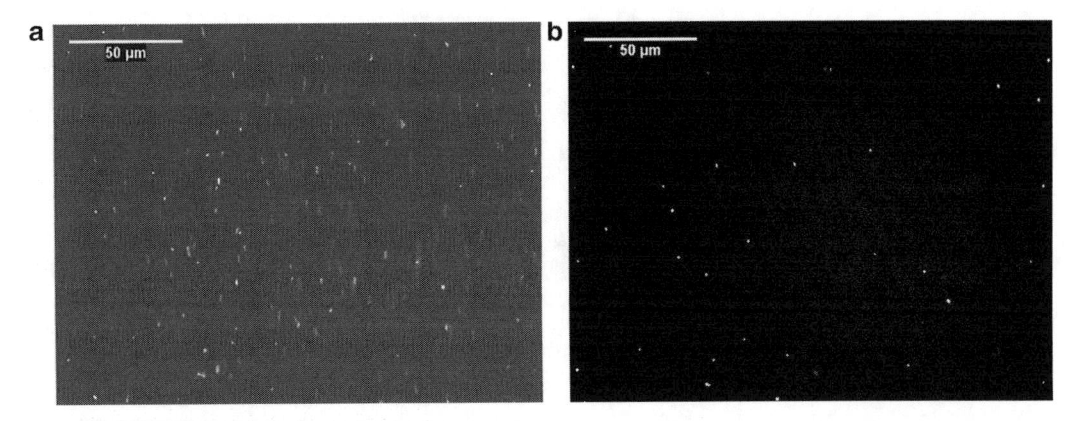

Fig. 2. Analysis of *N. meningitidis* initial adhesion on endothelial cells in the presence of flowing liquid and resistance to shear stress of individual bacteria. Typical images illustrating bacterial adhesion on the surface of endothelial cells are shown. (**a**) Initial adhesion: fluorescent bacteria circulating over the cells appear as *short lines*, whereas bacteria adherent on the cell monolayer appear as still *bright spots*. Some small aggregates of less than four bacteria are visible but are not included in the analysis step using ImageJ software. (**b**) Bacterial resistance to flow: bound fluorescent bacteria resistant to shear stress after application of the flow in the chamber. All unbound bacteria are washed away and only adherent bacteria remain on the cell surface.

3.2.2. Measuring the Resistance to Flow of Individual Bacteria Adherent to Host Cells

1. Following the initial adhesion experiment, the resistance of individual adherent bacteria is tested with the following protocol:
2. Program the syringe pump set-up to generate shear stress ranging from 3 to 100 dynes/cm^2 (see Note 13) for 5 min.
3. At the end of the 5 min, acquire ten fields randomly chosen (see Note 15) as described in step 9, Subheading 3.2.1, but with the flow stopped prior to the acquisition. An example of the acquired images is shown in Fig. 2b.
4. Analyze the acquired images as described in step 10, Subheading 3.2.1 and quantify the mean number of remaining bacteria per field.

3.2.3. Measuring the Growth of an Isolated Adherent Bacterium to a Micro-Colony

1. Prepare a bacterial suspension as described in step 1, Subheading 3.2.1.
2. Infect the host cells with 3×10^6 bacteria (multiplicity of infection, MOI = 100) (see Note 17) and allow adhesion to proceed for 30 min in static conditions.
3. Wash away any unbound bacteria in the µ-Slide by adding 120 µL of Endo-SFM containing 10% (v/v) FBS, removing the medium and repeating an additional 4 times. Place the washed µ-Slide onto the microscope stage, with the heated platform to maintain the temperature at 37°C (see Note 7).
4. Repeat step 3, Subheading 3.2.1, except using pre-warmed Endo-SFM supplemented with 10% (v/v) FBS.

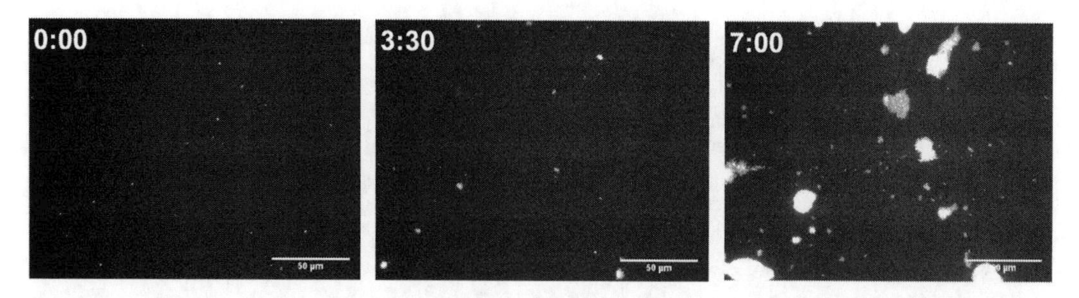

Fig. 3. Individual bacteria proliferate into micro-colonies in the presence of high shear stress. Individual video stills illustrating the growth of adherent bacteria on endothelial cells under flow conditions are shown. Compact bacterial colonies appear and increase in size during the course of the experiment (7-h time period with a 1 mL/h flow rate).

5. Affix the "entry" tubing to the μ-Slide with care to avoid introducing air into the chamber (see Note 10) and connect the "exit" tubing to the other end (see Note 18).
6. Acquire 2–3 images before flow application, in order to monitor both host cells (by phase contrast) and bacteria (by GFP-fluorescence). These captured images are used as controls for the experiment.
7. Introduce Endo-SFM containing 10% (v/v) FBS, maintained at 37°C with the heated platform, into the chamber using a syringe pump with low shear stress (see Note 11) for several hours. Record images for real-time video microscopy at one frame every 5 min. Examples of the video frames are shown in Fig. 3.
8. Following video microscopy, stop the shear stress by turning off the syringe pump, acquire 2–3 additional images and then stop the image acquisition sequence on the software.

3.2.4. Measuring the Resistance to Flow of Adherent Bacterial Micro-Colonies

1. Repeat steps 1–5 of Subheading 3.2.3 for this experiment, except for step 3 where after washing, the bacterial infection is allowed to proceed for an additional 2 h in the incubator.
2. Acquire 2–3 images of both cells (by phase contrast) and bacteria (by GFP-fluorescence) before flow application. For the rest of the experiment, the acquisition protocol will monitor only the fluorescence.
3. Repeat step 7, Subheading 3.2.3, except applying the shear stress for 5 min and recording one frame every 5 s.
4. Stop the shear stress by turning off the syringe pump, acquire 2–3 additional images and then stop the image acquisition sequence on the software.
5. Next, remove the "exit" tubing first and carefully, to avoid emptying the channel and then add pre-warmed fresh medium to fill the wells before removing the "entry" tubing.
6. To quantitatively determine the effect of shear stress on bacterial resistance, a plating assay can be performed as

follows: collect the remaining medium and wash the μ-Slide twice with 120 μL of PBS, which removes the remaining medium from the channel (both washes are also collected).

7. Detach the infected cells with the addition of 50 μL of trypsin–EDTA for 5 min at 37° and mix this detached sample with the suspension collected in the previous step.
8. Perform serial dilutions in PBS and plate a 10 μL fraction onto GCB agar plates, in triplicate, in order to determine the number of colony forming units (CFU) on the next day, after incubation of the plates at 37°C in an incubator with 5% (v/v) CO_2.

4. Notes

1. In addition to the μ-Slide VI$^{0.4}$ used in our studies, the BioValley Corporation offers a vast range of other designs with various numbers of channels, channel sizes and shapes. For more information, visit the IBIDI® website: http://www.ibidi.de/.
2. As an alternative to using the disposable μ-Slides, we have also used a commercial laminar-flow chamber from Immunetics, USA, that can be re-used. For such experiments, the HUVEC cells are cultured on 8-mm-diameter wells on glass slides coated with hydrophobic Teflon (CML). Prior to seeding at a density of 10^4 cells per well, we coat the slides for 1 h at room temperature with a solution of fibronectin (10 μg/mL; Sigma-Aldrich). The slides are then placed in the parallel flow chamber (3.3 × 0.6 cm × 250 μm) and sealed under a vacuum. In this design, the flow rates are calculated according to the formula $\tau = 6\eta/h^2b^*\phi$, where τ is the shear stress value (dynes/cm^2), η the fluid viscosity (dynes × s/cm^2), h the chamber height (cm), b the chamber width (cm) and ϕ the flow rate (mL/min).
3. Strains are frozen at −80°C in a GCB solution containing 20% (v/v) glycerol. A loop full of bacteria grown on a GCB-agar plate is collected and suspended in 1.5 mL of GCB-glycerol solution in suitable vials.
4. Alternative system control software can be used, including the OpenLab darkroom software from Improvision or MetaMorph® from Molecular Device.
5. The flow system described in our protocol can be used to study the interactions of many different bacterial species with many host cell types. For example, epithelial cells such as Hec1B or Caco-2 have been used successfully as models to study the commensal lifestyle of the meningococcus inside the human nasopharynx.

6. When introducing cells into the channels, the slide is leaned back and forth, and the tip is placed at the channel entry. During this procedure air should not be introduced into the chamber, but in case air bubbles do form, the slide should be tapped carefully to remove them.
7. Our parental strain was made to express the GFP by introducing the broad host range plasmid pAM239 (13) by conjugation. However, this conjugation experiment is poorly efficient and other methods might be considered to introduce the GFP protein into the bacterial strain, e.g., by transformation.
8. Since placing and removing the tubes allows movement of the μ-Slide, we recommend fixing it to the platform with tape.
9. We have seen that serum auto-fluorescence may overcome individual bacterium fluorescence. We tested various quantities and qualities of fetal bovine serum and found that at 2% (v/v) it is possible to distinguish between individual adherent bacteria without affecting the cells.
10. To avoid air bubbles it is best to have a drop of medium at the end of the tube that comes into contact with the medium in the well.
11. We usually do this step by introducing medium via the syringe pump, with a 5 mL/min setting.
12. The reservoir is made from a 2 mL syringe (Terumo) which is cut to size and volume.
13. Flow rates and corresponding shear stress are calculated according to the formula $\tau = 1.761\phi$, as provided by the manufacturer and where τ is the shear stress value and ϕ the flow rate.
14. Since we want to see only adherent bacteria, we recommend a visual check using the microscope to demonstrate that most of the non-adherent bacteria have been driven through the channel.
15. Since the liquid flow is not homogenous along the channel due to the forces generated on the chamber walls, the acquired fields are selected randomly along a line at the center of the chamber.
16. With a 1–2 s acquisition time, adhering bacteria are visualized as small dots on the acquired images while circulating bacteria are seen as lines.
17. In the μ-Slide, culture medium is removed from the wells but not from the channel, to avoid introducing air into the channel. For infection, a 30 μL volume of an $OD_{600} = 0.1$ bacterial culture is introduced into the channel.
18. In order to avoid shear stress when placing the tubes, the connection of the tubes to the chambers should be performed as slowly and carefully as possible.

Acknowledgments

The authors wish to thank Emilie Mairey for her initial participation in setting up the system and Anne-Flore Imhaus and Silke Silva for critical reading of the manuscript. The work was funded by INSERM (ATIP – AVENIR), ANR (Agence Nationale de la Recherche) and Université Paris Descartes.

References

1. McEver RP (2002) Selectins: lectins that initiate cell adhesion under flow. Curr Opin Cell Biol 14: 581–586.
2. Lehoux S, Tedgui A (2003) Cellular mechanics and gene expression in blood vessels. J Biomech 36: 631–643.
3. Janmey PA, Schliwa M (2008) Rheology. Curr Biol 18:R639–R641.
4. Valbuena G, Walker DH (2006) The endothelium as a target for infections. Annu Rev Pathol 1: 171–198.
5. Mairey E, Genovesio A, Donnadieu E et al (2006) Cerebral microcirculation shear stress levels determine *Neisseria meningitidis* attachment sites along the blood-brain barrier. J Exp Med 203: 1939–1950.
6. Mikaty G, Soyer M, Mairey E et al (2009) Extracellular bacterial pathogen induces host cell surface reorganization to resist shear stress. PLoS Pathog 5: e1000314.
7. Thomas WE, Trintchina E, Forero M et al (2002) Bacterial adhesion to target cells enhanced by shear force. Cell 109: 913–923.
8. Tchesnokova V, McVeigh AL, Kidd B et al (2010) Shear-enhanced binding of intestinal colonization factor antigen I of enterotoxigenic *Escherichia coli*. Mol Microbiol 76: 489–502.
9. Konto-Ghiorghi Y, Mairey E, Mallet A et al (2009) Dual role for pilus in adherence to epithelial cells and biofilm formation in *Streptococcus agalactiae*. PLoS Pathog 5: e1000422.
10. Kellogg DS, Jr. Cohen IR, Norins LC et al (1968) *Neisseria gonorrhoeae*. II. Colonial variation and pathogenicity during 35 months *in vitro*. Journal of bacteriology 96: 596–605.
11. Soyer M, Duménil G (2011) Introducing shear stress in the study of bacterial adhesion. J Vis Exp. doi: 10.3791/3241. In press.
12. Abramoff MD, Magelhaes PJ, Ram SJ (2004) Image Processing with ImageJ. Biophotonics International 11: 36–42.
13. Solomon JM, Rupper A, Cardelli JA et al (2000) Intracellular growth of *Legionella pneumophila* in *Dictyostelium discoideum*, a system for genetic analysis of host-pathogen interactions. Infect and Immun 68: 2939–2947.

Chapter 13

Techniques to Measure Pilus Retraction Forces

Nicolas Biais, Dustin Higashi, Magdalene So, and Benoit Ladoux

Abstract

The importance of physical forces in biology is becoming more appreciated. *Neisseria gonorrhoeae* has become a paradigm for the study of physical forces in the bacterial world. Cycles of elongations and retractions of Type IV pili enables *N. gonorrhoeae* bacteria to exert forces on its environment, forces that play major roles in the life cycle of this pathogen. In order to better understand the role of these forces, there is a need to fully characterize them. Here, we present two different techniques, optical tweezers and Polyacrylamide MicroPillars (PoMPs), for measuring pilus retraction forces. Initially designed for *N. gonorrhoeae*, these assays can be readily modified to study other pilus-bearing bacteria including *Neisseria meningitidis*.

Key words: Neisseria, Type IV pilus, Retraction force, Optical tweezer, PoMPs

1. Introduction

Type IV pili (Tfp) play a critical role in the life cycles of both *Neisseria meningitidis* and *Neisseria gonorrhoeae,* the two main pathogens of the *Neisseria* genus. These appendages are thin (6 nm in diameter) and long (up to 30 μm in length) dynamic polymers emanating from the surface of the bacteria. They are associated with a wide range of functions including DNA uptake, twitching motility, cell adhesion, and biofilm formation (1). An explanation for this wide spectrum of functions stems from the extraordinary diversity of Tfp filaments: the main subunit, pilin, can undergo high frequency antigenic variation, (2) harbors various post-translational modifications (3) and the filaments themselves possess different minor pilins (4). Alongside these chemical and genetic variabilities, force generation has also been implicated as another important function of Tfp biology.

Myron Christodoulides (ed.), *Neisseria meningitidis: Advanced Methods and Protocols*, Methods in Molecular Biology, vol. 799,
DOI 10.1007/978-1-61779-346-2_13, © Springer Science+Business Media, LLC 2012

During the last decade, the role of physical forces in biology has emerged as a central theme. From development to differentiation, from motility to signaling, physical forces modulate biological fates and the bacterial world is no exception (5–8). *N. gonorrhoeae* Tfp can undergo cycles of extension and retractions and they can generate force during retraction (9). A single Tfp fiber can exert forces of 100 picoNewtons (pN) (10) and bundles of 8–10 Tfp can be retracted in unison exerting forces of up to 1 nanoNewtons (nN) which can be maintained for hours (11). In the context of cellular infection, these forces elicit dramatic rearrangements in the cortex of eukaryotic cells (12) and have been shown to trigger cytoprotective pathways (13). The techniques designed to measure those forces in *N. gonorrhoeae* can readily be adapted to measure the force generation in *N. meningitidis* or other pili-bearing bacteria.

For this chapter, we present two complementary techniques to measure the forces generated by the retraction of pili. The first one utilizes optical tweezers and is geared toward the measurement of forces from a single bacterium. The second utilizes arrays of Polyacrylamide MicroPillars (PoMPs) and is geared toward the measurement of forces from multiple bacteria in "infection-like" conditions. The use of these techniques will increase our understanding of the role of physical forces in the biology of the pathogenic *Neisseriae* and other pilus-bearing bacteria.

2. Materials

2.1. Calibration of Optical Tweezers

1. A basic laser tweezers set-up with a fixed laser trap obtained from a 2 W neodymium-doped yttrium aluminum garnet neodymium (Nd:YAG) laser (14).
2. Carboxylated silica or latex beads (Polysciences or Bangs laboratories).
3. Eppendorf tubes.
4. Piezoelectric stage P-517 (Physik Instrumente).
5. #1 coverglass (Fisher Scientific).
6. Observation chamber: a metallic slide with a 20×20 mm opening (15) (see Fig. 1).
7. Vacuum grease.
8. A computer with image analysis software, e.g., Image J (see Note 1).

2.2. Optical Tweezers Assay

1. A basic laser tweezers set-up with a fixed laser trap obtained from a 2 W neodymium-doped yttrium aluminum garnet neodymium (Nd:YAG) laser (14).

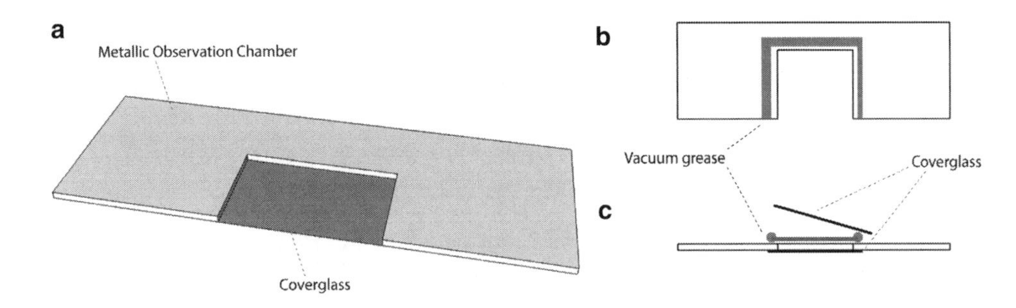

Fig. 1. Schematics of the observation chamber. (**a**) Schematic of an observation chamber with one coverglass at the bottom of it already in place. (**b**) Top view of an observation chamber with vacuum grease on the sides of the opening ready to be used to seal a coverglass on top. (**c**) Side view of an observation chamber with vacuum grease on the sides of the opening ready to be used to seal a coverglass on top. A coverglass is already at the bottom. After the liquid is added on the side, the chamber is sealed with more vacuum grease applied between the two coverglasses.

2. Carboxylated Silica or latex beads (Polysciences or Bangs laboratories).
3. Agar plates to grow bacteria: use GCB-agar for pathogenic *Neisseriae.*
4. Dulbecco's modified Eagle's medium (DMEM).
5. Spectrophotometer.
6. Humidified incubator with 5% (v/v) CO_2 for bacterial culture.
7. A computer with image analysis software, e.g., Image J (see Note 1).

2.3. Polyacrylamide MicroPillars Assay

The reagents and equipment listed below are required for the methods that describe activation of coverglasses, casting of PoMPs, coating of PoMPs, calibration of PoMPs using optical tweezers or magnetic tweezers, and PoMPs assays for measuring dynamic and static forces.

1. A silicon mould for the pillars, designed by usual photolithography techniques (see Note 2) (16, 17).
2. A plasma cleaner, e.g., PDC-32G (Harrick Plasma).
3. #1 coverglass.
4. 0.1 M NaOH solution, prepared in distilled water.
5. Parafilm.
6. 2 pairs of flat tipped tweezers, e.g., 2A model (SPI).
7. 3-Aminopropyl-trimethoxysilane (Fluka).
8. 70% (v/v) ethanol solution.
9. Solutions of 0.5% (v/v) and 4% (v/v) glutaraldehyde in phosphate buffered saline (PBS) pH 7.4.
10. Polyacrylamide solution 40% (Biorad) and bis-acrylamide solution 2% (Biorad). Mix to obtain the ratio of polyacrylamide and bis-acrylamide desired. For instance, to obtain 1 mL of 20%

Acrylamide–1% bis-Acrylamide, combine 500 μL of both solutions.

11. 10% (w/v) ammonium persulfate solution.
12. *N,N,N',N'*-Tetramethylethylenediamine (TEMED).
13. Sulfo-SANPAH (Thermo Scientific).
14. 50 mM HEPES buffer, pH 7.4.
15. Poly-L-Lysine solution: prepare a 30 μg/mL working concentration in PBS by diluting a stock solution of 1 mg/mL of Poly-L-Lysine hydrobromide (Sigma-Aldrich) (see Note 3).
16. Sterile tissue culture plates, 6-well.
17. An ultraviolet (UV) lamp: a germicidal lamp fitted within a laminar flow hood will suffice.
18. Piranha solution: mix three volumes of 1 M H_2SO_4 with one volume of 30% (v/v) hydrogen peroxide (H_2O_2).
19. Neodymium magnet (Applied Magnets).
20. Sewing needles: the size of the tip of the needle will control the size at which the magnetic gradient and thus the magnetic force is imposed (18).
21. Carboxylated magnetic beads, e.g., Dynabeads M270 carboxylic acid (Invitrogen Dynal).
22. Latex non-magnetic beads of a different diameter than the magnetic beads e.g., 1 μm beads (Polysciences or Bangs laboratories).
23. 12 mL pipettes.
24. Pipettor tips.
25. 3-Axis Micromanipulator e.g., Sutter instrument MP-285 (Sutter) (see Note 4).
26. Table-top centrifuge.
27. Dimethylpolysiloxane, with known viscosity of 12,500 cSt at 25°C (DPMS12M; Sigma-Aldrich).
28. An inverted microscope with a high magnification objective, preferably ×100 high numerical aperture.
29. A digital camera mounted on one of the ports of the microscope.
30. A computer with image analysis software e.g., Image J (see Note 1).

3. Methods

The principle behind optical tweezers is quite simple (19): when an intense light is focused on a point, the resulting three-dimensional gradient of electromagnetic energy will result in the ability to trap a dielectric object at that point. In other terms, if the light of a laser for example is concentrated in the focal plane of a microscope, all small transparent materials can be trapped. When the transparent object is exactly at the center of the concentrated light, no force is exerted on the object in the plane of focus. If the distance between the object and the center of the trap was to change, a restoring force would be exerted on the object. This force depends on several parameters including the shape, intensity, and wavelength of the focused light, the optical properties of both the object and the surrounding medium and the shape and size of the object. Formulae can be derived but calibration of the force is always preferred. We will assume that a basic optical tweezers set-up is readily available and outfitted with a fixed laser trap obtained from a 2 W neodymium-doped yttrium aluminum garnet neodymium (Nd:YAG) laser.

3.1. Calibration of Optical Tweezers

In order to be able to relate the displacement of the bead in the trap to a given force, the trap first needs to be calibrated. An easy and popular way to calibrate the trap is based on the fact that a bead of radius R moving in a medium of viscosity η at a speed v will be submitted to a force $F_{drag} = 6\pi\eta Rv$ (20) (see Fig 2a, b).

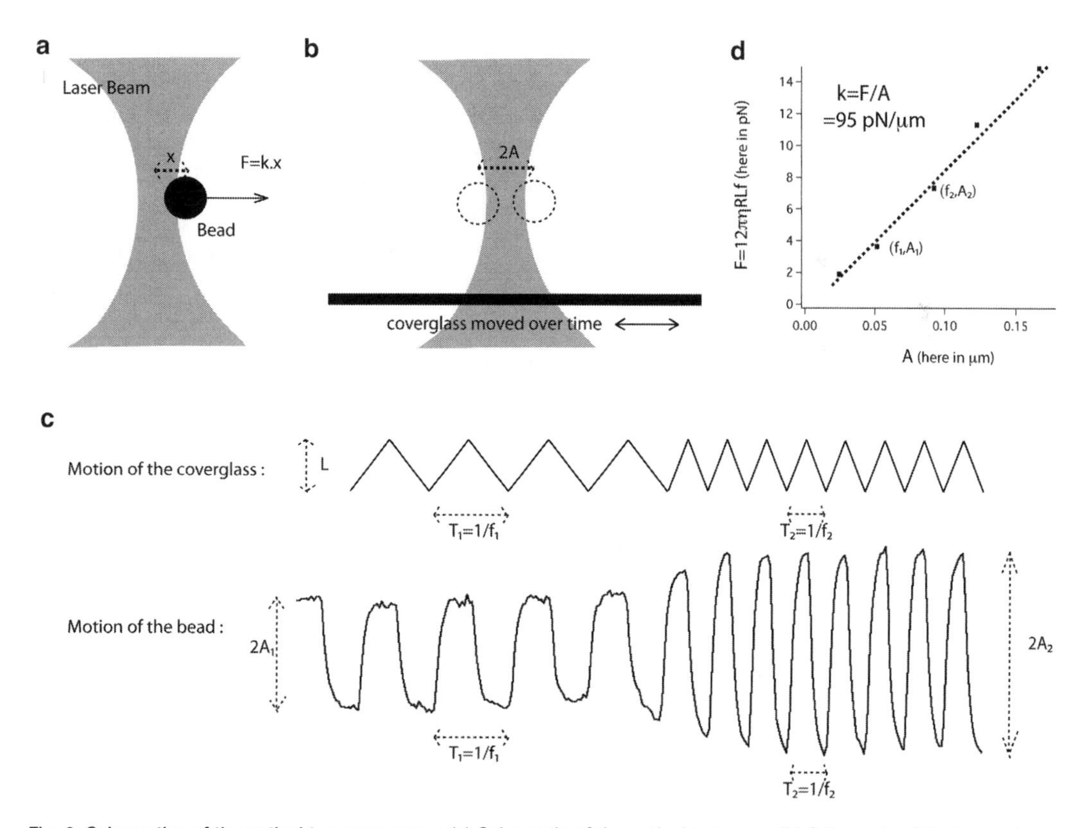

Fig. 2. Schematics of the optical tweezers assay. (**a**) Schematic of the optical tweezers. (**b**) Schematic of the two extreme positions of the bead during calibration. (**c**) Representation of the motion of the coverglass and the resulting motion of the bead at two different frequencies. Frequencies and amplitudes of the bead motion for each frequency can be measured on the trace of the bead motion. (**d**) An example of calibration of the trap: the slope of the graph of the viscous drag force vs. bead displacement gives the stiffness of the trap.

This method has the advantage of facile set-up, but other methods based on the analysis of the Brownian fluctuations of the bead exist and require more complicated set-up (20, 21). All the microscopy material is kept on an optical table.

1. Turn on the laser and ensure its stability by monitoring the output intensity over time. One hour of warming-up should be sufficient.
2. Prepare an observation chamber, which is a millimeter thick piece of aluminum with an opening of 20 × 20 mm with coverglasses on both sides maintained by vacuum grease (see Fig. 1) (15).
3. Place ~1 μL of the carboxylated silica or latex beads in 1 mL of distilled water. Add 0.2–1 μL of this bead suspension to 500 μL of water and place it into the observation chamber. Seal the chamber with vacuum grease (see Note 5 and Fig. 1).
4. Place the chamber on the piezoelectric stage. The beads should be spaced out far enough that you can trap an individual bead and move the stage by a full field of view without trapping other beads.
5. Measure the amplitude L of the movement of the stage by recording the motion of a bead stuck to the coverglass or another defect attached to the coverglass. The movement must be a periodic triangular signal so that the speed of the stage is constant during each half period. It is important to maximize L while utilizing a full field of view.
6. With the frequency of the signal set to zero, open the laser shutter to trap a single bead. Silica beads are easier to trap, whereas with latex beads, the bead must be as close as possible to the center of the trap when the shutter is opened.
7. Start recording a digital movie (see Note 6).
8. Slowly increase, by steps, the frequency of the signal (see Note 7). Each step should last at least ten periods of the current frequency. Figure 2c shows an example of two such steps.
9. Analyze the digital movie obtained using an implementation of cross-correlation algorithms (22) in your favorite image analysis software to retrieve the motion of the bead over time.
10. For each step, measure the frequency of the signal f and the maximum displacement in the trap A from the captured motion of the bead (see Fig. 2c). Assuming that the viscous drag force $F_{drag} = 12\pi\eta RLf$ and the elastic force $F_{elas} = kA$ are equal (see Note 8), the stiffness of the trap can be obtained as the slope of a graph representing force vs. A for all the frequency steps taken during the calibration (see Fig. 2d). As mentioned previously, the exact value to the trap's stiffness will depend on several parameters such as the laser used, the exact optical set-up

and the type of bead used, but as an approximate reference, a laser of a couple of watts will lead to stiffness of around 100 pN/μm for a silica bead of 1.5 μm diameter. Using latex beads will result in a trap approximately twice as stiff (see Notes 9 and 10).

3.2. Optical Tweezers Assay

Once the trap has been calibrated, the beads and bacteria are now placed in close vicinity. The method described below for using pathogenic *Neisseriae* can be readily adapted to investigate other pilus-bearing pathogens.

1. Streak out *Neisseria* bacteria from frozen stocks, maintained in 20% (v/v) glycerol GCB broth at –80°C, onto GCB-agar plates and grow in a humidified incubator with 5% (v/v) CO_2 for 16–20 h.
2. Suspend the bacteria at a concentration of ~10^8 bacteria/mL in DMEM by measuring the optical density (OD) at $\lambda_{600\,nm}$ in a spectrophotometer (1.4 OD_{600} ~10^9 CFU/mL) (see Note 11).
3. Place ~1 μL of the carboxylated silica or latex beads in 1 mL of water. The choice of bead is paramount to ensure that they will adhere to the pilus (see Note 12).
4. Add 10 μL of the bacterial suspension and 1 μL of the bead suspension in 500 μL of DMEM and add this volume to an observation chamber. Allow the bacteria to settle for 30 min.
5. In the meantime, turn on the laser and allow it to stabilize.
6. Place the observation chamber under the microscope.
7. Locate a freely diffusing bead and trap the bead with the optical tweezers as described in step 6 of Subheading 3.1 (see Note 13).
8. Bring the trapped bead in close vicinity to a bacterium tethered to the surface (see Fig. 3a, b). The distance between the two is commensurate with the length of the pilus being investigated, which is usually a few micrometer (see Note 14).
9. Start recording a digital movie (see Note 6).
10. Analyze the digital movie obtained using cross-correlation algorithms to retrieve the motion of the bead over time, as described in step 9 of Subheading 3.1.
11. The motion of the bead over time can be translated into the force exerted on the bead over time, now that the trap is calibrated (see Fig. 3c).

3.3. Polyacrylamide MicoPillars Assay

Both *N. gonorrhoeae* and *N. meningitidis* infect mucosal epithelia primarily. Many bacteria also interact directly with biological or abiotic surfaces as communities, forming micro-colonies or biofilms. In this context, the optical tweezer-based assay cannot be used reliably to measure retraction forces in the more relevant in vivo "infection-like" conditions found within micro-colonies or biofilms.

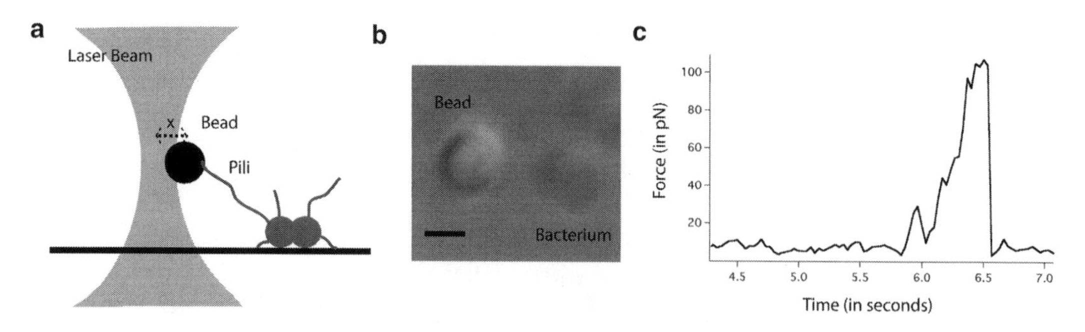

Fig. 3. Use of the optical tweezers assay. (**a**) Attachment between a trapped bead and a pilus-bearing bacterium. (**b**) Optical microscopy image of the attachment event. The *scale bar* is 1 μm. (**c**) Example of the measurement of a pulling event using the optical tweezers assay.

Recently, arrays of microscopic elastomeric pillars have been used to measure the forces exerted by eukaryotic cells on their environment (16, 23). In addition, previous methods based on the fabrication of substrate coated flexible gels have been extensively used to analyze the mechanical forces exerted by mammalian cells on their underlying support (24). The combination of these two techniques has permitted the manufacture of force measurement devices perfectly suited to analyze the retraction forces of *Neisseria* species and other pilus-bearing bacteria. The principle is to obtain a hexagonal array of identical micron-size elastomeric pillars. The motion of the pillars enables the measurement of the forces exerted by bacteria through retraction of their pili.

3.3.1. Activation of Coverglasses

Unless otherwise mentioned, all the following steps are performed on the bench.

1. Place a piece of Parafilm at the bottom of a flat clean glass dish.
2. Using tweezers, briefly pass the coverglasses through a flame and when cooled place them on the Parafilm. Prepare as many coverglasses as needed, as the activated coverglasses can be kept for up to a year.
3. With a pipette, add 0.1 M NaOH solution onto each coverglass. Add 1 mL to cover the entire surface of the coverglass and then remove the excess liquid with a pipette, leaving approximately 100–200 μL of the solution covering the full surface area of each coverglass.
4. Let the coverglasses dry at room temperature. Drying takes a few hours in a low humidity atmosphere, but overnight is usually recommended.
5. Add ~200 μL of 3-aminopropyl-trimethoxysilane on the top surface of each coverglass and incubate for 5 min at room temperature. This step should be done in a fume hood.

6. Slowly add water to immerse all of the coverglass(es) and incubate for 5 min at room temperature.
7. Use tweezers to remove the coverglasses individually and use a water squirt bottle to wash any residue from both sides. Keeping track of the activated side, put the coverglass on a rack immersed in water and agitate by slow rotation for 5 min.
8. Replace the water with a solution of 0.5% (v/v) glutaraldehyde in PBS. Ensure that all the coverglasses are immersed and incubate for 30 min at room temperature. This step should be done in a fume hood and the glutaraldehyde disposed of by following institutional health and safety recommendations.
9. Wash the coverglasses three times with water and dry overnight at room temperature.
10. The coverglasses, with their activated side maintained in an upward orientation, can be kept in Petri dishes sealed with Parafilm. Protected from dust, they can be used up to a year after preparation.

3.3.2. Casting of PoMPs

1. Take the desired number of activated coverglasses and place them in a clean plastic Petri dish with the activated side facing upward. The number chosen should be commensurate with the number of moulds to be used.
2. Prepare a sufficient volume of the polyacrylamide–bis-acrylamide solution at the concentration required (see Note 15). A volume of 15 μL is needed per coverglass.
3. Add 1/100th of the volume of 10% (w/v) ammonium persulfate and then 1/1,000th of the volume of TEMED to the acrylamide solution. Vortex for a few seconds.
4. Add 15 μL of the solution on top of each activated coverglass.
5. Using tweezers, delicately deposit the silicon mould with the holes facing the coverglass.
6. Allow the mixture to harden at room temperature, which usually takes from 15 to 45 min depending on the concentration of acrylamide and bis-acrylamide used. An easy way to follow the reticulation process is to prepare a greater volume of acrylamide solution and observe hardening of the remaining mixture after application to the coverglasses.
7. When the acrylamide gel has hardened, submerse the mould–gel–coverglass sandwich in 50 mM HEPES buffer, pH 7.4.
8. This step is critical and can be described as "shucking the mould." Using two pairs of tweezers, or a pair of tweezers and a razor blade, use one pair of tweezers to hold the coverglass and slowly push one side of the second pair of tweezers or the razor blade between the coverglass and the mould. Be cautious in the application of force; the principal concern should be to

go forward on the side of the mould. At some point during this process, the mould will be freed and the hydrogel micropillars on the coverglass will be exposed. These arrays of polyacrylamide micropillars over an activated coverglass are referred to as PoMPs.

9. Next, wash both sides of the mould with 70% (v/v) ethanol using a squirt bottle. Use a tissue to remove the excess ethanol and then dry the mould with clean air or preferably argon.
10. Store the mould in a plastic Petri dish sealed with Parafilm until next use.
11. Should some of the hydrogel remain stuck to the mould, apply a sufficient volume of 70% (v/v) ethanol to remove it. The hydrogel will turn white which will aid in identifying and removing all remaining gel. Use Kimwipes tissue to remove final traces. Finally, clean the mould using a plasma cleaner for 1 min, according to the manufacturer's instructions. If traces of hydrogel still cannot be removed using this method, the mould can be immersed in Piranha solution for a few hours to remove any organic debris, then washed with ample water and dried with clean air or Argon and subsequently cleansed in a plasma cleaner.

3.3.3. Coating of PoMPs

Another advantage to using polyacrylamide is that the surface of the micropillars can be readily modified with functional groups for protein binding. For example, bifunctional chemicals such as SulfoSANPAH (25) can be used to activate the surface of the PoMPs to allow binding of virtually any protein.

1. Cover the PoMPs with 80 μL of a solution of 1 mM SulfoSANPAH in 50 mM HEPES buffer, pH 7.4 and place them under a UV lamp for 5 min. The distance between the UV lamp and the sample should be approximately 10–15 cm.
2. Aspirate the solution, add 80 μL of fresh 1 mM SulfoSANPAH in 50 mM HEPES buffer, pH 7.4 and repeat the UV treatment.
3. Wash the PoMPs once with 50 mM HEPES buffer, pH 7.4 and add on top of the PoMPs a solution of the protein to be covalently linked. For instance, coating with a solution of 30 μg/mL of polylysine in PBS is recommended for investigating the interactions of *N. gonorrhoeae* pili.
4. Place the PoMPs in an incubator at 37°C for 1 h and then wash three times with PBS prior to using them for calibration and force assays.

3.3.4. Calibration of PoMPs Using Optical Tweezers

As with using optical tweezers, there is a need to know what the relationship is between the displacement of each pillar and the force exerted on them. One method of achieving this would be to rely on the equation from elastic theory in a linear approximation, $F = (3\pi E r^4 / 4L^3)x$, where E is the Young modulus of the material, r the radius of the beam, L the length of the beam and x the displacement. However, this equation is totally valid only if the pillars are exactly cylindrical. Moreover, the high power dependence on both the radius and the length means that the value of both parameters must be measured very precisely. This is usually accomplished by scanning electron microscopy, but this technique is not suitable for samples composed of aqueous materials such as hydrogels as it involves dehydration of the samples, which is impossible for hydrogels. We therefore prefer to measure directly the relationship between force applied and displacement of the pillars. A calibration is necessary only once for each type of mould-polyacrylamide–bis-acrylamide combination that is used. A calibration of the pillars can be done using optical tweezers as follows:

1. Take the polylysine-coated PoMPs to be calibrated and mount them on an observation chamber.
2. Add ~1 μL of carboxylated silica or latex beads to 1 mL of water. Add 2–10 μL of these beads to the observation chamber.
3. Start recording a digital movie (see Note 6).
4. Trap a bead and bring the side of a pillar tip to it until contact. Slowly pull the pillar away from the bead (see Fig. 4a, b).
5. Analyze the digital movie as described in step 9 of Subheading 3.1 and track the positions of the bead, the pillar and a reference point in the field of view. From these positions, obtain the displacement x of the bead from the center of the trap, which will give the force exerted on the pillar as the trap is calibrated. Obtain also the displacement Δ of the pillar over time. The graph of force vs. Δ will enable the measurement of the stiffness of the pillar (see Fig. 4c).

3.3.5. Calibration of PoMPs Using Magnetic Tweezers

If optical tweezers are unavailable, an alternative method that is cheaper and easier to implement relies on the use of magnetic tweezers.

1. Attach a small cylindrical neodymium magnet to the end of a handle so that it can be mounted to a 3-axis micromanipulator.
2. Attach a sewing needle to the magnet in a fixed position. The needle will become magnetized and provide the small tip of the magnetic tweezers.
3. Before calibrating the PoMPs, the magnetic tweezers first need to be calibrated. This is done once again by using the Stokes equation ($F_{drag} = 6\pi\eta R v$) (see Fig. 5a, b).
   (a) Add ~1 μL of magnetic beads as well as ~1 μL of non-magnetic beads of a different diameter into separate Eppendorf tubes containing 100 μL of water.

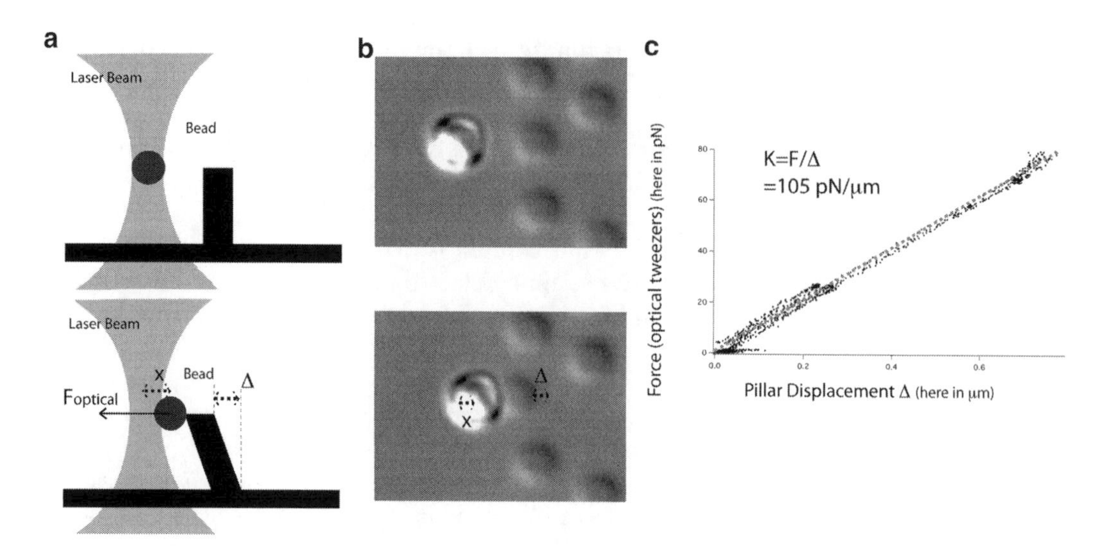

Fig. 4. Calibration of the pillars using optical tweezers. (**a**) Schematics of the calibration of the pillars using optical tweezers: at the beginning of the calibration when the bead is not in contact with the pillar and then in the middle of the calibration when force is exerted on the pillar. (**b**) Optical microscopy images of the calibration of the pillars using optical tweezers: at the beginning of the calibration when the bead is not in contact with the pillar and then in the middle of the calibration when force is exerted on the pillar. (**c**) An example of the calibration of the pillars with optical tweezers: the slope of the graph of the force of the optical tweezers vs. the displacement of the pillar gives the stiffness of the pillar.

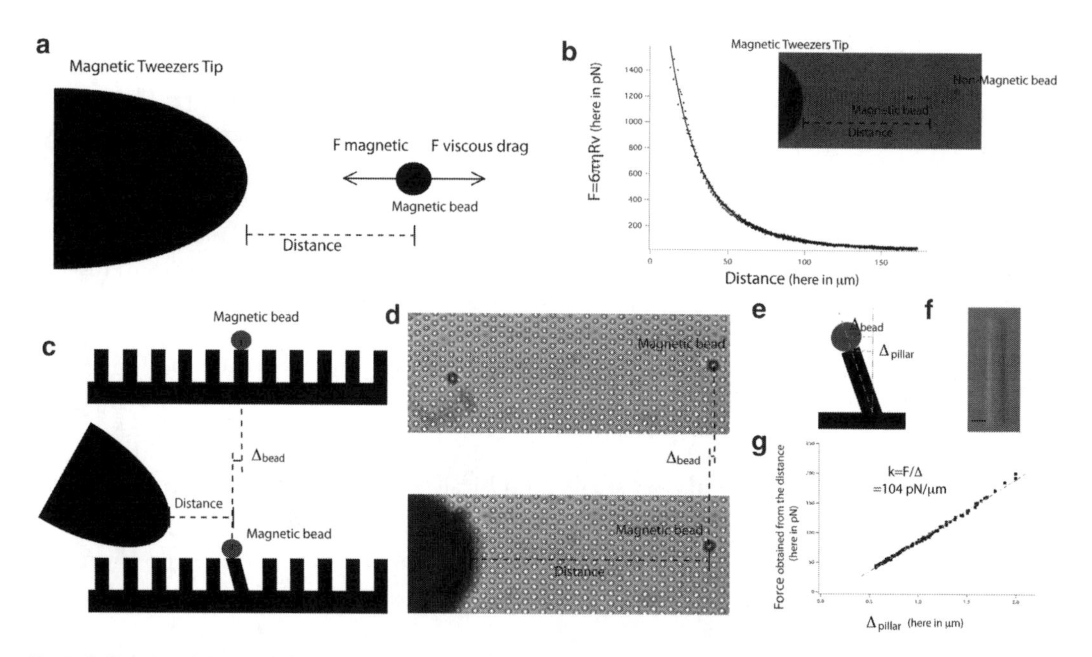

Fig. 5. Calibration of the pillars using magnetic tweezers. (**a**) Schematic of the calibration of the magnetic tweezers. (**b**) An example of the calibration curve of the magnetic force exerted by the magnetic tweezers vs. distance and an optical microscopy image of the setup. (**c**) Schematic of the calibration of the pillars by the magnetic tweezers. (**d**) Optical microscopy images of the calibration of the pillars by the magnetic tweezers. (**e**) A schematic that explains the difference between the bead's and pillar's displacements. (**f**) Optical microscopy image of a pillar enabling accurate measurement of the pillar's length. (**g**) An example of the calibration of the pillars with magnetic tweezers: the slope of the graph of the force of the magnetic tweezers vs. the displacement of the pillar gives the stiffness of the pillar.

(b) Remove all traces of water by using a magnet for the magnetic beads, or centrifugation at 13,000 × *g* for 5 min on a table-top centrifuge, for the non-magnetic beads.

(c) Add 300 μL of dimethylpolysiloxane, a medium of known viscosity (see Note 16).

(d) Stir well with a pipette tip to homogenize the mixture and incorporate the beads in the medium.

(e) Prepare a slide with a gasket a few millimeters high and add some of the viscous medium to it.

(f) Add a few droplets of both the magnetic and non-magnetic beads at the center of the gasket and mix again with a tip. A suspension of magnetic and non-magnetic beads in a viscous medium will be produced.

(g) Put the slide on a microscope stage and bring the tip into focus in the vicinity of the beads. Allow the system to settle for 10 min.

(h) Start recording a digital movie (see Note 6) and record the displacement of magnetic beads towards the tip.

(i) Analyze the digital movie as described in step 9 of Subheading 3.1 to determine the positions of the tip, magnetic beads, and non-magnetic beads.

(j) The movement of the non-magnetic beads enables the residual motion of the fluid itself to be tracked. The speed of the beads compared to the fluid can be computed and assuming that at all times the magnetic force and viscous drag force are equal, calibration of the magnetic tweezers will be obtained, i.e., the relationship between the distance between the bead and the tip and the force applied to the bead (see Fig. 5b).

4. Following calibration of the magnetic tweezers, the next step is to calibrate the PoMPs. Take the PoMPs, coated with polylysine as described in Subheading 3.3.3 and mount them onto an open observation chamber.
5. Make a dilute suspension of the carboxylated magnetic beads, e.g., 1 μL in 1 mL of water.
6. Add a few microliters of the suspension onto the pillars and wait for 20 min to allow most of the beads to settle before calibration can begin.
7. Select beads that are in contact with only a single pillar. These beads will be evenly surrounded by six pillars (see Fig. 5c, d).
8. Start recording a digital movie (see Note 6).
9. Bring the tip of the magnetic tweezers into focus and use the micromanipulator to change the force applied on the magnetic bead by slowly bringing the tip closer to the bead and then farther away. Repeat a couple of times.

10. Analyze the movie as described in step 9 of Subheading 3.1 to determine the position of the tip of the magnetic tweezers and the displacement of the bead.
11. The distance between the bead and the tip of the magnetic tweezers allows calculation of the force applied on the bead and the displacement. The stiffness of the pillar (force/displacement) can then be calculated, with one *caveat*. It is important to remember that the motion of the bead is followed, not the motion of the pillar. As such, the height of the pillar and the radius of the bead have to be measured in order to be able to calculate the stiffness of the pillar (see Fig. 5g). We will assume that the motion of the center of the pillar (Δ_{Pillar}) can be calculated as follows: $\Delta_{Pillar} = (L/(L+R))\Delta_{Bead}$ where R is the radius of the bead and L the length of the pillar (see Fig. 5e). An accurate measurement of the length of the pillar can be done by mechanically disrupting the PoMPs and imaging the resulting free floating pillars (see Fig. 5f).

3.3.6. PoMPs Assay

Once the PoMPs have been coated with polylysine or other protein of choice, the next step involves adding bacteria to the surface of the device at a known concentration. In order to mimic bacterial interactions with host cells, we recommend using the same surface density of bacteria to host cells that are used during infection assays. Depending on the type of data to be collected, the forces can be measured in two ways, termed dynamic or static force measurements (see Note 17).

1. Dynamic PoMPs force measurements are made as follows:
   (a) Mount the calibrated PoMPs onto an observation chamber.
   (b) Suspend ~10^6 *Neisseria* bacteria in 500 μL of DMEM (see Note 11).
   (c) Add this suspension to the observation chamber and put it under the microscope.
   (d) Start recording a digital movie (see Note 6).
   (e) Focus on the tip of the pillars.
   (f) The motion of the pillars over time provides a measurement of the dynamic forces exerted by the bacteria on the pillars (see Fig. 6a, b).
2. Static PoMPs force measurements are made as follows:
   (a) Place the calibrated PoMPs into the well(s) of 6-well tissue culture plates.
   (b) Suspend ~10^8 bacteria *Neisseria* bacteria in 2 mL of DMEM (see Note 11).

(c) Add the 2 mL of bacterial suspension onto the PoMPs and allow the bacteria to interact with the PoMPs over time.

(d) Fix the PoMPs with interacting bacteria by adding 2 mL of 4% (v/v) glutaraldehyde in PBS to the well for 1 h at room temperature.

(e) Transfer the fixed PoMPs to an observation chamber and take images under the microscope of both the tip and base of the pillars.

(f) The differences of position between the two images give the deflection of the pillars and thus the forces exerted by the bacteria, as the PoMPs are calibrated (see Fig. 6a, c). The differences of position between the two images can be obtained by using the implementation of cross-correlation algorithm as described in step 9 of Subheading 3.1 or by using a manual tracking software like the manual tracking plug-in from Image J (http://rsb.info.nih.gov/ij/plugins/manual-tracking.html).

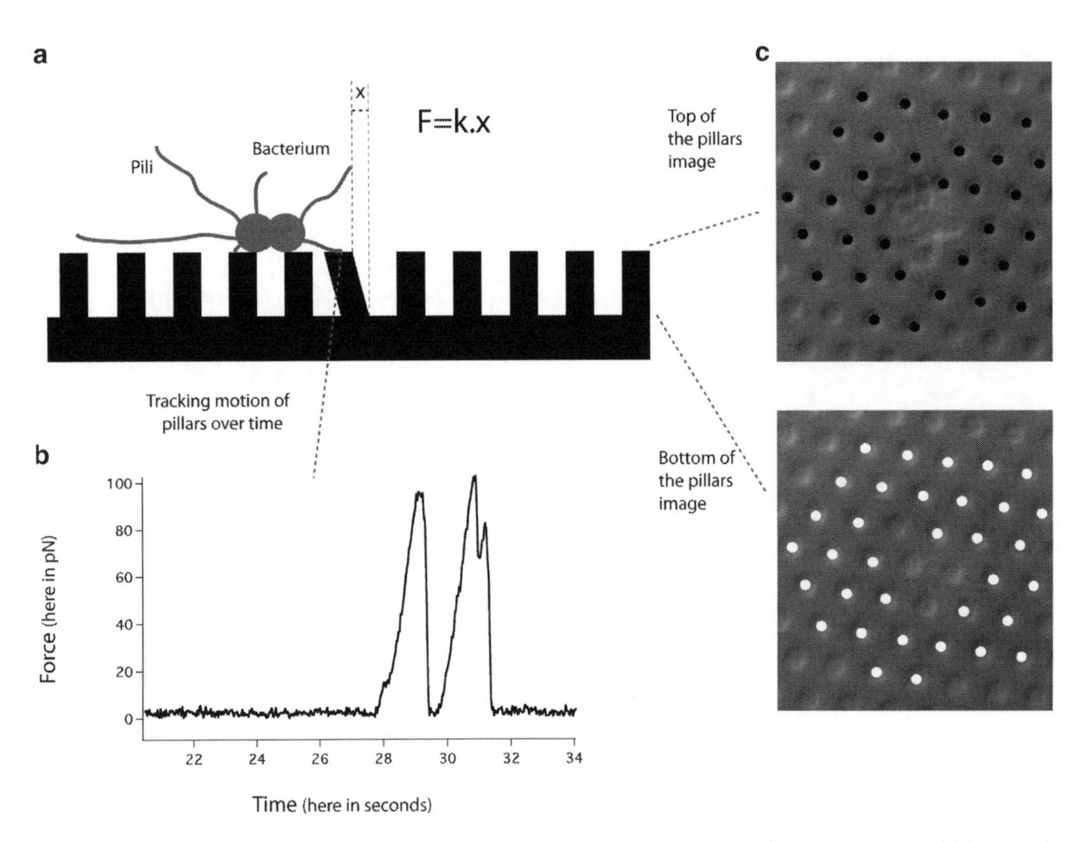

Fig. 6. The PoMPS assay. (**a**) Schematic of the PoMPs assay. (**b**) An example of a dynamic force measurement. (**c**) An example of a static force measurement.

4. Notes

1. Other popular and powerful image analysis software programs are Matlab, Metamorph, Igor and IDL.
2. The moulds must be etched in silicon. Moulds made in SU8 resins or polymethylmethacrylate (PMMA) resin, which can be demoulded easily with PDMS, will not work for multiple demouldings with polyacrylamide. After the silicon mould has been made, it needs to be cleaned with Piranha solution for 2 h and then plasma cleaned in an air plasma cleaner for 1 min. After this treatment, the mould can be used repeatedly to obtain arrays of PoMPs on an activated coverglass. For use, the mould needs to be cut into pieces smaller than the coverglass and we recommend areas of 1 cm^2.
3. Practically any protein can be used to coat the pillars. Trial and error with your protein of interest will be necessary to determine the concentration needed for the desired coating. When possible the use of fluorescently labeled protein helps in optimizing coating.
4. Other types of macromanipulators that can be used are the Burleigh PCS-5000, the Eppendorf Micromanipulator 5171 or the Injectman NI2.
5. The density of beads needs to be high enough that they will be found easily and low enough that no other bead is going to be trapped in the optical tweezers during the calibration.
6. The cost of both digital cameras and computer hard drives has dramatically decreased over recent years. A digital camera with capturing frequencies for at least part of their field of view, of 100 Hz or more is relatively inexpensive. The plethora of image analysis software available, e.g., Matlab, Metamorph, ImageJ, Igor, and IDL, combined with the availability of high computing power makes the use of recorded digital movies to analyze and to track microscopic movements a very attractive and affordable method. The implementation of cross-correlation algorithms (22) can lead to spatial resolution of less than 1/10th of a pixel, which is approximately 10–20 nm with the right magnification.
7. The frequency of the signal should not exceed a quarter of the recording frequency as this would hamper accurate analysis of the movie.
8. In order to make sure that the viscous drag force imposed on the bead is indeed equal to the elastic force of the optical trap, use only frequencies where a plateau in the bead displacement is attained at each half-cycle.

9. The viscous drag force equation (Stokes's law) is valid only if the bead is far away from the surface. Thus, perform the calibration at a distance from the surface between 5 and 10 bead diameters. However, this will slightly alter the profile of the light and potentially affect the stiffness of the trap. If calibration must be done near the surface, correction formulae exist to account for the distance between the bead and the surface (26).
10. An important parameter to record when performing a calibration is the temperature of the sample. The viscosity of water varies from 10^{-3} kg/s m at 20°C to 0.65×10^{-3} kg/s m at 40°C and thus the viscosity reading inserted into the equation might change by more than 30%, depending on the temperature (25 or 37°C) at which calibration is done. Taking these parameters into account, this method typically leads to calibration of the trap with greater than 20% precision.
11. The nature of the buffer could change the value of the forces measured as it could affect the interaction of the pili with the beads or the interaction between pili. We recommend using the same buffer that is used for other biological assays for the bacteria of interest. For example, the buffer often used for studies of the interactions between human host cells and *N. gonorrhoeae* or *N. meningitidis* is DMEM and this medium would be recommended for all force measurements.
12. With the optical tweezers assay, it is important that the beads used will attach to the pili of interest. In the case of the pathogenic *Neisseria,* pili adhere to many different surfaces. The corrugated surface of the *N. gonorrhoeae* pili presents helical alternative patches of positively charged and negatively charged surfaces. This peculiar configuration explains why *N. gonorrhoeae* pili will strongly bind both to negative surfaces such as carboxylated beads or positive surfaces such as poly-L-lysine-coated beads. Depending on the physico-chemical properties of the pili on your bacteria of interest, different beads may need to be used. As a starting point, we recommend the use of carboxylated beads that are readily available from commercial sources. Beads covalently coated with poly-L-lysine or antibodies against pili are valuable reagents for force measurement assays.
13. Depending on the type of bead and the coating used, it is possible that the beads will sediment quickly and bind to the surface of the coverglass. In this case, you should use an open observation chamber where the beads can be added at varying times.
14. If the bacteria do not readily adhere to the bottom of the coverglass, a single bacterium can be easily isolated by allowing its interaction with larger beads (adherent to the surface of the coverglass). This will ensure that the bacteria do not move.

15. Linear elastic theory for a beam leads to the following equation for the relationship between force and displacement: $F=(3\pi Er^4/4L^3)x$ where E is the Young modulus of the material, r the radius of the beam, L the length of the beam and x the displacement. From this equation, it is evident that the smaller the radius and the longer the beam, the softer the beam will be. However, if the geometrical parameters of the pillars can be adjusted to change their stiffness, there are still some technical constraints to take into account. With conventional approaches, the resolution of the features that can be patterned is around 1 μm, so the diameter for the pillars cannot be much smaller than this. Obtaining a ratio of length to radius greater than 20 is difficult and then demoulding becomes an issue. For this reason, we departed from the use of the widely popular PDMS (polydimethylsiloxane) polymer for moulding the features. The high Young modulus of the PDMS (~2 MPa) and the technological constraints mentioned above regarding the size and ratios of the features mean that PDMS pillars would have a stiffness too high to measure the lower forces that bacteria exerted, compared to eukaryotic cell forces. However, the use of polyacrylamide hydrogels considerably extends the flexibility of the micropillars system. Gels as soft as a 10% polyacrylamide – 0.1% bis-acrylamide and as stiff as 20% polyacrylamide – 1% bis-acrylamide have been used. The constraints in the choice of the percentages are the following: too low percentage gels will prevent demoulding and too high percentages will lead to brittle and opaque gels. Pillars of a diameter of 1 and 7 μm in length have a stiffness of around 100 pN/μm when casted with a 20% polyacrylamide – 0.2% bis-acrylamide mix. Given the high dependence of the stiffness with the geometrical characteristic of the pillars this number may vary for different moulds, e.g., a 10% increase in the radius would lead to almost a 50% increase in the stiffness.
16. Any medium of known viscosity can be used, as long as the viscosity is high enough to enable recording. The medium recommended (dimethylpolysiloxane, with known viscosity of 12,500 cSt at 25°C) allows movies to be analyzed at 1 Hz with the recommended beads.
17. Both methods have advantages and disadvantages. In the dynamic method, all the pillars that are not released during image acquisition cannot have their force properly measured leading to a selection bias towards short-lived pulling events. However, all the dynamic information of the forces is available. In the static method, there is no dynamic information, but the longer-lived events can be recorded as well and more data can be collected for statistical analysis. It should also be noted that for the static method a number of pillars may be distorted during

the demoulding process and thus care must be taken to avoid areas of the PoMPs that have pillars uniformly bent, as they are a tell-tale sign of a demoulding artifact rather than force generation. Depending on the data to be collected, add the bacteria at the desired concentration on top of the PoMPs and then either acquire a movie or fix and observe the differences in displacement of the pillars. Studies on the force generation of *N. gonorrhoeae* using the PoMPs technology led to the discovery that *N. gonorrhoeae* pili can bundle and exert pulling forces up to 1 nN. Such a technology can be used to study the retraction forces of *N. meningitidis* or other force-exerting bacteria. The ability to measure these forces may also be used to identify new bacterial phenotypes.

Acknowledgments

N. B. and M. S. acknowledge the award of NIH grant AI079030. This work was also supported by the Agence Nationale de la Recherche(ANR)(ProgrammeBlanc2010SVSE5“MECANOCAD”) and the CNRS (Program Prise de Risques “Interface physique, biologie et chimie”).

References

1. Mattick JS (2002) Type IV pili and twitching motility. Ann Rev of Microbiol 56: 289–314.
2. Hagblom P, Segal E, Billyard E et al (1985) Intragenic recombination leads to pilus antigenic variation in *Neisseria gonorrhoeae*. Nature 315: 156–158.
3. Aas FE, Egge-Jacobsen W, Winther-Larsen HC et al (2006) *Neisseria gonorrhoeae* type IV pili undergo multisite, hierarchical modifications with phosphoethanolamine and phosphocholine requiring an enzyme structurally related to lipopolysaccharide phosphoethanolamine transferases. J Biol Chem 281: 27712–27723.
4. Brown DR, Helaine S, Carbonnelle E et al (2010) Systematic functional analysis reveals that a set of seven genes is involved in fine-tuning of the multiple functions mediated by Type IV Pili in *Neisseria meningitidis*. Infect Immun 78: 3053–3063.
5. Farge E (2003) Mechanical induction of twist in the Drosophila foregut/stomodeal primordium. Current Biol 13: 1365–1377.
6. Gilbert PM, Havenstrite KL, Magnusson KEG et al (2010) Substrate elasticity regulates skeletal muscle stem cell self-renewal in culture. Science 329: 1078–1081.
7. Zhang XH, Halvorsen K, Zhang CZ et al (2009) Mechanoenzymatic cleavage of the ultralarge vascular protein von Willebrand Factor. Science 324: 1330–1334.
8. Engler AJ, Sen S, Sweeney HL et al (2006) Matrix elasticity directs stem cell lineage specification. Cell 126: 677–689.
9. Merz AJ, So M, Sheetz MP (2000) Pilus retraction powers bacterial twitching motility. Nature 407: 98–102.
10. Maier B, Potter L, So M et al (2002) Single pilus motor forces exceed 100 pN. Proc Nat Acad Sci USA 99: 16012–16017.
11. Biais N, Ladoux B, Higashi D et al (2008) Cooperative retraction of bundled type IV pili enables nanonewton force generation. Plos Biol 6: 907–913.
12. Higashi DL, Zhang GH, Biais N et al (2009) Influence of type IV pilus retraction on the architecture of the *Neisseria gonorrhoeae*-infected cell cortex. Microbiol-SGM 155: 4084–4092.

13. Howie HL, Glogauer M, So M (2005) The *N.gonorrhoeae* type IV pilus stimulates mechanosensitive pathways and cytoprotection through a pilT-dependent mechanism. Plos Biol 3: 627–637.
14. Sterba Re, Sheetz MP (1997) Basic laser tweezers. Meth Cell Biol 55: 29–41.
15. McGee-Russell SM, Allen RD (1971) Reversible stabilization of labile microtubules in the reticulopodial network of *Allogromia*. Adv Cell Molec Biol 1: 153.
16. du Roure O, Saez A, Buguin A et al (2005) Force mapping in epithelial cell migration. Proc Nat Acad Sci USA 102: 2390–2395.
17. Ghassemi S, Biais N, Maniura K et al (2008) Fabrication of elastomer pillar arrays with modulated stiffness for cellular force measurements. J Vac Sci Technol B 26: 2549–2553.
18. Tanase M, Biais N, Sheetz M (2007) Magnetic tweezers in cell biology. Meth Cell Biol 83: 473–493.
19. Ashkin A (1970) Acceleration and trapping of particles by radiation pressure. Phys Rev Lett 24: 156–&.
20. Simmons RM, Finer JT, Chu S et al (1996) Quantitative measurements of force and displacement using an optical trap. Biophys J 70: 1813–1822.
21. Clausen M, Koomey M, Maier B (2009) Dynamics of Type IV Pili is controlled by switching between multiple states. Biophys J 96: 1169–1177.
22. Gelles J, Schnapp BJ, Sheetz MP (1988) Tracking kinesin-driven movements with nanometre-scale precision. Nature 331: 450–453.
23. Tan JL, Tien J, Pirone DM et al (2003) Cells lying on a bed of microneedles: An approach to isolate mechanical force. Proc Nat Acad Sci USA 100: 1484–1489.
24. Wang YL, Pelham RJ (1998) Preparation of a flexible, porous polyacrylamide substrate for mechanical studies of cultured cells. Meth Enzymol 298: 489–496.
25. Kandow CE, Georges PC, Janmey PA et al (2007) Polyacrylamide hydrogels for cell mechanics: Steps toward optimization and alternative uses. Cell Mech 83: 29–46.
26. J.Happel, H.Brenner (1983) Low Reynolds number hydrodynamics with special applications to particulate media. Kluwer, Boston.

Chapter 14

Human Dendritic Cell Culture and Bacterial Infection

Hannah E. Jones, Nigel Klein, and Garth L.J. Dixon

Abstract

Dendritic cells (DC) play a key role in the development of natural immunity to microbes. The DC form a bridge between the innate and adaptive immune system by providing key instructions particularly to antigen naïve T-cells. The interaction of DC with T lymphocytes involves three signals: (1) antigen processing and presentation in context of MHC Class I and/or II, (2) expression of T cell co-stimulatory molecules, and (3) cytokine production. Studying the interactions of DCs with specific pathogens allows for better understanding of how protective immunity is generated, and may be particularly useful for assessing vaccine components. In this chapter, we describe methods to generate human monocyte-derived DCs and assess their maturation, activation, and function, using interaction with the gram-negative bacterial pathogen *Neisseria meningitidis* as a model.

Key words: Human dendritic cells, *Neisseria meningitidis*, Maturation, Phagocytosis, T cell polarization

1. Introduction

The immune system is constantly under microbial challenge, from both pathogenic and commensal organisms. How the immune system initially responds to these encounters appears to be critical for the induction of a protective immune response. Situated at the center of the immune system is the dendritic cell (DC). DCs are professional antigen-presenting cells that have high phagocytic capacity and act as sentinels and are often located in an immature form in submucosal or subcutaneous compartments. Upon encounter with a microbial stimulus the DC then undergoes a process of maturation, resulting in loss of phagocytic ability coupled with an increase in the expression of molecules involved in antigen

Myron Christodoulides (ed.), *Neisseria meningitidis: Advanced Methods and Protocols*, Methods in Molecular Biology, vol. 799, DOI 10.1007/978-1-61779-346-2_14,

presentation and T cell co-stimulation (1). Typically, these mature DCs migrate to secondary lymphoid tissue to initiate the immune response through stimulation of T cells.

The initial interaction of DCs with microbial stimuli is critical for shaping the outcome of the primary immune response. This response is mediated through the engagement of pathogen-associated molecular patterns (PAMPs), widely expressed by bacteria, fungi, and viruses, with pattern recognition receptors (PRR) such as toll-like receptors (TLR), NOD-like receptors, C-type lectins, and scavenger receptors (2). The consequence of ligand recognition by these receptors can result in antigen uptake, but also starts signaling mechanisms that drive DC maturation and cytokine production. The profile of cytokine production by DC appears to be crucial for inducing either T helper (Th1 and Th2), T regulatory, and Th17 type immune responses (3, 4).

In humans, two DC lineages have so far been described: (1) myeloid DCs and (2) plasmacytoid DCs. Myeloid DCs can be found in either peripheral tissue, secondary lymphoid tissue, or circulating in the blood. Plasmacytoid DCs commonly circulate through the blood into lymph nodes but are rarely found in peripheral tissue.

The isolation of human DC subsets can be difficult for a number of reasons: (1) enzymatic and mechanical processes used to isolate DCs from tissues can result in DC activation, (2) DCs in blood are rare, and (3) they are often heterogenous in nature. A major breakthrough in the understanding of human DC biology was made by the development of a method for in vitro generation of monocyte-derived dendritic cells (MDDC) by Sallusto and Lanzavecchia (5). This process allowed the differentiation of a homogenous population of immature DCs that could be produced in relatively large numbers from human blood and are potent stimulators of naïve T cells (5). DCs can also be generated from CD34+ hematopoietic stem cells following stimulation with GM-CSF and TNF-α, but this method gives rise to a more heterogenous population in a semimature state and requires either the use of umbilical cord or G-CSF mobilized peripheral blood (6–8).

A better understanding of how pathogens interact with DCs provides an insight into pathogenesis and gives valuable information to aid novel vaccine design (9, 10). Work from this laboratory and others has focused on the interaction of MDDC with *Neisseria meningitidis*, particularly with serogroup B (11–20). This work has provided valuable information about which *N. meningitidis* surface components interact with PRRs on the surface of DCs and how this affects DC maturation, cytokine production, and T cell stimulation. In this chapter, we describe methodologies that can be used to investigate the interaction of pathogens and microbial components with DCs using *N. meningitidis* as an example.

2. Materials

All chemicals are from Sigma-Aldrich unless otherwise stated.

2.1. In Vitro Generation of Human Immature Monocyte-Derived Dendritic Cells Using CD14 Magnetic Bead Separation

1. Sterile heparin 5,000 IU/mL.
2. Complete RPMI: Roswell Park Memorial Institute Medium (RPMI 1640) supplemented with 25 mM HEPES, 2.4 mM L-glutamine (all Invitrogen), and 10% (v/v) fetal calf serum. This can be stored at 4°C, but used at 37°C (see Note 1).
3. Lymphoprep (Axis-Shield). Store at room temperature (RT) and protect from light.
4. MACs buffer, Dulbecco phosphate buffered saline (DPBS) supplemented with 0.1% (w/v) low endotoxin bovine serum albumin (BSA) and 2 mM EDTA, sterile filtered through a 0.2-μm pore filter. Store and use this buffer at 4°C.
5. MACS human CD14 microbeads (Miltenyi Biotec).
6. Interleukin-4 (IL-4) and granulocyte-macrophage colony stimulating factor (GM-CSF) (R&D systems) (see Note 2).
7. Trypan blue.
8. Sterile tissue culture 6 well plates (Corning).
9. Hemocytometer.
10. 50 mL conical tubes (Corning).
11. Humidified CO_2 incubator.
12. Refrigerated Sorvall Legend RT centrifuge.

2.2. Characterization of the Surface Membrane Phenotype of Human MDDC

1. Monoclonal antibodies to the following surface proteins for characterization of immature MDDC: CD1a (HI149), CD3 (UCHT1), CD14 (TuK4), CD19 (SJ25-C1), CD25 (3G10), CD40 (HB14), CD83 (HB15e), CD86 (BU63), HLA-DR (TU36), HLA-Class 1 (TU149) (all Invitrogen), and DC-SIGN (R& D Systems).
2. Wash buffer: PBS, pH 7.4 containing 0.2% (w/v) BSA and 0.05% (w/v) sodium azide.
3. Paraformaldehyde (4% v/v) in PBS.
4. Fibronectin.
5. FACs tubes, 5 mL polystyrene round bottom tubes (BD Falcon).
6. Refrigerated Sorvall Legend RT centrifuge.
7. *FACS*calibur flow cytometer using CellQuest Pro software (BD Biosciences).

2.3. Infection of Human MDDC with N. meningitidis

1. *N. meningitidis* stored in Mueller-Hinton broth (Oxoid) with 15% (v/v) glycerol at −80°C.
2. GC agar supplemented with vitox (Oxoid).
3. Fluorescein isothiocyanate (FITC) in PBS.
4. RPMI 1640 with no phenol red (Invitrogen).
5. Polypropylene 5 mL tubes (VWR International).
6. Jenway 6300 spectrophotometer or equivalent.
7. Refrigerated Sorvall Legend RT Centrifuge (for pelleting cells).
8. Heraeus Biofuge Centrifuge (for pelleting bacteria).

2.4. Methods for Assessing Human MDDC Activation

2.4.1. Cytokine Production

1. Brefeldin A: solution of 1 mg/mL in 100% ethanol.
2. Paraformaldehyde (4% v/v) in PBS.
3. Permeabilization buffer: Hank's buffered saline solution (HBSS, Invitrogen) with calcium and magnesium containing 0.1% (w/v) saponin, 2 mM HEPES, and 0.05% (w/v) sodium azide.
4. Monoclonal antibodies used for intracellular cytokine staining of MDDC: specific for IL-1β (HIb-98), IL-6 (MQZ-13A5), IL-12p40 (C11.5) (Biolegend), and TNF-α (Sigma-Aldrich).
5. FACs tubes, 5 mL polystyrene round bottom tubes (BD Falcon).
6. Refrigerated Sorvall Legend RT Centrifuge (for pelleting cells).
7. Heraeus Biofuge Centrifuge (for pelleting cellular debris).
8. *FACS*calibur flow cytometer using CellQuest Pro software (BD Biosciences).
9. Cytokine ELISA kits using matched antibody pairs (eBioscience).

2.4.2. Determining Bacterial Internalization by Flow Cytometry

1. Trypan blue.
2. Wash buffer: PBS containing 0.2% (w/v) BSA and 0.05% (w/v) sodium azide.
3. Paraformaldehyde (4% v/v) in PBS.
4. FACs tubes, 5 mL polystyrene round bottom tubes (BD Falcon).
5. Refrigerated Sorvall Legend RT Centrifuge.
6. *FACS*calibur flow cytometer using CellQuest Pro software (BD Biosciences).

2.4.3. Determining Bacterial Internalization by Confocal Microscopy

1. Paraformaldehyde (4% v/v) in PBS.
2. Blocking buffer: PBS containing 1% (w/v) BSA.
3. Mouse antibody specific to the P1.7 epitope on *N. meningitidis* strain H44/76 PorA protein (NIBSC, UK).

4. Nuclear stain To-Pro3 (Invitrogen).
5. Alexa Fluor 568 goat antimouse IgG (Invitrogen).
6. Vectashield mountant (Vector laboratories).
7. Polylysine coated adhesion slides (Paul Marienfeld Gmbh).
8. Leica SP2 confocal laser scanning microscope system (Leica).

2.4.4. Determining Expression of Antigen Presentation and Activation Molecules on Human MDDC by Flow Cytometry

1. Monoclonal antibodies to the following DC activation and maturation molecules: specific for CD25 (3G10), CD40 (HB14), CD83 (HB15e), CD86 (BU63), HLA-DR (TU36), and HLA-Class 1 (TU149) (all Invitrogen).
2. All other reagents as itemized in Subheading 2.2.

2.5. Interactions of Human Naïve T Cells with Dendritic Cells

1. T cell medium: RPMI 1640 containing 10% (v/v) FCS, 2.4 mM of l-glutamine, 100 U/mL of penicillin/streptomycin (all Invitrogen), 10 U/mL of recombinant IL-2.
2. MACS CD3+ T cell isolation kit (Miltenyi Biotec). CD3 positive cells are isolated through negative selection.
3. MACS Naïve CD4+ T cell isolation kit II (Miltenyi Biotec). T cells are isolated using negative selection.
4. Trypan blue.
5. Carboxyfluorescein succinimidyl ester (CFSE).
6. Phytohemagglutinin (PHA).
7. Phorbol myristate acetate (PMA) and ionomycin.
8. Brefeldin A.
9. Monoclonal antibodies to CD4 T cells; CD45RA (MEM-56), CD45RO (UCHL1), and CD25 (CD25-3G10) (all Invitrogen).
10. Monoclonal antibodies used to assess T cell polarization: specific for IL-4 (8D4-8) and IFN-γ (25723.11) (ebiosciences).
11. *Escherichia coli* lipopolysaccharide (LPS).
12. Prostaglandin E2.
13. Poly I:C.
14. *Staphylococcus* enterotoxin B (SEB).
15. 15 mL conical tubes (Corning).
16. 96-well flat-bottomed sterile plates (Corning).
17. 24 well sterile plates (Corning).
18. Hemocytometer.
19. FACs tubes, 5 mL polystyrene round bottom tubes (BD Falcon).
20. Humidified CO_2 incubator.
21. Refrigerated Sorvall Legend RT Centrifuge.
22. *FACS*calibur flow cytometer using CellQuest Pro software (BD Biosciences).

3. Methods

3.1. In Vitro Generation of Human Immature Monocyte-Derived Dendritic Cells Using CD14 Magnetic Bead Separation

Human DCs can be generated in vitro by either culturing CD14+ cells (blood monocytes) with IL-4 and GM-CSF (5) or CD34+ cells either from cord blood (6) or G-CSF mobilized peripheral blood (7) with GM-CSF and TNF-α. Here we focus on MDDC as these can be generated from a homogenous population of monocytes isolated from blood taken from healthy donors. The MDDC preparation is carried out under sterile conditions using a Class II biosafety level laminar flow cabinet.

1. Take blood from a healthy human donor who has given informed consent (see Note 3).

   Heparin is added to the blood to a final concentration of 10 IU/mL. In general, 1×10^6 MDDC are expected from every 10 mL of blood. Dilute the blood sample 1:1 with RPMI with 5% (v/v) FCS.

2. Next, layer the blood carefully onto lymphoprep, typically 30 mL of diluted blood over 20 mL of lymphoprep in a 50 mL conical tube, and promptly centrifuge at $400 \times g$ for 30 min at RT without a break to prevent disruption of the gradient.
3. Following centrifugation, a layer of peripheral blood mononuclear cells (PBMCs) is found between the interface of the lymphoprep and blood plasma. Carefully remove the cells with a Pasteur pipette, making sure not to disturb the gradient and place the cells into a new 50 mL conical tube. Wash the cells with RPMI containing 5% (v/v) FCS and centrifuge at $200 \times g$ for 10 min at RT to remove platelets.
4. Count the cells with a hemocytometer, using trypan blue stain (mixed with the cell sample at a 1:1 ratio) to discriminate between live and dead cells. Dead cells appear blue under light microscope as the cell membranes are permeable to the blue dye.
5. Prepare the MACs buffer, which should be kept on ice or at 4°C at all times. Keep the PBMCs on ice for 30 min with the addition of 20 µL of MACS human CD14 beads and 80 µL of MACs buffer per 10^7 cells. An alternative method can be used that depends on adherence of cells to plastic and is therefore independent of the microbead technology (see Note 4).
6. Isolate CD14 positive cells using the magnetic microbead separation technology according to manufacturer's instructions (see Note 5).
7. Suspend the CD14 positive cells, at a concentration of 5×10^5 cells/mL, in complete RPMI supplemented with 100 ng/mL of GM-CSF and 50 ng/mL of IL-4. Normally, cells are seeded into 6 well plates with a typical volume of 4 mL in each well.

Incubate the cells in a humidified incubator at 37°C with 5% (v/v) CO_2 for 5–7 days.

8. After 5–7 days, harvest the MDDC into a 50 mL conical tube and then centrifuge at 300 × *g* for 10 min at 21°C. Next, count the cells and then suspend the cells at a concentration of 5×10^5 cells/mL.

3.2. Characterization of the Surface Membrane Phenotype of Human MDDC

There are a number ways to characterize DCs, including morphology, surface protein expression, and functional studies. DCs have a distinct morphology that is characterized by the presence of dendrites, which are branched projections that act to increase the surface area to maximize trapping of microbial products (see Fig. 1 and Note 6). Functionally, they possess the ability to readily take up and process antigen, migrate to secondary lymphoid tissue, and they are potent stimulators of T cells. However, DCs can also form many different subsets and the surface membrane phenotype can be useful in defining these subsets.

1. The cells generated in Subheading 3.1 are characterized by monoclonal antibody (mAb) labeling of surface membrane proteins to confirm that monocytes have differentiated into immature MDDC (Table 1).
2. Aliquot the cells into FACs tubes, add 2–3 mL of Wash buffer, and then centrifuge the cells at 300 × *g* for 10 min at 4°C. Pour off the supernatant gently. Typically, 2.5–5×10^4 cells in a 100 μL volume of Wash buffer are used for each stain.
3. Add the appropriate fluorescently labeled conjugated mAbs (2 μg/mL) to the cells and chill on ice for 30 min. At this stage

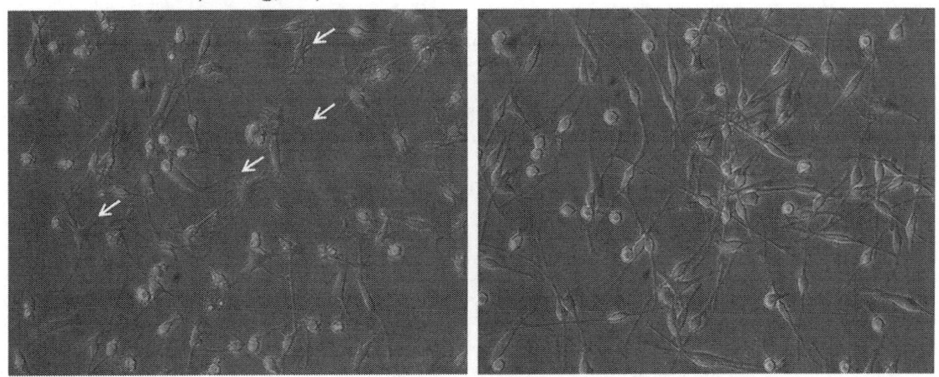

Fig. 1. Morphology of MDDC. Immature, unstimulated MDDC appear long and thin, but upon stimulation with meningococcal LPS (100 ng/mL) for 6 h they develop more projections (as indicated by *arrows*) known as dendrites which are the major characteristics of mature MDDC. MDDC in this instance are fixed to the plate using fibronectin.

Table 1
Phenotype of immature monocyte-derived dendritic cells (MDDC)

Molecule	Expression: yes or no?
CD1a	Yes
CD3	No
CD14	Yes, low
CD19	No
CD25	No
CD40	Yes
CD83	No
CD86	Yes, low
HLA-DR	Yes
HLA-DQ	Yes
HLA-Class 1	Yes

and for all subsequent steps, protect the cells from light in order to prevent loss of fluorescence (see Note 7).

4. Wash the cells with 2–3 mL of cold wash buffer and then centrifuge at 300×*g* for 5 min at 4°C. Repeat the wash step and then fix the cells with 100 μL of 4% (v/v) paraformaldehyde in PBS for 10 min at RT. Store the fixed cells at 4°C in the dark until ready to analyze by flow cytometry, following the equipment manufacturer's instructions (see Note 8).

3.3. Infection of Human MDDC with N. meningitidis (see Note 9)

1. Place a frozen aliquot of *N. meningitidis*, stored in Mueller-Hinton broth with 15% (v/v) glycerol, onto a GC agar plate supplemented with vitox. Grow the organism in a humidified incubator at 37°C with 5% (v/v) CO_2 for 8–24 h.
2. Harvest the meningococci from the plate using a sterile cotton swab and place the bacteria into 2 mL of RPMI (with no phenol red). Measure the optical density (OD) of the suspension with a spectrophotometer at an absorbance of λ600 nm.
3. Adjust the bacterial suspension to an OD of 1, which is equivalent to 1×10^9 bacteria/mL as determined by serial dilution and plating using standard microbiological techniques. For phagocytosis experiments, the meningococci are fluorescently labeled with FITC at this stage, by incubating the bacteria with

0.5 mg/mL of FITC for 15 min at 37°C with agitation (see Note 10).

4. Centrifuge the FITC-labeled bacteria at 3,500 × *g* for 2 min and a yellow pellet should be observed. Remove the supernatant and wash the bacteria with RPMI (no phenol red) and then centrifuge at 3,500 × *g* for 2 min. Repeat this wash step a further 2 times, in order to remove excess fluorescent dye. The bacteria can then be used at this stage or fixed with paraformaldehyde as described below. As an alternative to FITC labeling of the bacteria, intrinsically fluorescent meningococci can be used instead. These organisms express the green fluorescent protein (GFP) (see Note 11).
5. Fix the meningococci with a solution of 0.5% (v/v) paraformaldehyde in PBS for 10 min at RT. Wash the treated bacteria extensively with RPMI (no phenol red) with centrifugation for 2 min at 3,500 × *g*. Repeat this wash step for a further 3 times to remove excess paraformaldehyde.
6. Finally, suspend the bacteria to an OD of 1 in RPMI (no phenol red) ready for use. Fixed meningococci can be used either on the day of preparation or can be stored at −80°C for a few months, but repeated freeze–thaw cycles are not recommended as the bacteria will degrade (see Note 12).
7. Following 5–7 days of culture of MDDC as described in Subheading 3.1, harvest the immature MDDC from the 6 well plates by removing the cells carefully using a 5-mL pipette followed by gentle washing with 2–3 mL of complete RPMI to remove any loosely adhered MDDC (see Note 13).
8. Count the MDDC using a hemocytometer, as some loss is observed at this stage due to adherence of a small percentage of MDDC to the plastic. These adherent MDDC almost always have a mature DC phenotype and should not be used in the experiment.
9. Suspend the immature MDDC in polypropylene tubes at a concentration of 5×10^5 cells/mL in complete RPMI.
10. Add the fixed bacteria at an appropriate multiplicity of infection (e.g., a MOI of 100 is 100 bacteria to 1 DC) to the cells. FITC-labeled bacteria can be used to study bacterial internalization (see Notes 14 and 15). Alternatively, rather than using whole bacteria, microbial products such as purified LPS can be used to stimulate MDDC. Typically, meningococcal LPS is used at a concentration of 10–100 ng/mL and is a potent stimulator of DCs (Fig. 2).
11. For the phagocytosis assay, aliquots of cells at various time points may be required. In the case of fixed *N. meningitidis* used at a MOI of 100, a time course of between 1 and 6 h is

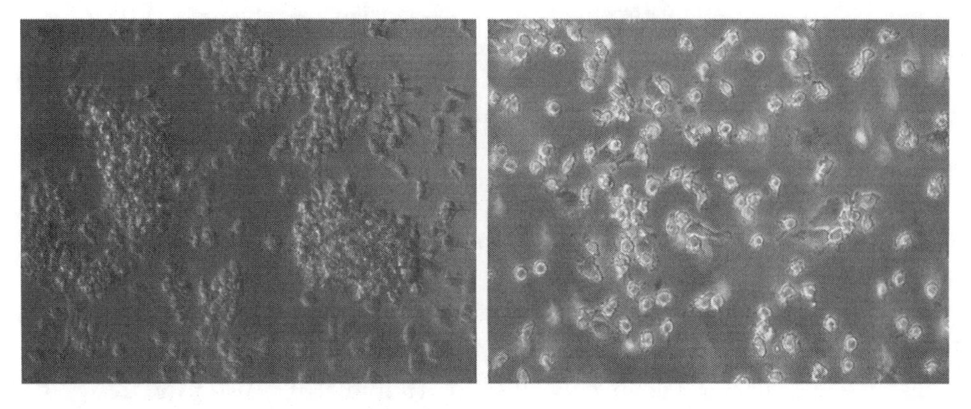

Fig. 2. Characteristic aggregation of MDDC following stimulation with LPS. Immature DCs following meningococcal LPS stimulation for 18 h begin to form aggregates in solution.

recommended. To assess MDDC maturation and cytokine production in the presence of fixed N. meningitidis (MOI = 10), 18–24 h is usually sufficient.

3.4. Methods for Assessing Human MDDC Activation

3.4.1. Cytokine Production

There are two methods that are generally used to assess cytokine production at the protein level: (1) intracellular cytokine staining using flow cytometry or (2) an ELISA-based method to assess cytokine release into the cell culture medium. It is important to consider both the advantages and limitations of each protocol before deciding which one to use. In the case of intracellular cytokine staining, the major benefit of this method is that if your cell population is heterogenous it is possible to determine exactly which cell is producing a particular cytokine by fluorescently labeling the cell type of interest with an antibody. The limitations of intracellular cytokine staining are that the use of brefeldin A, a protein transport inhibitor, not only prevents release of the cytokine of interest, but may affect other cytokines that are involved in regulation. For example, IL-10 is known to down regulate IL-12 production: therefore, preventing release of IL-10 can affect IL-12 levels (12). By contrast, assessing cytokine production by ELISA does reflect how cytokines may affect each other and an actual concentration can be determined using a standard curve, but for mixed cell populations it is difficult to determine which cell type is responsible for cytokine production.

1. In order to assess cytokine production by MDDC by intracellular cytokine staining, following co-culture of MDDC with the stimuli of choice as described in Subheading 3.3 add brefeldin

A (10 μg/mL) to MDDC for 18–24 h to prevent the release of cytokines from the cells into the supernatant (see Note 16).

2. Next, transfer the cell to FACs tubes. Fix the cells ($2.5–5 \times 10^4$) with 100 μL of 4% (v/v) paraformaldehyde in PBS at 4°C for 15 min. Then, add 2–3 mL of permeabilization buffer to the fixed cells and then promptly centrifuge at $300 \times g$ for 5 min at 4°C.
3. Remove the supernatant by tipping the tube and suspend the cells in 200 μL of permeabilization buffer. Incubate the samples with cytokine (IL-1β, IL-6, IL-12p40, TNF-α) specific mAbs and appropriate isotype matched controls, for 45 min at RT (see Note 17).
4. Wash the cells twice with permeabilization buffer as described in step 2 and then fix them with 100 μL of 4% (v/v) paraformaldehyde. Analyze the fixed cells by flow cytometry, following the equipment manufacturer's instructions.
5. In order to assess cytokine production by ELISA, following stimulation of MDDC, harvest the cell culture medium by pelleting the cells by centrifugation at $300 \times g$ for 10 min. Next, centrifuge the supernatant at $3{,}500 \times g$ for 5 min to remove any cell debris. In our studies, 18–24 h following stimulation represents the peak of cytokine accumulation in the MDDC medium, but it is important to carry out a time course for different microbial stimuli as the kinetics may vary. Store the supernatants at −80°C until required. It is important to avoid repeated freeze–thaw cycles, so freeze the supernatant samples in small aliquots that can be used as required. Cytokine protein production, including TNF-α, IL-1β, IL-6, IL-10, IL-12, and IL-23, is quantified using commercial ELISA kits according to the respective manufacturer's instructions (see Note 18).

3.4.2. Determining Bacterial Internalization by Flow Cytometry

1. Following infection of MDDC with FITC-labeled or GFP-expressing bacteria as described in step 10 of Subheading 3.3, take samples at various intervals to follow the course of phagocytosis. Typically, take $2.5–5 \times 10^4$ cells at each time point and transfer them to FACs tubes. Next, add 2–3 mL of Wash buffer and then centrifuge at $300 \times g$ for 5 min at 4°C. Repeat this process 2 times.
2. Add trypan blue to the cell pellet to a final concentration of 0.3% (v/v) in Wash buffer and incubate at RT for 10 min, in order to quench the FITC signal from extracellular bacteria on the surface of the MDDC (see Note 19).
3. Wash the cells twice with Wash buffer and then fix for 15 min at RT with 100 μL of 4% (v/v) paraformaldehyde Store fixed cells at 4°C until analyzed by flow cytometry, following the equipment manufacturer's instructions (see Note 20).

3.4.3. Determining Bacterial Internalization by Confocal Microscopy (see Note 21)

1. Following infection of MDDC with FITC-labeled or GFP-expressing bacteria as described in step 10 of Subheading 3.3, take cell samples at various time points. Add 2–3 mL of cold PBS to the cells and then centrifuge at 300×*g* for 10 min at 4°C. Add the cells to a 12-well adhesion slide and allow them to adhere for 10 min at RT. Typically, 5,000 cells are used in 10 μL of cold PBS.
2. Remove the PBS by gently touching a paper tissue to the edge of the slide.
3. Fix the cells with 10 μL of 4% (v/v) paraformaldehyde for 10 min at RT. Wash the cells 3× with 50 μL of PBS at RT, each time removing PBS with paper tissue. Next, stain the cells as follows or store them at −20°C to use at a later date.
4. Add 50 μL of Blocking buffer and incubate at RT for 10 min to reduce nonspecific binding. Then, incubate the cells with 10 μL of Blocking buffer containing 10 μg/mL of mouse antibodies specific for the P1.7 epitope on *N. meningitidis* strain H44/76 PorA protein.
5. Wash the cells twice by gently pipetting 50 μL of PBS onto the slide and then removing the PBS with paper tissue. Next, incubate the cells with 10 μL of blocking buffer containing 10 μg/mL of Alexa Fluor 568 goat antimouse IgG, to allow visualization of extracellular bacteria (see Note 22), and a 1/500 dilution of nuclear stain To-Pro3. Incubate the cells with the stains for 30 min at RT.
6. Wash the cells twice with PBS and then apply a coverslip over Vectashield mountant. Finally, analyze the sample slides using a confocal laser scanning microscope, following the equipment manufacturer's instructions (Fig. 3) (see Note 23).

3.4.4. Determining Expression of Antigen Presentation and Activation Molecules on Human MDDC by Flow Cytometry

Follow the protocol described in steps 1–4 of Subheading 3.2, except using antibodies to CD25, CD40, CD83, CD86, HLA-DR, and HLA-Class I to label the cells.

3.5. Interactions of Human Naïve T Cells with Dendritic Cells

A functional characteristic that defines DCs is their ability to stimulate T cells. Here, we describe methodologies to determine the proliferation and cytokine production from T cells upon stimulation with mature MDDC. We also describe a method to assess DC driven polarization of naïve T helper responses, adapted from a protocol originally described by de Jong et al. (21).

1. Stimulate MDDC with bacteria for 18–24 h as described in Subheading 3.3. Transfer DCs to 15 mL conical tubes, add 5–10 mL of RPMI, and then centrifuge at 300×*g* for 10 min at RT. Count the cells with a hemocytometer using trypan blue

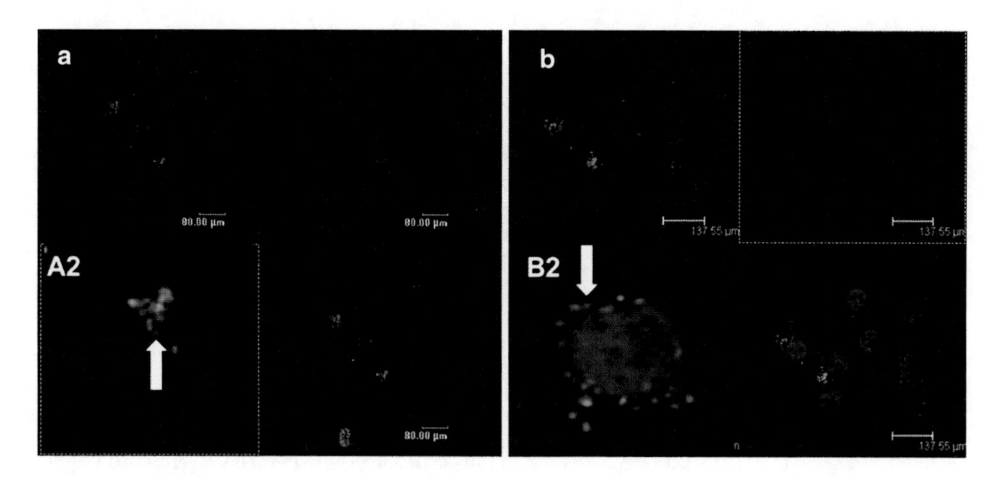

Fig. 3. Adherence and internalization of *N. meningitidis* into human MDDC studied by confocal microscopy. GFP (*green*) expressing bacteria (MOI 100) were incubated with MDDC for 4 h. Non-permeabilized cells were stained with mouse anti-PorA antibody and goat antimouse conjugated ALEXA fluor 568 (*red*) and nuclear/DNA stain To-Pro3 (*blue*). *Panel A* shows results with *live* bacteria and *Panel B* with *fixed* bacteria. Internalized bacteria are only stained *green* whereas external bacteria are double stained (*purple* in merged images, *bottom right panels*, *red* in single channel view, *top right panels*). Note that fixed organisms are readily internalized by MDDC, whereas very few live bacteria are found within the cells. *Panel A2* inset shows an expanded view of a single MDDC with a cluster of *purple* stained live bacteria (external). *Panel B2* inset shows a single MDDC with mainly internalized fixed bacteria (*green*) with occasional adherent external bacteria (*purple*).

as described in step 4 of Subheading 3.1, in order to discriminate live and dead cells. This is to take into account any cell loss during the incubation period. Next, suspend the cells in T cell medium supplemented with 10 IU/mL of IL-2.

2. Prepare a human PBMC suspension as described in steps 1–3 of Subheading 3.1 and then isolate pure T cells from these PBMC using the CD3 MACS beads separation kit according to the manufacturer's instructions. Co-culture the pure T cells with stimulated and unstimulated MDDC in 96-well flat-bottomed plates in a total volume of 200 μL: typically, 10^4 MDDC are included in co-culture with a range of T cell concentrations, depending on the T cell/MDDC ratio. It is important to optimize the T cell/MDDC ratio for each set of experiments.
3. Incubate the co-cultures for 5 days at 37°C in a humidified incubator with 5% (v/v) CO_2.
4. To assess for T cell proliferation, label the T cells with CFSE as follows:
   (a) Suspend the T cells in RPMI 1640 between 1 and 2×10^7 cells/mL in the presence of 10 μM CFSE. Incubate the cells for 5–10 min at 37°C (see Note 24).
   (b) Stop the labeling reaction by adding 20–30 mL of cold RPMI 1640 containing 10% (v/v) FCS. Centrifuge the cells at $300 \times g$ for 10 min and wash the cells a further 2 times to remove any excess CFSE (see Note 25).

(c) Count the cells and use them for the T cell/MDDC co-culture experiments. Include PHA (5 ug/mL) in co-culture as a positive control for T cell proliferation.

5. To assess for intracellular cytokine production from proliferating cells after 5 days of culture, follow the method described in Subheading 3.4.1 for intracellular staining using fluorescently labeled mAb (see Note 26), except that 6 h prior to staining, add 10 ng/mL of PMA and 1 μg/mL of ionomycin in the presence of 10 μg/mL of brefeldin A (see Note 27).
6. To assess the phenotype of proliferating T cells, use surface staining with fluorescently labeled mAbs as described in Subheading 3.2. The molecules of interest include CD4 (T helper cells), CD45RA (naïve T cells), CD45RO (memory T cells), and CD25 (T regulatory cells).
7. To assess T cell polarization:

(a) Treat MDDC with microbial components for 18–24 h as described in Subheading 3.3. Also, stimulate MDDC with a variety of control reagents known to induce T helper cell polarization: these include 100 ng/mL of *E. coli* LPS (mixed Th1 and Th2), 100 ng/mL of *E. coli* LPS + 10 μg/mL prostaglandin E2 (Th2), and 20 μg/mL of Poly I:C (Th1).

(b) Following stimulation for 24 h, transfer the cells to 15 mL conical tubes and add 5–10 mL of RPMI and then centrifuge at 300 × *g* for 10 min. Count the cells as described in step 4 of Subheading 3.1 (see Note 28).

(c) Isolate naïve T cells (CD4+ CD45RA+) from human PBMCs through negative selection by MACs magnetic bead separation, according to the manufacturer's instructions. The population of naïve CD4 cells should be at least 95% pure and this should be confirmed by flow cytometry with antibody staining to CD45RA and CD4. Follow the surface membrane protein staining method described in Subheading 3.2.

(d) Then, co-culture MDDC with naïve T cells in T cell medium in 200 μL total volumes in 96-well flat-bottom plates, for 5 days at 37°C in a humidified incubator with 5% (v/v) CO_2. Typically, use 10^4 MDDC with T cells numbers that vary depending on the MDDC/T cell ratio. For example, a MDDC/T cell ratio of 1:10 requires 10^5 T cells.

(e) In the case of autologous naïve T cell stimulation, it is common practice to include super antigens, such as SEB at this stage. Typically, use 100 pg/mL of SEB (see Note 29).

(f) On the fifth day, expand the cells into 24 well plates by suspending the cells present in each well of a 96 well plate into 1 mL of T cell medium and transferring this volume to 24 well plates. Incubate the cultures for a further 9 days at 37°C in a humidified incubator with 5% (v/v) CO_2.

(g) On the final day of culture, stimulate the T cells for 6 h with 10 ng/mL of PMA and 1 μg/mL of ionomycin in the presence of 10 μg/mL of brefeldin A.

(h) Assess the intracellular cytokine staining pattern as described in Subheading 3.4.1 to determine T cell polarization. The cytokines of interest include IFNγ and IL-4, which typically characterize Th1 and Th2 responses, respectively.

4. Notes

1. The quality of the FCS used can be critical, as some batches of FCS can result in premature activation of MDDC. It is therefore important to culture immature MDDC with different FCS batches to check that no activation is attributed to the FCS itself. This variability is probably due to the endotoxin content of different FCS batches. If in doubt endotoxin content can be assayed using the Limulus Amebocyte Lysate assay (Charles River).
2. It is critical to batch test IL-4 and GM-CSF as these cytokines are essential for inducing the correct phenotype of the MDDC.
3. In our experience it is often easier to interpret data from MDDC derived from a single human donor rather than using Buffy coats prepared from multiple donors. MDDC generated from Buffy coats are often less responsive or can give a mixed response, which is likely due to the natural variation observed with human donors. It is therefore important to repeat experiments with different donors to take into account this natural variation.
4. For this alternative method, PBMCs are suspended at 3×10^6 cells/mL in complete RPMI and then cultured in 6 well plates for 3 h at 37°C in an atmosphere of 5% (v/v) CO_2. A volume of between 3 and 4 mL of cells is added to each well. After 3 h, nonadherent cells, which include B and T cells, are removed from each well using a Pasteur pipette, taking extra care not to disrupt those cells that are adherent. Complete RPMI (3–4 mL per well) supplemented with 100 ng/mL of GM-CSF and 50 ng/mL of IL-4 is added to each well. Cells are then incubated for 5–7 days at 37°C in 5% (v/v) CO_2.
5. It is important to check the purity of the CD14 cell purification at this stage. This can be done by labeling an aliquot of cells with a fluorescent anti-CD14 antibody, before and after CD14 bead separation. Cells can then be analyzed by flow cytometry. Ideally, a purity of >97% is required to ensure that there is minimal lymphocyte contamination.

6. In order to see DC morphology (Fig. 1) it is often easier to immobilize the DC using fibronectin. Sterile plastic 24 well plates are coated with 50 μg/mL of fibronectin in PBS for 24 h at 37°C. DCs (1 mL, 5×10^6cell/mL) are allowed to adhere to the well for 30 min.
7. It is essential that the cells are kept on ice or at 4°C to prevent internalization of antibodies by Fc receptors expressed on MDDC. It is also important to include isotype controls for each monoclonal antibody to assess Fc receptor-mediated antibody uptake.
8. Cells were analyzed using a *FACS*calibur flow cytometer with a single argon laser which allows the detection of three fluorophores simultaneously. Typical fluorophore combinations include FITC, PE, and PerCP. DCs are first separated by forward and side scatter where they form a discrete population and this forms the collection gate. At least 5,000 events within this gate are collected for analysis.
9. Although we have focused in this chapter specifically on *N. meningitidis*, the protocols can also be broadly applicable to infection of DC with other bacteria or bacterial components.
10. For *N. meningitidis* H44/76, an OD of 1.0 at λ600 nm is equal to 1×10^9 bacteria/mL as determined by serial dilution and plating.
11. If live bacteria are being used in a study, it is important to confirm that the bacteria remain viable following FITC labeling. An alternative approach used by us and others is to use bacteria transfected with GFP.
12. The use of fixed bacteria can be useful for a number of reasons: for example, the bacterial concentration does not alter throughout the experiment and using fixed bacteria may be necessary for certain organisms that require containment facilities of level 3 or more. During a natural infection the host is likely to be exposed to both and live and dead organisms, so the response to dead bacteria does have some relevance during a natural infection. Responses of live and killed bacteria can be different, which is the case for *N. meningitidis* (12), so ideally both states should be compared.
13. DCs should be treated gently and with minimal handling as this can initiate activation. Cells should also be phenotyped to confirm they are immature MDDC as described in Subheading 3.2.
14. It is important to include appropriate positive and negative controls at this stage. Stimulation for 24 h with TNF-α (50 ng/mL) can be used as a positive control for MDDC maturation and activation. For a negative control, the diluent used for bacterial suspension is appropriate.

15. When using live bacteria it is often necessary to use a range of multiplicities of infection as bacteria can grow during MDDC infection. For example, in 6 h, *N. meningitidis* numbers can increase tenfold and this should be taken into consideration when comparing live and killed bacteria.
16. The concentration and the time of addition of brefeldin A was optimized for MDDC and in our hands was nontoxic to MDDC stimulated with *N. meningitidis*. The concentration and time of addition of brefeldin A must be optimized for different stimuli.
17. The concentration of antibody is important. All antibodies should be titrated for optimal staining conditions. If nonspecific binding of an antibody is a problem, then reduce the antibody concentration or the incubation time.
18. We usually use human cytokine ELISA kits from eBioscience, but other manufacturer's reagents are also suitable for use in this instance.
19. In phagocytosis experiments using flow cytometry, trypan blue staining should be used to assess the proportion of extracellular bacteria compared to intracellular bacteria.
20. It is important to take aliquots at different time points as cells from different donors vary in their ability to phagocytose. It is also worth noting that poor phagocytosis can occur as a result of the greater presence of mature MDDC at the start of the experiment, compared to immature MDDC.
21. Bacterial internalization can be blocked by using an inhibitor of actin polymerization such as cytochalasin D. Add the inhibitor 30 min prior to co-culture with bacteria. In our studies, 1 μg/mL of cytochalasin D is optimal to prevent phagocytosis as shown by flow cytometry and confocal analysis. At this concentration, other DC functions such as maturation and cytokine production (other than IL-10 and IL-12) are unaffected.
22. Internalized and surface-bound bacteria are differentiated using this staining method. Surface bound bacteria are FITC (green: emission λ520 nm) positive and are also counter-stained with Alexa Fluor 568 specific antibody to P1.7 PorA (red: emission λ603 nm). Therefore, extracellular bacteria appear yellow (red and green overlay) whereas intracellular appear green only. DCs are visualized using To-Pro3 which stains the nucleus blue (emission λ661 nm).
23. In order to identify internalized bacteria, at least ten optical sections (0.2–0.5 μm) spanning the entire DC were visualized by Leica confocal imaging software.
24. CFSE diffuses into the T cell, where its acetate group is cleaved by intracellular esterases found in the cytoplasm. This cleavage

yields a highly fluorescent protein (green) that remains in the cell. Cell division results in a reduction of CFSE fluorescence intensity.

25. Before continuing with the T cell/MDDC co-culture it is good practice to confirm that the CFSE labeling has worked by checking the cells for CFSE incorporation by flow cytometry.
26. T cell cytokines of interest include IL-2 (T cell proliferation inducing cytokine), IFNγ (Th1), IL-4 (Th2), and IL-17 (Th17).
27. PMA and ionomycin are used to re-stimulate T cells. PMA is a phorbol ester that activates protein kinase C (PKC) which is important for many cellular processes. Phorbol esters activate PKC because they resemble diacylglycerol. Ionomycin, an ionophore, is used in conjunction with PMA because PKC requires the presence of calcium ions to function.
28. Washing MDDC at this stage is a necessary step to remove excess antigen. If live bacteria have been used to stimulate the cells, it may be important to treat the MDDC with antibiotics, especially as MDDC/T cell co-cultures can last up to 14 days.
29. It is difficult to stimulate naïve T cell responses to study T cell polarization, as the frequency of antigen-specific T cells is very low. This can be overcome by using either an allogeneic (different MDDC and T cell donor) MDDC/T cell co-culture or an autologous MDDC/T cell co-culture, followed by SEB stimulation to further enhance stimulation.

Acknowledgements

The authors would like to thank the Meningitis Research Foundation for their financial support. The authors would also like to thank Professor Robin Callard, Dr Heli Uronen-Hansson, and Dr Jenny Allen for their roles in the development of these assays in the Infectious Diseases and Microbiology and Immunobiology Units at the Institute of Child Health, London, UK.

References

1. Banchereau J, Briere F, Caux C et al (2000) Immunobiology of dendritic cells. Annu Rev Immunol 18: 767–811.
2. Takeuchi O, Akira A (2010) Pattern recognition receptors and inflammation. Cell 140: 805–20.
3. Murphy K M, Stockinger B (2010) Effector T cell plasticity: flexibility in the face of changing circumstances. Nature Immunol 11: 674–680.
4. Coquerelle C, Mosser M (2010) DC subsets in positive and negative regulation of immunity. Immunol Rev 234: 317–34.
5. Sallusto F, Lanzavecchia A (1994) Efficient presentation of soluble antigen by cultured human dendritic cells is maintained by granulocyte/macrophage colony-stimulating factor plus interleukin 4 and down-regulated by tumor necrosis factor-alpha. J Exp Med 179: 1109–1118.
6. Caux C, Dezutter-Dambuyant C, Schmitt D et al (1992) GM-CSF and TNF-alpha cooperate in the generation of dendritic Langerhans cells. Nature 360: 258 –261.

7. Bernard H, Disis ML, Heimfeld S et al (1995) Generation of immunostimulatory dendritic cells from human CD34+ hematopoietic progenitor cells of the bone marrow. Cancer Res 55: 1099–1104.
8. Paczesny S, Li Y, Li N et al (2007) Effcient generation of CD34+ progenitor-derived dendritic cells from G-CSF-mobilized peripheral mononuclear cells does not require hematopoietic stem cell enrichment. J Leuk Biol 81: 957–67.
9. Ueno H, Schmitt H, Klechevsky E et al (2010) Harnessing human dendritic cell subsets for medicine. Immunol Rev 234: 199–212.
10. Connolly NC, Whiteside TL, Wilson C et al (2008) Therapeutic immunization with human immunodeficiency virus type 1 (HIV-1) peptide-loaded dendritic cell is safe and induces immunogenicity in HIV-1 infected individuals. Clin Vac Immunol 15: 284–292.
11. Villcock A, Schmitt C, Schielke S et al (2008) Recognition via the class A scavenger receptor modulates cytokines secretion by human dendritic cells after contact with *Neisseria meningitidis.* Microbes Infect 10: 10–11.
12. Jones H E, Uronen-Hansson H, Callard RE et al (2008) The differential response of human dendritic cells to live and killed *Neisseria meningitidis.* Cell Microbiol 9: 2856–2869.
13. Steeghs L, van Vliet SJ, Uronen-Hansson H et al (2006) *Neisseria meningitidis* expressing lgtB lipopolysacccharide targets DC-SIGN and modulates dendritic cell function. Cell Microbiol 8: 316–25.
14. Kurzai O, Schmitt C, Claus H et al (2005) Carbohydrate composition of meningococcal lipopolysaccharide modulates the interaction of *Neisseria meningitidis* with human dendritic cells. Cell Microbiol 7: 1319–1334.
15. Al Bader T, Jolley KA, Humphries HE et al (2004) Activation of human dendritic cells by the PorA protein of *Neisseria meningitidis.* Cell Microbiol 6: 651–662.
16. Uronen-Hansson H, Steeghs L, Allen J et al (2004) Human dendritic cell activation by *Neisseria meningitidis*: phagocytosis depends on expression of lipooligosaccharide (LOS) by the bacteria and is required for optimal cytokine production. Cell Microbiol 6: 625–637.
17. Al Bader T, Christodoulides M, Heckels JE et al (2003) Activation of human dendritic cells is modulated by components of the outer membranes of *Neisseria meningitidis.* Infect Immun 71: 5590–5597.
18. Unkmeir A, Kammerer U, Stade A et al (2002) Lipooligosaccharide and polysaccharide capsule: virulence factors of *Neisseria meningitidis* that determine meningococcal interaction with human dendritic cells. Infect Immun 70: 2454–2462.
19. Kolb-Maurer A, Unkmeir A, Kammerer U et al (2001) Interaction of *Neisseria meningitidis* with human dendritic cells. *Infect. Immun.* 69, 6912–6922.
20. Dixon GL, Newton PJ, Chain BM et al (2001) Dendritic cell activation and cytokine production induced by group B *Neisseria meningitidis*: interleukin-12 production depends on lipopolysaccharide expression in intact bacteria. Infect Immun 69: 4351–4357.
21. de Jong EC, Vieira PJ, Kalinski P et al (2002) Microbial compounds selectively induce Th1 cell-promoting or Th2 cell-promoting dendritic cells in vitro with diverse Th cell-polarising signals. J Immunol 168: 1704–1709.

Chapter 15

Hydrogen–Deuterium Exchange Coupled to Mass Spectrometry to Investigate Ligand–Receptor Interactions

Jessmi M.L. Ling, Leslie Silva, David C. Schriemer, and Anthony B. Schryvers

Abstract

A method for exploring protein–protein interactions using hydrogen/deuterium exchange coupled to mass spectrometry is described. The method monitors the exchange of backbone (amide) hydrogens in solutions of deuterated water that primarily occur on portions of the protein exposed to solvent. In the presence of a protein binding partner, regions that experience reduced exchange are either part of the protein–protein interaction interface or undergo conformational changes to reduce accessibility to solvent. This method has the advantage of being used under physiological conditions with unmodified proteins. In this chapter, we describe an approach suitable for probing interactions among relatively large proteins using conventional mass spectrometry systems. The interaction between human transferrin and the *Neisseria meningitidis* receptor protein, transferrin binding protein B, provides a challenging system as an example.

Key words: Hydrogen/deuterium exchange, Mass spectrometry, Human transferrin, Transferrin binding protein B, *Neisseria meningitidis*

1. Introduction

Hydrogen/deuterium exchange (H/DX) involves the exposure of protein to a solution containing deuterated water (D_2O) such that there is measurable exchange on accessible, labile hydrogens. The exchange rates for hydrogen bonded to carbon are effectively zero, while hydrogen in the side-chains exchanges too rapidly. The hydrogen of backbone amides exchange at a rate that is detectable in H/DX experiments (1). Only residues that are accessible to solvent, such as those on the surface of the protein or in solvent filled channels, will be subject to H/DX in the time course of an H/DX experiment. Proteolytic digestion of a protein exposed to

Myron Christodoulides (ed.), *Neisseria meningitidis: Advanced Methods and Protocols*, Methods in Molecular Biology, vol. 799, DOI 10.1007/978-1-61779-346-2_15,

H/DX followed by separation and analysis of the peptides by mass spectrometry allows identification of deuterated peptides when compared to analysis of peptides from the protein not exposed to D_2O. Thus, H/DX coupled to mass spectrometry (H/DX-MS) identifies peptides that are solvent exposed and is capable of monitoring changes in solvent exposure upon perturbations to the system. This feature is especially useful in analyzing protein–protein or protein–ligand interactions (2). Comparative H/DX-MS analysis of the protein in solution, with or without its binding partner, can result in identification of the sites of interaction.

The technical challenges associated with H/DX-MS relate to rapid generation and separation of peptides after H/DX in order to minimize the degree of back-exchange, exchanging hydrogen for deuterium in the absence of excess D_2O. These steps are normally carried out at low temperature and pH, thus providing a challenge for analysis by LC/MS. Conventional experiments in "bottom-up" protein MS benefit from long digestion periods (overnight) and extended chromatographic separation times (>60 min). In order to retain sufficient deuterium for analysis, digestion times are usually less than 5 min in H/DX-MS experiments, with chromatography complete in 10 min. This results in very complex sets of peptides entering the mass spectrometer at any given time. Large protein complexes can easily "overwhelm" conventional LC/MS systems, leading to poor sequence coverage and a reduced ability to localize binding sites.

The approach that was used for dealing with the large size of the partner proteins, human transferrin (hTf) and transferrin binding protein B (TbpB) from *Neisseria meningitidis* in this example, was to provide a rapid means of separating the proteins after H/DX. This was accomplished by engineering a site for enzymatic biotinylation on the N-terminus of one partner protein (TbpB) so that it could be rapidly separated from the non-tagged partner protein (hTf) under the low pH and temperature of the quenching conditions. Recombinant proteins that are expressed in the *Escherichia coli* cytoplasm such as TbpB will be endogenously biotinylated, whereas exported or secreted proteins can be biotinylated by recombinant BirA (3). Two additional challenges faced with the H/DX-MS analysis of the hTf used in this study are its resistance to protease due to the high number of disulfide bridges present and the presence of oligosaccharides. These were dealt with by optimizing the composition of the protease digestion buffer and enzymatic removal of the oligosaccharide chains that are described in the methods below. In spite of the challenges, the regions of hTf involved in binding meningococcal TbpB were successfully identified (4) and one might expect higher resolution with proteins that are more susceptible to proteolytic cleavage.

Improved resolution of the binding interface and maps of conformational changes may be achieved with enhancements in

LC/MS instrumentation and methods, but the relatively simple approach outlined in this chapter should be suitable for investigators with access to a standard mass spectrometry laboratory.

2. Materials

Unless otherwise stated, chemicals are purchased from Sigma-Aldrich.

2.1. Bacterial Strains, Culture Conditions and Expression of Recombinant TbpB

1. *E. coli* Top10 carrying a plasmid expressing full-length *tbpB* from *N. meningitidis* strain B16B6 (4). *E. coli* strain C41(DE3) for expression of *tbpB* (see Note 1). Bacterial strains are stored in 10% (v/v) glycerol in proteose peptone broth at −80°C.
2. GenElute™ Plasmid Miniprep Kit (Sigma-Aldrich).
3. Luria–Bertani (LB) base medium (Invitrogen): 10 g of peptone, 5 g of yeast extract, and 5 g of sodium chloride per liter of distilled water (dH_2O). Sterilize using an autoclave at 121°C for 15 min with 2.68 kg/cm^2 pressure.
4. LB agar medium (Invitrogen) contains LB base and 15 g per liter of agar. Sterilize using an autoclave at 121°C for 15 min with 2.68 kg/cm^2 pressure.
5. LB medium containing 100 μg/mL of ampicillin: add 1 mL of 1,000× ampicillin stock to 1 L of medium.
6. Recovery medium: LB medium containing 20 mM of glucose.
7. 1,000× ampicillin stock (100 mg/mL): dissolve ampicillin in dH_2O and sterilize through a 0.2 μm pore filter. Store 1 mL aliquots at −20°C until use.
8. Transformation buffer: 100 mM $CaCl_2$ in dH_2O. Sterilize using an autoclave at 121°C, for 15 min with 2.68 kg/cm^2 pressure and store the solution at 4°C.
9. Water bath set at 42°C.
10. Cell lysis buffer: 50 mM Tris–HCl, pH 8.0 containing 50 mM NaCl. Store at 4°C.
11. Centrifuge with refrigeration capability.

2.2. Enzymatic Deglycosylation of Human Serum Transferrin

1. Deglycosylating enzymes: Peptide-*N*-glycosidase (PNGase), and endoglycosidases (Endo-) F1, F2, and F3 (4–7).
2. Deglycosylation buffer 1: 100 mM of sodium phosphate (Na_2PO_4), pH 8.5.
3. Deglycosylation buffer 2: 100 mM of sodium acetate ($C_2H_3NaO_2$), pH 4.5.

4. Concanavalin A (ConA) sepharose (GE Healthcare).
5. ConA equilibration buffer: 50 mM sodium acetate, pH 6.9 containing 1 mM $CaCl_2$, 1 mM $MgCl_2$, and 1 mM $MnCl_2$.
6. Transferrin storage buffer: 50 mM Tris–HCl, pH 8.0 containing 1 M NaCl.
7. Human serum transferrin (1 mg/mL) in deglycosylation buffer 1.
8. 0.1 N NaOH.
9. Poly-Prep® chromatography columns, 2 mL bed volume, 0.8 × 4 cm (Bio-Rad).
10. Amicon Ultra-15 centrifugal filter device with Ultracel-10 membrane (Millipore).
11. Centrifuge with refrigeration feature that can accommodate 50 mL capacity tubes. A swinging bucket rotor allows more efficient application of the Amicon Ultra-15 centrifugal filter device.

2.3. Preparation of Immobilized TbpB-Transferrin Complex

1. Cell lysis buffer: 50 mM Tris–HCl, pH 8.0 containing 50 mM NaCl. Store at 4°C.
2. High salt wash buffer: 50 mM Tris–HCl, pH 8.0 containing 1 M NaCl. Store at 4°C.
3. Equilibration buffer: 10 mM Tris–HCl, pH 8.0, stored at 4°C.
4. Enzymatically deglycosylated human serum transferrin (40 mg/mL) prepared in cell lysis buffer and stored at 4°C.
5. Streptavidin-agarose (Invitrogen).
6. French pressure cell press (Thermo Spectronic).
7. Centrifuge with refrigeration feature, and capable of higher speeds to pellet cell debris. A suitable benchtop micro-centrifuge for volumes less than 2 mL.

2.4. Peptic Digestion and Generation of Peptide Identification List for Human Serum Transferrin

1. Immobilized pepsin (Thermo Fisher Scientific).
2. Transferrin dilution buffer: 10 mM Tris–HCl, pH 8.0.
3. Pepsin activation buffer: 0.1 M Glycine, pH 2.3 containing 250 mM tris(2-carboxyethyl)phosphine (TCEP)-HCl, and 30% (v/v) methanol. Prepare fresh buffer and chill on ice.
4. A micro-centrifuge.
5. Liquid chromatography (LC) columns. For example, C_{18} (Magic 200 Å, Michrom BioResources) beads packed in-house in fused silica capillaries (6.5 cm, 360 μm external diameter, 200 μm internal diameter). Columns of similar dimensions may be purchased commercially.
6. LC mobile phase A: 3% (v/v) acetonitrile, 0.05% (v/v) trifluoroacetic acid (TFA).
7. LC mobile phase B: 90% (v/v) acetonitrile, 0.05% (v/v) TFA.

8. A quadrupole time-of-flight mass spectrometer: for example, a QSTAR® Pulsar *i* quadrupole time-of-flight (QqTOF) mass spectrometer (Applied Biosystems), or later model.
9. An liquid chromatography (LC) system capable of delivering reproducible, low flow rates: for example, an Exigent 1D-LC with a loading pump, where gradients can be delivered at <10 μL/min and column loading at <30 μL/min.
10. Data-base search tool for identification of pepsin-generated peptides: for example, a site-license to Mascot (version 2.1 or later) (8) allowing the creation of customized small databases containing only the proteins present in the sample.

2.5. Hydrogen/Deuterium Exchange Coupled to Mass Spectrometry (H/DX-MS)

1. Materials for peptic digestion, liquid chromatography, and mass spectrometry as described in Subheading 2.4.
2. Enzymatically deglycosylated hTf (40 mg/mL stock) (see Note 2).
3. Streptavidin-agarose (Invitrogen).
4. Deuterium oxide (D_2O).
5. Quench buffer: 0.2 M Glycine, pH 2.3 containing 250 mM TCEP-HCl and 55% (v/v) methanol. Prepare fresh buffer and chill on ice (see Note 3).

2.6. Analysis of H/DX-MS Data

1. Software program to extract deuterium levels from peptides detected in the LC/MS data. Options include Hydra (9), the Deuterator (10), HeXicon (11), and HX-Express (12). Some approaches are dependent on the instrument platform, and thus available options should be considered carefully.
2. Microsoft Excel or other software, offering statistical functions and plotting.
3. Molecular visualization software, for mapping deuterium levels, changes in deuterium levels and regions of sequence coverage for protein systems under study. For example, the PyMOL molecular visualization system (version 1.3, Schrödinger, LLC.) and Swiss PDB Viewer (http://www.expasy.org/spdbv/).

3. Methods

H/DX-MS for protein interface detection involves determining the amount of deuterium incorporation for a given protein, in the presence and absence of its binding partner. The protein or a protein complex is exposed to D_2O-containing buffer to allow exchange with labile hydrogens. The exchange is rapidly suppressed by quenching with a low pH buffer at low temperature. The quenching step also removes the binding protein and the proteins digested.

The resulting peptides are analyzed by mass spectrometry for changes in mass due to deuteration (13). As outlined in the introduction, the analysis of the interaction between hTf and meningococcal TbpBs posed particular challenges due to the relatively large size of the proteins and the stability or protease-resistance of hTf. We have found that a particularly useful strategy for reducing the complexity of the resulting digest and optimizing digestion conditions for a particular component, involves engineering a tag/capture approach that is tolerant of the quenching conditions used in the work-up prior to digestion and LC/MS analysis.

One of the few tag/capture systems that are tolerant of the quenching conditions is the biotin–streptavidin interaction. A simple and convenient approach involves engineering an optimized biotinylation sequence at the N-terminus of the recombinant protein of interest. This results in enzymatic biotinylation of the recombinant protein during expression in *E. coli* so that it can be captured by covalently immobilized streptavidin. The method involves labeling of the non-biotinylated protein (free, or in the presence of its biotinylated binding partner), followed by quenching of the labeling and retention of the biotinylated protein on covalently immobilized streptavidin. As this interaction is resistant to release in the low pH of the quenching solution, and the protein–protein interaction under study typically will not be, the biotinylated protein can be effectively removed from the analysis. This strategy could be alternately applied to different protein partners if they are expressed in the *E. coli* cytoplasm or to proteins expressed in other systems, by enzymatic biotinylation with recombinant BirA.

To demonstrate this strategy, we describe an application involving the TbpB-transferrin interaction. Both transferrin and TbpB are large proteins (~80 kDa each) that challenge digestion strategies and analysis by conventional LC/MS systems. We addressed these challenges by focusing on transferrin, *i.e.* to identify the effects of TbpB binding. We have developed a method to capture transferrin with immobilized TbpB, and then rapidly release transferrin during the quench stage for digestion and peptide analysis.

3.1. Bacterial Strains, Culture Conditions, and Expression of Recombinant TbpB

1. Purify the expression plasmid encoding the full-length TbpB (*N. meningitidis* B16B6) from *E. coli* Top10 using the plasmid DNA extraction kit, according to the manufacturer's instructions. Elute the plasmid in sterile distilled water and this will be used to transform *E. coli* C41(DE3).
2. Prior to transformation, prepare competent *E. coli* C41(DE3) (see Note 4) as follows:
   (a) Grow *E. coli* C41(DE3) from frozen stock onto a LB agar plate overnight at 37°C and on the following day inoculate 3 mL of LB broth with a colony of bacterium. Grow this culture overnight at 37°C and on the following day inoculate 100 mL of LB broth with 1 mL of the overnight culture.

(b) Incubate the 100 mL culture at 37°C with shaking at 200 rpm for 2.5 h (to mid-log growth phase).

(c) Cool the culture on ice for 10 min and then transfer it to a chilled centrifuge tube.

(d) Centrifuge at 600 × *g* for 5 min in a refrigerated centrifuge to collect the cells. Carefully decant the supernatant (this is waste).

(e) Suspend the cell pellet gently in 50 mL of cold transformation buffer. When the cells are fully suspended, chill the suspension on ice for 5 min.

(f) Repeat the centrifugation (step 2d) and suspend the cell pellet gently in 25 mL of cold transformation buffer. When the cells are fully suspended, chill the suspension again on ice for 5 min.

(g) Repeat the centrifugation (step 2d) and finally suspend the cell pellet gently in 2.5 mL of cold transformation buffer. Transfer 50 μL aliquots of this competent cell suspension to chilled 1.5 mL micro-centrifuge tubes and store these on ice.

3. Transform *E. coli* C41(DE3) as follows:

(a) Add 1 μL of plasmid DNA (50–100 ng) prepared as described in step 1 to an aliquot of competent cells. Mix well.

(b) Chill the mixture on ice for 15–30 min.

(c) Quickly transfer the tube to a 42°C water bath without shaking the tube. Heat shock the cells for 45 s and return the tube back on ice for 2 min.

(d) Add 250 μL of recovery medium and incubate the cells in a shaking incubator at 37°C and 200 rpm for 45 min.

(e) Plate 50 μL aliquot onto LB agar containing 100 μg/mL of ampicillin and incubate the plate overnight at 37°C. This step shows the efficiency of transformation (a lawn of cells) and the presence of any potential contaminant that is also ampicillin resistant.

(f) Transfer the remaining 250 μL of transformed cells to 25 mL of LB broth containing 100 μg/mL of ampicillin. Incubate the culture overnight in a shaking incubator at 37°C and 200 rpm.

4. Express recombinant TbpB protein as follows:

(a) Inoculate 1 L of LB broth containing 100 μg/mL of ampicillin with the 25 mL overnight culture of transformed *E. coli* C41(DE3).

(b) Incubate the culture for 20 h at 37°C with shaking at 200 rpm, which produces sufficient fusion protein for the H/DX-MS experiment (see Note 5).

(c) Harvest the cells by centrifugation at 7,000 × *g* for 10 min at 4°C, and then suspend the cell pellet in 30 mL of cell lysis buffer.

3.2. Enzymatic Deglycosylation of Human Serum Transferrin

hTf contains two N-linked carbohydrate chains that can vary in size (14). Possible heterogeneity of these chains and the nature of the carbohydrate side-chains can complicate the identification of peptides by mass spectrometry. As the carbohydrate chains are not involved in the binding with TbpB (15), we opted to enzymatically deglycosylate commercial preparations of hTf. The deglycosylating enzymes were prepared in-house from recombinant *E. coli* (4), but similar enzymes may be obtained commercially.

1. Add 0.5 mL each of PNGase F and EndoF1 enzymes to 40 mL of human serum transferrin (at a concentration of 1 mg/mL in Deglycosylation buffer 1). Incubate the mixture overnight at room temperature.
2. Exchange the buffer to Deglycosylation buffer 2 by three cycles of concentration and dilution with an Amicon Ultra-15 centrifugal filter device, according to the manufacturer's instructions (see Note 6).
3. Add 0.5 mL each of EndoF2 and EndoF3 enzymes to the treated transferrin solution and incubate the mixture overnight at room temperature.
4. Change the buffer of the digestion mixture to the ConA equilibration buffer using an Amicon Ultra-15 centrifugal filter, according to the manufacturer's instructions.
5. Add 8 mL of ConA equilibration buffer to the reservoir to dilute transferrin, and then transfer the digested transferrin mixture to a clean 15 mL tube.
6. Remove transferrin that still contains carbohydrate residues, as follows:
   (a) Hydrate 1.5 g of ConA sepharose in 5 mL of ConA equilibration buffer.
   (b) Pack a ConA sepharose column by applying the slurry to a Poly-Prep® chromatography column. Allow the buffer to drain by gravity until the meniscus of the buffer reaches the top of the column.
   (c) Equilibrate the column with 5 column volumes of ConA equilibration buffer.
   (d) When the meniscus of the buffer reaches the top of the column, apply the digested transferrin mixture.
   (e) Collect the follow through and 2 column volumes of ConA equilibration buffer to recover dghTf (see Note 7).
7. Change the ConA equilibration buffer containing dghTf to the transferrin storage buffer and concentrate to 40 mg/mL using an Amicon Ultra-15 centrifugal filter device. Store dghTf at 4°C (see Notes 2 and 6).

3.3. Preparation of Immobilized TbpB-Transferrin Complex

Immobilized TbpB-transferrin complex allows for the rapid release of transferrin, which is then retained for analysis. To prepare aliquots of this complex for storage:

1. Lyse the cells expressing the different TbpB species by passing the cell suspension through a French pressure cell. Repeat this step twice, or more, until the cell lysate runs clear and is not viscous.
2. Centrifuge the resulting cell lysate at 40,000 × *g* for 20 min at 4°C to remove cell debris. Divide the supernatant into 5 aliquots (~6–8 mL each), and store at −20°C until use. An aliquot of expressed TbpB species is used for immobilization by streptavidin-agarose.
3. Equilibrate 150 μL of streptavidin-agarose (300 μL of a 50% (v/v) slurry) by washing the resin three times with 500 μL of cell lysis buffer (see Note 8).
4. Mix the equilibrated streptavidin-agarose with a single aliquot of cell lysate containing expressed TbpB species. Incubate the mixture for 1 h with gentle rocking at 4°C.
5. Collect the streptavidin-agarose by centrifugation (1,000 × *g* for 5 min at 4°C). TbpB is captured by the resin via its biotinylated tag.
6. Remove unbound TbpB from the resin by washing three times with high salt wash buffer. Centrifuge (1,000 × *g* for 1 min at 4°C) to gently pellet the resin between each wash.
7. Add 2 μL of deglycosylated hTf (dghTf, 40 mg/mL) and chill the mixture on ice for 1 h with gentle agitation every 10 min.
8. Wash the resin eight times with 1 mL of high salt wash buffer for each wash, in order to remove unbound dghTf. Centrifuge (1,000 × *g* for 1 min at 4°C) to gently pellet the resin between each wash.
9. Change the buffer of the resin by washing it five times with 1 mL of equilibration buffer.
10. Remove as much of the buffer after the final wash and then restore the resin to 50% slurry in equilibration buffer.
11. Transfer 60 μL aliquots to 5 microfuge tubes (30 μL bed volume), and store on ice until use.
12. Prepare the equivalent 5 aliquots of streptavidin-agarose resin (30 μL bed volume) that have been washed three times with equilibration buffer. Add sufficient dghTf to match the concentration of immobilized dghTf preparations, which will provide comparable signal strengths following LC-MS. This amount was 200 μg of dghTf per aliquot (10 μL of a 20 mg/mL concentration of dghTf). These aliquots provide the samples for free, unbound transferrin for H/DX-MS.

3.4. Peptic Digestion and Generation of Peptide Identification List for Human Serum Transferrin

This step allows for the optimization of peptic digestion conditions, which will be used for both generating the peptide identification list as well as the H/DX-MS experiment. Any modification to the peptic digestion conditions must be monitored and assessed by tandem mass spectrometry. The main objective of this step is to generate a list of peptides that is representative of the sequence coverage attainable under labeled conditions. Recursive independent data acquisition (IDA) experiments are performed to obtain maximal sequence coverage.

1. Add 0.5 mL of pepsin activation buffer to 50 μL of immobilized pepsin (bed volume). Invert several times to gently mix the slurry.
2. Centrifuge at 600 × *g* for 15 s to pellet the pepsin resin. Remove as much supernatant as possible and repeat the wash with pepsin activation buffer twice more.
3. Next, add 50 μL of chilled dghTf (1 mg/mL in transferrin dilution buffer) to 50 μL of chilled pepsin activation buffer (see Note 9).
4. Transfer the dghTf solution to the pepsin resin and tap gently to mix the contents.
5. Maintain the mixture in ice-water (4°C) for 2 min, then spin the mixture at 600 × *g* for 15 s to pellet the pepsin resin and remove 90 μL of the digested dghTf, taking care not to include any resin.
6. Inject 15 μL of digested dghTf onto the C_{18} column, at a flow rate of <30 μL/min with mobile phase A.
7. Perform a reverse phase chromatography by separating the peptides on the column using a short gradient of 5–90% mobile phase B (<10 min.).
8. Collect MS/MS spectra on peptides eluted into the mass spectrometer, using settings for data-dependent acquisition. To maximize coverage, multiple LC-MS/MS runs may be required. For most instruments, peptide identification should be confirmed by such fragmentation spectra rather than mass determinations alone. Peptide assignments are readily achieved with conventional data-base search tools such as Mascot, using standard settings. Note, however, that singly charged peptides should also be included in the search – these are normally excluded in typical proteomics applications.
9. Assemble a list of confirmed peptide identifications, retaining both the peptide charge state and retention time. In Hydra, for example, this list permits rapid extraction of deuteration data from the multiple LC-MS datasets obtained during H/DX experiments.

3.5. Hydrogen/Deuterium Exchange Coupled to Mass Spectrometry

Each H/DX-MS experiment consists of two samples, the free and complexed protein, dghTf and TbpB-bound dghTf in this example. Replicate data collections of each sample are generated, in paired runs to avoid error arising from drift. Upon addition of D_2O, event timing becomes critical. If automation robotics are not available (such as from Leap Technologies, Inc.), care should be taken to avoid differences between runs greater than ~10 s. This would lead to back-exchange that differs between samples and the generation of meaningless data. Preliminary replicates of the free protein should be carried out, in order to establish the reproducibility of the workflow and to determine in advance the precision of the method.

1. Prepare all materials for all replicate samples in advance:
   (a) Wash a 100 μL bed volume of immobilized pepsin three times with 1 mL of pepsin activation buffer. Centrifuge (1,000×g for 1 min at 4°C) to gently pellet the resin between each wash. After the final wash, replace sufficient buffer to yield 10 aliquots of 20 μL, each containing 50% of pepsin slurry. Place the tubes on ice until use.
   (b) Prepare aliquots of free and bound dghTf samples, as described in Subheading 3.2.
2. Working with one sample at a time, expose each sample to 15–50% D_2O to label the proteins (see Note 10). If large percentages of D_2O are used, be mindful of the buffer dilution that results. When adding D_2O to the immobilized complex, part of the volume will be occupied by the beads. To ensure that an equivalent percentage of D_2O exists between free and complexed protein, one method involves the addition of an equivalent volume of bead slurry for the free protein sample. Since the free protein is not biotinylated, it will not be retained. Note that imprecise formulation of the percent D_2O solution represents the largest source of error in these experiments.
3. Incubate the sample in D_2O solution at room temperature for either a single time period or conduct a kinetics analysis by generating multiple samples at time-points spanning 1–1,000 min. A single time-point (e.g., 5 min.) will often be sufficient for interface detection, provided that measurable levels of deuteration are detected in all peptides. Upon completion of incubation, pellet the streptavidin resin by a 5 s pulse spin on a micro-centrifuge. Remove and discard the supernatant.
4. Transfer 30 μL of ice-cold quench buffer to the sample, and chill on ice for 90 s. This step simultaneously slows the hydrogen–deuterium back-exchange, while dissociating bound dghTf from immobilized TbpB. It is important that all subsequent steps are performed on ice to keep back-exchange of deuterium to hydrogen at a minimum.

5. Pellet the resin again with a 5 s spin on a micro-centrifuge and then transfer 15 μL of the supernatant containing dghTf to the prepared immobilized pepsin resin.
6. Incubate dghTf in pepsin on ice for 3 min. Separate the digested dghTf from the immobilized pepsin with a 5 s spin in a micro-centrifuge (see Note 11).
7. Immediately remove 10 μL of the supernatant containing dghTf peptides and inject it into the LC-MS system for analysis. Operate the LC-MS system as per the sequencing phase, except that the mass spectrometer is run in MS mode rather than MS/MS mode (see Note 12).
8. Export the raw LC-MS data in a format supported by the deuteration quantitation software. For example, Hydra supports .wiff and .xml file formats.

3.6. Analysis of H/DX-MS Data

The following method describes the use of Hydra in the analysis of dghTf mass spectral data. The masses of dghTf peptides are compared between the free and bound dghTf samples, whereby significant differences between the peptide masses suggest differences in deuteration that is attributable to protein–protein interaction or conformational changes.

1. Import the peptide identification list and all LC-MS datasets into Hydra and configure the peak detection parameters as described by Slysz and colleagues (9). Process all datasets and validate with the available viewer. Adjust parameters as necessary to improve the precision and/or accuracy of the deuterium levels extracted from the peptide sets and process again if necessary (see Note 13).
2. To identify significant differences between individual peptides across the two protein states (bound vs. free), a useful method permitting variable thresholding involves a determination of P values for each peptide, based on a simple Student's t test between the replicates from the bound and free states. However, to avoid over interpreting on the basis of P values, insignificant differences (based upon a threshold of $P > 0.05$, e.g.) are used to determine the noise distribution for deuterium differences. A confidence interval based on this distribution may then be set. Therefore, in order to identify a peptide having a significant difference between bound and free states, this interval must be exceeded and the P value must also be above a chosen threshold. Such data is conveniently displayed in two-dimensional plots of ΔD vs. $1 - P$ (Fig. 1).
3. Significant differences may then be mapped to existing structures (e.g., PDB 2HAU for apo-transferrin) using PyMOL or other molecular visualization software.

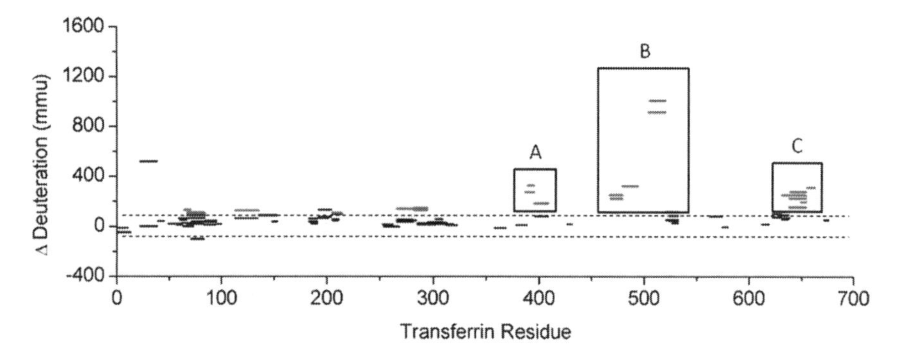

Fig. 1. Sequence plots demonstrating differential deuteration analysis of receptor-bound dghTf relative to free dghTf. The receptor in this dataset was a full-length TbpB derived from *N. meningitidis* B16B6. *Dashed lines* demarcate a 95% confidence, *bars* represent peptides with significant mass differences that are located above and below the *dashed lines*; *bars* that fall between the two *dashed lines* represent peptides with insignificant differences. *Boxes A, B,* and *C* represent discrete regions of contiguous sequence. Analysis was based on data from four replicates. (Reproduced from Ling et al. (4), with permission from Blackwell Publishing Ltd).

4. Notes

1. Sequences encoding mature TbpB from *N. meningitidis* B16B6 (leader sequence absent) was inserted downstream of a T7 promoter, biotinylation tag and maltose-binding protein in an expression plasmid, CV103 (an in-house vector). Expression of the gene requires *E. coli* strains with IPTG-inducible T7-RNA polymerase in the chromosome, such as BL21(DE3) (New England Biolabs) and C41(DE3) (16). The biotinylation tag allows the expressed protein to be biotinylated by the *E. coli* host, while the maltose-binding protein increases the stability and solubility of meningococcal TbpB that is expressed in high quantities.

   General methods for the construction of expression vector plasmids and for gene expression in heterologous hosts are outside the scope of the current protocol and can be found elsewhere in this book.

2. The high concentration for dghTf described here was more than sufficient as a stock solution from which to prepare all free dghTf and dghTf-TbpB samples for the H/DX-MS experiments. It is important to determine that your protein of interest at stock solution concentrations retains solubility and lacks aggregation or precipitation.

3. Via mass spectrometry, we have found that the composition of the quench buffer described here simultaneously removed dghTf from TbpB and denatured it to provide the highest sequence coverage of the protein. We recommend that you optimize the composition of the quench buffer for your protein.

As pH 2.3 is most favorable to reduce back-exchange of deuterium to hydrogen, other components of the quench buffer may be altered to optimize digestion. These include the choice and concentration of reducing agents (e.g., TCEP-HCl or urea), agents affecting protein structure (e.g., alcohols and salts) and by experimenting with different or combinations of proteases.

4. During the preparation of chemically ($CaCl_2$) competent *E. coli*, it is important to keep the cells chilled at all times. The culture volumes can be adapted depending on specific needs. At the final step, competent cells may be suspended in cold transformation buffer containing 10–15% (v/v) glycerol. This allows the cells to be stored in −80°C for up to 3 months, following a snap freeze step in either liquid nitrogen or ethanol-dry ice mixture. To use a previously frozen aliquot of cells for transformation, allow the cell suspension to thaw on ice prior to the transformation protocol.

5. Expression of any gene of interest in an expression plasmid should first be evaluated. This includes the addition or absence of an inducer, and assessing protein production by SDS–PAGE. In this study, we have found no difference in TbpB production from overnight cultures grown in the presence or absence of IPTG. Furthermore, affinity-binding assays showed that the TbpB produced were capable of binding to transferrin.

6. Buffers can also be exchanged by dialysis, followed by concentration of the transferrin solution using an Amicon Ultra-15 centrifugal filter device.

7. We would assess the deglycosylation process using SDS–PAGE and a western blot analysis. Loss of N-glycan residues results in a noticeable shift in molecular weight in SDS–PAGE. In a western blot analysis, ConA conjugated to horseradish peroxidase should not bind to dghTf compared to the undigested transferrin control (17).

8. Washing the resin includes gentle mixing of the resin in buffer, followed by centrifugation at 600 × *g* for 30 s, and removal of the supernatant. More buffer solution is then added for the second wash and the process repeated.

9. hTf contains 18 disulfide bonds (18). TCEP-HCl in the peptide activation buffer reduces dghTf, although not likely to completion given the time constraints of digestion.

10. The isotopic distributions become wider with increasing levels of deuterium (Fig. 2). Therefore, when working with large proteins (e.g., dghTf) or complexes, many peptides may overlap due to the wider isotopic distributions (19).

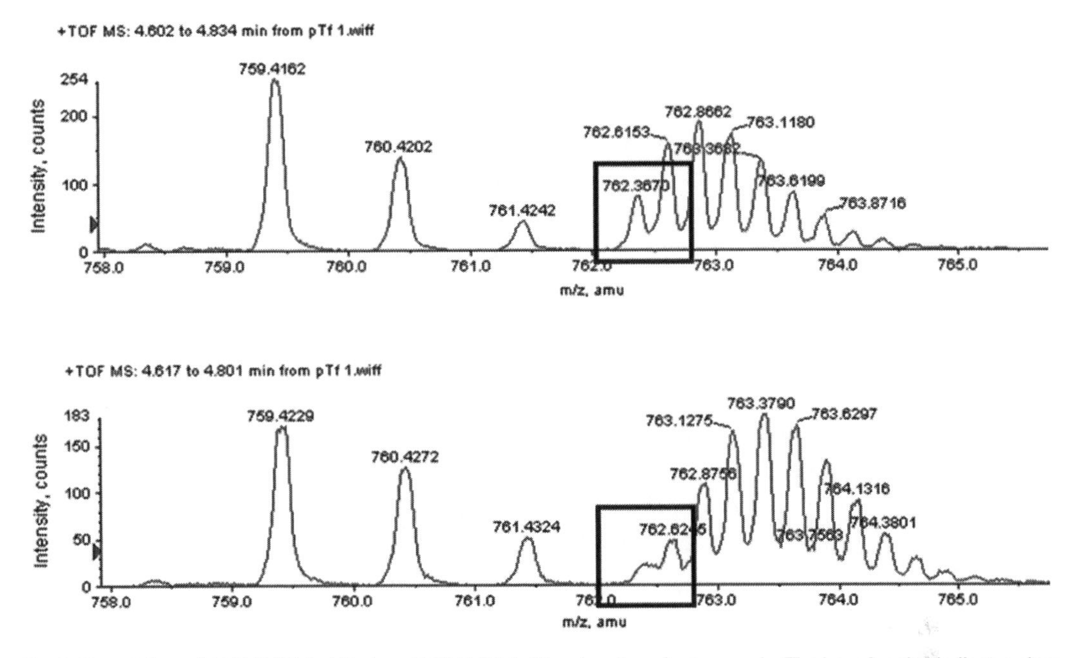

Fig. 2. Comparison of 10% D_2O label (*top*) vs. 25% D_2O label in a free transferrin sample. The boxed region indicates clear separation of the 759.4 *m/z* peptide from the 762.3 *m/z* peptide in the *top panel*, while the fourth isotope from the 759.4 *m/z* peptide interferes with the first isotope of the 762.3 *m/z* peptide in the *bottom panel*. The figure was generated using Analyst® LC/MS Software (Applied Biosystems).

11. An online pepsin digestion can be used in place of the slurry-based digestion. Pepsin, immobilized on POROS-AL20 beads (Applied Biosystems/SciEX), can be packed into a column and placed online prior to the C_{18} column. This removes the spin and transfer steps of the slurry-based approach and allows direct loading of the sample peptides onto the C_{18} column.
12. The LC system must be chilled and maintained at near 0°C. This is easily accomplished by submerging the LC column and connecting tubing in an ice-water slurry.
13. The initial configuration for the peak detection parameters in Hydra is determined from LC-MS data where the user will then define the *m/z* mass tolerance and width of a chromatographic peak in order to collate data from replicate LC-MS experiments. It is important to manually check the data as several factors, including variation in retention time, overlapping peaks, or undetectable peaks can cause the wrong peak (and mass) to be used in the analysis (9). Where necessary, peak detection parameters are adjusted and data re-processed.

Acknowledgements

The authors would like to thank Rong-hua Yu and Collin H. Shima for some of the sample preparation and data collection. This work was supported by a Canadian Institutes of Health Research Grant MOP 77558 (ABS).

References

1. Marcsisin SR, Engen JR (2010) Hydrogen exchange mass spectrometry: what is it and what can it tell us? Analyt Bioanalyt Chem 397: 967–972.
2. Engen JR (2009) Analysis of protein conformation and dynamics by Hydrogen/Deuterium Exchange MS. Analyt Chem 81: 7870–7875.
3. Cull MG, Schatz PJ (2000) Biotinylation of proteins in vivo and in vitro using small peptide tags. Applic Chimer Gene Hybrid Prot, Pt A 326: 430–440.
4. Ling JML, Shima CH, Schriemer DC et al (2010) Delineating the regions of human transferrin involved in interactions with transferrin binding protein B from *Neisseria meningitidis*. Mol Micro 77: 1301–1314.
5. Tarentino AL, Quinones G, Schrader WP et al (1992) Multiple endoglycosidase (Endo)-F activities expressed by *Flavobacterium meningosepticum* - Endo-F1 - molecular cloning, primary sequence, and structural relationship to Endo-H. J Biol Chem 267: 3868–3872.
6. Tarentino AL, Quinones G, Changchien LM et al (1993) Multiple endoglycosidase-F activities expressed by *Flavobacterium meningosepticum* Endoglycosidases-F2 and Endoglycosidases-F3 - molecular cloning, primary sequence, and enzyme expression. J Biol Chem 268: 9702–9708.
7. Kuhn P, Guan C, Cui T et al (1995) Active-site and oligosaccharide recognition residues of Peptide-N-4-(N-Acetyl-Beta-D-Glucosaminyl) Asparagine Amidase-F. J Biol Chem 270: 29493–29497.
8. Perkins DN, Pappin DJC, Creasy DM et al (1999) Probability-based protein identification by searching sequence databases using mass spectrometry data. Electrophoresis 20: 3551–3567.
9. Slysz GW, Baker CAH, Bozsa BM et al (2009) Hydra: software for tailored processing of H/D exchange data from MS or tandem MS analyses. BMC Bioinform 10.
10. Pascal BD, Chalmers MJ, Busby SA et al (2007) The Deuterator: software for the determination of backbone amide deuterium levels from H/D exchange MS data. BMC Bioinform 8.
11. Lou XH, Kirchner M, Renard BY et al (2010) Deuteration distribution estimation with improved sequence coverage for HX/MS experiments. Bioinform 26: 1535–1541.
12. Weis DD, Engen JR, Kass IJ (2006) Semi-automated data processing of hydrogen exchange mass spectra using HX-Express. J Am Soc Mass Spec 17: 1700–1703.
13. Zhang ZQ, Smith DL (1993) Determination of amide hydrogen-exchange by mass spectrometry - a new tool for protein-structure elucidation. Prot Sci 2: 522–531.
14. Fu DT, Vanhalbeek H (1992) N-Glycosylation site mapping of human serotransferrin by serial lectin affinity chromatography, fast atom bombardment mass spectrometry, and H-1 nuclear magnetic resonance spectroscopy. Analyt Biochem 206: 53–63.
15. Padda JS, Schryvers AB (1990) N-Linked oligosaccharides of human transferrin are not required for binding to bacterial transferrin receptors. Infect Immun 58: 2972–2976.
16. Miroux B, Walker JE (1996) Over-production of proteins in *Escherichia coli*: Mutant hosts that allow synthesis of some membrane proteins and globular proteins at high levels. J Mol Biol 260: 289–298.
17. Clegg JCS (1982) Glycoprotein detection in nitrocellulose transfer of electrophoretically separated protein mixtures using Concanavalin A and peroxidase - application to Arenavirus and Flavivirus Proteins. Analyt Biochem 127: 389–394.
18. Hall DR, Hadden JM, Leonard GA et al (2002) The crystal and molecular structures of diferric porcine and rabbit serum transferrins at resolutions of 2.15 and 2.60 angstrom, respectively. Acta Crystall Sec D-Biol Crystall 58: 70–80.
19. Slysz GW, Percy AJ, Schriemer DC (2008) Restraining expansion of the peak envelope in H/D exchange-MS and its application in detecting perturbations of protein structure/dynamics. Analyt Chem 80: 7004–7011.

Chapter 16

Visualising PAMP–PRR Interactions Using Nanoscale Imaging

Kathy Triantafilou and Martha Triantafilou

Abstract

The innate immune system utilises a set of receptors, called pattern recognition receptors (PRRs), in order to recognise specific molecular patterns or motifs called pathogen-associated molecular patterns (PAMPs) on invading pathogens. The toll-like receptor (TLR) family of proteins is an integral part of the mammalian innate immune system. We are now beginning to decipher which TLRs are involved in the recognition of particular microbial patterns, but questions remain as to the homo- and heterotypic associations that TLRs form and how these associations affect their activation. Technical advances in fluorescence microscopy has enabled us to investigate the functional associations of TLRs and other PPRs in living cells in response to different pathogens using non-invasive fluorescence imaging methods. In this chapter, we will describe some of the fluorescent imaging techniques, such as FRET and FRAP, that we employ in order to study PAMP–PRR associations.

Key words: Innate immunity, Pattern recognition receptors (PRRs), Toll-like receptors, Fluorescence resonance energy transfer (FRET), Fluorescence recovery after photobleaching (FRAP)

1. Introduction

Almost 20 years ago, Charles Janeway changed our view of the innate immune system, by publishing "Approaching the asymptote" as part of the Cold Spring Harbor Symposium on immune recognition (1). In this publication, he predicted that there would be molecules that were encoded in the *germ line* that would recognise the presence of molecules produced by broad classes of pathogens. He called these molecules pattern recognition receptors (PRRs) and the ligands that they recognise, pathogen-associated molecular patterns (PAMPs). Janeway's view was justified in the late 1990s with the discovery of the toll-like receptors (TLRs).

Myron Christodoulides (ed.), *Neisseria meningitidis: Advanced Methods and Protocols*, Methods in Molecular Biology, vol. 799, DOI 10.1007/978-1-61779-346-2_16,

The discovery of the TLRs proved that the innate immune system is actually highly specific in recognising microbial signatures. It has virtues that are equally specific and elaborate as the features of adaptive immunity, e.g., TLR4 recognises bacterial lipopolysaccharide (LPS) or endotoxin (2, 3), TLR2 recognise lipoteichoic acid (LTA) and peptidoglycan (4), TLR3 can sense double stranded viral RNA (5), TLR5 recognises bacterial flagellin (6), TLR7 (7) and TLR8 (8) sense single stranded viral RNA, and TLR9 can delicately distinguish between methylated DNA from host DNA and unmethylated DNA from microorganisms (9). Thus, the challenge has now changed from trying to identify the PRRs involved to trying to characterise the interactions of microbial PAMPs with PRRs, in order to understand the molecular mechanisms involved.

Technical advances in fluorescence microscopy techniques including laser scanning confocal microscopy, fluorescence resonance energy transfer (FRET), and fluorescence recovery after photobleaching (FRAP) have helped us in our understanding of PAMP–PRR associations as well as helped us decipher some of the protein–protein interactions that are involved in innate immune recognition. In this chapter, we will describe some of the fluorescence imaging techniques, such as FRET and FRAP, that we employ in order to study PAMP–PRR associations.

2. Materials

2.1. Preparation of Fluorescently tagged Antibodies

1. 0.5–1 mg of the appropriate antibodies against two molecules of interest, e.g., monoclonal antibody (mAb) against human TLR4 (HyCult) and mAb 26ic against CD14 (LGC Promochem).
2. Centricon Concentrators Y10 (Fisher Scientific).
3. 1 M sodium bicarbonate buffer, pH 8.0 (Sigma-Aldrich).
4. 1.5 M hydroxylamine, pH 8.5 (Sigma-Aldrich).
5. Disposable PD10 columns packed with Sephadex G25, column volume 13.5 mL (GE Healthcare).
6. Two fluorophores (5 mg/mL, dissolved in 100% dimethyl sulphoxide) for amine-specific labelling. NHS-ester or isothiocyanate efficiently labels antibodies and other purified proteins at primary amines (lysine side chains). Different fluorophore pairs that can be used include fluorescein isothiocyanate (FITC) with tetramethyl-rhodamine isothiocyanate (TRITC), Cyanin (Cy)3 with Cy5, and Alexa 488 with Alexa 555 (Molecular Probes Fluorescent Dyes and Probes, Invitrogen) (see Note 1).
7. Sterile phosphate buffered saline (PBS), pH 7.4.
8. UV-Visible Double Beam Spectrophotometer (Thermo Scientific) or similar.

9. Coomassie Plus (Bradford) Protein Assay Kit (Thermo Scientific).

2.2. Fluorescent Labelling of Cells Stimulated with PAMPs for FRET

1. Appropriate human cell line or primary cells of myeloid and non-myeloid lineage (see Note 2).
2. Cell-specific growth medium, e.g., Dulbecco's modified Eagle's medium or RPMI (Invitrogen) supplemented with 10% (v/v) foetal bovine serum (Invitrogen).
3. Haemocytometer for cell counting.
4. Serum-free medium (Invitrogen).
5. 8-well glass slides (Lab-Tek™ Chamber Slide™ System, Fisher Scientific) and glass coverslips.
6. Appropriate PAMP to be tested, e.g., LPS (100 ng/mL) (see Note 3).
7. PBS, pH 7.4.
8. PBS containing 0.02% (w/v) bovine serum albumin (BSA) and 0.02% (w/v) sodium azide (NaN_3).
9. 4% (v/v) Paraformaldehyde (PFA): prepare by dissolving 8 g of paraformaldehyde (Sigma-Aldrich) in 100 mL of distilled H_2O, heat to 60°C, and add a few drops of 1 M NaOH to aid dissolution of the fixative. Store at 4°C.
10. Antibodies against two different receptors conjugated to different fluorophore pairs, e.g., HTA-125-FITC (against TLR4) and 26ic-TRITC (against CD14).

2.3. Fluorescent Labelling of Cells Stimulated with PAMPs for FRAP

Use the same reagents as stated in Subheading 2.2 but with the omission of the PFA fixative.

2.4. FRET and FRAP Microscopy Measurements

1. Carl Zeiss LSM510 confocal microscope (with an Axiovert 200 fluorescent microscope) using a 1.4 NA 63× Zeiss objective and with a temperature-controlled stage; or equivalent.
2. Carl Zeiss LSM 2.5 analysis software.

3. Methods

The FRET technique allows the quantification of the association between fluorescently labelled molecules ≤10 nm apart. FRET is unique in providing signals sensitive to intra- and intermolecular distances in the 1–10 nm range. It works by capturing the interaction between labelled molecules when they are in close proximity, and then using this to calculate their distance relative to one another.

Since FRET only occurs if the fluorophore are ≤10 nm apart, its detection has the effect of extending the resolution of fluorescence microscopy to the molecular level. This uncovers cellular localisations that were previously undetectable. Due to its capabilities, FRET is increasingly used to visualise and quantify the dynamics of protein–protein interactions in living cells, with high spatio-temporal resolution.

FRET utilises the non-radiative energy transfer between two specifically selected different fluorophores that have been used to label two molecules of interest. These molecules could be present in a number of forms such as in solutions or in/on single living or fixed cells. Energy transfer occurs from an excited donor fluorophore to an appropriate acceptor fluorophore. The rate of energy transfer is inversely proportional to the sixth power of the distance between the donor and acceptor (10, 11). The efficiency of energy transfer (E) is defined with respect to r and R_0, the characteristic Förster distance by:

$$E = 1\left[1 + (r / R_0)6\right]$$

R_0 is the Förster distance of the donor/acceptor pair. This is a constant that represents the distance at which the energy transfer efficiency between the fluorophores is 50%. The Förster distance varies between different donor/acceptor pairs.

Due to the nature of FRET, the fluorophores have to be carefully selected in order to achieve an appropriate donor and acceptor fluorophore pair, referred to as a donor/acceptor pair. The emission wavelength from the donor has to be of the wavelength that excites the acceptor, the wavelength the acceptor absorbs, and therefore these have to overlap (Fig. 1).

The donor fluorophore is excited by a laser, which causes emission of light. If the acceptor fluorophore is ≥10 nm apart from the donor this energy will be released as light unaffected by the acceptor. However, if the acceptor is ≤10 nm in distance from the donor then two scenarios can present. Assuming a donor/acceptor pair, the emission wavelength of the excited donor fluorophore is sufficient to excitate the acceptor fluorophore, when these are ≤10 nm apart, causing emission of light from the acceptor. The emitted light of the acceptor will be of a set known band of wavelength and can be collected and quantified. Another result of FRET is a reduction in donor fluorophore emission on excitation. If donor and acceptor are ≤10 nm apart, a portion of the energy emitted from the excitated donor fluorophore will be absorbed by the acceptor fluorophore. This will have the effect also of reducing the light emission and excited state lifetime of the donor fluorophore for the energy has gone into acceptor excitation. The change in donor emission can also be used to quantify FRET.

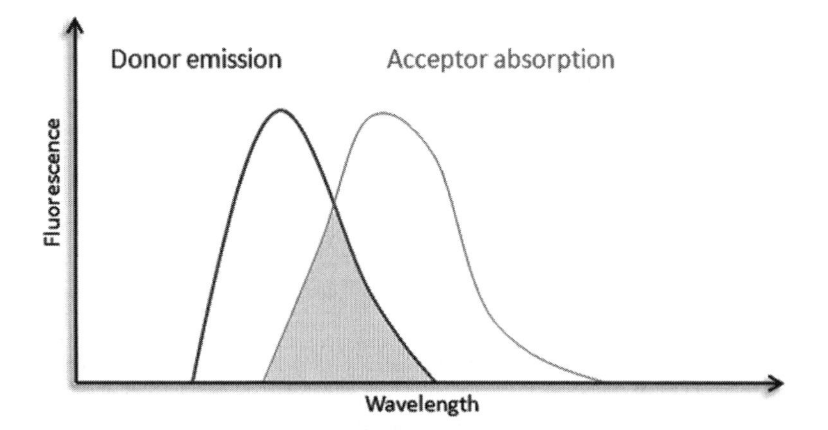

Fig. 1. Schematic diagram of donor emission and acceptor absorption spectra. There is overlapping (shown by light *gray area*) of donor emission (red channel) and acceptor absorption (blue channel) spectra indicating an appropriate fluorophore pair for fluorescence resonance energy transfer (FRET). The emission wavelength from the donor has to be of the wavelength that excites the acceptor, the wavelength the acceptor absorbs, and therefore these have to overlap. FRET will only occur when donor and acceptor fluorophores are ≤10 nm apart.

In our studies, FRET is calculated by measuring the change in donor (FITC or Cy3) emission following acceptor (TRITC or Cy5) bleaching (10, 12). By bleaching the acceptor, the fluorophore is effectively removed. If the acceptor is bleached (removed), then FRET will no longer occur between the pair and thus donor emission will no longer be sequestered. If the donor/acceptor pair are ≤10 nm apart then donor emission will increase after bleaching of the acceptor.

Although some technical complications must be carefully considered, FRET can be applied successfully to studying host–pathogen interactions. Care must be taken when choosing how to label the interacting molecules of interest. Two principal tools can be applied in labelling the molecules, either fluorophore-tagged antibodies or recombinant fusion proteins. The latter take advantage of the discovery of spontaneously fluorescent proteins, like green fluorescent protein (GFP). In this chapter, we describe a detailed protocol for labelling receptors of interest using fluorophore-tagged antibodies.

Another fluorescence imaging technique that can be used in order to study host–pathogen interactions is FRAP. FRAP is a biophysical method that measures the lateral diffusion of a population of molecules on a micrometre scale. FRAP is used to measure the dynamics of 2D molecular mobility, e.g., diffusion, transport, or any other kind of movement of fluorescently labelled molecules in membranes or in living cells. In this technique, a region of interest (ROI) on the cell membrane is bleached and then the recovery of fluorescence is monitored by taking a time series of images. The recovery of fluorescence results from the movement of unbleached fluorophores from the surroundings into the bleached area. The mean intensity in the ROI is plotted against time, where the recovery time (half-time) indicates the speed of this mobility, e.g., diffusion time. The diffusion coefficient gives us information as to how fast the molecules of interest are moving in the plasma membrane. In addition, FRAP provides us with a percentage of molecules that are recovering within the photobleached area. This gives us information on whether the molecules are diffusing freely and also on the existence of segregated plasma membrane regions or microdomains. Thus, FRAP provides us with the perfect tool in order to unravel the mechanism by which PRRs are engaged by their ligands. If, for example, TLR molecules associate with immobile or slowly diffusing proteins upon ligand stimulation, then their diffusion coefficient is expected to be lower than the one observed before stimulation. In addition, if the molecules accumulate in membrane microdomains then this should result in low percentage recovery or mobile fraction.

A related method is fluorescence loss in photobleaching (FLIP), which is the decrease/disappearance of fluorescence in a defined region adjacent to a repetitively bleached region. Thus, the same principle applies, which is to bleach a ROI, but the difference is that the ROI for fluorescence recovery is not monitored, but rather the adjacent region for loss in fluorescence. Like FRAP, FLIP is used to measure the dynamics of molecular mobility in membranes or in living cells, especially the ability of a molecule to move between different organelles in the cell.

3.1. Preparation of Fluorescently tagged Antibodies

1. Fluorescently tagged antibodies provide us with the flexibility of studying molecules of interest without the need of generating spontaneously fluorescent proteins. When investigating the associations between two molecules or their lateral mobility, choose antibodies that are directed against a non-functional domain of the protein of interest, which alleviates the possibility of the antibody inhibiting the interaction due to steric hindrance.
2. Once the antibodies of interest have been chosen, 0.5–1 mg of IgG for each molecule will be needed in order to conjugate it to the appropriate fluorophore. Concentrate the IgG to 1 mg/mL using Centricon concentrators and buffer exchange with sodium bicarbonate, as described by the manufacturer (see Note 4).
3. Prepare the fluorophores fresh in DMSO to a concentration of 5 mg/mL and mix with the concentrated IgG at a ratio of 5:1. Place the mixture in the dark for 1 h at room temperature.
4. Terminate the reaction by adding 10% by volume of 1.5 M hydroxylamine, pH 8.5 and incubate for a further hour in the dark (see Note 5).
5. Remove the unreacted succinimide by gel exclusion chromatography using a PD-10 column in the darkroom. Pre-equilibrate the PD-10 column with PBS, apply the reaction mixture with

gravity and elute in the same buffer. The labelled protein appears in the second or third 1 mL fraction, followed by the unlabelled fluorophore.

6. Determine the labelling ratio (fluorophore:protein ratio) by reading the absorbance of the fluorophore and determining the protein concentration of the IgG by Bradford assay, following the kit manufacturer's instructions or by measuring absorbance at $\lambda_{260\ nm}$ (see Note 6).

3.2. Fluorescent Labelling of Cells Stimulated with PAMPs for FRET

1. Seed human cells (either cell lines or primary cells) onto 8-well glass slides at a density of 10,000 cells per well, determined using a haemocytometer. Allow the cells to grow in cell-specific medium for 48 h to a confluency of ~60–80%.
2. Once the cells reach the required density, rinse with pre-warmed PBS (37°C) in order to remove all culture medium. Keep the buffer warm in order to avoid heat shocking the cells, which may cause them to detach.
3. Once the cells are washed and all the culture medium removed, add 100 μL of serum-free medium to the cells and the appropriate PAMP to be studied.
4. In order to study interactions between two molecules of interest, both donor and acceptor conjugated antibodies are added at the same time to each well. Add 100 μL of a mixture of donor conjugated antibody (FITC or Cy3) and acceptor conjugated antibody (TRITC or Cy5) to all wells.
5. Use one well as a control, with the cells labelled with the TRITC or Cy5 probe only, in order to determine the minimum time required to bleach the fluorophore. Bleach the fluorophore by continuous excitation using an appropriate filter set. Under these conditions, the donor (TRITC or Cy3) should not be bleached.
6. Rinse the cells twice with PBS containing 0.02% (w/v) BSA and 0.02% (v/v) NaN_3, prior to fixation with 4% (v/v) paraformaldehyde (PFA) for 15 min.
7. Following fixation with PFA, rinse the cells twice with PBS containing 0.02% (w/v) BSA and 0.02% (v/v) NaN_3 and remove the upper plastic housing of the chamber slide, leaving the gasket behind.
8. Gently remove the gasket using a razor blade and then add PBS followed by a glass coverslip. Seal the coverslip edges with nail varnish (see Note 7).

3.3. Fluorescent Labelling of Cells Stimulated with PAMPs for FRAP

1. Repeat steps 1–3 of Subheading 3.2.
2. In order to study the lateral mobility of a receptor of interest, add 20 μL of a fluorescently labelled antibody against the receptor to each well and incubate for 20 min at room temperature.

3. Rinse the cells twice with pre-warmed PBS (37°C) in order to remove any unbound fluorescently labelled antibody.
4. Remove the upper plastic housing of the chamber slide, leaving the gasket behind. Gently remove the gasket using a razor blade and then add PBS followed by a glass coverslip.

3.4. FRET Microscopy Measurements

1. Image the cells on a Carl Zeiss LSM510 confocal microscope (with an Axiovert 200 fluorescent microscope) using a 1.4 NA 63× Zeiss objective used in conjunction with Zeiss LSM 2.5 analysis software (see Note 8). Detect the donor and acceptor using the appropriate microscope filter sets. Using typical exposure times for image acquisition (less than 5 s), no fluorescence should be observed from a donor-labelled specimen using the acceptor filters, nor is acceptor fluorescence detected using the donor filter sets.
2. Collect an image of donor (FITC or Cy3) fluorescence in the presence of the acceptor ($FITC_{pre}$ or $Cy3_{pre}$), followed by an image of acceptor fluorescence ($TRITC_{pre}$ or $Cy5_{pre}$).
3. Next, irreversibly photobleach the acceptor by continuous excitation (typically requiring 1–2 min), and collect an image of the residual acceptor signal ($TRITC_{post}$ or $Cy5_{post}$) to ensure that complete photobleaching has occurred. This photobleaching step eliminates TRITC or Cy5 as an energy transfer acceptor.
4. Next, obtain a final image of the donor fluorescence ($FITC_{post}$ and $Cy3_{post}$). Subtract the dark-current contribution from each image and tabulate the fluorescence intensity from identical regions of interest (ROIs) on individual cells for each of these images, i.e., $Cy3_{pre}$, $Cy5_{pre}$, $Cy3_{post}$, and $Cy5_{post}$ (Fig. 2). The *E* of each ROI is then calculated as follows:

$$E(\%) \times 100 = 10{,}000 \times [(\text{Cy3postbleach} - \text{Cy3pre - bleach}) / \text{Cy3postbleach}]$$

5. Each set of FRET data shown is from a single experiment and is representative of two or more independent experiments. In a typical experiment, four to five fields of cells will be measured for each sample.
6. In order to determine whether the FRET measurements are due to random associations, carry out control experiments using a varying ratio of donor and acceptor as follows:
   (a) In all experiments, the concentration of the donor-labelled antibody is held constant (at 50 μg/mL) and the concentration of the acceptor-labelled antibody is varied to achieve the indicated donor-to-acceptor ratio (D:A).
   (b) Add the different antibody mixtures to different wells on the 8-well chamber slide and repeat the labelling procedure as described in step 4 of Subheading 3.2.

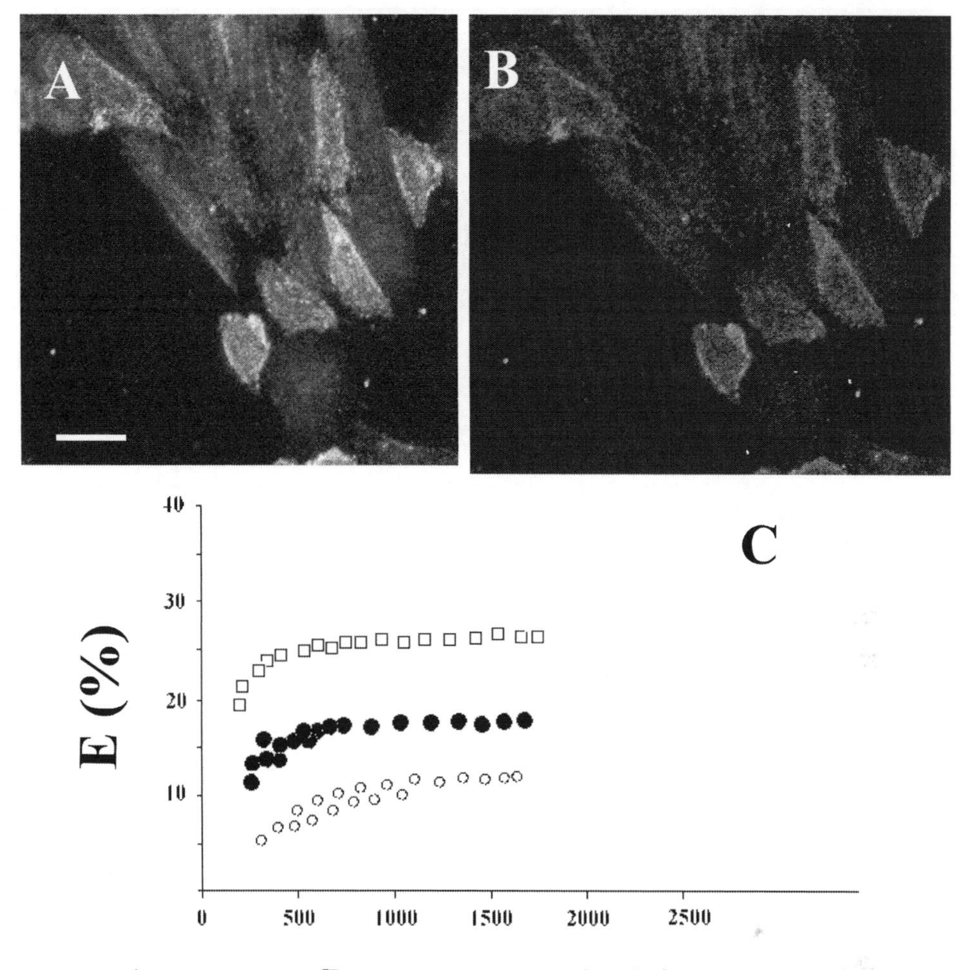

Fig. 2. Measurements of FRET. Energy transfer between receptors of interest can be detected from the increase in donor fluorescence after acceptor photobleaching. An image of the donor is acquired before (**a**), and after (**b**), bleaching the acceptor. In order to determine whether *E* is not due to random associations, the ratio of donor:acceptor (D:A) is varied and *E* is expressed as a function of fluorescence, for D:A of 1:1 (*closed circles*), 1:2 (*squares*), and 1:4 (*ellipse*) (**c**). When *E*% is found to be independent of acceptor surface density, to be sensitive to D:A ratio, and not to go to zero at low surface density, it suggests that the FRET values observed are due to clustered molecules and not random associations. The *scale bar* represents 10 μm.

(c) Following the labelling, image the cells using the confocal microscope, and the energy transfer (*E*%) being detected from each ratio is plotted against acceptor fluorescence. When *E*% is found to be independent of acceptor surface density, to be sensitive to donor:acceptor ratio and not to go to zero at low surface density, it suggests that the FRET values observed are due to clustered molecules and not random associations (Fig. 2).

3.5. FRAP Microscopy Measurements

1. Image the cells on a Carl Zeiss LSM510 confocal microscope (with an Axiovert 200 fluorescent microscope) using a 1.4 NA 63× Zeiss objective used in conjunction with Zeiss LSM 2.5 analysis software. Detect the fluorescently labelled molecule using the appropriate filter sets.
2. Place the slide containing labelled cells onto a temperature-controlled microscope stage and allow it to equilibrate for 5 min to the desired temperature (37°C) for FRAP measurements.
3. Focus the beam of the appropriate laser (e.g., if you are using a FITC-label then an argon ion laser will be used) onto the desired area on the cell, which we call the ROI.
4. Use one well as a control, with the cells labelled with the appropriate probe (i.e., FITC), in order to determine the minimum time required to bleach the fluorophore. Bleach the fluorophore by continuous excitation using the appropriate filter set.
5. Once the amount of time required to bleach the fluorophore is determined, use the laser to photobleach the ROI by continuous excitation for the desired amount of time.
6. Following photobleaching, obtain a time series of images in order to record the movement of new fluorescent molecules as they enter the photobleached area from the adjacent areas.
7. The raw data are collected by a detector, which measures the light through the microscope. These data are displayed on a computer screen and by the end of the experiment, a graph (recovery curve) will be produced that looks similar to Fig. 3. For each set of conditions, collect ten curves and average before analysis.
8. The percent recovery uses the formula: $(Y/X) \times 100 = \%$ recovery. In the curve, the percentage of fluorescence lost due to photobleaching is X and the amount of fluorescence that returned to the bleached area is Y.
9. The lateral mobility is determined by the slope of the curve. The steeper the curve is, the faster the recovery and therefore, the more mobile the molecules. The diffusion coefficient is calculated by the Zeiss LSM software using the non-linear least squares fitting to the equation:

$$F(t) = \frac{F(0) + F(\infty)(t / \beta\tau_D)}{1 + (t / \beta\tau_D)}$$

Where $F(t)$ is the measured fluorescence as a function of time t, $F(0)$ is the measured pre-bleach fluorescence and $F\%)$ is the value to which the fluorescence finally recovers. The parameter β is dependant on the depth of bleach and τ_D is the diffusion time. The diffusion coefficient, D is given by $D = \omega^2/4\tau_D$, where ω is the

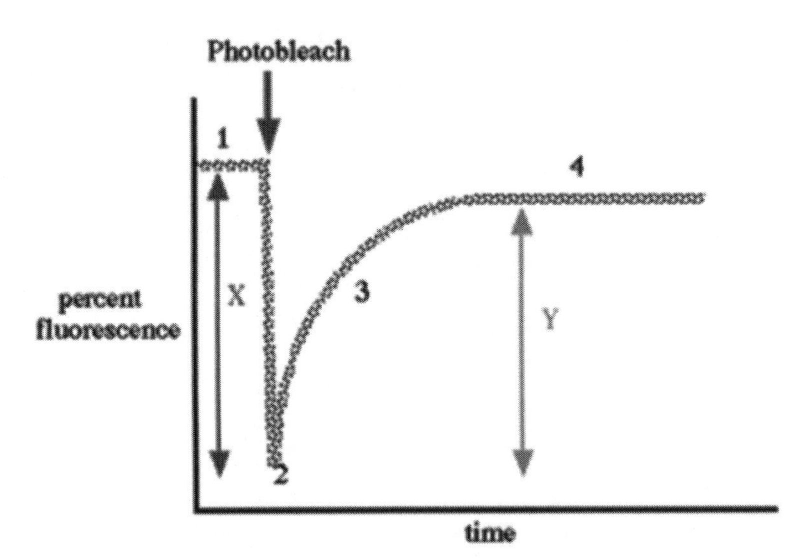

Fig. 3. Schematic diagram of FRAP measurements. Schematic diagram of the data collected during a FRAP experiment. A baseline of fluorescence is collected (1) before the photobleaching occurs (arrow) so that the amount of fluorescence is reduced significantly (2). Over time, the amount of fluorescence in the photobleached area increases as unbleached molecules diffuse into this area (3). Later, there is a stabilisation of the amount of fluorescence recovery (4) and a flat line is obtained. The percent recovery uses the formula: $(Y/X) \times 100 = \%$ recovery. In the diagram, the percentage of fluorescence lost due to photobleaching is *X* and the amount of fluorescence that returned to the bleached area is *Y*. In practice, the percent recovery almost never reaches 100%. The lateral mobility is determined by the slope of the curve. The steeper the curve is, the faster the recovery and therefore, the more mobile the molecules.

radius at $1/e^2$ the height of the illuminating laser spot. The analysis assumes the laser illumination to be circular with a Gaussian intensity profile.

4. Notes

1. FITC is the base fluorescein molecule functionalised with an isothiocyanate reactive group (–*N*=*C*=*S*) at one of two hydrogen atoms on the bottom ring of the structure. This derivative is reactive towards primary amine groups on proteins, peptides, and other biomolecules. NHS-fluorescein is activated with the *N*-hydroxy-succinimidyl-ester (NHS-ester) functional group. Compared to FITC, the NHS-ester derivative has greater specificity toward primary amines in the presence of other nucleophiles and results in a more stable linkage following labelling.
2. The technique can be applied to many cell types. It is preferred though that adherent, or semi-adherent, cell lines are used in order to eliminate any out of focus light that "rounded" cells will produce. In the past, we have used human monocytes, macrophages (12), meningeal cells (13), human embryonic kidney cells (HEK293), or Chinese Hamster ovary (CHO)

cells transfected with different receptors of interest (14–16), as well as fibroblasts and epithelial cells.

3. Different PAMPs that our group and others have studied in the past include bacterial LPS, LTA, lipoproteins, and toxins.
4. It is very important that the IgG is buffer-exchanged with sodium bicarbonate buffer of pH 8.5, as the reaction needs to take place at pH 8.5. We normally use centrifugal concentrators for buffer exchange in order to avoid lengthy dialysis steps. The IgG is added to the concentrator and during concentration, while the IgG is retained by an appropriate membrane, the dilution buffer can pass freely through the membrane. By diluting the concentrated IgG back to the original volume, the concentration of the original buffer is lowered, thereby exchanging the original buffering substance for sodium bicarbonate. If we cannot buffer exchange the IgG, then we should at least mix it with 20% by volume with 1.5 M sodium bicarbonate buffer, pH 8.5.
5. There is no need to incubate the mixture for 1 h in the dark when terminating the reaction with 1.5 M hydroxylamine pH 8.5. This step can be shortened to 15–30 min.
6. Quantitation of the protein:fluorophore ratio is essential for predicting the amount of probe necessary for an experiment and for controlling fluorescence intensity between experiments. The protein and the fluorophore content are determined separately, based on absorbance measurements and then these concentrations are expressed as a ratio in order to calculate the degree of labelling. The ratio is the number of dye molecules conjugated to each protein molecule. In order to determine the protein:fluorophore ratio you need to know the following: the extinction molar coefficient (ε) of the unlabelled protein, the A_{max} of the fluorophore which is the wavelength that the fluorophore absorbs maximally (λ_{max}), and the correction factor for the protein (CF), which is needed because absorbance at $\lambda_{280\ nm}$ (A_{280}) is used to determine protein concentration. However, the fluorophore also absorbs at that wavelength. Values for ε, λ_{max}, and CF of fluorophores that are routinely used for IgG labelling are listed in Table 1. The procedure for determining the labelling ratio is as follows:
   (a) Remove any unlabelled fluorophore by dialysis or PD-column.
   (b) Measure the absorbance of the conjugate at $\lambda_{280\ nm}$ using a spectrophotometer and 1 cm cuvette (dilute the conjugate if the absorbance is above a value of 2).
   (c) Measure the absorbance of the conjugate at the λ_{max} of the fluorophore (see Table 1).
   (d) Calculate the molarity of the protein using the formula below (taking $\varepsilon = 210{,}000\ M^{-1}\ cm^{-1}$ for IgG):

Table 1
Critical values for various fluorophores that are routinely used for labelling IgG

Fluorophore	Wavelength maximum (λ_{max}) (nm)	Extinction coefficient (ε) (M^{-1} cm^{-1})	Correction factor (CF)
Fluoroscein isothiocyanate (FITC)	494	68,000	0.3000
Tetramethyl-rhodamine-5 (and -6)-isothiocyanate (TRITC)	555	65,000	0.3400
Cy3	550	150,000	0.0800
Cy5	649	250,000	0.0500

$$\text{Protein concentration }(M) = \frac{A_{280} - (A_{\max} \times \text{CF})}{\varepsilon} \times \text{dilution factor}$$

$$\text{Moles dye per mole protein} = \frac{A_{\max} \text{ of the labeled protein}}{\varepsilon' \times \text{protein concentration }(M)} \times \text{dilution factor}$$

7. It is very important that for FRET experiments the cells should always be mounted in PBS and not in anti-fadant, as the goal of the experiment is to bleach the acceptor molecule and this is not achievable when mounted in anti-fade.
8. The choice of LSCM is not important and different manufacturers' machines can be substituted.

Acknowledgements

The authors would like to thank Prof. Michael Edidin for his advice while setting up the FRET technique in our lab. This work has been supported by the Wellcome Trust.

References

1. Janeway CA (1989) Approaching the Asymptote - Evolution and Revolution in Immunology. Cold Spring Harbour Symp Quant Biol: Immunological Recognition, Pts 1 and 2 54: 1–13.
2. Poltorak A, He XL, Smirnova I et al (1998) Defective LPS signaling in C3H/HeJ and C57BL/10ScCr mice: Mutations in Tlr4 gene. Science 282: 2085–2088.
3. Qureshi ST, Lariviere L, Leveque G et al (1999) Endotoxin-tolerant mice have mutations in toll-like receptor 4 (Tlr4). J Exp Med 189: 615–625.

4. Takeuchi O, Hoshino K, Kawai T et al (1999) Differential roles of TLR2 and TLR4 in recognition of gram-negative and gram-positive bacterial cell wall components. Immunity 11: 443–451.
5. Alexopoulou L, Holt AC, Medzhitov R et al (2001) Recognition of double-stranded RNA and activation of NF-kappa B by Toll-like receptor 3. Nature 413: 732–738.
6. Hayashi F, Smith KD, Ozinsky A et al (2001) The innate immune response to bacterial flagellin is mediated by Toll-like receptor 5. Nature 410: 1099–1103.
7. Lund JM, Alexopoulou L, Sato A et al (2004) Recognition of single-stranded RNA viruses by Toll-like receptor 7. Proc Nat Acad Sci USA 101: 5598–5603.
8. Heil F, Hemmi H, Hochrein H et al (2004) Species-specific recognition of single-stranded RNA via toll-like receptor 7 and 8. Science 303: 1526–1529.
9. Ahmad-Nejad P, Hacker H, Rutz M et al (2002) Bacterial CpG-DNA and lipopolysaccharides activate Toll-like receptors at distinct cellular compartments. Eur J Immunol 32: 1958–1968.
10. Kenworthy AK, Edidin M (1998) Distribution of a glycosylphosphatidylinositol-anchored protein at the apical surface of MDCK cells examined at a resolution of <100 angstrom using imaging fluorescence resonance energy transfer. J Cell Biol 142: 69–84.
11. Kenworthy AK, Edidin M (1998) Imaging fluorescence resonance energy transfer as probe of membrane organisation and molecular associations of GPI-anchored proteins, in: M.Gelb (Ed.), Meth Mol Biol., Humana Press, Towota, NJ, 1998, pp. 37–49.
12. Triantafilou K, Triantafilou M, Dedrick RL (2001) A CD14-independent LPS receptor cluster. Nature Immunol 2: 338–345.
13. Humphries HE, Triantafilou M, Makepeace BL et al (2005) Activation of human meningeal cells is modulated by lipopolysaccharide (LPS) and non-LPS components of *Neisseria meningitidis* and is independent of Toll-like receptor (TLR)4 and TLR2 signalling. Cell Micro 7: 415–430.
14. Triantafilou K, Triantafilou M, Ladha S et al (2001) Fluorescence recovery after photobleaching reveals that LPS rapidly transfers from CD14 to hsp70 and hsp90 on the cell membrane. J Cell Sci 114: 2535–2545.
15. Triantafilou M, Brandenburg K, Kusumoto S et al (2004) Combinational clustering of receptors following stimulation by bacterial products determines lipopolysaccharide responses. Biochem J 381: 527–536.
16. Manukyan M, Triantafilou K, Triantafilou M et al (2005) Binding of lipopeptide to CD14 induces physical proximity of CD14, TLR2 and TLR1. Eur J Immunol 35: 911–921.

Chapter 17

Transcriptome Analyses in the Interaction of *Neisseria meningitidis* with Mammalian Host Cells

Biju Joseph, Matthias Frosch, Christoph Schoen, and Alexandra Schubert-Unkmeir

Abstract

Infectious disease research has been revolutionized by two recent developments in the field of genome biology: (1) the sequencing of the human genome as well as many pathogen genomes and (2) the development of high-throughput technologies including microarray technology, proteomics, and metabolomics. Microarray studies enable a deeper understanding of the genetic evolution of pathogens and investigation of the determinants of pathogenicity on a whole-genome scale. Host studies, in turn, allow for an unprecedented holistic appreciation of the complexities of host cell responses at the molecular level. In combination, host–pathogen studies allow global analysis of gene expression in the infecting bacterium as well as in the infected host cell during pathogenesis, providing a comprehensive picture of the intricacies of pathogen–host interactions. In this chapter, we briefly explain the principles underlying DNA microarrays including the major points to consider when planning and analyzing microarray experiments and we describe in detail their practical application, using the interaction of *Neisseria meningitidis* with human endothelial or epithelial cells as examples.

Key words: *Neisseria meningitidis*, Infection biology, Cellular microbiology, Gene expression, Transcriptomics, Microarrays

1. Introduction

The large amounts of data from genome sequencing projects have led to a wealth of information about many organisms. In particular, the completion of the human genome project (1, 2), together with the complete genome sequence of around 1,250 bacteria (http://www.ncbi.nlm.nih.gov/genomes/lproks.cgi) including all major human pathogenic species, offers new possibilities for exploring the molecular pathogenesis of infectious diseases (3). In particular,

Myron Christodoulides (ed.), *Neisseria meningitidis: Advanced Methods and Protocols*, Methods in Molecular Biology, vol. 799, DOI 10.1007/978-1-61779-346-2_17, © Springer Science+Business Media, LLC 2012

the DNA microarray technology has made it possible to study the expression of every single gene within a genome at once (4). This genome-wide study of the complete mRNA expression profile (the transcriptome) has been termed accordingly "transcriptomics." Compared to conventional gene expression technologies such as reverse transcriptase (RT)-polymerase chain reaction (PCR) or Northern blotting, microarrays allow for the interaction between different molecular pathways in the cell to be studied – arguably one of the most important advantages of this novel technology. In addition, microarray-based approaches allow researchers to interrogate host and pathogen genomes without prior bias to which genes might be involved in a disease process.

Although newer technologies like the new generation sequencing methods have additional advantages to differential gene expression like identification of transcriptional start sites, identification of operon structures, identification of small noncoding RNAs, etc., microarrays are still a cost-effective method to study the gene expression of an organism on a whole genome scale. For this chapter, we describe very briefly the basic principles of DNA microarray technology and highlights in particular the application of microarrays to the analysis of the changes in gene expression in two *Neisseria meningitidis* strains upon adhesion to FaDu epithelial cells (5), as well as the transcriptional response in endothelial cells upon infection with the bacterium (6). The interested reader should refer to a review of eight published studies that analyzed mammalian gene transcription that occurs on infection with *N. meningitidis* (7). Several reports also describe microarray analyses of the infection biology of *N. meningitidis* (8, 9).

1.1. What Is a Microarray?

A microarray is a densely packed array of DNA probes which are fixed to a solid substrate in a predetermined matrix on a small surface area. The arrays are then amenable to hybridization with RNA or, more common cDNA nucleotide samples isolated from experimental cultures for gene expression analyses, or with chromosomal DNA for comparative genome hybridization studies. The nucleotides are labeled either fluorescently or radioactively before hybridization and afterwards the fluorescent intensity or radioactivity from each probe spot can be measured. When compared to a control, the spot intensities can be taken as a measure of the relative level of the complementary nucleotide in the sample.

Currently, there are two main types of microarrays: in situ synthesized oligonucleotide arrays (Affymetrix gene chips) and spotted DNA arrays. Affymetrix (http://affymetrix.com) uses a method that directly synthesizes an oligonucleotide sequence on a silicon slide/chip using a technique known as light-directed combinatorial chemical synthesis (10). Although still quite expensive, Affymetrix chips can achieve very high spot densities allowing the incorporation of, for example, up to 47,000 probes as in the latest available human oligonucleotide gene chip. However, for the whole-genome analysis of prokaryotes, which have a considerably smaller genome, the more easily obtainable spotted arrays are sufficient and are more commonly used for most projects in microbiology.

Spotted arrays most commonly use coated glass slides (microarray) or nylon membranes (macroarray) as the solid substrate. The coating chemistry of microarray slides can vary, but the common feature of the activated glass is that it allows cross-linking between the positively charged surface and the negatively charged nucleotide probes. The spotted array uses either PCR amplicons or presynthesized oligonucleotides of unique fragments from each of the target genes as probes. The oligonucleotide probes are in general melting-temperature normalized and are commonly 50–70 nucleotides long. As they are specifically designed to locations in each ORF, oligonucleotides have an increased specificity and a reduced cross-reactivity with other probes when compared with PCR-based microarrays. In addition, predesigned set of probes covering the entire genome of many bacterial pathogens are commercially available (e.g., Eurofins, MWG Operon, http://www.mwg-biotech.com/). The PCR product or oligonucleotide probes are then spotted onto the array surface by an automated robot that uses specially designed pins to pick up the sample from a well of a microtiter plate and deposit it onto the array surface in a prearranged grid formation. This allows the analysis software to link each spot with the gene it represents.

In this chapter, we will focus on spotted oligonucleotide arrays as they are the most common platform for microbial gene expression analysis using microarrays. More comprehensive details of microarray technology are available elsewhere (11).

1.2. Designing Microarray Experiments

Producing meaningful data from microarrays depends heavily on the manner in which the study is performed, so that potential sources of error are reduced as much as possible. There are two fundamental sources of error that occur during a microarray experiment: random errors and systematic errors (12). Random errors are minimized by obtaining an adequate number of technical as well as biological repeats to reduce extraneous influencing factors. Systematic error is reduced as far as possible by using an appropriate experimental design. With respect to the reference used, microarray experiments can in principle be divided into type 1 and 2 experiments (13). A type 1 experiment uses control and experimental cDNA labeled with two different fluorescent dyes to make a direct comparison between the two on one array (see below). The identification of genes up- or down-regulated at a certain time point during infection compared to in vitro culture would be an example of a type 1 experiment. Alternatively, radioactively labeled samples are hybridized separately to two different substrates and the results are compared to each other computationally. This design is ideal when directly comparing one or two different strains or experimental conditions. But if the candidates to be analyzed are more in number, then this design becomes tedious. Therefore, in a type 2 experiment an indirect comparison is made between two samples. In such a "common reference design" one strain or condition is a common reference and all other strains or conditions are indirectly compared to this common reference, which also allows comparisons between a large number of various experimental strains and conditions.

As an empirical rule of thumb, one should include at least three biological replicates of each experiment. In addition, a first technical replicate is to have at least two spots for each probe on each microarray. For fluorescent labeling with two different dyes (see below) a second technical replicate is to include a dye swap in each experiment to compensate for different labeling efficiencies of the dyes. Therefore, even a simple type 1 experiment comparing just two different samples/conditions already results in $3 \times 2 \times 2 = 12$ data sets for each probe that could be analyzed.

However, a detailed discussion of the bio-informatical and bio-statistical methods required for the proper planning and analysis of DNA microarray experiments is beyond the scope of this chapter and only rough guidelines will be given. For detailed information, we refer the reader to several excellent manuals (12, 14, 15). In addition, it is important to contact local biostatistics or bioinformatics experts for their advice on statistically appropriate, state-of-the-art design and analysis of the intended microarray-based experiment(s) (16).

1.3. Using Microarrays

There are four fundamental laboratory steps in using a microarray to measure gene expression in a sample: (1) sample preparation and labeling, (2) hybridization, (3) washing, and (4) image acquisition.

1.3.1. Sample Preparation and Labeling

There are a number of different ways in which a sample can be prepared and labeled for microarray experiments. In any case, rapid extraction of good quality RNA is always a crucial step if it aims to assess gene expression levels in a meaningful manner. Additionally, mRNA has a short half-life and the speed of isolation through to analysis is thus critical. A number of commercial kits have become available to be used for RNA extraction, but essentially the RNA is purified by precipitation and suspended in RNase-free water. To ensure that the prepared RNA is free of contaminating genomic DNA, which would give a false positive signal in any array experiment, the RNA is treated with DNase before it is used for further analysis.

Although it is possible to hybridize complementary RNA, it is much more common to hybridize complementary (cDNA) to the arrays. Therefore, total RNA is reverse transcribed to cDNA at which point it is labeled in a process known as "direct labeling," either by incorporation of radioactively labeled nucleotides such as (^{33}P)αATP or dCTP to which the fluorescent dyes Cy3 (excited by a green laser) or Cy5 (excited by a red laser) have been covalently attached.

1.3.2. Hybridization

Hybridization is the step in which the DNA probes on the glass and the labeled DNA target form heteroduplexes via Watson–Crick base pairing. Before hybridization, the cDNA/DNA samples have to be denatured to single-stranded molecules with a brief incubation at 95°C and added to the hybridization mixture, which guarantees a stringent environment for the binding reaction. In addition, to prevent nonspecific binding of nucleotides, the slides must also be prehybridized before use. During the hybridization reaction, the samples are hybridized to the array over night at a desired temperature (40–65°C) depending on the type of array used, the bacterial species, and the G/C content of the genome.

1.3.3. Washing

After hybridization, the slides are washed with solutions containing decreasing concentrations of saline sodium citrate (SSC), with and without sodium dodecyl sulfate (SDS) to remove excess hybridization solution and unbound labeled cDNA from the array.

1.3.4. Image Acquisition

The final step of the laboratory process is to produce an image of the surface of the hybridized array. For radioactively labeled samples, the hybridized and washed microarray is exposed with a low-energy phosphoimaging screen which is used in conjunction with a suitable detection system such as the PhosphorImager™ instrument. Phosphoimaging screens retain energy from β-particles, X-rays, and γ-rays generated by the radioactive decay of the labeled sample. Upon laser-induced stimulation in the PhosphorImager™ instrument, for each spot light is emitted from the storage phosphor screen in proportion to the amount of radioactivity in the sample. The emitted light is then quantified and further analyzed by image analysis softwares like the commercial AtlasImage™ (BD Biosciences Clontech, http://www.bdbiosciences.com).

For fluorescently labeled samples, the levels of fluorescence from each spot for both Cy3 and Cy5 are determined using a confocal scanner. The scanner contains two (or more) lasers that are focussed onto the array. After a prescan to check for the negative and positive controls and that the nonspecific background is not too imposing, the array is scanned at a gain just below the threshold where any of the spots become saturated. The array is scanned at a wavelength of 635 nm for Cy5 and 535 nm for Cy3 resulting in two monochrome images as a TIFF (tagged information file format) that reflects the fluorescent intensities on a black and white gradient. Consecutively, the monochromatic TIFF images are exported to image analysis software like the free Spotfinder by TIGR (http://www.tigr.org/software/tm4/spotfinder.html), or the commercially available softwares like ImaGene (http://www.biodiscovery.com/imagene.asp) or GenepixPro (http://www.axon.com) to create the familiar red-green false-color images of microarrays and to quantify the fluorescent intensities.

1.4. Computational Analysis and Storage of Microarray Data

For fluorescently labeled samples, the quantified spot intensities have to be further processed computationally by applying a data normalization method (both within slide and between slides) such as variance stabilization (VSN) (17) or locally weighted regression scatter plot smoothing (LOWESS) (18) to further reduce systematic errors. Useful software tools for microarray analysis are, amongst others, the freely available software packages like Limma which are part of the Bioconductor project (http://www.bioconductor.org/) (18) or the TIGR software MIDAS (http://www.tm4.org/midas.html). After normalization, statistical testing for differentially regulated genes, clustering of genes with similar expression patterns, and detailed database analyses are necessary before a well-founded biological interpretation of the data is possible (14, 15).

As microarrays provide a vast amount of complex data, the proper storage and interpretation of these data is a subject matter in its own right. A standardized storage format is pivotal for interpretation of these data and to enable other researchers to use them for meta-analyses. One attempt to facilitate the access and usability of microarray data has been set up in the form of MIAME (Minimum Information About a Microarray Experiment, http://www.mged.org/) microarray data file format (19) which enables researchers to compare results between various platforms.

1.5. Experimental Validation of Microarray Data

Microarray data are often heavily influenced by a large number of different parameters such as growth media used for cultivation, method of RNA extraction, type of microarray platform used, etc. As a result, there is a lot of uncertainty about using microarray data as concrete evidence of biologically relevant responses. Therefore, microarrays are often used as mainly exploratory tools that generate leads that can be followed up individually with simpler and robust low-throughput techniques such as real-time quantitative PCR, Northern blots, or other suitable phenotypic validation methods of the interesting candidates identified.

1.6. Application of Microarrays in the Study of N. meningitidis Host Cell Interaction

In the following sections, first the analysis of gene expression between *N. meningitidis* serogroup B strains upon adhesion to mammalian nasopharyngeal epithelial cells is presented, using spotted 70mer oligonucleotide microarrays comprising the genomes of the meningococcal strains α14, FAM18, MC58, and Z2491, respectively (20). Second, the analysis of the host-cell's transcriptome in response to infection with *N. meningitidis* MC58 will be described using radioactively labeled cDNA probes and plastic membranes containing more than 11,000 human oligonucleotides spotted in duplicate (BD atlas™ Plastic Microarrays; BD Bioscience/Clontech, http://www.clontech.com/).

2. Materials

2.1. Bacterial Strains and Growth Conditions

1. *N. meningitidis* serogroup B strain MC58 is a clinical isolate of the ST-32 complex, which was isolated in 1983 in United Kingdom (R. Borrow, personal communication) and was kindly provided by E.R. Moxon, University of Oxford. The *N. meningitidis* serogroup B strain α710 of the ST 41/44 complex was isolated from a healthy carrier under the Bavarian carriage study (NRZM, Würzburg, Germany). The strains are stored in a 20% (v/v) glycerol solution (prepared by adding 20% (v/v) glycerol, 25 g of Standard-I-nutrient broth (S-I-media, Merck) to 1 L of distilled H_2O) at –80°C.
2. GC agar for growth of meningococci: prepare by adding 36 g of Difco™ GC medium base to 1 L of distilled H_2O, autoclaving for 15 min at 121°C for 2.68 kg/cm^2, and adding Polyvitex (bioMérieux) to a final concentration of 1% (v/v). Alternatively, blood agar (Biomerieux) can be used (21).
3. Proteose peptone medium (PPM) for growth of meningococci: add 15 g of proteose peptone, 5 g of NaCl, 0.5 g of starch, and 20 mL of a stock KH_2PO_4/K_2HPO_4 solution to 1 L of distilled H_2O. The stock KH_2PO_4/K_2HPO_4 solution is prepared by adding 200 g of KH_2PO_4 and 50 g of K_2HPO_4 to 1 L of dH_2O. Polyvitex is then added to a final concentration of 1% (v/v).
4. Spectrophotometer and orbital shaker.
5. Humidified CO_2 incubator.
6. RPMI 1640 cell culture medium (with HEPES) supplemented with 10% (v/v) human serum (HS) or 10% (v/v) fetal calf serum (FCS).

2.2. Cell Culture and Bacterial Infection Assay

1. Human brain microvascular endothelial cells (HBMEC) have been described by Stins et al. (22, 23). The cells are positive for factor VIII-Rag, carbonic anhydrase IV, Ulex Europeus Agglutinin I, they take up fluorescently labeled acetylated low density lipoprotein and they express γ-glutamyl-transpeptidase, all of which are characteristic markers of brain endothelial cells.
2. FaDu: this epithelial cell line was established from a punch biopsy of a hypopharyngeal tumor and is designated as HTB-43™ at ATCC.
3. HBMEC Complete Medium: RPMI 1640 medium supplemented with 10% (v/v) FCS, 10% (v/v) Nu Serum IV (Becton Dickinson), 1% (w/v) vitamins (Biochrom), 1% (w/v) nonessential amino acids (Biochrom), 1 mM sodium pyruvate, 2 mM L-glutamine, 5 U/mL heparin, and 30 μg/mL of endothelial cell growth supplement (ECGS, Cell Systems Clonetics) (see Note 1). HS, FCS, and Nu Serum IV are decomplemented by

heat inactivation at 56°C for 30 min and stored at –20°C. HS is prepared from a serum pool (voluntary staff) (see Note 2).

4. FaDu Complete Medium: Eagle's Minimum Essential Medium (Lonza) supplemented with 10% (v/v) FCS, 1 mM sodium pyruvate, and 1× nonessential amino acid mix (Lonza).
5. Solution of trypsin and ethylenediaminetetraacetic acid (EDTA).
6. Sterile T75 and T175 cell culture flasks.
7. Phosphate buffered saline, pH 7.4 (PBS): it is useful to make a 10× stock solution for frequent use. Prepare a 1× PBS solution for infection assays.
8. Centrifuge (e.g. Heraeus Megafuge 1.0R), 15 and 50 mL polypropylene centrifuge tubes.
9. Sterile Pasteur pipettes.
10. RLT beta-mercaptoethanol (RLT-βME) solution (RNeasy kit, Qiagen).

2.3. The *N. meningitidis* Transcriptome

2.3.1. Isolation of RNA from Adherent *N. meningitidis*

1. RNeasy Mini Kit (Qiagen) (see Note 3).
2. Lysing tubes containing Lysing Matrix B (Q-BIOgene).
3. Fast Prep machine (Q-BIOgene).
4. Micro centrifuge (with rotor for 2 mL tubes).
5. RNAse-free tubes.
6. Ethanol (100%).

2.3.2. Quality Control Analysis of *N. meningitidis* RNA

1. DNA-free™ kit (Ambion).
2. PCR reaction with primer pairs for one of the seven housekeeping genes used for Multi Locus Sequence Typing (MLST) to confirm absence of chromosomal DNA from http://pubmlst.org/neisseria/info/primers.shtml.
3. Other components for a PCR reaction: dNTP mix, Taq Polymerase, appropriate buffer, and distilled water.
4. Positive control: genomic DNA from *N. meningitidis* serogroup B strain MC58.
5. Thermocycler.
6. Electrophoresis grade agarose.
7. 10× Tris borate EDTA (TBE) buffer: 890 mM tris base, 890 mM boric acid, and 20 mM EDTA pH 8.0.
8. Gel casting moulds, suitable power pack.
9. 6× sample loading buffer: 0.25% (w/v) bromophenol blue, 0.25% (v/v) xylene xyanol, and 40% (w/v) sucrose.
10. Nanodrop 2000 spectrophotometer (Thermo Scientific).
11. Ultraviolet (UV) transilluminator.
12. 10× 3-morpholinopropane-1-sulfonic acid (MOPS) buffer: 0.2M MOPS, pH 7.0 containing 0.1M sodium acetate (NaOAc),

pH 7.0 and 10 mM EDTA, pH 7.0. Dissolve all components in RNase-free water, store at room temperature, and protect from light. A working solution of 1× MOPS is prepared in RNase-free water.

13. Ethidium bromide solution (10 mg/mL).
14. Agilent RNA 6000 Nano Kit for quality analysis of the RNA.
15. 2× RNA-sample solution: 50% (w/v) formamide, 2.2M formaldehyde, 1× MOPS-buffer, 4% (w/v) Ficoll 400, 0.001% (v/v) bromophenol blue in RNase-free water. Store in aliquots at –20°C.
16. RNA-MOPS-agarose-gel: Prepare a 1% (w/v) agarose solution in 1× MOPS by boiling. Cool the solution to 50–60°C and add 40 μL of 1M guanidine thiocyanate (see Note 4) and 7.5 μL of 30% (w/v) ethidium bromide solution. Mix the gel and pour onto the gel support.

2.3.3. Reverse Transcription Reaction with Direct Cy-Labeling of N. meningitidis RNA

1. CyDye-dNTP: Cy3-dCTP and Cy5-dCTP (GE Healthcare).
2. 1 mg/mL luciferase control RNA (Promega): prepare a 10 ng/μL solution in RNase-free water and store in aliquots of 20 μL at –80°C.
3. 100 mM dNTP Set (Invitrogen), containing the four deoxynucleotides, each at 100 mM stock concentration: prepare a 20 mM dATP, a 20 mM dTTP, a 20 mM dGTP, and a 10 mM dCTP stock in RNase-free water and store at –20°C.
4. SuperScript™ II RT (200 U/μL, Invitrogen), 0.1M dithiothreitol (DTT), and 5× First-Strand buffer.
5. RNaseOut™ Recombinant Ribonuclease Inhibitor (40 U/μL, Invitrogen).
6. 0.1% (v/v) Diethylpyrocarbonate (DEPC) (Sigma-Aldrich) treated distilled water.
7. NONA-Random-Oligo (5′-NNNNNNNNN-3′; MWG Operon) is suspended to a final concentration of 5 μg/μL in RNase-free water.
8. RNase, DNase-free (500 μg/mL, Roche Applied Science).
9. Illustra ProbeQuant™ Sephadex G-50 Micro Columns (GE Healthcare).

2.3.4. Hybridization of a N. meningitidis Microarray

1. *N. meningitidis* whole-genome microarray slides. Due to space limitations, only a very brief description of the fabrication of spotted microarrays can be given and we refer the reader elsewhere for more information (11). As a convenient alternative, custom-made microarrays can be used instead, as these will be offered for most micro-organisms by different vendors in the near future. *N. meningitidis* whole-genome slides were obtained by robotically spotting (Omnigrid 100, GenMachines) 70mer

oligonucleotides (70mer-Oligos including an aminolinker, Operon) in triplicate, including luciferase-specific oligonucleotides as specificity and stringency controls, onto pretreated epoxy coated glass slides (Schott Nexterion) as described by the manufacturer (Omnigrid 100, GenMachines). The oligonucleotide-based microarray was constructed in collaboration with Eurofins MWG Operon containing 2,872 oligonucleotides representing 2,098 ORFs from *N. meningitidis* MC58 (NCBI Ref_Seq NC_003112), 2,119 ORFs from *N. meningitidis* Z2491 (NCBI Ref_Seq NC_003116), 2,131 ORFs from *N. meningitidis* FAM18 (NCBI Ref_Seq NC_008767), and 2,067 ORFs from *N. meningitidis* α14 (Genbank accession no. AM889136), respectively. The oligonucleotides were designed according to the method recently described by Li and Stormo (24). Accordingly, 2,078 oligonucleotides were directly designed from the primary source *N. meningitidis* MC58, but the probes were selected such that a larg- number of *N. meningitidis* Z2491, *N. meningitidis* FAM18, and *N. meningitidis* α14 ORFs were also represented. The printed slides were stored protected from dust in the dark under dry storage conditions and at room temperature.

2. 1% (w/v) SDS, prepared in distilled water.
3. 20× SSC: 3M NaCl in 0.3M sodium citrate buffer. Prepare this buffer by adding 175.3 g of NaCl, 88.2 g of Na_3 Citrate · H_2O to 900 mL of H_2O and adjust the pH to 7.0 with 1M HCl. Make the volume up to 1 L with H_2O and store the buffer at room temperature.
4. Prehybridization solution: 0.1% (v/v) Triton X, 1 mM HCl, 100 mM KCl.
5. Blocking solution: 50 mM ethanolamine and 0.1% (w/v) SDS in 0.1M Tris–HCl buffer, pH 9.0.
6. Microarray hybridization chamber (Corning) for manual hybridization.
7. Washing solution I: 2× SSC, 0.1% (w/v) SDS.
8. Washing solution II: 2× SSC.
9. Washing solution III: 0.2× SSC.
10. Washing solution IV: 0.1× SSC.
11. Tecan HS 4800™ Pro Hybridization Station for automated hybridization.
12. Speed Vac concentrator (Genevac).
13. Microcentrifuge (with rotor for 2 mL Eppendorf tubes).
14. Dyesaver solution (Genisphere).

2.3.5. Scanning of the N. meningitidis Microarray Slide and Image Acquisition

1. Microarray scanner Genepix Professional 4200A scanner and data acquisition software Genepix Pro 6.0 (MDS Analytical Technologies), or similar.

2.4. The Host Cell Transcriptome

2.4.1. Isolation of RNA from Infected and Noninfected HBMEC

1. BD Atlas™ Pure Total RNA labeling system (BD Biosciences Clontech).
2. Saturated phenol: use 100 g of phenol for 160 mL solution. In a fume hood, heat phenol to 70°C in a water bath for 30 min until it has melted. Add 95 mL of phenol directly to the saturation buffer and mix well.
3. Dry ice and a refrigerated centrifuge capable of 15,000 × *g*.
4. Polypropylene centrifuge tubes: 15 and 50 mL tubes.
5. Chloroform (100%), isopropanol (100%), and ethanol (80% v/v).

2.4.2. DNase Treatment of Total HBMEC-RNA

1. RNase-free DNase I (10 U/μL, Roche).
2. RNase-free H_2O.
3. 10× DNase I buffer: 400 mM Tris–HCl, pH 7.5 containing 100 mM NaCl and 60 mM $MgCl_2$.
4. Saturated phenol: chloroform (Sigma).
5. 95% (v/v) ethanol.
6. 2M NaOAc, pH 4.5.
7. 10× Termination Mix.

2.4.3. Quantifying the Yield and Purity of Total Mammalian RNA

Please refer to reagents for RNA-MOPS–agarose gel in Subheading 2.3.2.

2.4.4. cDNA Probe Synthesis from Host Cell RNA

1. BD Atlas™ total RNA labeling system (BD Bioscience Clontech).
2. dATP (10 μCi/μL); >2,500 Ci/mmol (Amersham).
3. Reagents included with the BD Atlas™ total RNA labeling system: 10× dNTP, 10× Random Primer Mix, PowerScript™ RT, 5× PowerScript™ Reaction Buffer, DTT (100 mM), 10× Termination Mix.
4. Magnetic particle separator (Promega). It is important to use a separator designed for 0.5 mL tubes.
5. Polypropylene centrifuge tubes: 1.5, 2, 15, and 50 mL (Greiner).

2.4.5. Column Chromatography for cDNA Purification

1. BD Atlas NucleoSpin® Extraction Kit (BD Bioscience Clontech).
2. Centrifuge (Centrifuge 5417, Eppendorf) and polypropylene centrifuge tubes of 1.5 and 2 mL volume.

3. 95% (v/v) ethanol.
4. LS 6500 Liquid Scintillation counter (Beckman Coulter), scintillation-counter vials, and a container for radioactive waste.

2.4.6. Hybridization of the BD Atlas™ Plastic Microarrays

1. BD Atlas™ Plastic Microarrays (BD Bioscience Clontech, http://www.clontech.com). The methods described for host cell transcriptome analysis are based on the use of this plastic microarray, which consists of more than 11,000 long oligonucleotides immobilized by UV radiation on a rigid, translucent plastic support. They contain 384 repeating blocks of 64 spots. Furthermore, there are three blocks (these blocks are labeled with "C") with identical sets of control spots plus experimental genes. The control spots comprise negative controls (various λ phage specific oligonucleotides) and housekeeping genes.
2. BD Atlas™ Plastic Array Hybridization Box.
3. Rocker platform.
4. 10× Denaturation solution: 1M NaOH, 10 mM EDTA.
5. 2× Neutralization solution: 1M $Na_2H_2PO_4$.
6. 20× SSC.
7. 20% (w/v) SDS.
8. High salt wash solution: 2× SSC, 0.1% (w/v) SDS.
9. Low salt wash solution 1: 0.1× SSC, 0.1% (w/v) SDS.
10. Low salt wash solution 2: 0.1× SSC.
11. Phosphorimaging screen (Molecular Dynamins), Imager Analyzer FLA-3000 with Image Reader software (Fuji).
12. Stripping solution: 0.1M Na_2CO_3 for stripping cDNA probes from the plastic microarray.

2.4.7. Analysis of the BD Atlas™ Plastic Microarrays

1. BD AtlasImage 2.7 software.
2. BD AtlasImage Grid Template.

3. Methods

3.1. Bacterial Strains and Growth Conditions

1. Streak out the *N. meningitidis* strain MC58 and α710, stored frozen in glycerol solution at –80°C, onto fresh GC agar plates and incubate at 37°C for 18 h in an atmosphere of 5% (v/v) CO_2.
2. Suspend the bacterial culture using cotton swabs into 10 mL of PPM + 1% (v/v) Polyvitex and incubate at 37°C with shaking at 200 rpm for 1.5–2 h.

3. After 1.5–2 h, prepare a 1/10 dilution of the culture and measure the absorbance in a spectrophotometer at λ600 nm. An absorbance at λ600 nm of 0.1 is equivalent to 1×10^8 meningococci. Prepare the bacterial inoculum using the formula:

$$\text{Total in oculum required (mL)} \times \frac{0.001}{A_{600\,\text{nm}}\ \text{of}\ 1:10\text{dilution}} = \text{mL of neat culture required.}$$

   This will give ~1×10^8 colony forming units (cfu) in the final inoculum, which corresponds to a multiplicity of infection (MOI) of 10 (the number of HBMEC in a T75 or T175 tissue flask grown to confluence is around 1×10^7 or 5×10^7 cells, respectively). For the comparison of the transcriptome profiles of the two serogroup B meninogococcal strains (MC58 and α710), an MOI of between 100 and 200 should be used to obtain sufficient adherent bacteria.

4. Suspend the bacteria in 10 mL of RPMI cell culture medium with 10% (v/v) HS or 10% (v/v) FCS (see Note 2).

3.2. Cell Culture and Bacterial Infection Assay

1. Culture HBMEC in HBMEC Complete Medium and FaDU cells in FaDu Complete Medium in cell culture flasks at 37°C in an incubator with a humid atmosphere and 5% (v/v) CO_2.
2. Passage the HBMEC and FaDu cell cultures when they approach confluence. Wash the cultures twice with prewarmed 1× PBS and treat with trypsin/EDTA solution for 10 min at 37°C to detach the cells from the flasks. Pellet the cells by centrifugation for 2 min at 292 g and suspend the cells into T75 or T175 tissue culture flasks with fresh culture medium. Prepare two T75 confluent flasks (for HBMEC transcriptome analysis) to six T175 confluent flasks (for *N. meningitidis* transcriptome analysis) for each experimental data point. Seed the cells into the appropriate numbers of T75 or T175 tissue culture flasks with an inoculum of between 5×10^6 cells and 1×10^7 cells for T75 and T175 flasks, respectively, and incubate the cultures for 48 h at 37°C with 5% (v/v) CO_2. This procedure will provide confluent cell monolayers.
3. Aspirate the medium from the flasks and carefully wash the cells with prewarmed 1× PBS. Repeat this washing step an additional two times.
4. Add the 10 mL of bacterial suspension, prepared as described in steps 3 and 4 of Subheading 3.1, in RPMI cell culture medium containing 10% (v/v) HS, to each HBMEC monolayer at a MOI of 10. For analyzing gene expression profiles in *N. meningitidis* strains following adhesion to FaDu cells, suspend the bacteria in RPMI cell culture medium containing 10% (v/v) FCS.

5. Incubate the cells with bacteria at 37°C with 5% (v/v) CO_2 for 4- and 8-h periods for analysis of the HBMEC transcriptome or for a 3-h period for analysis of the adherent *N. meningitidis* transcriptome (5, 6). In parallel, incubate flasks of uninfected HBMEC cultures for control samples.
6. At each time point, aspirate the infection media from the monolayers using a sterile Pasteur pipette.
7. Carefully wash the cell monolayers with 1× PBS and repeat this washing procedure an additional two times.
8. Add 10 mL of trypsin/EDTA solution to the washed monolayers and incubate at 37°C for 10 min with gentle rocking on a shaker to detach the bacteria and host cells (25) (see Note 5).
9. Transfer the detached bacteria and host cells to a 15 mL centrifuge tube and centrifuge at 1,600 × *g* for 10 min at 4°C.
10. Transfer the supernatant containing bacteria only to a fresh 15 mL centrifuge tube and centrifuge at 2,770 × *g* for 10 min at 4°C. Suspend the bacterial pellet immediately in 1.0 mL of RLT-βME buffer.
11. Suspend the pellet remaining in the tube in step 10, which contains HBMEC, in ice-cold PBS, centrifuge at 1,000 × *g* for 10 min at 4°C, and "shock" freeze the pellets at −70°C for isolation of RNA.

Figure 1 shows a flow diagram of the steps to follow for analysis of the *N. meningitidis* transcriptome upon adhesion to FaDu cells (or indeed other host cells) and of the HBMEC host cell transcriptome in response to infection with the bacteria. We begin with the *N. meningitidis* transcriptome.

3.3. The *N. meningitidis* Transcriptome

3.3.1. Isolation of RNA from Adherent *N. meningitidis*

1. Transfer the complete bacteria-RLT-β-ME mixture of one strain, prepared in step 11 of Subheading 3.2, into a lysing tube containing Lysing Matrix B. Lyse the cells by centrifugation for 30 s at a speed of 6.5 in the Fast Prep machine, following the instrument manufacturer's instructions. Chill the lysing tubes for 30 s on ice and then repeat the centrifugation for another 20 s (see Note 6).
2. Centrifuge the lysing tubes for 2 min at 16,000 × *g*, which sediments the Lysing Matrix B.
3. Transfer the supernatant to an RNase-free tube, add 550 μL of ethanol (100%), and mix the contents by shaking (see Note 7).
4. Perform the further purification steps according to the manufacturer's instructions with an on-column DNAse treatment supplied with the RNeasy Mini Kit. Transfer the purified RNA to an RNase-free tube.

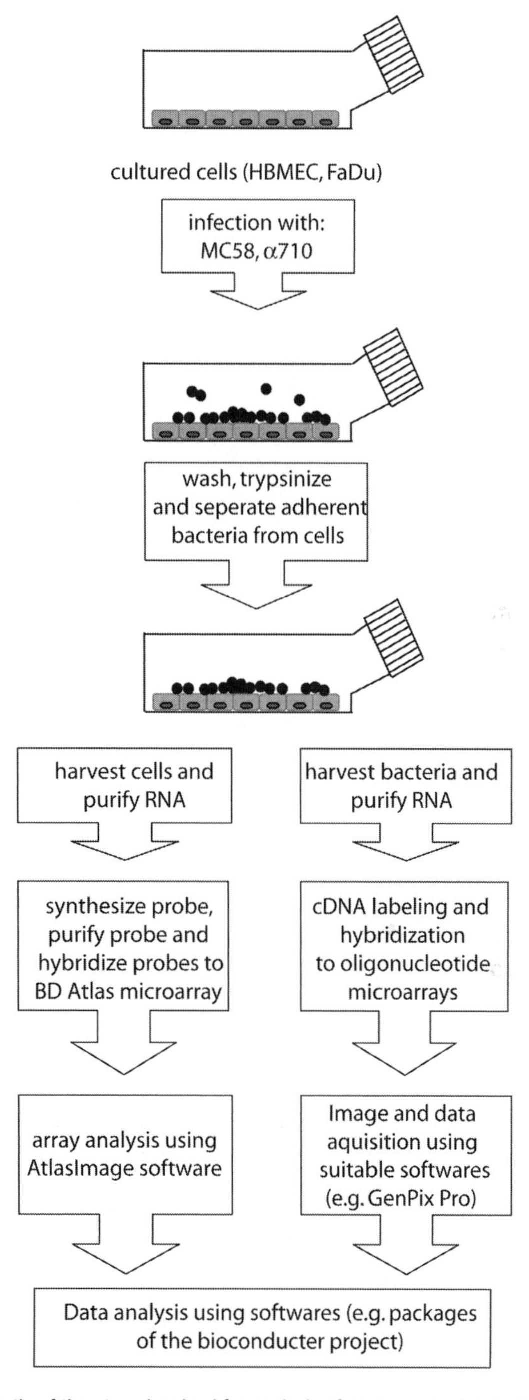

Fig. 1. Schematic of the steps involved for analysis of gene expression in *Neisseria meningitidis* upon adhesion to FaDu cells and the host cell transcriptome (HBMEC) in response to meningococcal infection.

3.3.2. Quality Control Analysis of N. meningitidis RNA

1. To test for traces of contaminating genomic DNA in the isolated RNA, perform the following PCR reaction. For a 50 μL reaction volume, combine the following reagents in the following order:

5 μL	10× Thermopol buffer
2.5 μL	4 mM dNTP
0.25 μL	100 μM Primer 1
0.25 μL	100 μM Primer 2
0.25 μL	Taq DNA polymerase (5 U/μL)
1 μL	Template (RNA, positive/negative control) (see Note 8)
50 μL	ddH_2O

Place the reaction tube in a thermal cycler and subject to PCR with the following conditions: an initial denaturation step at 95°C for 10 min, followed by 32 cycles each at 95°C for 60 s, 58°C for 45 s, and 72°C for 45 s, and an additional 72°C for 5 min. Store the PCR product(s) at 4°C.

2. Mix 10 μL of the PCR product with 6× sample loading buffer (end concentration of 1×) and load onto a 1% TBE-agarose gel. The gel is prepared by boiling, cooling, and casting 1% (w/v) agarose in 1× TBE buffer into gel cast moulds. Run the electrophoresis at 5 V/cm for 1 h and then stain the gels with ethidium bromide and visualize using a UV transilluminator.
3. If chromosomal DNA is detected on the agarose gel, subject the isolated RNA preparation to an additional DNAse treatment using the DNA-free™ kit according to the manufacturer's protocol. Then repeat the PCR and agarose gel electrophoresis to confirm that the RNA is not contaminated with DNA. Presence of PCR amplified product of the expected size using RNA as template indicates DNA contamination in the preparation.
4. Check the quantity of the RNA using a Nanodrop 2000 spectrophotometer, according to the manufacturer's instructions (see Note 9).
5. Dilute the RNA to a concentration of between 50 and 500 ng/μL and subject 1 μL of the RNA sample(s) through an Agilent RNA 6000 Nano Kit, according to the manufacturer's instructions to evaluate the quality of the isolated RNA. An example of intact and degraded RNA is shown in Fig. 2. Alternatively, use a 1% (w/v) RNA-MOPS-agarose-gel as follows: prepare a 1% (w/v) MOPS agarose gel in 1× MOPS buffer and subject 5 μL of the RNA mixed with 2× RNA sample buffer to electrophoresis at 2.5 V/cm for 1.5 h and visualize the RNA using a UV transilluminator.

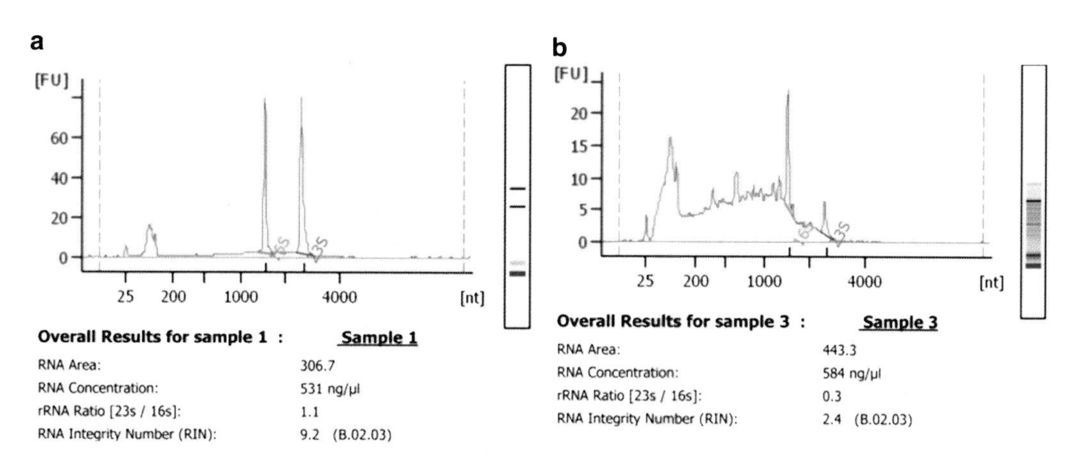

Fig. 2. Electropherogram of an intact bacterial RNA (**a**) and a partially degraded RNA (**b**).

3.3.3. Reverse Transcription Reaction with Direct Cy-Labeling of N. meningitidis RNA

In order to get reproducible results it is important that pipetting is accurate and that RNase-free working conditions are observed during all the following steps.

1. For the labeling reaction, use 5–20 µg of total RNA for each strain, i.e., adherent MC58 and adherent α710 in a volume of 14.1 µL of RNase-free water (see Note 10).
2. Add 1 µL of 10 ng/µL of luciferase-RNA (Promega) to each labeling reaction as a “spike-in” control.
3. Add 0.9 µL of a 5 µg/µL NONA random oligo-primer and heat the probes to 70°C for 5 min to dissolve the secondary and tertiary structures of the RNA and primer dimers. Chill on ice for 5 min.
4. Prepare the master mix by mixing in this order, 20 µL of 5× First-Strand buffer with 1 µL of 0.1M DTT, 2 µL of 10 mM dCTP, 2.5 µL of 20 mM dATP, 2.5 µL of 20 mM dTTP, 2.5 µL of 20 mM dGTP, 14.5 µL of DEPC-H_2O, 2.5 µL of 40 U/µL of RNase Out, and 2.5 µL of 200 U/µL RT.
5. Add 20 µL of the master mix to both reactions and 2 µL of Cy3-dCTP to one of the labeling reactions and 2 µL of Cy5-dCTP to the other (see Note 11).
6. Incubate the probes at room temperature for 10 min and then at 42°C for 2 h to complete the reverse transcription reaction.
7. Inactivate the reaction, especially the RNase inhibitor, by heating the sample to 70°C for 15 min. Then, add 2 µL of 500 µg/mL RNase, DNase-free, to the samples and incubate the mixture at 37°C for 45 min to degrade the RNA.
8. Purify the Cy3 and Cy5 labeled cDNA by Sephadex-G-50 chromatography using illustra ProbeQuant™ G-50 Micro

Columns. First of all, centrifuge the columns for 1 min at 2,500 × *g* and discard the flow-through. Place the columns in new collection tubes and pipette the complete labeling reaction directly onto the Sephadex matrix. Centrifuge for 1 min at 2,500 × *g* to collect the purified labeled cDNA.

3.3.4. Hybridization of a N. meningitidis Microarray

Hybridization of fluorescently labeled samples can be done either manually or by using an automated hybridization station.

1. For manual hybridization, pool both purified fluorescent labeled samples and concentrate the resulting mixture to a final volume of 30 μL in a Speed Vac. Add 6 μL of 20× SSC buffer and 4 μL of 1% (w/v) SDS.
   For the automated hybridization station method, pool both purified fluorescent labeled samples and make up the final volume to 130 μL with 20× SSC and 1% (v/v) SDS to yield a final concentration of 3× SSC and 0.1% (v/v) SDS.
2. For both hybridization methods, denature the mixture of the labeled samples by heating at 95°C for 3 min, perform a quick spin in a micro-centrifuge, and chill the sample on ice.
3. Next, pipette the mixture, avoiding the introduction of air bubbles, onto a preprocessed *N. meningitidis* whole genome microarray slide (see Note 12), and then place a glass coverslip onto the solution.
4. For the manual hybridization method, place the microarray slide into a hybridization chamber with a humid atmosphere. Heat the chamber to 60°C for 16 h for hybridization. After hybridization, wash the slides according to the manufacturer's protocol (see Note 13). Briefly, wash the hybridized slides in wash solutions I–IV for 10 min each and then dry the slides by centrifugation at 200 × *g* for 5 min. At this point, dip the slide for 10 s in the Dyesaver solution to prevent degradation of the fluorescent Cy5.
5. For the automated hybridization method, carry out step 3 above but hybridize the slides using the Tecan HS 4800™ Pro Hybridization Station, according to the manufacturer's protocols. The principle of the hybridization station is identical to the manual hybridization method except that all the hybridization and washing steps are automated using defined programs and with the slides dried subsequently with liquid N_2 and then stored away from light to prevent bleaching of the dyes.

3.3.5. Scanning of the N. meningitidis Microarray Slide and Image Acquisition

1. Scan the slide with a microarray scanner (e.g., Genepix Professional 4200A) at λ535 nm for Cy3 and λ647 nm for Cy5 to yield two TIFF formatted images. To achieve two comparable images (low background, similar spot intensity of most spots (~80%) in both channels and a signal maximum of 80%

of the brightest spots), adjust the photo multiplier tube gain and the laser settings.

2. Analyze the images using the Genepix Pro 6.0 analysis software, following the manufacturer's instructions. First of all, examine the spots manually and if necessary flag the spots that have high background, stray fluorescence, and nonuniform spot morphologies. Next, quantify the images using an appropriate array description file (gal file) by measuring the mean or median signal intensity of each spot and the local background of the spots. Finally, analyze these raw data files using specialized software such as the Limma package (26) implemented in the R language to identify statistically significant genes that are differentially regulated between the two samples.

3.4. The Host Cell Transcriptome

The following protocol describes how to generate a transcriptome from HBMEC infected with *N. meningitidis*, but the methodology is transferable to other types of human host cells infected with meningococci or other bacteria.

3.4.1. Isolation of RNA from Infected and Noninfected HBMEC

1. Isolate total RNA using the BD Atlas™ Pure Total RNA labeling System, following the manufacturer's instructions. In this system, the key components are streptavidin-coated magnetic beads and biotinylated oligo(dT), which allow both poly A^+ RNA enrichment and probe synthesis to be done in a single procedure.
2. Add 3 mL of denaturation solution, supplied with the BD Atlas™ Pure Total RNA labeling System, to $1–3 \times 10^7$ HBMEC. Pipette up and down vigorously and vortex well until the cell pellet is completely suspended.
3. Chill the denatured cell suspension on ice for 5–10 min.
4. Centrifuge the homogenate at $15{,}000 \times g$ for 5 min at 4°C. This step removes cellular debris.
5. Transfer the entire supernatant to new 50 mL centrifuge tubes and add 6 mL of saturated phenol.
6. Vortex for 1 min and chill the sample on ice for 5 min.
7. Next, add 1.8 mL of chloroform (100%) to the sample and shake and vortex vigorously for 2 min. Again, chill the sample on ice for 5 min.
8. Centrifuge the homogenate at $15{,}000 \times g$ for 10 min at 4°C.
9. Transfer the upper aqueous phase containing the RNA to a new 50 mL centrifuge tube. Do not pipette any material from the white interface or the lower organic phase.
10. Perform a second round of phenol:chloroform extraction.
11. Remove the upper aqueous phase and add 1.8 mL of isopropanol (100%). Mix the solution thoroughly and chill on ice for 10 min.

12. Centrifuge the samples at 15,000 × *g* for 5 min at 4°C.
13. Remove the supernatant without disturbing the RNA pellet.
14. Add 3 mL of 80% (v/v) ethanol to the RNA pellet.
15. Centrifuge at 15,000 × *g* for 5 min at 4°C. Quickly and carefully discard the supernatant.
16. Air-dry the pellet and suspend it in RNase-free H_2O to an RNA concentration of 1–2 μg/mL.
17. Store the RNA samples at –70°C until ready to proceed with DNase treatment.

3.4.2. DNase Treatment of Total HBMEC-RNA

The following protocol describes DNase I treatment of 0.5 mg of total RNA prior to purification of poly A^+ RNA.

1. Treat the isolated RNA with DNase I enzyme using following recipe: 500 μL of total RNA (1 mg/mL), 100 μL of 10× DNase I buffer, 50 μL of DNase I (1 U/μL), and 350 μL of deionized H_2O to make a final total volume of 1 mL. Mix well by pipetting.
2. Incubate the reaction at 37°C for 30 min.
3. Add 100 μL of 10× termination mix and mix well by pipetting.
4. Split each reaction into two 1.5 mL micro-centrifuge tubes (500 μL per tube).
5. Add 500 μL of saturated phenol and 300 μL of chloroform to each tube and vortex thoroughly.
6. Centrifuge at 16,000 × *g* for 10 min at 4°C to separate the phases.
7. Carefully transfer the top aqueous layer to a fresh 1.5 mL micro-centrifuge tube. Avoid pipetting any material from the interface or the lower phase.
8. Add 500 μL of chloroform to the aqueous layer and vortex thoroughly.
9. Centrifuge at 16,000 × *g* for 10 min at 4°C to separate the phases.
10. Carefully transfer the top aqueous layer to a fresh 2 mL micro-centrifuge tube.
11. Add 1/10 of the volume (i.e., 50 μL) of 2M NaOAc and 2.5 volumes (i.e., 1.5 mL) of 95% (v/v) ethanol. If treating <20 μg of total RNA, add 20 μg of glycogen.
12. Vortex the mixture thoroughly and chill on ice for 10 min.
13. Centrifuge at 16,000 × *g* for 15 min at 4°C.
14. Carefully remove the supernatant and any traces of ethanol.
15. Gently overlay the pellet with 80% (v/v) ethanol.

16. Centrifuge at 16,000 × *g* for 5 min at 4°C and carefully remove the supernatant.
17. Air-dry the pellet for approximately 10 min.
18. Dissolve the cell precipitate in 250 μL of RNase-free H_2O an assess the yield and purity of RNA as described in Subheading 3.4.3. Alternatively, store the RNA at −70°C.

3.4.3. Quantifying the Yield and Purity of Total RNA

The yield of total RNA will vary depending on the cells from which it is obtained. 1–3 × 10^7 HBMEC will yield around 50 μg of total RNA. Determine the quality and quantity of RNA as described in Subheading 3.3.2. On a RNA-MOPS-agarose gel, total RNA from mammalian cells should appear as two bands at approximately 4.5 and 1.9 kb (28S and 18S ribosomal RNA). The ratio of intensities of 28S and 18S rRNA should be 1.5–2.5:1. Lower ratios indicate degradation.

3.4.4. cDNA Probe Synthesis from Host Cell RNA

1. Carry out poly A^+ RNA enrichment using the streptavidin-coated magnetic beads and biotinylated oligo(dT), following the instructions supplied with the BD Atlas™ Pure Total RNA labeling System.
2. Prepare a Master Mix for all labeling reactions plus one extra reaction using the following reagents and volumes: 4 μL of 5× PowerScript Reaction Buffer; 2 μL of 10× dNTP Mix (for dATP label); 5 μL of (α-33P) dATP (10 μCi/μL); >2,500 Ci/mmol; 0.5 μL of DTT (100 mM) (total volume of 11.5 μL).
3. Preheat a PCR thermal cycler to 65°C.
4. For each experimental poly A^+ RNA, combine the following in a labeled 0.5 mL PCR tube: 1 μg (1 μg/μL) of Poly A^+ RNA sample and 1 μL of Random Primer Mix.
5. Mix well by pipetting and spin the tubes briefly in a microcentrifuge.
6. Incubate tubes in the preheated PCR thermal cycler at 65°C for 2 min.
7. Reduce the temperature of the thermal cycler to 42°C and incubate the tubes for 2 min. During this incubation, add 2 μL of PowerScript RT per reaction to the Master Mix by pipetting and keep the Master Mix at room temperature.
8. After completion of the 2-min incubation at 42°C, add 13.5 μL of Master Mix to each reaction tube. Mix the contents of the tubes thoroughly by pipetting and immediately return them to the thermal cycler.
9. Incubate the tubes at 42°C for 30 min.
10. Add 2 μL of 10× Termination Mix and mix well.
11. Proceed with the column chromatography for cDNA purification.

3.4.5. Column Chromatography for cDNA Purification

This protocol separates the labeled cDNA from unincorporated ^{33}P-labeled nucleotides and small (<0.1 kb) cDNA fragments with Nucleospin extraction spin columns. All reagents required are provided with the BD Atlas NucleoSpin® Extraction Kit.

1. Add 180 µL of Buffer NT2 to dilute the probe synthesis reaction to a total volume of 200 µL and mix well by pipetting.
2. Place a Nucleospin Extraction Spin Column into a 2 mL centrifuge tube and pipette the sample onto the column. Centrifuge at 18,000 × *g* for 1 min. Reserve the flow through for estimation of the reverse transciptase reaction efficiency. Discard the centrifuge tube into an appropriate container for radioactive waste.
3. Insert the NucleoSpin column into a fresh 2 mL centrifuge tube. Wash the column with 95% (v/v) ethanol before use and then add 400 µL of Buffer NT3. Centrifuge at 18,000 × *g* for 1 min. Discard the centrifuge tube and the flow through. Repeat this step an additional two times.
4. Transfer the NucleoSpin column to a clean 1.5 mL micro-centrifuge tube. Add 100 µL of Buffer NE and allow the column to soak for 2 min.
5. Centrifuge at 18,000 × *g* for 1 min to elute the purified probe.
6. Check the radioactivity incorporation in the probe by scintillation counting. Add 5 µL of each purified probe to 5 mL of scintillation fluid in separate scintillation-counter vials. In addition, add 5 µL of the flow through from each purified probe to 5 mL of scintillation fluid in separate scintillation-counter vials. Count samples on the ^{33}P channel and multiply the counts by a dilution factor of 20. The probes should yield a minimum of $5–25 \times 10^6$ counts per minute.
7. Discard the flow through fractions, columns, and elution tubes in the appropriate container for radioactive waste.

3.4.6. Hybridization of the BD Atlas™ Plastic Microarrays

For hybridization, four procedures are followed in this order: (1) prerinsing the plastic microarray, (2) hybridizing cDNA probes to the plastic microarray, (3) washing of the plastic microarray, and (4) detection of the signals.

1. Prerinsing the plastic microarray.
   (a) Fill a hybridization box with deionized H2O to 80% of its volume and warm to 55–60°C. Also, prewarm 30 mL of the BD PlasticHyb hybridization solution at 60°C in a separate container (see Note 14).
   (b) Carefully place the microarray into the hybridization box containing prewarmed H_2O, with the printed surface facing upwards. During the hybridization procedure,

ensure that the printed surface of the microarray is always facing upwards. Furthermore, ensure that the hybridization box is well sealed and shaking continuously.

(c) Pour off the H_2O and replace with 15 mL of prewarmed hybridization solution. Firmly attach the lid.

(d) Rock the microarray at 60°C for 10–30 min.

2. Hybridizing cDNA probes to the plastic microarray.

(a) To prepare the probe for hybridization, incubate the probe in a boiling (95–100°C) water bath for 2 min (see Note 15). Then chill on ice for 2 min.

(b) Combine the denatured probe with 15 mL of prewarmed hybridization solution in a disposable 50 mL plastic tube. Make sure that the two solutions are thoroughly mixed together.

(c) Carefully pour off the prerinsing solution from the microarray, and add the mixture prepared in Step 2a to the microarray. Ensure that the printed surface is facing upwards and firmly attach the lid.

(d) Hybridize overnight at 60°C with continuous rocking. Ensure that all regions of the plastic are in contact with the hybridization solution at all times.

3. Washing the plastic microarray.

(a) The next day, prewarm 300 mL of high salt wash solution and 300 mL of low salt wash solution 1 to 58–60°C. Fill a 500 mL beaker with low salt wash solution 2 maintained at room temperature.

(b) Carefully remove the hybridization solution and discard into an appropriate container for radioactive waste. Replace with 40–50 mL of prewarmed High Salt Wash Solution. Attach the lid and rock for 5 min at 58°C to remove residual radioactive probe. Discard the Wash Solution and repeat this wash procedure.

(c) After pouring off the second high salt wash, fill the hybridization box with 40–50 mL of prewarmed low salt wash solution 1 and wash the microarray in a 58°C incubator for 5 min. Repeat this low salt wash procedure.

(d) Reduce the temperature of the hybridization oven to 25–30°C. Discard the low salt wash solution, and fill the box to approximately 80% capacity (40–50 mL) with room-temperature low salt wash solution 2. Rock the microarray for 5 min.

(e) Remove the microarray from the hybridization box using forceps and immediately transfer the microarray to the beaker of room-temperature Low Salt Wash Solution 2.

Rinse the microarray by dipping it several times into this wash solution.

(f) Remove the microarray from the beaker of Low Salt Wash Solution very slowly, allowing the Wash Solution to drain off the surface.

(g) Allow the microarray to air-dry completely, which takes around 15 min.

4. Detection of the signals.

   (a) Expose the printed surface of the microarray to a phosphorimaging screen suitable for ^{33}P detection, following the manufacturer's instructions. We recommend exposure times of 12 and 72 h and if necessary of 1 week.

   (b) Scan the phosphoimager screen at a resolution of 50 μm.

3.4.7. Analysis of the BD Atlas™ Plastic Microarrays

For cDNA microarray analysis, use the BD AtlasImage 2.7 software according to the manufacturer's guidelines. The arrays are aligned with BD AtlasImage Grid Template automatically and then fine-tuned for each gene using manual adjustment options. Calculate the background using the median intensity of the "blank spaces" between different panels of the array (default method of calculation) and then measure the raw signal intensity of each spot. A raw intensity (before normalization) of 2.0-fold over background is taken as an indication that a gene is expressed at a significant level.

For comparing the expression patterns of infected cells and uninfected cells, normalize the signal intensities by the global normalization-sum method, which is best suited for the comparison of two similar tissue samples. Signal values in arrays hybridized with meningococci-infected cDNA should be normalized with respect to those from arrays reprobed with cDNA of uninfected cells. The adjusted signal intensities of each of the individual cDNA spots should be compared for infected and uninfected cells. The ratio is calculated as the adjusted intensity of array 2 (infected HBMEC): adjusted intensity of array 1 (uninfected cells), according to the manufacturer's guidelines. Differences are estimated when a gene signal either in array 1 or 2 is at background level.

Instead of using numerical values, we prefer to indicate these genes as up-regulated (↑) or down-regulated (↓) in infected HBMEC (6) and we save the data in a tab-delimited format for further statistical analysis using Microsoft Excel.

4. Notes

1. The splitting of cells, changing of media, preparation of infection conditions, and infection of cells must be done using sterile techniques in a Class II biological safety cabinet to prevent contamination.

2. The recommendation for using RPMI cell culture medium supplemented with 10% (v/v) HS is based on the observation that meningococcal entry is supported by binding of the outer membrane protein Opc via fibronectin to integrins on HBMEC (27). HS should be pooled and decomplemented by heat inactivation at 56°C for 30 min. For analyzing gene expression profiles of adherent *N. meningitidis* strains upon adhesion to FaDu epithelial cells, supplementation with 10% (v/v) FCS is recommended instead.
3. Standard procedures to minimize RNase contamination should be used for all solutions and glassware (28). Whenever possible, disposable sterile plasticware should be used for any steps prior to the RNA being converted to single-stranded cDNA. Wear disposable gloves during preparation of materials and solutions that will be used for isolation and analysis of RNA and during any manipulations involving RNA. Change gloves frequently.
4. Guanidine thiocyanate denatures proteins and inhibits RNases in the presence of reducing agents. It is used as a substitute for toxic formaldehyde.
5. Cells should not be incubated for longer than 10 min with trypsin/EDTA, because of cell-surface protein release and toxic effects.
6. The cells are disrupted and simultaneously homogenized by a combination of turbulence and mechanical shearing. After the lysis of the cells there should be no foam on the sample. It is important to ensure that no lysing matrix is transferred.
7. To elute RNA from the column, pipette 100 μL of RNase-free water directly onto the membrane of the column and let it stand for at least 10 min, then centrifuge for 1 min at ≥8,000 × *g*.
8. The application of a Bioanalyzer (Agilent) to check the integrity and purity of the RNA preparation is more accurate than the analysis using conventional agarose gel electrophoresis.
9. The quality of the RNA is the most important factor influencing the sensitivity and reproducibility of the experiments. Contamination by genomic DNA is a critical point; therefore all RNA samples must be treated with RNase-free DNase I. We recommend considering the guidelines for working with RNA (28). Organic solvents, salt, and protein are detected at λ230 nm, whereas only proteins are detected at λ280 nm. The ratios of the absorbance values at λ260 and λ280 nm, and at λ260 and λ230 nm, respectively, indicate the purity of the RNA. The results of both ratios should be over 1.8. For exact determination of the ratios, the measurements should be done in 10 mM Tris–HCl buffer, pH 7.5, or in buffered deionized water as the pH has a strong effect on these values at 280 nm.

10. If the 5–20 μg of RNA is in a volume greater than 14.1 μL, concentrate the sample in a Speed Vac.
11. Exposure to light should be minimized during all working steps involving these dyes, as both dyes are photosensitive.
12. The preprocessing of the slide was done according to the manufacturer's instructions for epoxy coated slides (Schott Nexterion). Briefly, rinse the spotted microarray slides once for 5 min in 0.1% (v/v) Triton X, twice in 1 mM HCl for 2 min each, and once in 100 mM KCl for 10 min. Block the slides for 15 min at 50°C in blocking solution consisting of 50 mM ethanolamine, 0.1% (w/v) SDS in 0.1M Tris–HCl buffer, pH 9.0. Finally, rinse in distilled water for 1 min and dry by centrifugation (200 × *g* for 5 min).
13. Washing of the slides follows the protocol recommended by the manufacturer of the glass slides, Schott Technical Glass Solutions, GMBH.
14. The BD Atlas™ Plastic array Hybridization Box is recommended by the manufacturer. We also strongly recommend using this hybridization box because plastic microarrays fit precisely and they are kept flat during the hybridization and washing steps. By contrast, roller bottles curl the plastic film and lead to poor images after postprocessing of the arrays.
15. We find that boiling is adequate to denature probes; however, if you prefer an alkaline denaturing procedure, you may use the following procedure
    (a) Mix approximately 100 μL of the labeled probe (entire sample) and 1/10th of the total volume of 10× denaturing solution for a total volume of ~111 μL.
    (b) Incubate at 68°C for 20 min.
    (c) Add the following to your denatured probe: approximately 115 μL of 2× neutralizing solution for a total volume ~230 μL.
    (d) Continue incubating at 68°C for 10 min.

Acknowledgments

Analysis of the host cell transcriptome was funded by the German Research Council priority program SPP1130, Infection of the Endothelium (grant Unk-135/1-1). The work on the meningococcal transcriptome was funded by the German Bundesministerium für Bildung und Forschung (BMBF) in the context of the PathoGenoMik and PathoGenoMik-Plus funding initiatives (grant 0313801A).

References

1. Lander ES, Linton LM, Birren B et al (2001) Initial sequencing and analysis of the human genome. Nature 409: 860–921.
2. Venter JC, Adams MD, Myers EW et al (2001) The sequence of the human genome. Science 291: 1304–1351.
3. Hacker J, Dobrindt U (2006) Pathogenomics. Wiley-VCH, Weinheim.
4. Schena M, Shalon D, Davis RW et al (1995) Quantitative monitoring of gene expression patterns with a complementary DNA microarray. Science 270: 467–470.
5. Joseph B, Schneiker-Bekel S, Schramm-Gluck A et al (2010) Comparative genome biology of a serogroup B carriage and disease strain supports a polygenic nature of meningococcal virulence. J Bacteriol 192: 5363–5377.
6. Schubert-Unkmeir A, Sokolova O, Panzner U et al (2007) Gene expression pattern in human brain endothelial cells in response to *Neisseria meningitidis*. Infect Immun 75: 899–914.
7. Schubert-Unkmeir A, Slanina H, Frosch M (2009) Mammalian cell transcriptome in response to meningitis-causing pathogens. Expert Rev Mol Diagn 9: 833–842.
8. Claus H, Vogel U, Swiderek H et al (2007) Microarray analyses of meningococcal genome composition and gene regulation: a review of the recent literature. FEMS Microbiol Rev 31: 43–51.
9. Schoen C, Joseph B, Claus H et al (2007) Living in a changing environment: insights into host adaptation in *Neisseria meningitidis* from comparative genomics. Int J Med Microbiol 297: 601–613.
10. Lipshutz RJ, Fodor SP, Gingeras TR et al (1999) High density synthetic oligonucleotide arrays. Nat Genet 21: 20–24.
11. Bowtell D, Sambrook J (2002) DNA Microarrays. Cold Spring Harbor Laboratory Press, Cold Spring Harbor.
12. Churchill GA (2002) Fundamentals of experimental design for cDNA microarrays. Nat Genet 32: Suppl. 490–495.
13. Yang YH, Speed T (2002) Design issues for cDNA microarray experiments. Nat Rev Genet 3: 579–588.
14. Stekel D (2003) Microarray bioinformatics. Cambridge University Press, Cambridge.
15. Parmigiani G, Garret ES, Irizarry RA et al (2003) The analysis of gene expression data. Springer, Berlin.
16. Tilstone C (2003) DNA microarrays: vital statistics. Nature 424: 610–612.
17. Huber W, von Heydebreck A, Sultmann H et al (2002) Variance stabilization applied to microarray data calibration and to the quantification of differential expression. Bioinformatics 18:Suppl 1, S96–104.
18. Cleveland WS (1979) Robust locally weighted regression and smoothing scatterplots. J Am Statist Assoc 74: 829–836.
19. Brazma A, Hingamp P, Quackenbush J et al (2001) Minimum information about a microarray experiment (MIAME)-toward standards for microarray data. Nat Genet 29: 365–371.
20. Schwarz R, Joseph B, Gerlach G et al (2010) Evaluation of one- and two-color gene expression arrays for microbial comparative genome hybridization analyses in routine applications. J Clin Microbiol 48: 3105–3110.
21. Tinsley CR, Heckels JE (1986) Variation in the expression of pili and outer membrane protein by *Neisseria meningitidis* during the course of meningococcal infection. J Gen Microbiol 132: 2483–2490.
22. Stins MF, Gilles F, Kim KS (1997) Selective expression of adhesion molecules on human brain microvascular endothelial cells. J Neuroimmunol 76: 81–90.
23. Stins MF, Badger J, Kim KS (2001) Bacterial invasion and transcytosis in transfected human brain microvascular endothelial cells. Microb Pathog 30: 19–28.
24. Li F, Stormo GD (2001) Selection of optimal DNA oligos for gene expression arrays. Bioinformatics 17: 1067–1076.
25. Dietrich G, Kurz S, Hubner C et al (2003) Transcriptome analysis of Neisseria meningitidis during infection. J Bacteriol 185: 155–164.
26. Smyth G (2005) Limma: linear models for microarray data. Springer, New York.
27. Unkmeir A, Latsch, K, Dietrich G et al (2002) Fibronectin mediates Opc-dependent internalization of *Neisseria meningitidis* in human brain microvascular endothelial cells. Mol Microbiol 46: 933–946.
28. Sambrook J, Fritsch EF, Maniatis T (1989) Molecular cloning: A Laboratory Handbook, 2nd Ed. Cold Spring Harbour Laboratory, Cold Spring Harbour NY.

Chapter 18

The Use of the Pan-*Neisseria* Microarray and Experimental Design for Transcriptomics Studies of *Neisseria*

Nigel J. Saunders and John K. Davies

Abstract

The pan-*Neisseria* microarray was the first bacterial microarray to address multiple strains and species, and is a tool specifically developed for the performance of comparative studies within and between species. To achieve this, its design was based upon a detailed comparison of multiple genomes, prior to probe selection, and serial triage to optimize sensitivity and specificity. While this tool can be used for transcriptional comparisons of the same species, such as isogenic mutants, or strains exposed to different environmental conditions, its features are also particularly suited to population and functional studies of unrelated strains. The optimal use of these tools, including the use of single-channel labeling for genomic studies, the biological replication needed to perform robust transcription studies, and key aspects of data analysis such as the use of cross-channel correction and Bayesian analytical approaches, is discussed.

Key words: Microarray, *Neisseria*, Transcriptomics, Comparative genomics

1. Introduction

1.1. The History and Nature of the Pan-*Neisseria* Microarray

To understand the proper function and use of the pan-*Neisseria* array, it is necessary to appreciate its particular properties and the purpose for which it was originally designed. Following the publication of the first two meningococcal genome sequences (1, 2), and the near completion of the first gonococcal genome sequence, the Wellcome Trust organized a postgenome meeting with group leaders from a range of laboratories working on *Neisseria meningitidis* and *Neisseria gonorrhoeae* to explore and promote the use of the genome information by the research community.

It was broadly recognized that microarrays were a tool that the community wanted to have access to and be able to use. One of the facets of *Neisseria* research is that we work with organisms with

Myron Christodoulides (ed.), *Neisseria meningitidis: Advanced Methods and Protocols*, Methods in Molecular Biology, vol. 799, DOI 10.1007/978-1-61779-346-2_18, © Springer Science+Business Media, LLC 2012

challenging population biology, reflecting its highly panmictic/recombinatorial properties, and there are several different laboratory strains that are used in different centers; only some of which have been sequenced. In addition, we work on two closely related but behaviorally contrasting species. Thus, there were issues related to the extrapolation of data from the genomes available to the other strains that people would choose to use experimentally and of generating data that would be comparable and cross-interpretable between different laboratories working with different model strains.

The consensus from this meeting was that the need to generate comparable data would be greatly facilitated by the production of a single microarray, with common probes, which the community as a whole would have access to. This meant that it should use common probes for common genes and be able to address multiple strains and both species. In addition, it was considered important that the same tool should be useful for both comparative genomics as well as transcriptomic studies of manipulated experimental strains. This tool would be produced and made available "at cost"/as cheaply as possible, and without restriction on its use or with requirements for co-authorship. At the conclusion of this discussion, John Davies, the co-author, was selected to lead this project, with the author taking the lead in array probe design.

Unfortunately, this arrangement rapidly fell apart after the withdrawal of several key investigators from the original project. Subsequently, at the biennial *International Pathogenic Neisseria Conference* in 2000, four US groups signed up to contribute resources and to assist in cross-center validation of the arrays. These groups were primarily focused upon *N. gonorrhoeae* research, and this determined the probe selection/design strategy that was adopted.

So, the final "brief" was to generate a microarray that could be used for each of the strains of both species that had been sequenced, and as far as possible could be used to investigate unrelated un-sequenced strains as well. To assist in this, since *N. gonorrhoeae* strain FA1090 does not possess the gonococcal genetic island (GGI), Joseph Dillard from the University of Wisconsin-Madison was generous enough to provide us with the then unpublished sequences for the strain MS11 GGI and two other strain-specific islands from this strain (3).

1.2. The Probe Design Strategy

The goal was to generate a microarray that could generate as consistent a signal as possible for each gene, and thus for which the hybridization properties of each probe were broadly similar. The aim was to use probe regions that were optimal for binding to orthologous (the same) gene in different genomes and were maximally able to discriminate between paralogous (similar, but functionally different) genes within each genome. The focus was upon the use of common probes for orthologous genes as much as possible, for the experimental reason that this would generate the most comparable data, and for the practical reason that this would minimize the number of probes required on the microarray.

To achieve this we adopted a novel design strategy. The genome sequence information for all three genomes were combined and the annotations overlayed in a series of three ACE databases, each indexed from one of the three genome sequences, with the unique *N. gonorrhoeae* strain MS11 sequences appended as an additional contig in the *N. gonorrhoeae* strain FA1090 database. In these comparative databases, the regions of greatest DNA identity were displayed, using width settings that highlighted regions of greater than 90% identity between the genomes. At the same time, translated sequence similarities were displayed so that we could identify orthologous and paralogous genes between and within the genomes. Regions that were of high identity within orthologous genes were extracted, initially working from *N. gonorrhoeae* strain FA1090. These extracted regions were then used as the templates for primer design using Primer3 with length optima of 250 bp.

This was an extremely slow manual task, largely performed by Lori Snyder, in the author's group. The process of sequence comparison, probe design, in silico probe testing, selection of revised regions to optimize probes within genomes and re-testing, and then the analysis of each probe within each genome were very time consuming, and ultimately took five iterations before the design phase was completed. It is the labor-intensive nature of this strategy, which is difficult to automate, which is probably why this approach has not subsequently been used by other groups for similar projects in other species. But, it is also the reason why this microarray remains the most complete polymerase chain reaction (PCR) product-based bacterial microarray for any species that we are aware of, and why it has particularly good gene discriminatory and detection properties.

Subsequently, the genome sequence of the serogroup C *N. meningitidis* strain FAM18 was released (4), and the pan-*Neisseria* version 1 (v1) microarray was revised (to generate the pan-*Neisseria* version 2 (v2) microarray) to include the additional genes from this sequence and annotation, as well as with genes identified in ongoing studies of neisserial minimal mobile elements (5, 6). This required the addition of several hundred additional probes but only ten were required to probe for divergent orthologues, and we also added long oligo probes for two genes for which PCR products could not be generated, so that the array was 100% inclusive. We also added probes for experimental markers, including antibiotic resistance cassettes and reporter genes, which can be of value in confirming the orientation of datasets being analyzed. In this version, the array contains 2,835 different probes.

1.3. Microarray Printing

The way in which a microarray is printed directly influences the quality of the resulting data and the number of replicates required for an experiment. Early microarray printers, including that used for the original production of the pan-*Neisseria* microarray were only capable of printing spot replicates in adjacent positions. Because processing artifacts tends to be physically regional within a microarray, adjacent replication is the functional equivalent of no replication, but where the probe has been divided into two or more components. To overcome this, and to provide a second source of production, we transferred the primers and primary PCR products for the original (v1) microarray to the University of Oxford and set up an additional printing facility using printers that could generate nonadjacent replicates within subgrids. All of the UK produced arrays, therefore, included threefold nonadjacent replicates: threefold replication is a functional minimum for triage and quality control, because of the need to be able to identify which spot feature is the "outlier." We have done experiments on the influence of replication on data quality, and find that there is continued improvement up to five- and eightfold replication, and that with eightfold replication data extraction can be reliably automated, but that is not normally experimentally practical.

1.4. Not All Pan-Arrays Are the Same

To distinguish our array we called it the pan-*Neisseria* array. However, another array was subsequently generated by the Bacterial Microarray Group at St. George's Hospital (http://bugs.sgul.ac.uk), which, unfortunately, was referred to by exactly the same name. However, these arrays are not equivalent, and the data from them are not necessarily directly comparable. The difference stems from a fundamental difference in design strategy.

The more usual design of a pan-array follows one of two strategies, both of which differ from ours, and generates probes with different sensitivity/specificity characteristics. In the most widely used approach, one genome is used to generate a set of primers/probes that addresses that single genome. It does this without reference to the regions within the annotated genes that are most conserved in different strains/related species, because it does not refer to the information from these genomes at this stage in the design process. It does, however, have a huge advantage of speed, and therefore cost, because it can be run automatically. In the next phase, the probes from the index strain are assessed for their binding to the second and subsequent genomes. If a hybridization threshold is passed, then the probe is considered good for both genes in both genomes. If that threshold is not passed, then the gene from the second genome is used to design an additional/new probe, which will be generated from this template. The process is then repeated for subsequent genomes.

There is an alternative approach, which is likely to have been adopted for the St George's *Neisseria* microarray. In this strategy, unique genes are identified by BLAST searching, by comparing each annotated gene with all of those present within and then between the genomes using a bit score (usually half that of the "self" vs. "self" match) to identify each "unique gene." This is much quicker and easier to do than working with percentage matches over specific regions, because it works with the type of data directly generated by the BLAST search engine, without further analysis, and is readily automated. Then, for each "unique gene" a primer pair and probe is designed and generated. However, this approach has the potential to be even less specific or stringent than the previous approach, particularly because some genes have very variable similarity and differences along the whole length of the gene, and the regions of similarity and difference are not specifically addressed.

The limitation of this approach, when designing a tool to address multiple strains and unrelated strains, is that it uses thresholds of adequacy, rather than seeking optimal solutions. This is partly because it does not compare the genes between the genomes to identify the most similar regions that exist between the different orthologues, and does not identify the regions to avoid that are most similar between paralogues within genomes. Only by using all of the available genome sequence information simultaneously can probe design be optimized in this way. But, pragmatically, this approach is not always practicable or affordable.

These two arrays also differ in the length of the probes. There is a direct relationship between the length of a microarray probe and its performance in hybridization assays. There is a progressive increase in binding affinity with probe length, such that the coefficient of variability of the signal intensity and measurement bias are best at 500 bp and above. Short oligos (less than 60 bp) are particularly difficult in this regard. Ideally, for stringency, probes need to be 70 bp or greater, and if they are to tolerate minor polymorphisms, as is necessary for comparative studies, or use on unrelated strains, they should ideally be over 150 bp in length. For a theoretical analysis of this, see the article from Chou et al. (7).

However, as the length of the probe increases, the specificity of the probe decreases. If the probe regions are selected at random, there is an asymptote between 60 and 70 bp at which the stringency of a probe is progressively lost, that becomes a progressively greater problem the longer the probe, with probes of greater than 150 bp being particularly problematic (7). To address this in our probe design, we tried to keep a relatively narrow window for probe length. Also, the process of selection for the regions of the genes that would be used for probe design included the comparisons against the index and other genomes, so regions with significant identity to other targets were specifically excluded. For genes greater than 150 bp in length, probes were sought between 150 and 300 bp, and the final mean probe length was 230 bp. For shorter genes, shorter probes were accepted. And, once the probes were designed they were serially rejected and re-designed if any probe contained a region of greater than 90% identity over any 20 bp, which is even more stringent than was used in oligo arrays (such as the 70 mer *Neisseria* array from Operon). The great majority of this work of design was done by Lori Snyder, and the quality of the resulting array reflects considerable effort on her part. By contrast, for the

St. George's pan-*Neisseria* microarray the size range used for primer/probe design was 100–800 bp, with an optimum of 500 bp, and it did not involve the same degree of subsequent filtering and re-design. This range spans a much wider range of hybridization characteristics and cannot be optimally specific. This approach is pragmatic and not unreasonable, and is far quicker and cheaper, but it is important to appreciate that there are qualitative differences in microarrays that depend upon this type of design process and the parameters used, and the two so-called pan-*Neisseria* arrays are not equivalent tools, because of the differences in their design and probe content, and this almost certainly accounts for some of the apparent discrepancies between the studies that have used them.

1.5. The General and Specific Features and Appropriate Uses of the Different Neisseria Microarrays, and the Experimental Settings for Which the Pan-Neisseria Array Is Specifically Suited

The best microarray to use depends upon the type of experiment undertaken. Our purpose was to generate a single tool that could be used for multiple neisserial species and strains, and it was the first microarray that did this. This was for pragmatic reasons reflecting the nature of the strains used in the research community, and because we wanted to be able to perform comparative genomic and transcriptomics studies. With regard to transcription experiments, if the strain to which the probes were designed is used, then there is no particular *a priori* advantage to the use of the pan-*Neisseria* microarray over any other well designed and optimally printed microarray with long-oligos or PCR product probes (for example, those from Operon or TIGR). This is especially true if the intention is to perform singular experiments, and do not intend to directly compare the data with other studies generated on different platforms.

The particular advantages of the pan-*Neisseria* microarray are in experiments on strains other than those that were used as the index strain for design/probe generation, if the wish is to address a strain that has not been sequenced, or to directly relate/compare data with other studies that have used it.

1.6. Preliminary Experiments for Use with an Un-sequenced Strain

It is wise to confirm which probes are applicable to an experiment, and thus which probes should be subsequently analyzed before proceeding to transcriptional profiling studies. This is for two essential reasons. First of all, the interpretation of the absence of a change for a gene that is not present represents a false negative. Second, data from probes that are negative tend to have greater noise than those that are positive. So, only the probes that address the strain under study should be included in the subsequent analysis, and the other probes should be filtered out (prior to normalization). This will improve the distribution of the data and the quality and power of the statistical tests.

In this chapter, transcriptomics is addressed primarily, rather than comparative genomics, but aspects of the latter are relevant to this preliminary work. There are two approaches that can be used for this purpose in both settings: dual channel (comparing the strain of interest against one of those represented on the microarray) or single channel. The better option is single channel, and by preference this should use the red/Cyanin (Cy)5 wavelength.

The inherent fluorescence from DNA probes on microarrays and the predominant background fluorescence from buffers and washing artifacts are almost exclusively in the "green/Cy3" channel. Thus, if one color is to be used, "red/Cy5" is almost always the better choice.

Also, although a well-performed microarray will generate reasonably uniform hybridization signals with labeled genomic DNA, not all probes hybridize similarly. These differences are partially controlled for by using comparative/dual channel approaches and relying upon the fold ratio more than the raw intensity. But, differences in copy number can be problematic with this approach, and the gene complement may not be identical between the strain of interest for expression profiling and the strain in the "control" channel. And, the copy number problem is actually compounded when using pools of strain DNA in the control channel.

Dual channel hybridizations are also subject to something we call "probe walking." Since we have not seen this described elsewhere, it will be discussed briefly here. In the process of pan-*Neisseria* array validation, we were primarily focused upon confirming the presence of functional probes, and optimizing hybridization conditions to ensure that the signals were specific. This was primarily done by performing comparisons using the three (subsequently four) genomes used for probe design and templating. Once optimized, we were able to obtain good fold-ratio data, consistent with what we were expecting in these comparisons, but it was notable that some probes for genes present in neither strain would sometimes give weak signals. In the following series of experiments we studied *N. gonorrhoeae* strains, which are divided into two broad groups that either do, or do not, possess an F-plasmid related element called the GGI, which is typically inserted at the *dif* site. This is an island with well-defined ends, to which we specifically included probes based upon *N. gonorrhoeae* strain MS11. We noted that when strains that did, and did not, contain the island were directly compared there was a "tailing" signal from the GGI-negative strain into the ends of the probes for the island. This troubled us for some time, and was one reason why we adopted a "triangulated experimental design," such that each strain was compared against two other unrelated strains, to ensure that the data generated for each was consistent. We investigated this artifact and found that the GGI-negative strains only generated the anomalous result when hybridized with a GGI-positive partner. Also, a similar process could not be demonstrated when transcripts from two such strains were compared. So the process is dependent upon the presence of continuous DNA and a mixture of positive and negative samples.

Our interpretation is that while the DNA selected for the probes is carefully designed and triaged for specificity, a complex process occurs during hybridization to double-stranded DNA samples, such that the bound DNA extends beyond the probe, into adjacent sequences and into less specific regions. As a result the probe bound sample effectively extends the probe, in a way that can generate an artifactual signal for adjacent sequences.

1.6.1. Klenow-Based Labeling of DNA Using Cy-5

In our experience the easiest and cheapest method for labeling DNA is to use the Klenow-based protocol with Cy-5, as follows:

1. Purify DNA using the CTAB method, as described in ref. (8).
2. Take 20 μg of purified *Neisseria* chromosomal DNA.
3. Incubate the purified DNA with a labeling mixture of dNTPs including 5 units of the Klenow fragment of DNA polymerase I, and 3 μg of random hexamer primers, in a 50 μL reaction volume (made up with pure water), for 3 h at 37°C.
4. Remove unincorporated nucleotides and random primers using a suitable column (e.g., a QIAquick nucleotide removal column) following the manufacturer's instructions (Qiagen).
5. Hybridize the probe to the microarray, in a 4× SSC, 0.3% (w/v) SDS buffer, or with commercial SDS-based microarray buffer, overnight at 65°C. Do NOT use so-called enhanced buffers, formamide-based buffers, or those containing polyethylene glycol to improve hybridization or use a lower hybridization temperature, because each of these alters stringency and can result in false positive signals. Any suitable hybridization chamber can be used, and we typically recommend using LifterSlips (Eerie Scientific). This can be done in a sealed hybridization unit in a water bath, but ideally should be done on a slide hybridization system that mixes the sample during hybridization.
6. Wash the slide thoroughly in 1× SSC, 0.05% (w/v) SDS, preheated to 65°C for 2–5 min with shaking (being careful that the slide does not break the meniscus of the wash solution, because this affects the uniformity of washing).
7. Wash the slide thoroughly in 0.06× SSC at room temperature for 2–5 min. Dry using an airbrush. If an airbrush is not available, dry by placing the slide in a 50 mL Falcon tube and centrifuging at 1,300 × *g* for 2 min.
8. Scan the microarray at 5 μm resolution, adjusting the PMTs so that the brightest spots achieve around 90–95% saturation.
9. Extract, triage, and analyze the data. The best tool to extract and triage this type of array data is BlueFuse (BlueGnome). BlueFuse contains a tool that was developed specifically for array CGH applications. This overcomes the need to perform dual channel/two-sample hybridizations with analysis based

upon fold ratios, but instead enables the analysis of two channels independently, and in single sample/color datasets. This method and its theoretical basis are described in detail in an Appendix in ref. (9). The key elements are that the shape, location, and other quality scores are used to determine a probability that the gene is present, which are then used to generate the "pON" score for each channel independently. This can be used alone or with quality measures such as "offset" or "circularity." These should be adjusted until the automatic scoring/ranking is identifying around 90% of the true positives, and then the remaining "boundary" should be addressed manually, to generate a list of genes that are present, which can then be used to filter the data used in the subsequent transcriptional studies.

Since confusion in this regard has been created in the literature, it should be said that this method of calling gene presence/absence (or, in the context of what will be discussed later in this chapter, detectable transcript presence/absence) is not primarily based upon signal intensity. This was raised in a discussion of why different neisserial array-CGH studies generated different findings/conclusions (10), so should be addressed. Clearly, the presence of detectable signal and a minimum of hybridization intensity above background is a prerequisite for subsequent quality scoring, so no microarray analysis is completely independent of signal intensity. However, after that threshold is passed, the pON data, especially when it is combined from spot replicates, is far more dependent upon the quality scores than the raw intensity data, and this is a key aspect of its reliability. We believe it to be the best method for calling presence/absence, and that the differences in the CGH projects are due to a combined effect of differing probe design strategies, and the issues of pair-wise dual channel comparison.

2. Experimental Design and Use of Any Microarray for Transcriptional Analysis of *Neisseria*

No matter what the method, expense, or the scale of the data generated, all experiments addressing biological systems are subject to the same general considerations with regard to experimental design. Specifically, the observations must be technically robust, *and they must be reproducible by the biological system.* The former can be ensured by using well-tested methods and tools, and gaining initial training from someone with practical experience. The latter can only be ensured by using adequate biological replication.

Different biological systems exhibit differing behavioral consistency. This reflects differences in their regulatory systems and responses to differences in their microenvironments, and the influence of stochastic processes, which are increasingly recognized even in systems that do not make use of random gene switching through phase variation. We have worked with plants, cell lines, animal samples, and multiple bacterial species, and *Neisseria* are amongst the most biologically variable of all that we have studied. In a series of direct comparisons, we found that liquid cultures of *N. gonorrhoeae* are more variable than solid. And, mid-log phase liquid cultures can be very different – even if grown in parallel, from common seed stocks, and sampled at the same time and optical density. This has several practical implications for microarray-based transcriptional profiling.

Based upon our observations, we recommend when working with *Neisseria* species that (a) comparisons should include at least six biological replicates, of which no more than two should be grown/prepared simultaneously and (b) biological replicates performed on different days should be obtained independently from the seed stock cultures, rather than sequentially from serially passaged cultures. However, these recommendations should be approached pragmatically, and the approach should be adjusted to the specific goals of the experiment.

2.1. Technical Replicates, Dye-Balanced Experiments, and Pooling

Experimental planning requires a consideration of the following:

1. If the technique is used properly, the primary source of differences between biological replicates is biological (and not technical) in origin. If it was not, it would not be a robust or appropriate experimental assay.
2. All technical noise will be present in the biological replicate data; thus, in a properly replicated experiment this is addressed in analysis.
3. Technical replicates are repeated measurements of the same biological samples, and thus for analytical purposes they must always be combined to generate a single dataset/observation, i.e., a single slide has an "$n = 1$," and three technical replicates of the same biological materials also has an "$n = 1$!" Essentially, given the cost and nature of microarray experiments, it is always better to generate more biological information than to perform technical replicates that will be reduced to a single dataset. So, unless the aim is to validate the technique in your hands (which is a distinct and separate objective) the focus should be on biological replicates.
4. The experimental design should avoid systematic technical errors and biases. The most important source of biases is related to dye- and detection-related biases that affect the two labeled samples differently. This is usually attributed to the different influences of Cy-3 and Cy-5 on directly incorporated labeled nucleotides during reverse transcription, when using a "direct labeling technique." However, there are other sources of "dye bias" such that this remains a consideration when working with

secondary labeling methods, such as aminoallyl-dUTP based methods, or with direct detection methods, such as using 3DNA dendrimers. The influence of dye-biased labeling on detection is often manifest by what is technically known as "a banana" in the dataset when the image is acquired, which describes the curve in the data distribution which usually tails with a green bias at low and/or high intensities. One of the reasons that it remains even when incorporation bias has been eliminated is that the inherent fluorescence of DNA (or protein) on microarrays tends to be in the green channel, as do washing artifacts (especially related to SDS). In addition, there is a phenomenon we refer to as "dendrimer spots," the exact basis for which is unknown, in which some spots will repeatedly report in one color when using 3DNA detection reagents. This is more frequently a problem when using very high density (eukaryotic) microarray probe sets, but there is one probe (for XNG1546/pilF) on the pan-*Neisseria* microarray that is influenced by this phenomenon and should be excluded from analysis. Notably, it only occurs when performing dual channel experiments, and does not occur when the detection reagent is added to the microarray in the absence of the tailed cDNA, so we hypothesize that it may reflect a "probe walking" phenomenon, similar to that described previously.

5. To avoid labeling and detection-related biases, the experimental design should be "dye balanced." Dye balanced means that, as far as possible, experiments should be equally labeled in each direction, and any changes should be verified that they occur in at least one microarray labeled in each orientation. This is not the same meaning as "dye-swap," which means a technical replicate pair, one of each of which has been labeled in each direction, but which must be reduced to a single biological observation/datapoint. For this reason, even when performing a relatively small experiment, or when screening data, an $n=4$ experiment considering genes reported in three or more of the assays is considerably more robust than an $n=3$ experiment.
6. Pooling is an inherently information-poor form of experimental design, and should be avoided. The statistical reason for this is that all of the replicates are performed with the same combination of biological materials, and are thus all technical replicates, and the experiment is essentially $n=1$. There are examples in the literature (including the *Neisseria* literature) of groups analyzing this type of data as if the technical pooled-sample replicates were independent, but this is not sound and is essentially performing an analysis of the reproducibility of the assay rather than the biological system. The biological reason for this is that a pooled experiment is unable to distinguish between common and inconsistent changes in the individual cultures. For example, if the samples from three test cultures are combined in equal

proportions, and compared to three similarly pooled controls, a twofold increase in the test pooled sample in the hybridizations might represent a twofold change in all of the test samples, or a fourfold increase in just one of them, while the majority of samples were no different. If these two "issues" are combined the problem is compounded, because three replicates in which the technique reliably reports a twofold change three times, if incorrectly analyzed as $n=3$, gives a robust statistical significance to the observation, while the true situation of two with no change, and a single replicate with a fourfold change would not be determined as statistically significant in the analysis.

2.2. The Influence of Labeling Strategy on the Final Data

When the pan-*Neisseria* microarray was first generated, samples were sent for a comparative validation experiment to four different laboratories to perform a combined experiment between the different groups that had contributed to the resources to generate it. At that time the predominant thinking was that microarrays tended to be technically noisy, and that the details of the labeling and processing had significant influences on the resulting data. This was a matter of concern, since one of our objectives was to be able to have a tool that would facilitate data exchange and meaningful comparisons, but the array infrastructures available to each group were different. Some groups used direct labeling, others aminoallyl-dUTP-based indirect labeling, some used Cy dyes, others Alexafluors; hybridization buffers and available systems differed, and at least three different types of scanner were used. The bottom line was that these differences did not make substantial differences to the final data, and that each method (if used properly) was broadly and similarly capable of generating similar data from common RNA samples.

2.2.1. Considerations for the Selection of Labeling Strategies

The individual researcher can and should feel free to use whatever labeling strategy is available. However, this does not mean that all methods are equal, although adequate replication will compensate for most weaknesses. When selecting a method the investigator should be aware of the following considerations:

1. Methods using direct incorporation of labeled fluorophores are prone to the influence of the dyes on the dNTPs affecting the reverse transcription efficiency. This can be both length- and sequence composition-dependent.
2. Indirect incorporation experiments are influenced by suboptimal incorporation of modified bases, but the effects are similar (at this stage) in both channels.
3. Dyes are differently sensitive to light, ozone, and hydrolysis. Cy-5 is the best dye in terms of its properties as a dye, but it is more labile than Cy-3. In fact, it is possible to judge the technical expertise of the operator by the final brightness of these two dyes.

An expert will have a slightly brighter Cy-5 signal, while a novice will typically have much weaker Cy-5 (or almost none at all). Alexafluors can be more reliable in inexperienced hands, but they are more expensive and not quite as inherently bright. Other more modern alternatives can also be considered, e.g., modified Cy dyes and Oysters.

4. Usually, the best way to reduce microarray artifactual noise and to obtain optimal signal-to-noise ratios is to generate more signal.
5. No matter whether you use direct or indirect methods in which a subset of bases in the reverse transcript are labeled, the signal will always be dependent upon the length of the original mRNA template. This is also a function of the location of the probe within the message. On average (assuming an even distribution of priming and extension to the end of the message) there will be twice as much of the 5′ region of the transcript reverse transcribed than the 3′ region. Thus, a probe at the 3′ end of a gene will be less sensitive. And, a 100-bp product will be a quarter as bright as a 400-bp product. This means that the probes on an array will have different sensitivities depending upon the length of the transcript and the position of the probe. The only solution to this is a unitary tag based system.

2.2.2. 3DNA Dendrimer Labeling

For most purposes, in recent years, our preferred method of labeling has been to use the 3DNA dendrimer labeling methodology (predominantly with Cy dyes, although the more recently available alternatives may have advantages). The reasons for this are:

1. The reverse transcription step involves only natural/unmodified bases, which we believe to be the most effective and least subject to bias.
2. The initial hybridization of DNA reverse transcript to the probes is a simple unmodified DNA-to-DNA hybridization, the properties of which are well understood.
3. Each reverse transcript is labeled with a DNA target sequence which is detected in a second hybridization step, and this is the first time that the fluorophores are used, so they are minimally exposed to situations in which they differentially fade.
4. Each "tag" is detected with a uniform number of fluorophores, so the strength of the signal is not influenced by incorporation differences (although there is still an influence from the length of transcript available for priming).
5. The sensitivity of the assay can be enhanced using highly labeled dendrimers. Thus, our standard protocol using the 3DNA 900 MPX kit (Genisphere) will use between 2 and 5 μg of total RNA in the initial labeling reactions, although we have successfully used 500 ng (and in eukaryotic experiments, in

which poly-A labeling is used, 100 ng). But, if you consistently have more RNA than this and wish to save resources, the "50" and "350" MPX kits can be used.

6. The method, when used optimally, gives a very high signal and thus good signal to noise, with saturating signals at intermediate laser/PMT settings, which reduces the influence of background noise and artifacts.
7. We recommend using the MPX kits with the following modifications and locally determined conditions:
   (a) Use Superscript III from Invitrogen, rather than the supplied reverse transcriptase, because we find that this gives consistently better yields.
   (b) Perform the reverse transcription step for 3 h.
   (c) Only use the SDS buffer (Vial 6) for both the first and the second hybridization. The alternative, the so-called EB, or enhanced buffer, does have a beneficial effect on hybridization, but it comes at the expense (in our hands) of increased drying and other artifacts. The formamide buffer should never be used because it is associated with false positive signals, which in control CGH experiments can be shown to indicate the presence of genes that are not present within known/sequenced genomes. Originally, formamide was included in Northern Blots because it reduces background from nitrocellulose membranes. This comes at the cost of altered DNA conformation, which reduces binding and also stringency, which is why the hybridization temperature is lower. Its use in glass-bound microarray experiments is an historical anachronism and should be avoided.
   (d) Use LifterSlips (Eerie Scientific) and SlideBoosters (Advalytix) for the hybridizations. The temperatures used are 60°C for the first hybridization (hybridization of probe and sample) and 55°C for the second hybridization (hybridization of DNA tags with the target sequences on the dendrimers). The first hybridization is typically done for 16 h, and there are clearly measurable improvements in the signal up to 12 h. The second hybridization is typically done for 4 h. The first wash after each hybridization step should be performed at the hybridization temperature. Subsequent washes should be at room temperature.
   (e) Use SlideBooster settings of 7:3 and a power of 25.
   (f) When washing the slides, it is important to keep the slides under the surface/meniscus of the wash solution. If part of the slide is brought through the meniscus repeatedly then this will cause differential washing and nonuniform signal and background that cannot be normalized, i.e., the data will not be safely extractable/usable.

(g) Scan the slides as soon as possible after the second hybridization. Slides are scanned at 20-μm resolution to optimize the scanning. Optimal scanning will gain interpretable (nonsaturating) signals from the largest number of probes. If none of the probes are saturating then a full power/sensitivity scan should be performed. Typically, the ribosomal RNA probes and the probes for an island of rRNA-associated proteins are the most abundant. It is possible to saturate these spots in exchange for better coverage of other probes, but the fold ratios for these probes will then be inaccurate. Once the laser/PMT settings are optimized, then a final scan at 5 μm resolution should be acquired for subsequent extraction and analysis.

2.2.3. Drawbacks to the 3DNA Dendrimer Labeling Method

However, there are some drawbacks to the 3DNA dendrimer labeling method, as there are to all of the alternatives. These are:

1. The method is comparatively slow, because it involves multiple steps that effectively take 2 days, including the two hybridization steps.
2. The column cleanup steps in this protocol are not infallible, and we occasionally lose a whole channel of data as a result.
3. The binding of the larger dendrimers is diffusion limited, and labeling is not optimal without on-slide mixing. (Actually, this is also true for DNA:DNA hybridization, but the effect is not as great.) This is also true for all labelling methods, so ideally hybridization should always involve agitation or mixing to generate the brightest, most consistent and most reliable data, but this is an additional factor, particularly when using the "900" dendrimers. The "50" kits perform reasonably in standard "water bath" and similar incubation methods, although they still give better results when performed with on-slide mixing.
4. It is comparatively expensive, although the "50" kits are priced to be close to the use of more traditional methods.
5. However, on balance, the 3DNA dendrimer method is preferred, particularly because of its sensitivity, and because it is comparatively easy to obtain 2–5 μg of RNA, or to have sufficient RNA to be able to use a single sample in more than one comparison.

2.3. Image Extraction and Triage

As mentioned in step 9 of Subheading 1.6.1, our preferred tool for image extraction is BlueFuse for microarrays (BlueGnome). It is comparatively expensive (especially compared to the free alternatives from TIGR), but in a high-throughput environment this is justifiable given its particular additional functions, that can be used to assist in the triage of the dataset for spots that have not reported a genuine signal, and to exclude hybridization and other artifacts.

1. First, inspect the image for overall quality. It is better to discard a poor quality dataset than to attempt to include it in the analysis, even if this reduces the number of experimental replicates. Ideally, there will be sufficient material to repeat a low quality hybridization. The most important thing to look for is areas of the image that have much brighter spots associated with higher background. This typically indicates regions that will have different properties such that the data will not be robustly normalized, because the median values will be different for spots extracted from regions of the image that differ in this way. The severity of the problem can often be determined by looking at the fold ratios of replicate spots that span the edges of this type of artifact. However, if the SDS-based buffers are used, and care is taken with washing, this type of artifact is uncommon.
2. Once it is decided that the data from the array are suitable for analysis, the next step is a manual exclusion of two things: obvious physical artifacts (e.g., dust, streaks from washing, smaller regions of the slide with untypical hybridization) and extracted data from probe spots that have not reliably reported in either channel. How the "absent" data is excluded is largely a matter of taste and experience. Some people simply filter on a particular intensity, but the background and suitable threshold does vary from image to image, so the use of a single value for all slides in a dataset at a later stage of analysis is not an optimal solution. Our preferred approach is to use the presence of a reasonable pON value in one or other channel, combined with a visual triage of spots with markedly off-center placement (usually not a true spot), followed by a subgrid-by-subgrid manual exclusion. The method chosen is not important, but because the fold-ratio data extracted from low intensity (absent/undetected) features is wide and statistical analysis is based upon the fold-ratio distribution, taking the time to do this properly pays significant dividends in subsequent data analysis.
3. Finally, an important point to consider is that is it essential that the researcher can always return to the slide images to check that the conclusions from the analysis match with what the image shows. This is critical, especially if just a small proportion of hybridizations show a change which is potentially important, it must always be possible to refer back to the image, because the images are the primary data, not the numbers in the analysis spreadsheets.

2.4. Data Analysis

The keys to microarray data analysis in any setting are (1) only analyze the real data, (2) keep it simple, and remember that every data transformation is associated with a loss of information, and (3) understand what you are doing and why. We prefer to use the BASE microarray LIMS and analysis platform, which is free to academic users. We actually use the last revision of BASE version 1, because BASE version 2 is somewhat harder to navigate and the first version is stable, contains a lot of historical data, and does everything that we need. Something we particularly like about this system is that it keeps a clear record of all steps in the analysis of the data, and it can readily be used to compare data from a large number of experiments. It also keeps things simple!

1. Once the data is loaded, filter the data to only address the probes that address genes within the genome under analysis. This should be done before any other normalization or analysis. Collaborators usually analyze their data within our BASE installation (which can be done remotely over the internet) which has the dual advantages that we can assist them with their analysis, and the system includes progressively updated tables with up-to-date annotations and columns that enable the data to be filtered for each of the five genomes currently indexed (i.e., *N. meningitidis* strains Z2491, MC58, and FAM18, and *N. gonorrhoeae* strains FA1090 and MS11). If someone is analyzing their data in a different system, a spreadsheet identifying which probes address each of these genomes can be provided. If a different genome is being used, then the data should be filtered on the basis of the preliminary CGH data obtained as described earlier in this chapter.
2. After filtering, carry out any correction for channel cross-talk. It is well known that there is a degree of stimulation and emission cross-talk between fluorophores, and this is routinely corrected for in other settings (for example, in flow cytometry). In a (currently unpublished) study comparing microarray and serial analysis of gene expression (SAGE) data, we found that correcting for channel cross-talk could generate microarray data that much more closely reflected the fold ratios from SAGE and from quantitative Reverse Transcriptase-Polymerase Chain Reaction (qRT-PCR) assays. There is an additional aspect of using 3DNA dendrimer labeling, in that a small proportion of the detection reagents can detach and reattach to the targeting sequences of the other channel. This can also be corrected for and we have a bespoke cross-channel correction plug-in for BASE, which is available on request. Because this cross-talk is a function of absolute intensity, it should be addressed before any other scaling or normalization. Empirical testing and comparisons with parallel data prepared using SAGE (unpublished) indicate that good default values are to subtract 2% of the other channel's intensity and to place a floor of 50, so that subtraction does not lead to values significantly below background.
3. Normalization is the next step of data processing and this should involve the minimum manipulation necessary to generate

comparable datasets within the experiment. Typically, the preferred normalization will be to use the Lowess algorithm. This addresses any "banana" in the data due to labeling or image acquisition, by tracing a median value through the curve in the data and using this variable value to normalize and straighten the data. If the data is not "curved," as can be the case when you have very high signals from dendrimer labeling with very low background, then only a global median normalization is necessary, and if appropriate this lesser manipulation is preferable. But, if any dataset within the experiment requires Lowess normalization, then it is important that all datasets are processed similarly.

Normalization of array data is based upon an underlying assumption that only a small minority of the transcripts differ, and if they are changed the net difference between those that are induced and repressed is similar. This assumption is badly flawed in bacterial studies, especially those comparing profiles in very different experimental conditions, and in studies of responses controlled by global regulators. We first identified this issue when comparing data from experiments addressing contrasting atmospheric conditions, in which the substantial changes in expression profiles resulted in clearly changed and adaptive genes not being reported to have significantly altered fold ratios, while unchanged, or marginally changed genes in the less common direction were being falsely identified. To address this, we use an alternative normalization method, in which it is possible to exclude the 10% (or other user-defined) most increased and decreased fold ratios, in addition to the traditionally excluded ratios from noise-prone high and low intensity data. This is particularly important if the data has a "wide spread," and the differing transcripts have a clearly visible bias in one direction. Again, a bespoke plug-in for BASE to do this is available on request.

4. Determine the fold ratio using the mean value. Most people determine the fold ratio using a median value, or a trimmed mean in which the least consistent data is excluded. When working with high quality data, which has been manually triaged so that only real data from probes addressing measurable transcripts within experiment have been included, then the use of median values is to fail to optimally make use of the data that has been collected. When working with properly triaged data, it is much better to use the mean value, which is also more coherent if subsequently working with confidence intervals. The mean, standard deviation, and confidence interval will then provide meaningful information with respect to the nature of the consistency of the system behavior.
5. The choice of statistical test for significance is important. We have tried multiple alternatives and the best is Cyber-T (11), which

was developed originally by Pierre Baldi at the University of California in Irvine, and has been used to analyze data reported from 100 of microarray studies. The traditional Student's *t*-test is a very poor tool for the analysis of biological data as collected by microarrays. This is because while the direction of change and relative rank position of a fold ratio change tend to be consistent, the absolute numerical value can vary quite widely. Nobody would question that a gene was induced if it was changed two-, four-, sevenfold in a triplicated experiment, but the Student's *t*-test would not call it significant, while putting a robust *p*-value on something altered 1.3-, 1.4-, 1.5-fold close to the technical resolution of the assay (or within it, if performed poorly). We struggled with this for a long time, when we could see within the data changes that made biological and functional sense that were not deemed significant, while other "noise" was identified. Eventually, we found Cyber-T, and these issues were clearly resolved, especially when combined with the modified normalization described in step 3 above. We now use Cyber-T as our primary test and, again, a bespoke BASE plug-in is available on request. If an experiment is well replicated, with six or more replicates, then a Student's *t*-test can also be used, and because the underlying models of the two tests (rank order of change and distribution, respectively) this can be more informative than simply lowering the threshold of significance using a single test. When using cyber-T, adjust the parameters to the dataset. The size of the window should not typically be larger than 5% of the number of genes being addressed (it also needs to be an odd number, at least in the BASE implementation): thus, 101 is a good typical value for an experiment addressing *Neisseria*. In addition, the "Bayes confidence estimate value" (BCEV) should be no more than three times the number of biological replicates, but with a typical maximum of 10. Thus, a preliminary study using two replicates would use a BCEV of 6, while a full dataset of six replicates would use a BCEV of 10.

6. Finally, it is valuable to formally address consistency within a dataset. Confidence intervals are a particularly good tool for this purpose, and it is also useful to be able to determine whether any biological replicates have made no change, or changed in the opposite direction. This can be screened for within Excel (or we have a "confidence overview" plug-in for BASE available on request). The reason that this is valuable, especially if an outlier is clearly confirmed by re-inspecting the image, is that it distinguishes direct and indirect effects, since if the regulation of a gene is directly and mechanistically linked to the response, it can never "go the other way," but it might still be part of an integrated set of secondary changes.

2.5. Interpretation of the Results

In the end, a judgment will always have to be made with respect to what changes are biologically important. This is a separate issue from the technical resolution of the assay, and from statistical significance, and should be recognized and formally considered. If the methods outlined in this chapter are used, and have similar technical variability to us, the result will be a technical resolution that will enable interpretation of differences in expression of ±0.2-fold, i.e., if the mean fold ratio is below 0.83 or above 1.2 then you can be confident that the abundance of the transcript differs. If other labeling methods, different hybridization conditions, or different protocols are used, then you may lose some of this resolution, and will have to determine what this technical variability is before the data can be interpreted robustly at these lower fold ratios. The best way to do this is to take the same RNA sample, label it as if it were two independent samples, and compare it against itself. This experiment should be carried out on three different days (with the same RNA sample, with the same number of freeze-thaw cycles), and the data should be analyzed as if it were an experiment with three biological replicates. This will give specific data about the technical reproducibility and resolution of the method in a particular laboratory.

Critically, the fold ratio that you choose to interpret as biologically important must be greater than the technical resolution of the assay *in a particular laboratory*. For example, even if the technical variability was twice what we normally expect, it would still be possible to rely on fold ratios outside of 0.71–1.4, and thus if the focus was upon transcripts differing by 1.5-fold or greater (outside of 0.67–1.5) the interpretation would be reasonably secure.

3. Do We Still Need Microarrays for Transcriptomics?

This is a valid and timely question, and is worth brief discussion. Now that RNA sequencing (or rather reverse transcribed DNA sequencing) is a practical proposition should we be using this approach instead of using microarrays? There is no simple yes or no answer to this and the answer, at least currently, is that there is a role for both methodologies and that one, the other, or a combination of the two may be the best experimental solution depending upon the question to be addressed.

1. The primary issues are of cost and experimental replication, and there are additional practical considerations related to processing and analysis. Currently, and admittedly figures vary widely, one can prepare a library for around £400, run a sample using Illumina technology (currently the cheapest option) for £1,200 per lane (for up to 100 bp reads), plus money for

informatics. Assuming tagged libraries were used, to increase the efficiency of processing for bacteria and one experiment with six tagged libraries was run in one lane, this would cost £3,600 for six channels of data, which is the minimum necessary to perform three biological replicates (compared to £9,600 if run in separate channels without tagging/multiplexing). The same experiment, using a dual channel approach addressing the same three pairs of biological replicates would cost between £600 and £1,000 using a microarray approach. Currently, this additional cost, and the pressures that it brings to performing fewer studies and fewer biological replicates, is difficult to justify.

2. Microarray analysis tools are well established and the software is free if tools such as BASE are used. Therefore, the researcher can usually, after modest training, perform their own analysis in multiple ways and really explore their data and findings rapidly. This can be done on the same day as the data has been triaged, which can be the day after performing the hybridizations. The analysis of high-throughput sequencing data is more complex and currently more specialized, which makes this slower, more expensive, and usually more remote from the researcher. And, finally, there is a real issue of sequencing capacity, which leads to weeks or months of delay before samples are processed, and since the informatics services are often also overburdened this potentially adds further delays at this point. If the RNA samples are at hand, a microarray experiment can be completed in a week, for a fraction of the cost of a comparable RNA-seq. Especially, if a study wants to look at multiple comparisons or conditions, and wants to incorporate transcriptional studies into an ongoing reciprocal functional study of regulation, at present the array approach remains an attractive option.
3. By contrast, where RNA sequencing really comes into its own is its ability to generate data that is not dependent upon the annotation, or where the investigation wishes to address the transcript pool at a resolution greater than the probe size or probe density. It is also useful if one wants to address the presence/abundance of RNA from intergenic regions, including antisense and small noncoding RNA. It can also map transcription start points and promoters across the genome. Some of these tasks can be addressed by tiling arrays, but these have their own complications, and considerable effort is needed to design a tiling array with optimized probe density and hybridization stringency properties, and this always represents a compromise that can be problematic. In these contexts, direct sequencing can be the best technique and the additional information obtained can clearly justify the additional expense.

4. While the debate typically focuses on "which" of the two methods should be used now and in the future, the pragmatic answer is that sometimes a hybrid approach is best. If you need data with no background signal, or to address regions that are not, or not sufficiently and specifically probed by a microarray, then preliminary studies using a small number of RNA sequencing datasets may be necessary to ensure that the microarray analysis addresses what is necessary. Then the microarray approach can be used to perform the biological replication and to test the generality of the observations. If necessary, the arrays can be supplemented with additional probes, as indicated from the RNA sequencing data, and this is relatively simple to do. It is also increasingly possible since bespoke oligo arrays can be generated for such purposes. This approach may be particularly useful in studies of un-sequenced strains, when preliminary experiments might include a combination of genome and transcript sequencing.
5. But, to date, the single greatest failing of microarray-based transcript analyses, in *Neisseria* and more generally, has been inadequate independent biological replication. The primary "excuse" for this has been cost, combined with some poorer experimental design using pooling and technical, rather than biological, replication. This "excuse" will be all the worse with an even more costly alternative method. Microarrays are the most time- and cost-efficient tool with which to generate properly replicated data, and will remain so for the immediate and foreseeable future. For some applications, particularly if data needs to be comparable between strains and/or with large existing datasets, a tool like the pan-*Neisseria* microarray remains the best tool for comparative studies, particularly when compared to (non-tiling) oligo arrays that generate signals that are much more susceptible to minor polymorphisms when using strains other than the strain used for design.

Acknowledgments

The original manufacture of the pan-*Neisseria* microarray v1 was supported in part by a grant from the Royal Society. Subsequent studies developing and using the microarray were supported by the Wellcome Trust, and microarray infrastructure was supported by the EPA Cephalosporin Trust. The input from the Computational Biology Research Group (formerly the Oxford University Bioinformatics Center), including from John Peden, Sarah Butcher, Simon McGowan, and Steve Taylor has been invaluable to work in this area. Bespoke plug-ins for BASE to perform cross-channel correction, a modification of global median ratio normalization (if a significant proportion of transcripts differ), and to perform the Cyber-T analysis are available directly from the Oxford Computational Biology Research Group. Further detailed protocols on the use of these arrays, which are outside the scope of this technical review, can be obtained from the authors on request.

References

1. Parkhill J, Achtman M, James KD et al (2000) Complete DNA sequence of a serogroup A strain of *Neisseria meningitidis* Z2491. Nature 404: 502–506.
2. Tettelin H, Saunders NJ, Heidelberg J et al (2000) Complete genome sequence of *Neisseria meningitidis* serogroup B strain MC58. Science 287: 1809–15.
3. Dillard JP, Seifert HS (2001) A variable genetic island specific for *Neisseria gonorrhoeae* is involved in providing DNA for natural transformation and is found more often in disseminated infection isolates. Mol Microbiol 41: 263–277.
4. Bentley SD, Vernikos GS, Snyder LA et al (2007) Meningococcal genetic variation mechanisms viewed through comparative analysis of serogroup C strain FAM18. PLoS Genet 3: article e23.
5. Snyder LA, Davies JK, Saunders NJ (2004) Microarray genomotyping of key experimental strains of *Neisseria gonorrhoeae* reveals gene complement diversity and five new neisserial genes associated with Minimal Mobile Elements. BMC Genomics 5: article 23.
6. Snyder LA, McGowan S, Rogers M et al (2007) The repertoire of minimal mobile elements in the *Neisseria* species and evidence that these are involved in horizontal gene transfer in other bacteria. Mol Biol Evol 24: 2802–2815.
7. Chou CC, Chen CH, Lee TT et al (2004) Optimization of probe length and the number of probes per gene for optimal microarray analysis of gene expression. Nucleic Acids Res 32: article e99.
8. Wilson K (1987) Current protocols in molecular biology. Wiley, New York.
9. Snyder LAS, Snudden G, Haan N et al (2007) Comparative genomics. The nature of CGH analysis and data interpretation. In: Falciani F (Ed) Microarray technology through applications. Taylor and Francis, UK.
10. Stabler R, Hinds J (2006) The majority of genes in the pathogenic Neisseria species are present in non-pathogenic *Neisseria lactamica* including those designated as virulence genes: response. BMC genomics. 7:article 129
11. Long AD, Mangalam HJ, Chan BY et al (2001) Improved statistical inference from DNA microarray data using analysis of variance and a Bayesian statistical framework. Analysis of global gene expression in *Escherichia coli* K12. J Biol Chem 276: 19937–19944.

Chapter 19

Analysis of Parameters Associated with Prevention of Cellular Apoptosis by Pathogenic *Neisseriae* and Purified Porins

Paola Massari and Lee M. Wetzler

Abstract

The process of cellular apoptosis is mediated by a number of microbial pathogens to modulate host defense mechanisms. Inhibition of apoptosis is thought to favor microbial survival, replication or immune evasion, while induction of apoptosis is likely to promote escape of the organisms from host cells. Several studies have reported that infection with *Neisseria* spp. can inhibit or reduce apoptotic cell death, thus allowing adaptation, intracellular replication, and immune evasion, events that are likely to spread infection. In this chapter, various techniques are described for direct measurement of host cell responses to infection with *Neisseria meningitidis* and to treatment with pure Neisseria porins, the major proteins found in the outer membrane of the pathogen.

Key words: *Neisseriae*, Human epithelial cells, Apoptosis, Porin, Mitochondria

1. Introduction

Apoptosis is a process of programmed cell death that is regulated by different intracellular and extracellular events, including excessive intracellular calcium, cell death receptor activation, exposure to chemical substances, or bacterial infections (1). It has been established that several bacterial organisms can influence cellular apoptosis by either inducing cell death or preventing it following infection. Some of the organisms that favor host cell survival and prevent or reduce cellular apoptosis are *Chlamydia* (2–4), *Shigella flexneri* (5), *Brucella suis* (6), *Porphyromonas gingivalis* (7, 8), and *Neisseria meningitidis* (9–12). Despite initial studies suggesting a potential pro-apoptotic effect of *Neisseria gonorrhoeae* and gonococcal porin (13–15), a general anti-apoptotic effect has also been established

Myron Christodoulides (ed.), *Neisseria meningitidis: Advanced Methods and Protocols*, Methods in Molecular Biology, vol. 799,
DOI 10.1007/978-1-61779-346-2_19, © Springer Science+Business Media, LLC 2012

for this organism and for purified gonococcal porin, PIB (11, 16–22), similar to what has been shown previously for the meningococcal porin, PorB (9, 11, 23). A role for neisserial porins has been identified as likely due to a direct effect on mitochondrial homeostasis via interaction with the mitochondrial porin VDAC (9, 24).

Mitochondria are among the key control points of cellular apoptosis (1). A typical consequence of apoptosis is the disruption of mitochondrial membrane potential, which can occur upon opening of the mitochondrial permeability transition pore (25). This results in the efflux of ions and small molecules followed by the release of pro-apoptotic mitochondrial factors into the cell cytosol (for example, cytochrome c (26), apoptosis-inducing factor (AIF) (27), or Smac/DIABLO (28)). This is then followed by cytosolic recruitment and/or activation of intracellular pro-apoptotic proteins, such as caspases (29) and members of the Bcl-2 family (30). In particular, pro apoptotic Bcl-2 proteins enhance the release of mitochondrial factors, thus amplifying activation of caspase enzymes. Apoptosis culminates in cell shrinkage, membrane blebbing, chromatin condensation, and DNA degradation.

Several methods have been employed to follow the mechanisms of protection from cellular apoptosis induced by *Neisseriae* and neisserial products, particularly in human epithelial cells. The purpose of this chapter is to highlight a variety of methods to measure the effect of live *Neisseriae* organisms and purified neisserial porin on several parameters associated with cellular apoptosis. These range from assessment of cell viability by trypan blue exclusion, determination of mitochondrial potential changes, measurement of DNA degradation, mitochondrial cytochrome c release, and activation of intracellular caspases. A variety of different techniques will be utilized, including flow cytometry, fluorescence and confocal microscopy, and biochemical methods. Representative figures are included describing experiments performed with purified meningococcal PorB or with live *Neisseria* organisms.

2. Materials

Unless otherwise stated, all buffers and reagent solutions are prepared with double-distilled water (ddH_2O). For cell culture, sterile ddH_2O is mandatory.

2.1. Eukaryotic Cell Culture

1. HeLa human epithelial cells (human adenocarcinoma) (ATCC).
2. Complete medium: Dulbecco's Modified Eagle's Medium (DMEM) (Gibco) containing 2 mMl-glutamine, 100 U/mL of penicillin, 100 mg/mL of streptomycin (Bio-Whittaker), and 10% (v/v) fetal bovine serum (FBS). Heat-inactivate FBS

at 56°C for 1 h to decomplement, prior to addition to Complete Medium.

3. Trypsin-Versene ethylenediamine tetraacetic acid (EDTA) mixture containing 0.25% (v/v) trypsin and 1 mM EDTA.
4. Phosphate buffered saline, pH 7.4 (PBS).
5. Trypan blue solution in 0.81% (w/v) NaCl and 0.09% (w/v) KCl (Fluka).
6. Dimethylsulfoxide (DMSO) supplied in 100 mL bottles (Sigma-Aldrich).
7. 75 cm^2 tissue culture-coated plastic flasks (Costar).
8. Neubauer hemocytometer.
9. Sorvall benchtop centrifuge (Thermo Scientific).

2.2. Culture of Neisseriae

1. *N. meningitidis* serogroup B strain H44/76 (31, 32) and *N. gonorrhoeae* strain F62 (18).
2. GC-agar base (Difco).
3. Sterile plastic Petri dishes (90 × 15 mm) (Fisher).
4. GC liquid medium containing 10% (v/v) IsoVitaleX supplement (BD).
5. Screw-cap glass Erlenmeyer flasks (250 mL) and Fernbach flasks (2 L) (Corning).
6. Equipment: Class II biosafety cabinet (BSL2+ NuAire), shaking incubator, spectrophotometer (Fisher).

2.3. Purification of Neisserial Porin by Column Chromatography

1. Stock solutions: prepare 1M sodium acetate (NaAc), pH 4.0; 10% (w/v) Zwittergent 3-14 (Z3-14) (Calbiochem); 1M calcium chloride; 1M Tris–HCl, pH 8.0; 0.5M ethylene diamine tetraacetic acid (EDTA) (Sigma-Aldrich); 20% (w/v) sodium azide (NaN_3) (see Note 1).
2. 1M HEPES buffer, pH 7.2 (Sigma-Aldrich).
3. 2,3-,dimercapto-1-propanol (BAL) (Sigma-Aldrich) dissolved in 100% ethanol.
4. Ethanol.
5. 10% (w/v) d-octyl-glucoside (Sigma-Aldrich) prepared in 10 mM HEPES, pH 7.2.
6. Loading buffer pH 8.0: 50 mM Tris–HCl, pH 8.0 containing 10 mM EDTA, 5% (w/v) Zwittergent 3-14, and 0.02% (w/v) NaN_3.
7. Buffer A: 50 mM Tris–HCl, pH 8.0 containing 10 mM EDTA, 0.05% (w/v) Zwittergent 3-14, and 0.02% (w/v) NaN_3.
8. Buffer B: 50 mM Tris–HCl, pH 8.0 containing 10 mM EDTA, 0.05% (w/v) Zwittergent 3-14, 0.8M NaCl, and 0.02% (w/v) NaN_3.

9. Buffer C: 100 mM Tris, pH 8.0 containing 10 mM EDTA, 0.2M NaCl, 0.05% (w/v) Zwittergent, and 0.02% (w/v) NaN_3.
10. Loading buffer pH 7.5: 50 mM Tris–HCl, pH 7.5 containing 10 mM EDTA, 5% (w/v) Zwittergent 3-14, and 0.02% (w/v) NaN_3.
11. Buffer D: 50 mM Tris–HCl, pH 7.5 containing 10 mM EDTA, 0.05% (w/v) Zwittergent 3-14, and 0.02% (w/v) NaN_3.
12. Buffer E: 50 mM Tris–HCl, pH 7.5 containing 10 mM EDTA, 0.05% (w/v) Zwittergent 3-14, 0.8M NaCl, and 0.02% (w/v) NaN_3.
13. Glass chromatography columns: 3 Econo columns, 2.5 × 10 cm (Bio-Rad) and 1 column, 2.6 × 180 cm (Spectrachrom).
14. DEAE Sepharose CL-6B resin, 50 mL (Amersham).
15. CM-Sepharose resin, 50 mL (Amersham).
16. Sephacryl S-300 resin 1 L (Amersham).
17. Matrex cellufine sulfate resin (Millipore).
18. AKTAPrime protein chromatography system (GE Healthcare Life Sciences).
19. Sorvall floor centrifuge (Thermo Scientific).
20. Protein concentration assay: Bicinchoninic Acid Kit for Protein Determination (PIERCE) using a 2 mg/mL solution of bovine serum albumin (BSA) fraction IV (Sigma-Aldrich), in 0.9% NaCl and 0.02% (w/v) NaN_3, to prepare dilutions for standard concentration curves.
21. Household bleach (50% v/v) solution.

2.4. Sodium Dodecyl Sulfate-PolyAcrylamide Gel Electrophoresis (SDS-PAGE)

1. Stock solutions: prepare 1.5M Tris–HCl, pH 8.8; 1M Tris–HCl, pH 6.8; 10% (w/v) ammonium persulfate (Biorad); 20% (w/v) SDS; water-saturated butanol.
2. 30% acrylamide/bis solution (37.5:1) (Protogel, National Diagnostics) (see Note 2).
3. N,N,N,N′-Tetramethyl-ethylenediamine (TEMED) (BioRad).
4. 12% separating gel: combine 3.3 mL of water, 4 mL of 30% acrylamide/bis solution, 2.5 mL of 1.5M Tris–HCl, pH 8.8, 100 μL of 20% (w/v) SDS solution, 100 μL of 10% (w/v) ammonium persulfate solution, and 20 μL of TEMED.
5. Stacking gel: combine 1.4 mL of water, 0.33 mL of 30% acrylamide/bis solution, 0.25 mL of 1M Tris–HCl, pH 6.8, 20 μL of 20% (w/v) SDS solution, 20 μL of 10% (w/v) ammonium persulfate solution, and 4 μL of TEMED.
6. SDS-PAGE running buffer: prepare a 10× stock solution containing 144 g of glycine, 30 g of Tris, and 10 g of SDS per liter of ddH_2O.

7. 4× SDS-PAGE protein loading buffer: 200 mM Tris–HCl, pH 6.8 containing 400 mM 1,4-dithiothreitol (DTT) (Sigma-Aldrich), 8% (w/v) SDS, 50% (v/v) glycerol, and 0.4% (w/v) bromophenol blue.
8. Minigel cell system (BioRad) and power supply.
9. Microcentrifuge (Eppendorf).
10. Acrylamide gel destaining solution: 10% (v/v) acetic acid and 40% (v/v) ethanol.
11. Coomassie brilliant blue (Sigma-Aldrich) staining solution: prepare a 0.4% (w/v) concentration of the dye in gel destaining solution.
12. Kaleidoscope prestained molecular weight protein standard (Biorad).

2.5. Induction of Apoptosis

1. HeLa human epithelial cells in Complete Medium.
2. GC liquid cultures of *N. meningitidis* or *N. gonorrhoeae* or purified *Neisseria* porins.
3. Trypsin-Versene EDTA mixture.
4. Phosphate buffered saline (PBS), pH 7.4.
5. Six-well sterile tissue culture plates (Costar).
6. Staurosporine (STS) (Sigma-Aldrich): prepare a 1 mM stock solution in sterile dimethyl sulfoxide (DMSO).
7. Sorvall benchtop centrifuge, or equivalent.

2.6. Assays for Host Cell Apoptosis

2.6.1. Trypan Blue Exclusion Assay

1. Trypan blue.
2. Neubauer hemocytometer.
3. Nikon Eclipse E400 microscope equipped with transmitted light halogen illuminator, blue, green, and red filters, 4×, 10×, 40×, 100× (oil) objectives; or similar.
4. Sorvall benchtop centrifuge, or equivalent.

2.6.2. Integrity of Cellular DNA Content Measured by Propidium Iodide Staining of Whole Cells and Flow Cytometry

1. Propidium iodide (PI) supplied as a 10-mL stock solution (Invitrogen). Store at 4°C.
2. Stock solution of RNase A (Sigma-Aldrich), 10 mg/mL. Store at −20°C.
3. Flow cytometry (FACS) wash buffer: PBS containing 2% (v/v) FBS.
4. PI staining solution: 50 μg/mL of PI and 0.5 mg/mL of RNaseA in FACS wash buffer.
5. Ethanol (100%), ice-cold.
6. *FACSCalibur* Flow Cytometer (Becton Dickinson).
7. Sorvall benchtop centrifuge, or equivalent.

2.6.3. Integrity of Cellular DNA Content Measured by Agarose Gel Electrophoresis and DNA Laddering

1. Stock solutions: prepare 20% (w/v) SDS; 20 mg/mL of Proteinase K, stored at –20°C; 10 mg/mL of RNase A, stored at –20°C; 5 mg/mL of ethidium bromide (see Note 3).
2. DNA cell lysis solution: 1% (w/v) SDS, 0.5 mg/mL of RNAse A, and 5 μg/mL of proteinase K.
3. TBE buffer: prepare a 10× stock solution containing 108 g of Tris base, 55 g of boric acid, and 8.3 g of EDTA per liter of ddH_2O.
4. Agarose powder (Biorad).
5. Agarose gel loading dye: prepare a 10× stock solution containing 15% (v/v) Ficoll, 0.2% (w/v) bromophenol blue, 0.2% (v/v) xylene cyanol FF (Kodak), and 0.1% (w/v) SDS. Store the dye solution at –20°C.
6. DNA ladder standard (Fisher).
7. Electrophoresis chamber, gel casting tray and combs (Bio Rad), and power supply.
8. Benchtop microcentrifuge.
9. UV transilluminator.

2.6.4. Chromosomal DNA Condensation Visualized by DAPI Staining Followed by Fluorescence Microscopy

1. Stock solutions: 10 μM of DAPI (4,6-diamidino-2-phenylindole) (Sigma-Aldrich) stored at 4°C (see Note 4); 16% (w/v) paraformaldehyde (see Note 5); 20% (w/v) sucrose.
2. Glass coverslips (22 × 40 × 0.15 mm) and glass slides (Fisher).
3. DAPI Fix buffer: 3.7% (v/v) paraformaldehyde solution in PBS containing 5% (w/v) sucrose (see Note 6).
4. Vectashield™ mounting medium (Vector Laboratories).
5. Nikon Eclipse E400 microscope, or similar.

2.6.5. Mitochondrial Membrane Integrity Assessment by Nonyl Acridine Orange (NAO) Staining of Intact Cells and Flow Cytometry

1. Nonyl acridine orange (NAO) (Molecular Probes, Invitrogen) dissolved in DMSO and stored in aliquots at –20°C (see Note 7).
2. Ethanol.
3. Flow cytometry (FACS) wash buffer: PBS containing 2% (v/v) FBS.
4. Sorvall benchtop centrifuge.
5. *FACSCalibur* Flow Cytometer (Becton Dickinson).

2.6.6. Mitochondrial Membrane Integrity and Potential Assessment by Rhodamine 123 Staining of Intact Cells and Flow Cytometry

1. Rhodamine 123 (Molecular Probes, Invitrogen): prepare a 1 mM stock solution in sterile DMSO and store in aliquots at –20°C (see Note 8).
2. Flow cytometry (FACS) wash buffer: PBS containing 2% (v/v) FBS.
3. Sorvall benchtop centrifuge.
4. *FACSCalibur* Flow Cytometer (Becton Dickinson).

2.6.7. Mitochondrial Membrane Integrity and Potential Assessment by JC-1 Staining of Intact Cells and Fluorescence Microscopy

1. JC-1 (5,5′,6,6′-tetrachloro-1,1′,3,3′-tetraethylbenzimidazolylcarbocyanine iodide) (Molecular Probes, Invitrogen): prepare a 1-mM stock solution in sterile DMSO and store in aliquots at –20°C (see Note 9).
2. Glass coverslips (22 × 40 × 0.15 mm) and glass slides (Fisher).
3. JC-1 wash buffer: PBS containing 1% (w/v) BSA.
4. Vectashield™ mounting medium (Vector Laboratories).
5. Nikon Eclipse E400 microscope, or similar.

2.6.8. Isolation of Mitochondria

1. Stock solutions: prepare 1M sucrose; 0.5M potassium chloride; 0.5M EDTA; 0.5M ethylene glycol tetraacetic acid (EGTA); 1M magnesium chloride; 1M HEPES, pH 7.4; 100 mM of phenylmethylsulfonyl fluoride (PMSF) in 100% ethanol; Protease inhibitor cocktail (Sigma-Aldrich) dissolved in DMSO.
2. Mitochondria isolation buffer: 20 mM HEPES, pH 7.4 containing 250 mM sucrose, 10 mM potassium chloride, 1 mM EDTA, 1 mM EGTA, 1.5 mM magnesium chloride, 0.1 mM PMSF, and 1/1,000 (v/v) of protease inhibitor cocktail.
3. Biorad Kit for protein concentration determination (BioRad).
4. 1-mL Glass Dounce tissue homogenizer (Kontes).
5. Benchtop microcentrifuge.
6. 4× SDS-PAGE protein loading buffer.

2.6.9. Preparation of Whole Cell Lysates

1. Stock solutions: prepare 1M Tris–HCl, pH 8.0; 0.5M EDTA; Protease inhibitor cocktail dissolved in DMSO.
2. Cell lysis buffer: 10 mM Tris–HCl, pH 8.0 containing 10 mM EDTA, 0.5% (v/v) Tween-20, and 1/1,000 (v/v) protease inhibitor cocktail.
3. Bio-Rad Kit for protein concentration determination (Bio-Rad).
4. Benchtop microcentrifuge.
5. 4× SDS-PAGE protein loading buffer.

2.6.10. Western Blotting for Intracellular Markers of Apoptosis

1. Mini Trans-Blot Cell system (Bio-Rad).
2. Transfer buffer: prepare a 10× stock solution containing 144 g of glycine and 30 g of Tris per liter of ddH_2O. Prior to use, dilute to a 1× solution containing 20% (v/v) methanol and chill at 4°C.
3. Tris-buffered saline (TBS), pH 7.6: prepare a 10× stock solution containing 88 g of NaCl, 2 g of KCl, and 30 g of Tris base per liter of ddH_2O.
4. TBST buffer: TBS containing 0.1% (v/v) Tween-20.
5. Polyvinylidene fluoride (PVDF) Immobilon membrane (Millipore) (see Note 10).

6. Methanol.
7. Whatman 3MM blotting paper (Fisher).
8. Solution of Ponceau S (0.1% w/v) in 5% (v/v) acetic acid.
9. Blocking buffer: TBST containing 5% (w/v) powdered nonfat milk.
10. Incubation buffer: TBST containing 5% (w/v) BSA.
11. Primary antibodies: anti-cytochrome c monoclonal antibody (clone 7H8.2C12, Pharmingen), rabbit anti-procaspase 3, anti-procaspase 7, anti-caspase 9, anti-caspase 8 and anti-AIF antibodies (Stressgen), rabbit anti-HSP70 antibody (Cell Signaling).
12. Secondary antibodies: goat anti-rabbit IgG or rat anti-mouse IgG conjugated to horseradish peroxidase (HRP) (Cell Signaling).
13. Chemiluminescent reagents: ECL Plus Western Blotting Detection Systems (GE Healthcare).
14. Hyperfilm ECL film (GE Healthcare).

3. Methods

3.1. Eukaryotic Cell Culture

1. Thaw an aliquot of frozen HeLa cells by transferring a cryovial from a liquid nitrogen storage Dewar in a 37°C water bath for 5 min. Wash the thawed cell suspension (1 mL) once with 10 mL of warm Complete Medium to remove the freezing solution (a solution of 10% (v/v) DMSO and 90% (v/v) FBS). Suspend the cells in 10 mL of warm Complete Medium.
2. Grow HeLa cells in Complete Medium at 37°C in a humidified incubator with 5% (v/v) CO_2 in 75 cm^2 tissue culture-coated plastic flasks. Maintain the cell cultures at 70% confluence.
3. For passaging the cells, remove the medium from the flask, wash the adherent cells once with 3 mL of cold PBS (see Note 11), and add 1.5 mL of Trypsin-Versene/EDTA solution to the flask to initiate cell detachment from the plastic surface. Return the flask to the incubator and incubate for 5 min until the cells detach. Then, add 5 mL of PBS, collect the cells using a 5 mL sterile pipette, and centrifuge for 5 min at 800 × *g* at 4°C in sterile tubes. Remove the supernatant and suspend the cell pellet in 5 mL of fresh PBS.
4. Transfer 10 μL of the cell suspension with a sterile pipette into an Eppendorf tube and mix with 10 μL of Trypan blue solution (1:1 v/v dilution). Allow the cells to incorporate the dye for 5 min at room temperature and then place 10 μL of the suspension onto a glass hemocytometer for cell counting.

3.2. Culture of Neisseriae

Handling of neisserial liquid cultures must be done in bio-safety 2 level plus (BSL2+) sterile cabinets. Laboratory personnel should wear appropriate personal protective equipment (PPE) including gloves, goggles, laboratory coat, and mask.

1. Plate *Neisseriae* organisms from frozen stock cultures in 50% (v/v) glycerol/50% (v/v) GC liquid medium on GC-agar plates containing 1% (v/v) Isovitalex. Grow overnight at 37°C in a humidified incubator with 5% (v/v) CO_2. For bacterial culture, we dedicate an incubator for this purpose.
2. The next day, collect colonies from the plates with a sterile cotton swab and inoculate 50 mL of liquid GC medium. Grow the bacteria in liquid culture to exponential phase for approximately 4–6 h in a dedicated shaker incubator at 37°C to an optical density (OD) at λ600 nm of 0.1 (approximately 10^8 colony forming units [CFU]/mL) as determined by spectrophotometer measurement. Aliquots of these cultures can be used for cell infection experiments.
3. For large-scale bacterial cultures for PorB isolation, transfer 12.5 mL of bacterial culture into each of four sterile 250-mL glass Erlenmeyer flasks containing 125 mL of medium each and grow for an additional 6 h. Then, transfer each culture into 1.5 L of medium using four 2-L glass Fernbach flasks and culture overnight at 37°C in a shaker incubator (33).

3.3. Purification of Neisserial Porin by Column Chromatography

1. In a dedicated BSL2+ safety cabinet, divide the bacterial culture prepared in step 3 of Subheading 3.2, into aliquots of equal volume. Place the bacterial suspension into 250 mL plastic screw-cap bottles and centrifuge at 3,900 ×*g* for 15 min at 4°C (see Note 12). Combine the pellets into one centrifuge bottle with cold 0.9% (w/v) NaCl and centrifuge again. Add 50% (v/v) bleach solution to the culture supernatants, in order to disinfect the samples prior to discard.
2. Transfer the bacterial pellet into a 1-L glass beaker and add 32 mL of 1.0M sodium acetate (NaAc) (see Note 13). Add 32 μL of 2,3-,dimercapto-1-propanol (BAL) as an anti-protease (see Note 14).
3. Place the bacterial cell suspension in a water bath sonicator and slowly add 112 mL of 1M $CaCl_2$ and 112 mL of 10% (w/v) Zwittergent 3-14. Mix for about 30–60 min with a glass rod until the suspension is homogeneous. Then, add 64 mL of 100% ethanol (to a final concentration of 20% v/v) to precipitate DNA, lipooligosaccharide, and cellular debris.
4. Centrifuge the suspension in polypropylene screw-cap centrifuge bottles for 15 min at 3,975 ×*g* at 4°C. Collect the supernatant and discard the pellet. Add 100% ethanol (to a final concentration of 80% v/v) to the supernatant and store the sample overnight

at 4°C in order to precipitate the total protein content. The next day, remove as much as possible of the upper clear portion of the solution without disturbing the precipitated proteins. Transfer the remaining suspension to 150 mL centrifuge glass bottles and centrifuge at 3,975 × *g* for 15 min at 4°C.

5. Suspend the pellet in approximately 100 mL of Loading Buffer pH 8.0 and centrifuge the suspension at 12,000 × *g* for 15 min at 4°C; collect the supernatant, and repeat the extraction procedure two more times (see Note 15).
6. Using an AktaPrime protein chromatography machine or if unavailable, a simple peristaltic pump, load the clear protein solution onto a DEAE Sepharose CL-6B column and a CM-Sepharose column in tandem equilibrated with Buffer A at a flow rate of 1.8 mL/min. Collect the flow through and wash the columns with Buffer A until the λ280 nm OD is back to baseline (see Note 16). Wash the columns with Buffer B to elute residual proteins from the columns prior to re-equilibration in Buffer A. Wash the columns with 20% (v/v) ethanol for long-term storage.
7. Precipitate the protein content in the flow through with 80% (v/v) ethanol as described in step 4 above, centrifuge for 15 min at 3,975 × *g*, and suspend in approximately 10 mL of Loading Buffer pH 8.0.
8. Load the protein solution onto a Sephacryl S-300 gel filtration chromatography column previously equilibrated with Buffer C at a flow rate of 0.3 mL/min. Collect 8 mL fractions; from the chromatogram, identify the fractions corresponding to the eluted protein peaks. Place 200 μL aliquots of selected fractions into Eppendorf tubes and add ethanol (80% v/v) and keep at 4°C overnight to precipitate total protein content and to remove the zwittergent-containing solution (see Note 17). Centrifuge the solution at 13,000 × *g* for 10 min, discard the supernatant, and air-dry the pellet.
9. Suspend the pellet in 1× SDS-PAGE protein loading buffer and identify the porin-containing fractions by SDS-PAGE and Coomassie staining as described in Subheading 3.4. Based on the SDS-PAGE analysis, pool the porin-containing fractions, precipitate with ethanol (80% v/v), and suspend the precipitate in Loading Buffer pH 7.5.
10. Load the protein solution onto a Matrex Cellufine Sulfate column previously equilibrated with Buffer D at a flow rate of 6 mL/min (33). Wash the column with Buffer D as described in step 6 above and apply a linear gradient of 0.2–0.5M NaCl with Buffer E. The porin-containing fractions will elute between 0.24 and 0.4M NaCl. Pool the PorB containing fractions, identified by SDS-PAGE and Coomassie staining, and precipitate the protein content with 80% (v/v) ethanol (see Note 18).

11. Suspend the purified PorB precipitate in 1 mL of 10 mM HEPES buffer, pH 7.2 containing the dialyzable detergent d-octyl-glucoside (10% w/v) and dialyze extensively (>5 × 10^{10} times the original volume of the sample for 36 h) against PBS containing 0.02% (w/v) NaN_3, to remove the zwittergent and to form the porin proteosomes. Store the dialyzed porin proteosomes at a temperature no lower than 4°C (33) (see Note 19).
12. Measure the protein concentration by the BCA protein assay reagent following the manufacturer's instructions.

3.4. Sodium Dodecyl Sulfate-Polyacrylamide Gel Electrophoresis (SDS-PAGE)

These instructions assume the use of a Bio-Rad minigel system. Clean glass plates with detergent after each use, rinse with distilled water, follow with 95% (v/v) ethanol, and air-dry.

1. Pour a 1-mm thick, 12% polyacrylamide separating gel, leaving approximately 1 cm of space for the stacking gel, overlay with water-saturated isobutanol, and let the gel polymerize for approximately 30 min.
2. Blot off the isobutanol and rinse the gel with ddH_2O. Pour a stacking gel above the polymerized separating gel and insert the comb. The stacking gel should polymerize in 30 min.
3. Prepare 1 L of SDS-PAGE Running buffer by diluting 100 mL of the 10× running buffer with 900 mL of ddH_2O.
4. Carefully remove the comb and wash the wells with running buffer using a 3 mL syringe fitted with a 22-gauge needle. Add the running buffer to the upper and lower chambers of the gel unit.
5. Boil the test samples in protein loading buffer for 5 min, centrifuge at 17,000 × g for 5 min, and place on ice. Load a maximum of 30 μL of each sample per well and include 10 μL of a prestained molecular weight marker to one well.
6. Assembly the gel unit and apply a current of 50 V using a power supply. The gel can be run overnight at room temperature. If cooling is available for the gel unit, then run during the day at a maximum of 100 V through the stacking gel and 200 V through the separating gel. The blue dye front will help to identify the positions of the lanes.
7. Disconnect the gel unit from the power supply and extract the gel. Using a metal spatula, pry open the glass plates. The gel will lie on one glass. Very carefully, lift the gel and place it into a plastic container (i.e., a disposable plastic Petri dish) containing approximately 15 mL of Coomassie blue staining solution. Cover the container with the dish lid or with Saran wrap, place on a rocker, and stain for a minimum of 2 h at room temperature. Discard the Coomassie stain in a dedicated waste container and add approximately 20 mL of destaining solution. Destain the gel for a minimum of 4 h, changing the destaining solution four times, or until the desired level of destaining has been reached.

3.5. Induction of Apoptosis

1. Seed HeLa cells in Complete Medium in six-well tissue culture plates (5×10^4 cells/mL, 2 mL/well) and culture the cells overnight at 37°C in a humidified incubator with 5% (v/v) CO_2 to allow them to adhere. One well is required for each experimental data point.
2. On the same day, inoculate liquid cultures of *N. meningitidis* or *N. gonorrhoeae* in 10 mL of GC medium containing 1% (v/v) Isovitalex and grow overnight at 37°C as described in Subheading 3.2.
3. On the next day, dilute the overnight *Neisseria* culture in Complete Medium without antibiotics to an O.D. at λ600 nm of 0.1, approximately 1×10^8 CFU/mL.
4. Aspirate the medium from each cell culture well using a sterile Pasteur pipette, wash the cells gently three times with 3 mL of cold PBS and add 2 mL of fresh Complete Medium without antibiotics, containing appropriate bacterial dilutions to obtain the desired multiplicity of infection (MOI). Keep the infected cultures for 24 h at 37°C in a humidified incubator with 5% (v/v) CO_2.
5. As an alternative to infection with live bacteria, incubate HeLa cells for 24 h in the presence of 10 μg/mL of purified *Neisseria* porins.
6. On the next day, induce apoptosis by adding 1 μM STS. Include positive controls of cells treated with STS alone and negative controls of cells incubated with medium alone alongside cells infected with *Neisseriae* or treated with purified porin. Incubate for additional time points ranging from 4 h up to 24 h.
7. To measure extracellular bacterial survival and to determine the number of viable bacteria in the co-cultures, collect 0.1 mL aliquots of culture medium at various time points postinfection and plate onto GC-agar plates overnight for counting CFU.
8. At desired time points after induction of apoptosis, collect individual cell culture supernatants. Wash the wells with PBS and add this wash to the culture supernatants. Add a 0.5–1 mL volume of Trypsin-Versene/EDTA solution to the cell monolayers and incubate for 2–3 min at 37°C to detach the cells. Collect the cells and add them to the supernatants and washes for each condition. Wash the wells with 1 mL of PBS and collect again (see Note 20).
9. Centrifuge the cells at $800 \times g$ for 5 min at 4°C and wash the cell pellet once with 5 mL of cold PBS.

3.6. Assays for Host Cell Apoptosis

In this section, we describe a series of assays for investigating host cell apoptosis and the effect of infection with live *Neisseriae* and/or treatment with purified neisserial porins on cellular apoptotic responses. For these types of experiments, we suggest a variety of assays to analyze cellular apoptosis, ranging from cell membrane integrity, mitochondrial physiology, DNA degradation to intracellular activation of apoptosis-related proteins.

3.6.1. Trypan Blue Exclusion Assay

1. Suspend the cells obtained in step 9 of Subheading 3.5 in 1 mL of PBS.
2. Stain 10 μL of the cell suspension with Trypan blue as described in step 4 of Subheading 3.1 and place 10 μL of this suspension onto a glass hemacytometer for cell counting. Count a minimum of 100 cells for each sample in triplicate.
3. Calculate the percentage of dead cells as a ratio between the number of dye-retaining cells (blue cells, nonviable) and the total number of cells per mL.

3.6.2. Integrity of Cellular DNA Content Measured by Propidium Iodide Staining of Whole Cells and Flow Cytometry

1. Suspend the cells obtained in step 9 of Subheading 3.5 in 300 μL of FACS wash buffer and add 700 μL of ice-cold ethanol for 15 min at 4°C in order to permeabilize the cells.
2. Centrifuge the cells at 800 ×*g* for 5 min, discard the supernatant, and wash the pellet with cold PBS. Add 300 μL of PI staining solution to the pellet and keep at 4°C for 15 min in the dark.
3. Centrifuge the stained cells, wash with 1 mL of PBS once, and then suspend the pellet in 300 μL of FACS wash buffer. Analyze the samples by flow cytometry according to the instrument manufacturer's instructions, with gating to exclude cell debris associated with necrosis and nonapoptotic cells. An example of STS-induced DNA degradation detected by PI staining of permeabilized HeLa cells compared to medium-incubated control cells is shown in Fig. 1a, b, indicated as the percentage (%) of hypodiploid DNA content. The protective effect of live *N. meningitidis* co-incubation on STS-induced DNA degradation is shown in Fig. 1c. Similar results have been obtained with purified meningococcal PorB porin (see Fig. 1d) (9, 11) as well as with live *N. gonorrhoeae* and with purified gonococcal porin (10, 18).

3.6.3. Integrity of Cellular DNA Content Measured by Agarose Gel Electrophoresis and DNA Laddering

1. Suspend the cells obtained in step 9 of Subheading 3.5 in 50 μL of DNA cell lysis solution.
2. Freeze the cells at –20°C for a minimum of 24 h (see Note 21).
3. Mix 1.25 g of agarose powder with 125 mL of 1× TBE buffer in a 500-mL glass flask. Melt the agarose in a microwave or hot water bath until the solution becomes clear (see Note 22). Cool the solution to about 50–55°C, swirling the flask occasionally.
4. Seal the ends of a gel casting tray with tape. Place the comb in the tray and pour the melted agarose solution into the casting

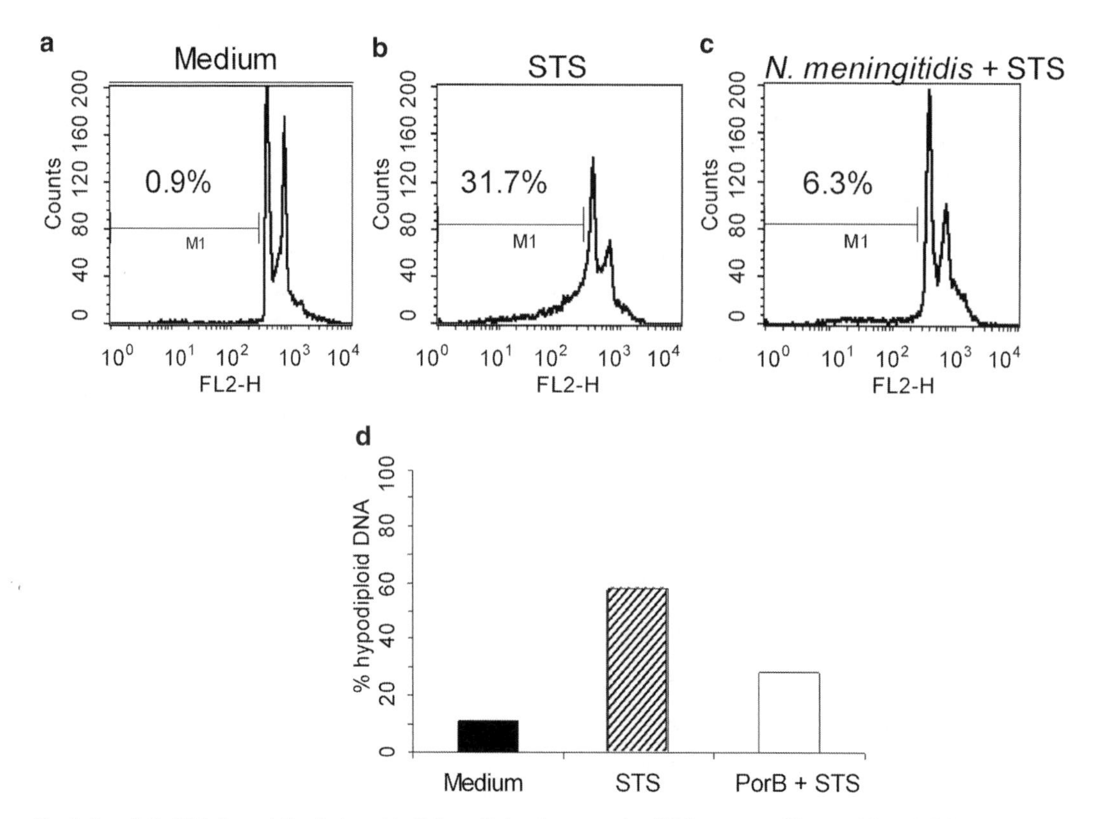

Fig. 1. Apoptotic DNA degradation induced in HeLa cells by staurosporine (STS) measured by propidium iodide staining and flow cytometry. Percent of hypo-diploid DNA content in the gated cell population is indicated. HeLa cells incubated for 24 h with (**a**) medium alone; (**b**) 1 μM STS; (**c**) live *Neisseria meningitidis* organisms MOI 10 for 24 h followed by STS treatment for further 24 h. Reproduced from Massari et al. (10), with permission from publisher. Copyright (2003) Wiley. (**d**) Representative bar graph relative to HeLa cells incubated for 24 h with medium alone (*black bar*), STS alone (*dashed bar*) or purified *N. meningitidis* PorB for 24 h followed by STS for 24 h (*white bar*). Percent of hypodiploid DNA content is shown.

tray. The gel should polymerize in about 30 min, turning milky white. Pull out the comb, remove the tape, and place the gel in the electrophoresis chamber. Add TBE buffer so that the gel is submerged by 2–3 mm.

5. Add 10× agarose gel loading dye to the samples (final concentration of 1×) and carefully load approximately 20 μL of each sample per well. Include 10 μL of DNA ladder standard in one well. Close the gel box, connecting the electrodes to the power supply making sure the positive (red) and negative (black) are correctly connected and apply a current of about 100 V.
6. Run the gel until the blue dye approaches the end of the gel. Turn off the power and disconnect the wires from the power supply before removing the lid of the electrophoresis chamber.
7. Using gloves, carefully remove the tray and gel. Visualize DNA by ethidium bromide fluorescence with an UV transilluminator (9).

3.6.4. Chromosomal DNA Condensation Visualized by DAPI Staining Followed by Fluorescence Microscopy

1. Before plating of the HeLa cells as described in step 1 of Subheading 3.5, place a sterile glass cover slip into each well of a six-well tissue culture plate. Use sterile tweezers to handle the coverslips.
2. Induce host cell apoptosis as described in step 6 of Subheading 3.5.
3. Rinse the cells carefully twice with ice-cold PBS to avoid loss of nonattached cells. Add 1 mL of DAPI Fix solution to the cell monolayers and leave at room temperature for 10 min.
4. Gently wash the wells twice for 5 min each with cold PBS. Discard the DAPI Fix solution into a hazardous waste container.
5. Add 1 mL of PBS containing 300 nM of DAPI to the cells and incubate for 10 min at room temperature in order to stain the DNA and identify the nuclei.
6. Wash the cells carefully with 5 mL of cold PBS and aspirate the wash solution from the edge of the wells to avoid disturbing the cell monolayer.
7. Air dry and mount the coverslips, carefully inverted, onto microscope glass slides using 20–25 μL of Vectashield™ mounting medium. A volume of 25 μL is sufficient for a 22 × 22 mm coverslip (see Note 23). For prolonged storage, coverslips can be permanently sealed around the perimeter with nail polish or a plastic sealant. Let the polish dry and store the glass slides in the dark at 4°C for up to a month.
8. Using phase contrast microscopy, identify the cells and the focal plane. Excitation at λ364 nm induces DAPI fluorescence (observed as blue emission). Overlay the phase contrast and fluorescence images to obtain merged images. An example of the protective effect of *N. meningitidis* on STS-induced DNA nuclear condensation is shown in Fig. 2 by fluorescence microscopy using a Nikon Eclipse E400 microscope (10).

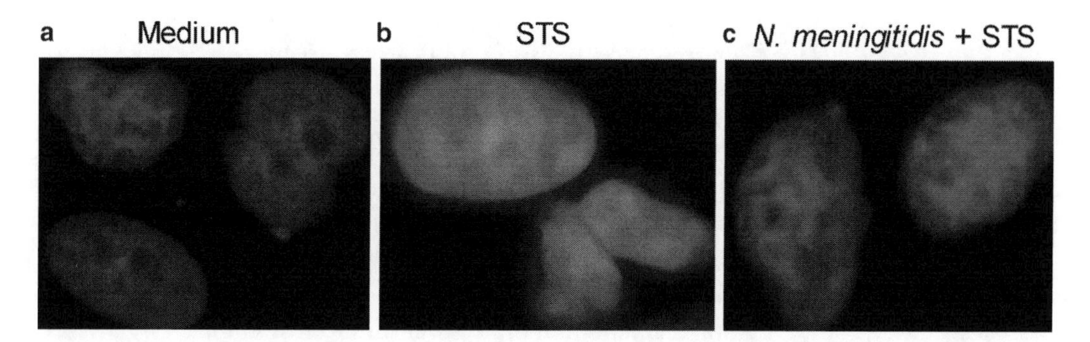

Fig. 2. Apoptotic DNA nuclear condensation measured by DAPI staining and fluorescence microscopy. HeLa cells are incubated for 24 h with (**a**) medium alone; (**b**) 1 μM STS, and (**c**) live *N. meningitidis* MOI 10 for 24 h followed by STS for 24 h.

3.6.5. Mitochondrial Membrane Integrity Assessment by Nonyl Acridine Orange (NAO) Staining of Intact Cells and Flow Cytometry

1. After washing twice in ice-cold PBS, fix the cells obtained in step 9 of Subheading 3.5 with 80% (v/v) ethanol at −20°C overnight.
2. Incubate the fixed cells with FACS wash buffer containing 10 mM NAO for 15 min at room temperature. Wash the cells once with FACS wash buffer and analyze by flow cytometry, following the instrument manufacturer's instructions. Cells containing intact mitochondria will display higher fluorescence than cells containing mitochondria with decreased mass due to apoptosis (9) (see Note 24).

3.6.6. Mitochondrial Membrane Integrity and Potential Assessment by Rhodamine 123 Staining of Intact Cells and Flow Cytometry

1. Incubate HeLa cells with bacteria and/or purified porins as described in steps 4 and 5 of Subheading 3.5. Then, induce apoptosis with STS as described in step 6 of Subheading 3.5.
2. Next, add 1 μM of rhodamine 123 to the medium in the wells and incubate for 30 min at 37°C.
3. Collect cells as described in step 9 of Subheading 3.5.
4. Suspend the cell pellet in 300 μL of FACS wash buffer and immediately analyze by flow cytometry. Store on ice for a maximum of 1 h after staining. Cells containing depolarized mitochondria will display lower fluorescence levels than cells containing mitochondria with intact membrane potential (9). An example of protection from STS-induced mitochondrial potential loss by purified PorB is shown in Fig. 3a while the effect of *Neisseria* organisms is shown in Fig. 3b (see Note 8).

3.6.7. Mitochondrial Membrane Integrity and Potential Assessment by JC-1 Staining of Intact Cells and Fluorescence Microscopy

1. Incubate HeLa cells with bacteria and/or purified porins as described in steps 4 and 5 of Subheading 3.5. Then, induce apoptosis with STS as described in step 6 of Subheading 3.5.
2. Next, add 5 μM of JC-1 (34) to the culture medium and incubate for 10 min at 37°C.
3. Gently wash the glass coverslips twice with cold PBS for 5 min each and analyze the slides by microscopy as described in step 8 of Subheading 3.6.4. Excitation at λ488 nm induces JC-1 fluorescence emission shift from green (~529 nm) to red (~590 nm). Overlay the phase contrast and fluorescence images to obtain merged images (see Note 9).

3.6.8. Isolation of Mitochondria

1. Incubate HeLa cells with bacteria and/or purified porins as described in steps 4 and 5 of Subheading 3.5. Then, induce apoptosis with STS as described in step 6 of Subheading 3.5.
2. Next, incubate the cells in 0.5 mL of mitochondria isolation buffer for 30 min on ice to induce cell swelling.

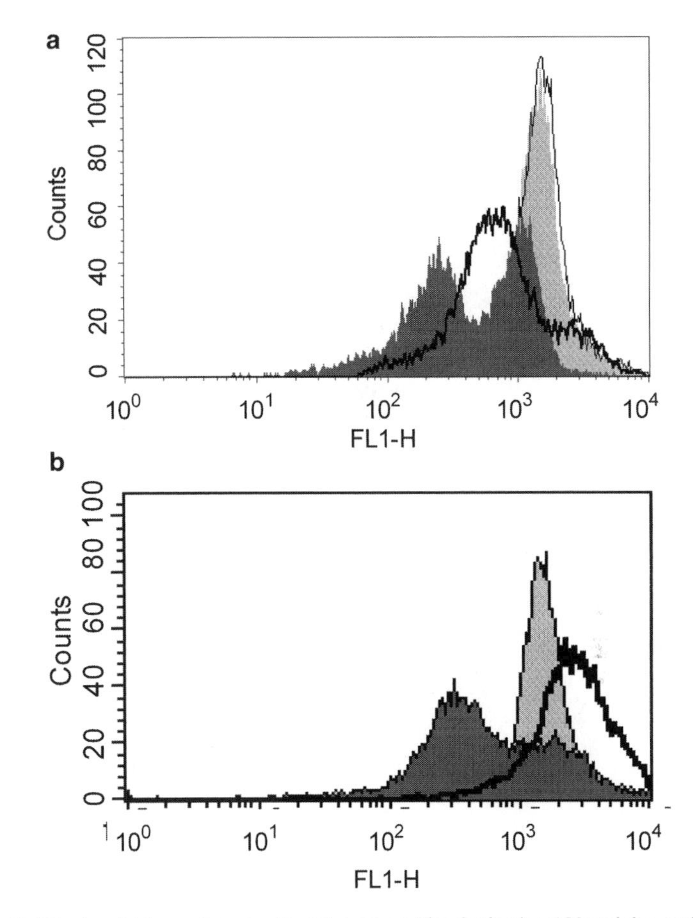

Fig. 3. Mitochondrial membrane potential measured by rhodamine 123 staining and flow cytometry. HeLa cells are incubated for 24 h with medium alone (**a** and **b**, *light gray area*) or 1 μM STS (**a** and **b**, *dark gray area*). (**a**) Cells are incubated with 10 μg/mL of purified PorB alone (*thin line*) or PorB followed by STS for further 24 h (*thick line*). (**b**) Cells are incubated with *N. meningitidis* MOI 10, alone (*thin line*) or *N. meningitidis* followed by STS for further 24 h (*thick line*). Mitochondrial membrane depolarization is determined by loss of rh123 fluorescence. Reproduced from Massari et al. (10), with permission from publisher. Copyright (2003) Wiley.

3. Homogenize the cells with a glass ball homogenizer placed on ice. Count 40 strokes. Transfer the cell homogenates to clean 1.5-mL Eppendorf tubes.
4. Centrifuge at a low speed (350 × *g*) for 10 min at 4°C to pellet the nuclear fractions and cell debris. Transfer the supernatant to a clean Eppendorf tube. Suspend the nuclear fractions in 100 μL of mitochondria isolation buffer and store a 20 μL aliquot at −80°C for determination of total protein concentration using the Bio-Rad protein assay kit according to the manufacturer's instructions. Add 26 μL of 4× SDS-PAGE loading buffer (final concentration of 1×) to the remaining 80 μL volume and store this at −20°C until ready to use.

5. Centrifuge the supernatants at high speed (13,000 × *g*) for 30 min at 4°C to separate the mitochondrial and cytosolic fractions. Transfer the supernatants into clean Eppendorf tubes. Store the cytosolic fraction (approximately 0.5 mL) at −80°C until ready to measure the total protein concentration. The cytosolic fractions might have low protein content and can be concentrated by ethanol precipitation as described in step 8 of Subheading 3.3 prior to electrophoresis, if required. Store at −20°C until ready to use.
6. Suspend the mitochondrial pellet in 50 μL of mitochondria isolation buffer. Store a 10 μL aliquot at −80°C for determination of total protein concentration. Add 12 μL of 4× SDS-PAGE loading buffer to the remaining 40 μL and store at −20°C until ready to use.

3.6.9. Preparation of Whole Cell Lysates

1. Incubate HeLa cells with bacteria and/or purified porins as described in steps 4 and 5 of Subheading 3.5. Then, induce apoptosis with STS as described in step 6 of Subheading 3.5.
2. Suspend the cells in 0.5 mL of cell lysis buffer and place on ice for 30 min.
3. Centrifuge the cell lysates in 1.5-mL Eppendorf tubes for 5 min at 13,000 × *g* at 4°C. Collect the supernatants and store a 20 μL aliquot at −80°C for determination of total protein concentration. Add 4× SDS-PAGE loading buffer to the remaining 80 μL and store at −20°C until ready to use.

3.6.10. Western Blotting for Intracellular Markers of Apoptosis

The following instructions assume the use of a Bio-Rad transfer tank system for electrophoretic transfer of proteins, separated on polyacrylamide gels prepared as described in Subheading 3.4, to a PVDF membrane.

1. Prepare a large Pyrex tray filled with 1× transfer buffer and place a transfer cassette system in the tray submerged in the buffer. Place a fitted sponge and two sheets of Whatman 3MM blotting paper on each side.
2. Wet a sheet of PVDF membrane, slightly larger than the size of the gel, in 100% methanol. Transfer the PVDF membrane to the tray (see Note 25). Disconnect the gel unit from the power supply and extract the gel. Using a metal spatula, pry open the glass plates. The gel will lie on one glass, which will be placed into the tray and submerged in transfer buffer. Very carefully, lift the gel and place it on top of the PVDF membrane, with the molecular marker on the left side. Gently smooth the gel on top of the PVDF membrane, removing any air bubbles. Mark the wells using a sharp pencil. Follow with two further sheets of wet 3MM paper on top of the gel, again ensuring that no bubbles are trapped in the resulting sandwich. Place the second wet sponge on top and close the transfer cassette.

3. Place the cassette into the transfer tank, making sure that the PVDF membrane is between the gel and the anode (see Note 26).
4. Turn on the refrigerated/circulating water bath to avoid overheating of the transfer system. Alternatively, the transfer can be placed into a large glass-door, moisture-free refrigerator equipped with internal electric plugs. If this option is chosen, pre-equilibrate the power supply at 4°C for several hours (preferably overnight). In the absence of such refrigeration, the transfer apparatus can be placed into a large tray filled with ice. Close the tank and apply a current of 20 mA overnight or a maximum of 200 mA for 2 h.
5. Carefully disassemble the transfer cassette and remove the membrane. The prestained molecular weight markers will be clearly visible on the membrane.
6. In a clean plastic Petri dish, stain the PVDF membrane with Ponceau S solution to control for transfer efficiency. Rinse extensively with ddH_2O followed by PBS until the red staining has disappeared (see Note 27).
7. Block the PVDF membrane in 20 mL of blocking buffer for 1 h at room temperature on a rocking platform. Discard the blocking buffer, rinse the membrane in TBST twice for 10 min followed by the addition of 10–20 mL of incubation buffer containing the appropriate dilution of primary antibody as specified by the manufacturer. Incubate overnight at 4°C on an orbital shaker or a rocking platform.
8. Remove the primary antibody-containing solution and wash the membrane three times for 5 min each with 50 mL of TBST.
9. Add HRP-labeled anti-mouse or anti-rabbit secondary antibody in blocking buffer at concentrations specified by the manufacturer. Incubate for 2 h at room temperature on a rocking platform.
10. Warm up the chemiluminescent reagents (ECL) to room temperature.
11. Discard the antibody-containing solution and wash the membrane three times for 10 min each with TBST.
12. After the final TBST wash, prepare a 2 mL aliquot of ECL reagent A containing 50 μL of reagent B per each blot. Place the PVDF membrane, face up, on a large sheet of Saran wrap and cover with the ECL solution. Ensure that the membrane is entirely covered by the ECL solution. Incubate for 5 min at room temperature in the dark (see Note 28).
13. Blot the ECL reagents off from the PVDF membrane and fold the Saran wrap to cover the membrane.

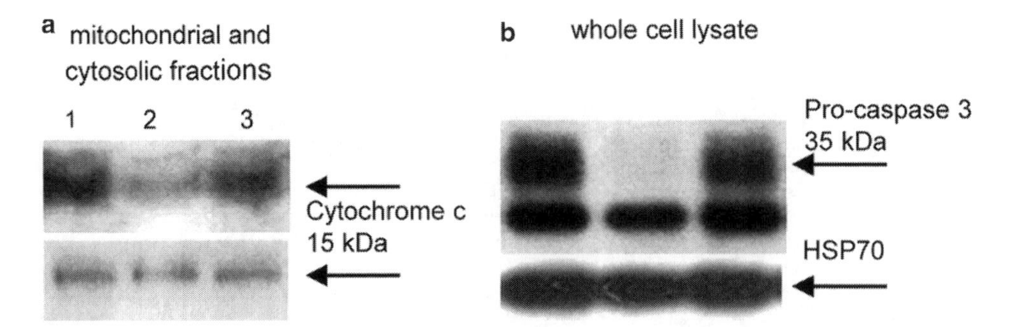

Fig. 4. Western blot of (**a**) mitochondrial and cytosolic fractions and of (**b**) whole cell lysates. (**a**) Cytochrome c release into the cell cytosol following induction of apoptosis with STS is prevented by incubation with 10 μg/mL of purified PorB. *Lane 1* medium; *lane 2* STS; *lane 3* PorB + STS. (**b**) Pro-caspase 3 cleavage induced by STS is inhibited by incubation of HeLa cells with 10 μg/mL of purified PorB, determined by Western blot of whole cell lysates with anti-caspase 3 antibody. *Lane 1* medium; *lane 2* STS; *lane 3* PorB + STS. Anti-HSP70 antibody is used as a loading control.

14. Place the wrapped membrane into an X-ray film cassette and in a dark room under safe light conditions, cover with film for a variable exposure time, ranging from a minimum of 10 s to a maximum of 2 min. Figure 4 shows examples of the protective effect of meningococcal PorB on STS-induced mitochondrial release of cytochrome c (see Fig. 4a) or cytosolic cleavage of pro-caspase 3 (see Fig. 4b). Protection from cleavage of pro-caspases 7 and 9 by both purified porins and *Neisseria* organisms has also been previously shown using this method (9–11).

4. Notes

1. Sodium azide is used to prevent bacterial growth in bulk reagents. It is highly toxic and potentially explosive, and exposure to this reagent should be avoided.
2. Acrylamide is a neurotoxic agent in liquid un-polymerized form and skin exposure to this reagent should be limited.
3. Ethidium bromide is considered a mutagen, carcinogen, and teratogen, and skin exposure to this reagent should be avoided.
4. DAPI is a light-sensitive dye and should be stored away from light in aluminum foil-wrapped tubes.
5. Paraformaldehyde is a toxic agent and skin exposure should be avoided. For preparation of a high concentration solution of paraformaldehyde, the solution needs to be heated on a heat/stir plate in a fume hood. The solution needs to be cooled to room temperature prior to use.
6. Addition of 5% (w/v) sucrose to the DAPI fix buffer allows for a better preservation of the cellular membrane.

7. NAO fluorescence is a light-sensitive indicator of changes in mitochondrial mass during apoptosis. Store in the dark.
8. Rhodamine 123 is a mitochondrial membrane potential-dependent fluorescent dye retained only by mitochondria with intact membrane potential. Store in the dark. Analysis of mitochondrial potential using Rh123 can only be performed on live cells.
9. JC-1 is a potential-dependent mitochondrial dye. The JC-1 monomeric, green fluorescent form accumulates in the cytosol of apoptotic cells following mitochondrial membrane depolarization, while red-fluorescent J-aggregates stain mitochondria with intact membrane potential. Store in the dark. Analysis of mitochondrial potential using JC-1 can only be performed on live cells.
10. PVDF membrane is preferable to nitrocellulose, as the latter shows nonspecific affinity for amino acids.
11. Trypsin is inhibited in the presence of FBS. Wash carefully the cells prior to the addition of trypsin solution.
12. Do not use over 150 mL per centrifuge bottle to avoid spills of bacterial culture.
13. Add the NaAc slowly and also suspend the organisms slowly to avoid clumping.
14. At this point the bacteria are lysed and the suspension can be removed from the safety cabinet.
15. All the following steps are performed at room temperature (25°C) unless otherwise indicated.
16. The porin will be found in the flow through.
17. It is necessary to remove the zwittergent from the samples before electrophoresis to ensure migration of the protein bands.
18. PorB purified following this procedure presents minimal LPS contamination (less than 0.01% as demonstrated by Limulus lysate assays or silver staining of gels) (33).
19. To avoid low temperature-induced degradation and loss of activity, avoid storage of purified PorB below 4°C (33).
20. As a consequence of STS treatment, cell rounding and detachment is often observed. Collecting all the steps will allow recovery of all the cells in each culture.
21. It is advisable to freeze the samples at –20°C followed by high-speed centrifugation for 10 min before agarose gel electrophoresis, due to the viscosity of the whole cell preparations.
22. If using a microwave, heat the solution for several short intervals to avoid boiling the solution out of the flask.

23. Slides mounted with Vectashield™ mounting medium will not dry out and can be viewed without sealing.
24. NAO fluorescence is an indicator of changes in mitochondrial mass during apoptosis.
25. PVDF membranes are hydrophobic and need to be properly hydrated prior to protein transfer. To allow PVDF membranes to become wet, they need to be soaked in 100% methanol for a few minutes.
26. Make sure that the sandwich is correctly positioned, otherwise the proteins might transfer into the buffer rather than onto the PVDF membrane.
27. Carefully rinse the Ponceau S solution off the membrane and the Western blot box to avoid formation of milk clots due to residual acetic acid.
28. This step can be also performed in the light without dramatically affecting the efficiency of the reaction.

Acknowledgments

The authors thank NIH/NIAID for funding grant RO1 AI40944-01.

References

1. Green DR (2005) Apoptotic Pathways: Ten Minutes to Dead. Cell 121: 671–674.
2. Fan T, Lu H, Hu H et al (1998) Inhibition of apoptosis in *Chlamydia*-infected cells: blockade of mitochondrial cytochrome c release and caspase activation. J Exp Med 187: 487–496.
3. Fischer SF, Schwarz C, Vier J et al (2001) Characterization of antiapoptotic activities of *Chlamydia pneumoniae* in human cells. Infect Immun 69: 7121–7129.
4. Xiao Y, Zhong Y, Su H et al (2005) *Chlamydia trachomatis* infection inhibits both Bax and Bak activation induced by staurosporine J Immunol 174: 1701–1708.
5. Clark CS, Maurelli AT (2007) *Shigella flexneri* inhibits staurosporine-induced apoptosis in epithelial cells. Infect Immun 75: 2531–2539.
6. Gross A, Terraza A, Ouahrani-Bettache S et al (2000) *In vitro Brucella suis* infection prevents the programmed cell death of human monocytic cells. Infect Immun 68: 342–351.
7. Murray DA, Wilton JM (2003) Lipopolysaccharide from the periodontal pathogen *Porphyromonas gingivalis* prevents apoptosis of HL60-derived neutrophils *in vitro*. Infect Immun 71: 7232–7235.
8. Nakhjiri SF, Park Y, Yilmaz O et al (2001) Inhibition of epithelial cell apoptosis by *Porphyromonas gingivalis*. FEMS Microbiol Lett 200: 145–149.
9. Massari P, Ho Y, Wetzler LM (2000) *Neisseria meningitidis* porin PorB interacts with mitochondria and protects cells from apoptosis. Proc Natl Acad Sci USA 97: 9070–9075.
10. Massari P, King CA, Ho AY et al (2003) Neisserial PorB is translocated to the mitochondria of HeLa cells infected with *Neisseria meningitidis* and protects cells from apoptosis. Cell Microbiol 5: 99–109.
11. Massari P, Gunawardana J, Liu X et al (2010). Meningococcal porin PorB prevents cellular apoptosis in a toll-like receptor 2- and NF-kappaB-independent manner. Infect Immun 78: 994–1003.
12. Linhartova I, Basler M, Ichikawa J et al (2006) Meningococcal adhesion suppresses proapoptotic

gene expression and promotes expression of genes supporting early embryonic and cytoprotective signaling of human endothelial cells FEMS Microbiol Lett 263: 109–118.
13. Muller A, Gunther D, Naumann M et al (1999) Neisserial porin (PorB) causes rapid calcium influx in target cells and induces apoptosis by the activation of cysteine proteases. EMBO J 18: 339–352.
14. Muller A, Rassow J, Grimm J et al (2002) VDAC and the bacterial porin PorB of *Neisseria gonorrhoeae* share mitochondrial import pathways. EMBO J 21: 1916–1929.
15. Kozjak-Pavlovic V, Dian-Lothrop EA, Meinecke M et al (2009). Bacterial porin disrupts mitochondrial membrane potential and sensitizes host cells to apoptosis. PLoS Pathog 5: e1000629.
16. Binnicker MJ, Williams RD, Apicella MA (2003) Infection of human urethral epithelium with *Neisseria gonorrhoeae* elicits an upregulation of host anti-apoptotic factors and protects cells from staurosporine-induced apoptosis. Cell Microbiol 5: 549–560.
17. Binnicker MJ, Williams RD, Apicella MA (2004) Gonococcal Porin IB activates NF-κB in human urethral epithelium and increases the expression of host antiapoptotic factors. Infect Immun 72: 6408–6417.
18. Follows SA, Murlidharan J, Massari P et al (2009) *Neisseria gonorrhoeae* infection protects human endocervical epithelial cells from apoptosis via expression of host antiapoptotic proteins. Infect Immun 77: 3602–3610.
19. Morales P, Reyes P, Vargas M et al (2006) Infection of human fallopian tube epithelial cells with *Neisseria gonorrhoeae* protects cells from tumor necrosis factor alpha-induced apoptosis. Infect Immun 74: 3643–3650.
20. Simons MP, Nauseef WM, Griffith TS et al (2006) *Neisseria gonorrhoeae* delays the onset of apoptosis in polymorphonuclear leukocytes. Cell Microbiol 8: 1780–1790.
21. Howie HL, Shiflett SL, So M (2008) ERK mediated downregulation of Bim and Bad during infection with *N.gonorrhoeae*. Infect Immun 76: 2715–2721.
22. Higashi DL, Lee SW, Snyder A et al (2007) Dynamics of *Neisseria gonorrhoeae* attachment: microcolony development, cortical plaque formation and cytoprotection. Infect Immun 75: 4743–4753.
23. Liu X, Wetzler LM, Massari P (2008) The PorB porin from commensal *Neisseria lactamica* induces Th1 and Th2 immune responses to ovalbumin in mice and is a potential immune adjuvant. Vaccine 26: 786–796.
24. Colombini M (2004) VDAC: The channel at the interface between mitochondria and the cytosol. Mol Cell Biochem 256-257: 107–115.
25. Marzo I, Brenner C, Zamzami N et al (1998) The permeability transition pore complex: a target for apoptosis regulation by caspases and bcl-2-related proteins. J Exp Med 187: 1261–1271.
26. Kluck RM, Bossy-Wetzel E, Green DR et al (1997) The release of cytochrome c from mitochondria: a primary site for Bcl-2 regulation of apoptosis. Science 275: 1132–1136.
27. Susin SA, Lorenzo HK, Zamzami N et al (1999) Molecular characterization of mitochondrial apoptosis-inducing factor. Nature 397: 441–446.
28. Du C, Fang M, Li Y et al (2000) Smac, a mitochondrial protein that promotes cytochrome c-dependent caspase activation by eliminating IAP inhibition. Cell 102: 33–42.
29. Budihardjo I, Oliver H, Lutter M et al (1999) Biochemical pathways of caspase activation during apoptosis. Annu Rev Cell Dev Biol 15: 269–290.
30. Levine B, Sinha S, Kroemer G (2008) Bcl-2 family members: dual regulators of apoptosis and autophagy. Autophagy 4: 600–606.
31. Holten E (1979) Serotypes of *Neisseria meningitidis* isolated from patients in Norway during the first six months of 1978. J Clin Microbiol 9: 186–188.
32. Tommassen J, Vermeij P, Struyve M et al (1990) Isolation of *Neisseria meningitidis* mutants deficient in class 1 (porA) and class 3 (porB) outer membrane proteins. Infect Immun 58: 1355–1359.
33. Massari P, King CA, Macleod H et al (2005) Improved purification of native meningococcal porin PorB and studies on its structure/function. Prot Expr Purif 44:136–46.
34. Cossarizza A, Baccarani-Contri M, Kalashnikova G et al (1993) A new method for the cytofluorimetric analysis of mitochondrial membrane potential using the J-aggregate forming lipophilic cation 5,5′,6,6′-tetrachloro-1,1′,3,3′-tetraethylbenzimidazolcarbocyanine iodide (JC-1). Biochem Biophys Res Comm 197: 40–45.

Chapter 20

Analysis of the Immune Response to *Neisseria meningitidis* Using a Proteomics Approach

Jeannette N. Williams, Myron Christodoulides, and John E. Heckels

Abstract

The availability of Neisseria genome sequences together with improvements in proteomic technologies provide the opportunity to study at high resolution the immune response to *Neisseria meningitidis*. In this chapter, we describe a protocol that combines two-dimensional (2D) SDS-PAGE of meningococcal outer membranes with western blotting of human antisera to identify proteins associated with the development of protective antibody responses. This methodology can identify putative vaccine candidates for incorporation in a multi-component serogroup B meningococcal vaccine.

Key words: Two-dimensional gel electrophoresis, Western blotting, Immuno-proteomics, Meningococcal antigens, Vaccine

1. Introduction

Neisseria meningitidis releases outer membrane (OM) blebs from the cell wall surface during growth and depletion of the toxic lipooligosaccharide gives rise to outer membrane vesicles (OMV) (1). Such vesicles have been exploited as vaccines because they contain the antigenic surface components of the donor meningococcal strain. However, a full understanding of the basis of the human immune response to meningococci has been hampered by a lack of detailed knowledge of the antigenic composition of the organism and the released OM blebs. The traditional approach to studying the humoral immune response to OM and OMV vaccines has been one-dimensional (1D) SDS-PAGE in which proteins are separated according to molecular weight followed by western blot analysis of serum reactivity. However, the availability of Neisseria genome sequences (2–5) and improved proteomic techniques (6)

Myron Christodoulides (ed.), *Neisseria meningitidis: Advanced Methods and Protocols*, Methods in Molecular Biology, vol. 799, DOI 10.1007/978-1-61779-346-2_20, © Springer Science+Business Media, LLC 2012

have revealed that both meningococcal OM and OMV preparations contain a larger number of proteins than previously elucidated by conventional 1D SDS-PAGE separation (7). Moreover, mass spectrometric analyses of OM/OMVs resolved on 1D SDS-PAGE gels clearly show that the resulting antigen bands contain multiple proteins (8, 9).

In order to overcome the poor resolution of 1D SDS-PAGE, we describe in this chapter a protocol for two-dimensional (2D) SDS-PAGE, a technique that separates proteins according to their isoelectric pH in the first dimension followed by molecular weight in the second dimension. An additional advantage of 2D SDS-PAGE is that using a combination of specific detergents and thoiurea has improved the solubility of membrane proteins (6), therefore enabling the study of the protein composition of meningococcal OM and OMV (10–13). We then describe how the utility of this method can be extended by combining it with western blot analysis to study the immune response to individual meningococcal antigens. Identification of immunogenic proteins is achieved by excision of protein spots in matching stained 2D gels followed by mass spectrometry fingerprinting analysis. This methodology has been used to analyse the human immune response to both meningococcal colonisation (13) and infection (14) and the potential of proteins from meningococci and the non-pathogen *Neisseria lactamica* to induce immunologically cross-reactive antibodies (15).

New strategies are needed to produce a broadly effective vaccine against serogroup B meningococcal strains: the described protocol that follows enables a more detailed analysis of the immune response to meningococcal proteins in order to identify potential cross-protective vaccine candidates for incorporation in a multi-component serogroup B meningococcal vaccine (13).

2. Materials

2.1. Sample Preparation for Isoelectric Focusing

1. OM preparation (7).
2. Rehydration/solubilisation buffer (R/SB): 7M urea, 2M thiourea, 2% (w/v) 3-((3-Cholamidopropyl)-dimethyl-ammonio)-1-proane sulphonate (CHAPS), 2% (w/v) tetradecanoylamid opropyl-dimethylammonio-butanesulphonate (ASB-14), 1% (w/v) dithiothreitol (DTT), 2% (w/v) 3–10 nonlinear (NL) immobilised pH gradient (IPG) buffer (GE Healthcare Biosciences) (see Note 1), 2 mM tributylphosphine solution (TBP), 2 mM $MgSO_4$, 1,000 U Benzonase solution (Sigma-Aldrich) and trace amounts of Orange G dissolved in ultra high quality (UHQ) water (13, 16) (see Notes 2 and 3).
3. Screw-capped 1.5 mL microtubes.

2.2. Rehydration of Immobiline DryStrips

1. IPGphor Strip Holders (GE Healthcare Biosciences) (see Note 4).
2. Immobiline DryStrips 7 cm (3–10 NL) (GE Healthcare Biosciences).
3. Immobiline DryStrip Cover Fluid (GE Healthcare Biosciences).
4. Ettan IPGphor II Isoelectric System (Pharmacia Biotech) (see Note 5).
5. R/SB.
6. Strip Holder Cleaning Solution (GE Healthcare Biosciences) (see Note 6).
7. Chromatography Whatman filter paper 3MM Chr (Schleicher and Schuell).

2.3. Isoelectric Focusing

1. Ettan IPGphor II isoelectric system (Pharmacia Biotech).
2. IPGphor cup-loading strip holder (GE Healthcare Biosciences).
3. Sample cups (GE Healthcare Biosciences).
4. Isoelectric focusing (IEF) electrode strips (GE Healthcare Biosciences) (see Note 7).
5. Immobiline DryStrip cover fluid (GE Healthcare Biosciences).
6. Strip Holder Cleaning Solution (see Note 6).
7. Ethanol.

2.4. Equilibration of Immobiline DryStrips

1. 0.5M Tris–HCl Buffer, pH 6.8.
2. Equilibration stock solution: 50 mM Tris–HCl buffer, pH 6.8 containing 6M urea, 1% (w/v) SDS, 30% (v/v) glycerol. To make 50 mL, add 18 g urea, 0.5 g SDS, 15 mL glycerol to 5 mL of 0.5M Tris–HCl buffer; dissolve and adjust the volume to 50 mL with UHQ water. This buffer can be stored at room temperature for up to 2 weeks.
3. Equilibration solution A: 100 mg DTT added to 10 mL equilibration stock solution. This solution must be made and used immediately.
4. Equilibration solution B: 250 mg iodoacetamide and a few grains of bromophenol blue added to 10 mL equilibration stock solution. This solution must be made and used immediately.

2.5. Second Dimension Electrophoresis

1. Multiphor II Flatbed electrophoresis unit (Pharmacia Biotech).
2. MultiTemp III thermostatic circulator (Pharmacia Biotech).
3. EPS 3501 XL high-voltage power supply (Pharmacia Biotech).
4. Paraffin liquid for oil baths.
5. ExcelGel SDS gradient XL 12–14 (GE Healthcare Biosciences).
6. ExcelGel SDS buffer strips, anode and cathode (GE Healthcare Biosciences).

7. IEF electrode strips (GE Healthcare Biosciences) (see Note 7).
8. LMW-SDS Marker Kit, Mr range 14,400–97,000 Da (GE Healthcare Biosciences).
9. Dual Colour Precision Plus Protein Prestained Standards (Bio-Rad) (see Note 8).

2.6. Staining of Second Dimension Gel

1. Fixing solution: 50% (v/v) ethanol, 10% (v/v) acetic acid.
2. Ethanol solution: 30% (v/v) ethanol.
3. ProteoSilver™ Plus Stain Kit (Sigma-Aldrich) containing ProteoSilver Sensitizer, ProteoSilver Silver Solution, ProteoSilver Developer 1, ProteoSilver Developer 2, and ProteoSilver Stop Solution. All solutions should be prepared just prior to use in 110 mL total volume (enough to cover a 2D gel) with UHQ water. The sensitisation and silver solutions are diluted to 1% (v/v) concentration; whilst the Developer Solution is prepared with 5% (v/v) ProteoSilver Developer 1 and 0.1% (v/v) ProteoSilver Developer 2. A volume of 5.5 mL of undiluted Stop Solution added to 110 mL of Developer Solution is required to stop the developing reaction.
4. Sodium azide. Prepare a working stock of 5% (w/v) in UHQ and store at 4°C.
5. Scanning equipment e.g. HP Scanjet G2410 or similar.

2.7. Electrotransfer of Proteins from Gel to Membrane

1. Trans-Blot Cell (Bio-Rad).
2. Power Supply Model 200/2.0 (Bio-Rad) or similar.
3. 2D gel of separated proteins.
4. Blotting buffer: prepare 10× stocks with 25 mM Tris, 192 mM glycine, and 0.1% (w/v) SDS. Adjust pH to 8.3 and store at room temperature. Dilute 100 mL with 700 mL water and 200 mL methanol.
5. Hybond-P polyvinylidene fluoride (PVDF) membrane, cut a little larger than the gel to allow for gel swelling (see Note 9).
6. 3MM Chr chromatography paper (see Note 10).
7. Foam pads (come as part of Trans-Blot Cell equipment).
8. FilmRemover (GE Healthcare).

2.8. Staining of Proteins Transferred to PVDF Membrane

1. MemCode Protein Stain Kit (Perbio Science) containing MemCode™ Sensitizer, MemCode™ Reversible Stain, MemCode™ Destain, and MemCode™ Stain Eraser. The Destain and Eraser solutions are prepared 1/1 with methanol just before use.
2. Methanol.
3. UHQ water.

2.9. Immunoreaction of Antisera and Outer Membrane Proteins

1. PVDF membrane incorporating OM proteins.
2. Serum samples (see Note 11).
3. Tris-buffered saline (TBS): prepare 10× stock with 0.5M NaCl, 20 mM Tris. Adjust pH to 7.5 with HCl and dilute 100 mL with 900 mL water for use.
4. Tween-Tris-buffered saline (TTBS) is made by adding 0.05% (v/v) Tween 20 to TBS (see Note 12).
5. Blocking buffer: 5% (w/v) non-fat milk powder in TTBS.
6. Anti-human IgG antibody conjugated to alkaline phosphatase, diluted 1/1,000 in blocking buffer (see Note 13).
7. Substrate components:
   (a) Buffer: 100 mM Tris–HCl, pH 9.5 containing 100 mM NaCl and 2 mM $MgCl_2$.
   (b) Nitro blue tetrazolium (NBT): 30 mg/mL in 70% (v/v) dimethyl formamide (DMF).
   (c) 5-Bromo-4-chloro-3-indolyl phosphate (BCIP): 15 mg/mL in 100% DMF.
   (d) Just before use, NBT and BCIP solutions are added at a concentration of 1% (v/v) to the substrate buffer (see Note 14).

2.10. Alignment of Images and Spot Excision

1. Graphics software, e.g. Corel PHOTO-PAINT.
2. Microtitre plates, 96 well sterile.
3. VersaDoc™ Imaging System (Bio-Rad) and GelPix robotic spot excision tool (Genetix).

3. Methods

3.1. Sample Preparation for Isoelectric Focusing

Protein samples can be applied to the Immobiline DryStrip gel by in-gel rehydration (16) or cup-loading (13, 17). In the method described below, the DryStrips are rehydrated in R/SB without sample, and then the OM sample is applied in a cup placed at the anodic end of the strip.

1. Suspend OM preparation (7) (containing 225 μg protein) to a total volume of 310 μL in R/SB in sterile screw-capped microtubes and vortex (see Note 15).
2. Incubate for 3 h at room temperature (see Note 2).
3. Centrifuge sample at room temperature for 2 h at 16,000×*g* to remove any insoluble material.
4. Carefully remove the supernatant and proceed directly with IEF, or freeze at −70°C in single use aliquots in sterile screw-capped microtubes for later use.

3.2. Rehydration of Immobiline DryStrips

Prior to IEF, the Immobiline DryStrips must be rehydrated to their original thickness of 0.5 cm in R/SB.

1. Pipette 125 μL of R/SB into an IPGphor Strip Holder (see Notes 16 and 17).
2. Remove the plastic backing from the DryStrip Gel with clean forceps and place gel (gel facing down) in the R/SB solution.
3. To ensure even hydration, remove any trapped bubbles under the DryStrip gel by gently stroking the plastic backing with a pair of forceps.
4. Overlay DryStrip with approximately 2 mL Immobiline DryStrip Cover Fluid to prevent evaporation and urea crystallisation.
5. Place cover over Reswelling Tray and incubate at room temperature overnight (10–20 h) (see Note 5).
6. After rehydration, rinse each strip briefly in UHQ water and blot gently on filter paper (see Note 18).

3.3. Isoelectric Focusing

This method describes IEF using the Ettan IPGphor II Isoelectric System and anodic cup-loading of protein samples. The IPGphor consists of a Peltier cooling plate divided into cathode and anode areas and an integral programmable high-voltage power supply.

1. Position the cup-loading strip holder on the Ettan IPGphor with the pointed end of the strip holder in contact with the anode (+), and the blunt end in contact with the cathode (−) electrode areas of the separation platform.
2. Using clean forceps, place the rehydrated DryStrip in the IPGphor cup-loading strip holder, gel side uppermost. The strip must be centred in the strip holder channel and extend into both plated electrical contact regions of the strip holder with the anodic part of the strip, indicated by a "+" sign on the DryStrip placed towards the pointed end of the holder.
3. Cut two electrode strips of approximately 5 mm long per Immobiline DryStrip. Dampen with UHQ water, blot until almost dry with filter paper and then apply to the ends of the DryStrip.
4. Position the movable electrodes above the filter paper pads and press down gently (see Note 19).
5. Position the movable sample cup onto the surface of the Immobiline DryStrip as close to the electrode at the anodic end of the strip as possible and gently apply pressure. The sample cup must form a seal with the DryStrip gel without damaging its surface.
6. Check sample cup placement by pipetting a 100 μL volume of R/SB into each cup. If leakage occurs, this will be apparent by seepage of the R/SB into the anodic positioned filter paper

pad. If necessary, remove the R/SB, reposition the sample cup and check again. When the cups are correctly placed, remove the R/SB and load 100 μL of sample per cup.

7. Cover the entire length of the Immobiline DryStrip with 2–4 mL of cover fluid (see Note 20).
8. Position the IPGphor cup-loading strip holder cover over the strip holder and close the lid of the IPGphor unit.
9. Program the Ettan IPGphor with the following run parameters using the up and down arrows:

Step 1	500 V	0.30 h	Step n Hold
Step 2	1,000 V	0.30 h	Step n Hold
Step 3	8,000 V	4.30 h	Step n Hold

   Limit the current to 50 μA per strip and program in the number of DryStrips in the focusing run (18). As IEF proceeds, the Orange G tracking dye (included in the R/SB) migrates towards the anode.

10. On completion of IEF, place each DryStrip into a sterile universal plastic container or other suitable tube. Next, proceed with equilibration and Second Dimension Electrophoresis or store the strips at –70°C (see Notes 21–23).
11. Gently clean the strip holders after use with the IPGphor Strip Holder Cleaning Solution, by applying a few drops of detergent directly from the bottle and spreading with a gloved finger. Rinse thoroughly with tap water, then deionised water and allow to air dry. Rinse electrodes, cup and strip holder cover with deionised water and air dry. The cup can also be briefly rinsed with 100% ethanol.
12. Wipe any excess Immobiline DryStrip Cover Fluid gently off the surface of the IPGphor platform with paper towels (see Note 24).

3.4. Equilibration of Immobiline DryStrips

Equilibration improves the protein transfer from the first to the second dimension by immersing the DryStrips in an SDS buffer system compatible with Second Dimension Electrophoresis. The strong binding of focused proteins to the fixed charges on the DryStrip can interfere with subsequent protein transfer from the immobiline gel matrix to the Second Dimension gel. The addition of urea and glycerol to the Equilibration Buffer increases the viscosity of the solution thereby reducing electro-endosmosis (18). In this protocol, two of the three DryStrips of separated OM proteins focused in the same IEF run are equilibrated simultaneously so that they can be run on the same Second Dimension Electrophoresis gel. The gel must be ready for the application of the DryStrips once they have been equilibrated.

1. Place each focused DryStrip in a sterile universal container with the plastic backing towards the container wall.
2. Add 5 mL of equilibration solution A to each container, place lengthways on a rotary shaker and gently rock for 15 min at room temperature. Ensure that the strips can move freely (see Note 25).
3. During equilibration, the Second Dimension Electrophoresis gel must be prepared (see Subheading 3.5).
4. Drain off Equilibration Solution A without losing the strips and replace with 5 mL of Equilibration Solution B.
5. Rock for an additional 15 min at room temperature.
6. Remove Equilibration Solution B and briefly rinse strips with UHQ water.
7. Drain the DryStrips by placing them on their sides on damp filter paper for a minimum of 3 min and a maximum of 20 min.

3.5. Second Dimension Electrophoresis

The following instructions assume the use of the Multiphor II Electrophoresis System together with the MultiTemp III Thermostatic Circulator and the EPS 3501 XL power supply for Second Dimension Electrophoresis. The Multiphor II is a horizontal electrophoretic apparatus consisting of a buffer tank, ceramic cooling plate, electrode holder with movable electrodes, and a polycarbonate safety lid.

1. Switch on the MultiTemp III Thermostatic Circulator at least 15 min prior to starting electrophoresis to allow the temperature of the Multiphor II cooling plate to reach 15°C.
2. Pour light paraffin oil onto the centre of the cooling plate (see Note 26).
3. Remove the ExcelGel XL SDS 12–14 from the foil package and place the gel on the Multiphor II cooling plate, gel side up, with the notched corner orientated to the right corner of the anodic region of the cooling plate (indicated by a "+" sign on the cooling plate). Straighten the gel so that the anodic edge of the gel is aligned uniformly with the anodic edge of the printed grid on the cooling plate.
4. Leave the plastic gel cover on and smooth out trapped air bubbles between the gel and cooling tray using paper towels or a roller. Work from the middle of the gel towards the outer edges.
5. Remove the plastic cover from the gel with forceps (see Note 27).
6. Allow the gel surface to dry for 5–15 min before proceeding with the next step.
7. Remove the SDS buffer strips from their foil packaging and place vertically, narrow side down, on the ExcelGel using the

printed grid on the cooling plate for alignment (see Note 28). Place the clear cathode SDS buffer strip in row 1 and the orange anode strip in row 17. Remove any trapped bubbles between the strips and gel by gently stroking the surface of the buffer strips with the back of a forceps.

8. Apply the two equilibrated Immobiline DryStrips approximately 2 cm apart and approximately 5 mm away from the cathode (white) buffer strip. The strips must be placed gel side down with the acidic part of the DryStrip (marked by a "+" sign) facing towards the bottom of the ExcelGel. Place small squares of water dampened filter paper (see Note 29) underneath either end of each Immobiline DryStrip, close to the gel to prevent "smiling." Then, stroke the plastic backing of each DryStrip with forceps to exclude any air bubbles.
9. Molecular weight markers can be applied to water dampened filter paper squares (see Note 29) positioned above and/or below the DryStrips on the ExcelGel. Pipette 5 μL of LMW molecular weight standards onto each square. If the gel is to be blotted, also include a damp filter square between the DryStrips with 15 μL of a prestained molecular weight marker.
10. Align the movable electrodes of the Multiphor II until they are above the buffer strips, connect to the unit and close the safety lid.
11. Program the EPS 3501 XL power supply using the "Set" button and up and down arrows with the following parameters: 1,000 V, 20 mA per ExcelGel, 40 W per ExcelGel, 45 min. Press Run.
12. Open the Multiphor II lid and remove the Immobiline DryStrips and filter paper pieces. Replace the lid and set the EPS 3501 XL power supply to 1,000 V, 40 mA per ExcelGel, 40 W per ExcelGel, 5 min. Press Run.
13. Open the Multiphor II lid and move the cathode buffer strip (white) to the area of the removed Immobiline DryStrip (lane 2) and adjust the position of the cathodic electrode so that it is above the buffer strip again.
14. Replace the lid and set the EPS 3501 XL power supply to 1,000 V, 40 mA per ExcelGel, 40 W per ExcelGel, 2.5 h. Press Run.
15. On completion, lift the Multiphor II lid and discard the buffer strips. Remove the gel from the electrophoresis unit and blot the plastic backing with paper towels to remove excess paraffin oil. Any residue oil can be cleaned off with methanol. Wipe the cooling plate of the Multiphor II with paper towels only.
16. Proceed to either Western blotting or silver staining of the gel.

3.6. Staining of Second Dimension Gel

The following protocol describes the staining of 2D gels using the ProteoSilver™ Plus Stain Kit as described in the manufacturer's instructions. It is compatible with downstream mass spectrometry analysis and it is able to detect low abundance proteins. The staining procedure is conducted at room temperature in successive solutions using a rotary shaker for constant gentle agitation.

1. Place the gel in a clean glass dish and immerse in fixing solution for 20–40 min (see Note 30).
2. Wash the gel in Ethanol Solution for 10 min, followed by 10 min incubation in Sensitisation Solution.
3. Subject the gel to two successive 10 min UHQ water washes followed by immersion in Silver Equilibration Solution for 10 min.
4. Wash the gel in UHQ water for 1 min (see Note 31), then immerse in Developer Solution until the desired staining intensity is observed (see Note 32).
5. Stop the reaction by adding 5.5 mL of Stop Solution to the gel Developer Solution and mixing for 5 min.
6. Finally, decant the Developer/ProteoSilver Stop Solution and wash the gel in numerous changes of UHQ water.
7. Store the gel in water with the plastic backing uppermost. For long-term storage, add sodium azide solution to the water at a final concentration of 0.1% (w/v) (see Note 33). The gel can be labelled with a permanent marker on the plastic backing.
8. Scan the gel to obtain an electronic image of the separated proteins (see Fig. 1a).

3.7. Electrotransfer of Proteins from Gel to Membrane

The following instructions describe electrophoretic transfer of proteins using a Trans-Blot Cell (Bio-Rad). In this system, the gel and membrane are held within a sandwich of blotting paper and foam pads in a cassette entirely submerged in transfer buffer.

1. Once gel electrophoresis is complete, use a scalpel blade and ruler to cut the area of each electrophoresed DryStrip and its accompanying molecular weight marker/s (see Note 34). Use the dye front of the DryStrips and markers as a guideline and exclude sections of the ExcelGel on either side of the positions formerly occupied by the SDS buffer strips. Cut through the gel to the plastic backing, but leave the plastic intact. Measure the area of each gel.
2. Immerse the gel in blotting buffer for 30 min to equilibrate.
3. Cut nine pieces of chromatography blotting paper and two sheets of PVDF a little larger than the dimensions of the gels.

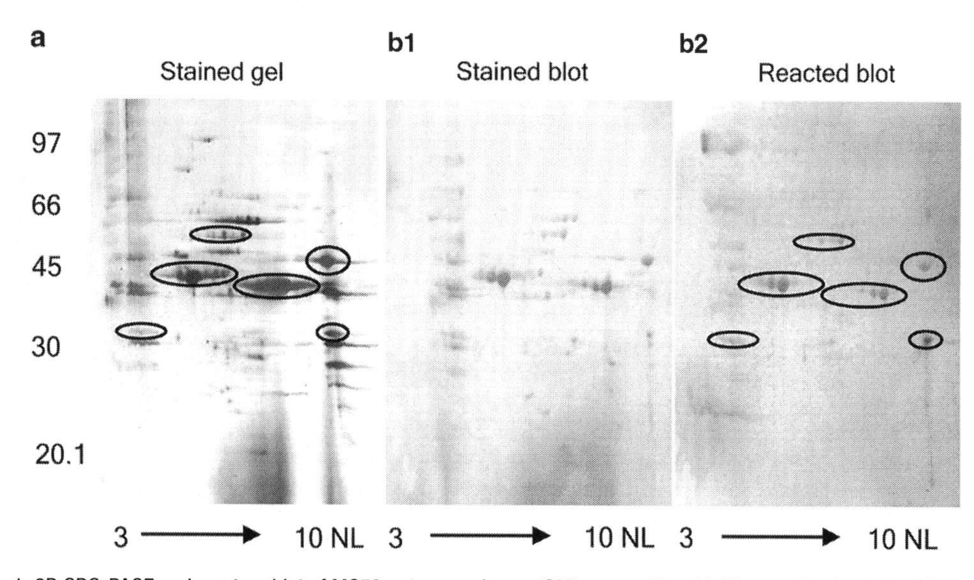

Fig. 1. 2D SDS-PAGE and western blot of MC58 outer membrane (OM) preparation. (**a**) Silver-stained gel. (**b**) Western blot of a duplicate gel transferred to a polyvinylidene difluoride (PVDF) membrane (1) stained for total protein with a reversible protein stain and (2) probed with serum obtained post-meningococcal colonisation.

4. Wet the PVDF sheets in 100% methanol and then immerse in blotting buffer.
5. Open the gel cassette and place it in a tray containing blotting buffer and construct a gel sandwich as described below. Dampen all components of the sandwich with blotting buffer and remove air bubbles after the addition of each layer.
   (a) Position a foam pad in the cassette.
   (b) Overlay the foam pad with three successive pieces of blotting paper.
   (c) Remove the plastic backing from the gel with the aid of a FilmRemover. Place the gel on the FilmRemover with the gel uppermost and clamp in place. Slide the wire gently between the gel and the plastic backing until the first precut gel area is removed. Remove the gel and position it on the blotting paper.
   (d) Position a PVDF sheet over the gel.
   (e) Overlay with a further three sheets of blotting paper.
   (f) Place the second gel on the blotting paper and overlay with the remaining PVDF membrane, blotting paper and foam pad.
6. Close the cassette and immerse in blotting buffer in the Trans-Blot Cell with the PVDF membranes positioned between the gels and the anode.

7. Perform electrophoresis at 4°C overnight with the current limited to 0.15 A.
8. The following day, rinse the membrane in several changes of UHQ water.

3.8. Staining of Proteins Transferred to PVDF Membrane

PVDF membranes can be stained using the MemCode™ Protein Stain, in order to verify transfer of proteins and to create a reference map of the electrophoretic location of proteins.

1. All steps are done at room temperature using a rotary shaker.
2. Immerse membrane in sensitizer solution for 2 min.
3. Decant the sensitizer solution and submerge the membrane in MemCode™ Reversible Stain for 1 min.
4. Remove background staining by rinsing in destain/methanol solution and then water.
5. Allow the membrane to dry (see Note 35).
6. Use a pencil to mark the bottom of the membrane with the IEF date and a unique number for the membrane (see Note 36).
7. Scan the membrane to obtain a reference image (see Fig. 1b).

3.9. Immunoreaction of Antisera and Outer Membrane Proteins

In this protocol, paired sera (e.g. pre- and post-colonisation) are reacted against membranes of OM proteins focused in the same IEF run to facilitate subsequent comparison of immunoreactivity. The method utilises a colorimetric substrate incorporating NBT and BCIP as described previously (19). All steps take place at room temperature using a rotary shaker unless otherwise stated.

1. Wet the stained PVDF membrane (see Subheading 3.8) with 100% methanol.
2. Reverse stain by immersion of the membrane in MemCode™ Stain Eraser/methanol solution for a maximum of 20 min, followed by rinsing in UHQ water (see Note 37).
3. Block non-specific immunoreactivity by immersing the membrane in blocking buffer and rock for 1 h.
4. Remove blocking buffer by washing the membrane in TTBS for 10 min. Repeat twice more.
5. Immerse the membrane in serum diluted (e.g. 1/400) in TTBS and rock overnight at 4°C.
6. Remove unbound antibody by washing the membrane in TTBS (3×10 min).
7. Incubate the membrane in blocking buffer (1/1,000) containing anti-human IgG antibody alkaline phosphatase conjugate.
8. Remove unbound conjugated antibody by washing in TTBS as described above, followed by washing in TBS (3×10 min) to remove Tween 20.

9. Detect immunoreactivity by the addition of substrate solution. Monitor the developing colour reaction and stop by washing in UHQ water once the spots have reached the desired intensity with minimal background staining (see Note 38).
10. Dry the membrane between sheets of filter paper and scan the image (see Fig. 1c).

3.10. Alignment of Images and Spot Excision

Identification of the immunogenic proteins on the membranes can be achieved by mass spectrometry of the corresponding protein spot on the matching silver-stained gel. The stained gel and immunoreacted membranes are scanned and the electronic images aligned. However, detachment of the 2D gel from its plastic backing prior to electrotransfer (see Subheading 3.7) of the resolved proteins onto PVDF causes the gel to swell and distort. Consequently, it is useful to stain the membrane with a reversible protein stain after electrophoresis to obtain an image of the overall complement of proteins to compare to the immunogenic proteins on the same membrane.

1. Open the separate protein stained and immunogically reacted membrane image files for each of the sera to be compared and the corresponding silver-stained gel using graphics software e.g. Corel PHOTO-PAINT.
2. Optimise the images by adjusting the contrast, intensity, and brightness. Increasing both the contrast and intensity of the image tonal ranges generally work well (see Note 39).
3. Crop each image using the Crop tool. Input the dimensions (width, height) of the area to be cropped and drag the resulting rectangle to select the area on the image to be cropped. If required, the cropping area can be enlarged or reduced by dragging the cropping handles. Standardise for each image. Now, double-click inside the cropping area. Save the new images.
4. For labelling of optimised images, we find that copying and pasting into Microsoft Word or PowerPoint programs is easiest. For each membrane compare the immunologically reactive image with the protein stained image and match the immunogenic spots to the protein spots by visual inspection of spot size, shape, and electrophoretic position. Similarly identify the selected protein spots to the image of the stained gel.
5. Excise the matching spots from the stained gel with separate clean scalpel blades or cut pipette tips and transfer to microtitre wells for mass spectrometry analysis (see Notes 40–42). Make a note of the well number and corresponding position of each spot to be identified on the gel. Cover the gel pieces with 100 μL of UHQ water and seal the microtitre plate. Once cut, the gel pieces can be stored frozen at –20°C until submitting for mass spectrometry analysis.

6. The alternative method is to acquire the images of the stained gels and membranes using the VersaDoc™ Imaging System, which can be linked directly to the GelPix robotic spot excision tool.
7. Once the images are aligned (see Fig. 1), it is possible to manually score serum antibody reactivity quantitatively as follows: 0, no reactivity; 1, very weak reaction; 2–4, progressively stronger reactions; and 5, very strong reaction (20). Thus, a level of immunoreactivity can be obtained facilitating the selection of protein spots showing increased immunoreactivity for further analysis.

4. Notes

1. The pH range of the IPG buffer must match that of the Immobiline DryStrip.
2. Care should be taken to ensure that urea containing solutions are kept at room temperature. Urea crystallises at low temperatures (<10°C) and carbamylates at higher temperatures (37°C) (18).
3. The R/SB must be made up fresh on the day it is to be used, or frozen at −70°C in single-use aliquots. The pH is not adjusted.
4. The ceramic IPGphor strip holders and IPGphor cup-loading strip holders are fragile. Handle with care.
5. Rehydration can take place on the bench if the room temperature is ≥20°C. Alternatively, using the Peltier cooling plate of the IPGphor unit with voltage set to zero ensures a constant 20°C temperature throughout rehydration (18).
6. The ceramic IPGphor strip holders and IPGphor cup-loading strip holders are coated to minimise protein adsorption. Washing with detergents other than the Strip Holder Cleaning Solution might compromise the coating (18).
7. These are strips of blotting paper approximately 3 mm wide and 20 cm long. Blotting paper cut to the right dimensions can be substituted.
8. Application of a pre-stained marker to the gel can be used to monitor electrophoresis and to assess subsequent electroblotting efficiency.
9. Nitrocellulose membrane can be used instead of PVDF; however, in our experience the transfer of proteins to PVDF is more efficient.

10. Whatman filter paper (Fisher Scientific) precut to commonly sized gels can be purchased to save time.
11. The primary antibody can be re-used for subsequent experiments if sodium azide is added to a final concentration of 0.02% (w/v). Sodium azide inhibits horseradish peroxidise, therefore it should not be used when this enzyme is conjugated to the secondary antibody. Sodium azide must be handled with caution in a fume hood as it is extremely toxic and potentially explosive. Refer to the appropriate Control of Substances Hazardous to Health Regulations (COSHH) and the Health and Safety Executive (HSE) guidance before disposal of sodium azide containing solutions.
12. In the current method, TTBS is used to dilute the human serum (primary antibody) and TTBS containing 5% (w/v) non-fat milk powder is used to dilute the alkaline phosphatase conjugated secondary antibody. When using alkaline phosphatase conjugates, do not substitute TTBS with a buffer containing phosphate buffered saline as a diluent, as phosphate inhibits this enzyme.
13. Ensure that the IgG species in the conjugate preparation matches that of the primary antibody.
14. BCIP and NBT solutions must be made up in glass containers since DMF erodes plastic. Wrap the glass vessels in foil since both reagents are sensitive to light.
15. This mixture is sufficient for loading three sample cups with a 100 μL volume of R/SB containing 75 μg of OM preparation.
16. The volume of R/SB required for rehydration is dependent on the length of the DryStrip gel. The manufacturer recommends 125, 200, 250, 340 and 450 μL volumes of R/SB for rehydration of 7, 11, 13, 18 and 24 cm strips respectively (18).
17. Do not use the cup-loading strip holder for rehydration of the strips because the groove is too wide to allow absorption of the entire volume of R/SB solution (18).
18. This step removes excess R/SB and prevents urea crystallisation on the surface of the gel during IEF.
19. At this point, the electrodes may not be in complete electrical contact with the surface of the IPGphor; however, closing the lid of the unit will apply pressure ensuring contact (18).
20. Do not add more than the recommended volume of cover fluid as the excess may capillary wick out of the strip holder, resulting in the gels drying out during IEF.
21. Focused DryStrips can be stored frozen for months.
22. Never equilibrate the Immobiline DryStrips prior to storage (18).

23. It is useful to make a note of the unique identifying number of the DryStrip for record keeping.
24. Using abrasive solutions to clean the IPGphor would damage the gold-plated copper surface of the electrode areas.
25. Shorter equilibration steps risk some proteins not transferring out of the DryStrip into the ExcelGel (18).
26. The manufacturer recommends a volume of approximately 3 mL of oil; however, a larger volume makes it easier to remove bubbles trapped between the gel and the cooling plate.
27. Take care to minimise transfer of oil to the gel surface as it may result in slippage of the buffer strips during electrophoresis.
28. Wear non-latex gloves when applying buffer strips, as latex sticks to the strips.
29. The paper must be just damp and not wet through. Blot with absorbent paper before applying to the gel.
30. Fixing for 40 min may decrease background staining caused by the presence of carrier ampholytes. However, care must be taken not to fix for any longer as the acid in the fixative causes the gel to degrade.
31. Washing for longer than 1.5 min decreases sensitivity of the stain.
32. Usually takes 3–7 min.
33. Even stored as described, the gel will eventually degrade when it comes away from its plastic backing.
34. Although the working area of the Trans-Blot Cell is 16 × 20 cm, handling such a large gel once it has been separated from its' plastic backing is difficult because it is 0.5 mm thick. It is easiest to cut the gels to size whilst still supported by a plastic backing.
35. Inclusion of this step is not recommended by the manufacturers of the staining kit; however, we have noticed no deleterious effects. The dry membrane is easier to scan and the background is lighter.
36. Marking the membrane is useful for orientation of the membrane image after it has been scanned.
37. Although the protein spots are no longer visible after 20 min of erasing the stain, a very pale blue background remains. This appears to have no effect on the subsequent immunoreaction.
38. React paired sera at the same time to facilitate comparison of intensity of immunoreactivity.
39. The images of the paired sera membranes should be manipulated to the same degree to facilitate comparison of intensity of immunoreaction.

40. Try not to cut too far around the protein spot as the more concentrated the protein sample the better for subsequent analysis.
41. Excise protein spots for mass spectrometry analysis as soon as practically possible as proteins will oxidise and degrade with time.
42. The identification of low abundance proteins may necessitate running 3–4 duplicate 2D gels and pooling the spot to be analysed in a single microtitre well.

Acknowledgements

This work was supported by Meningitis UK and Wessex Medical Research.

References

1. Beveridge TJ (1999) Structures of gram-negative cell walls and their derived membrane vesicles. J Bacteriol 181: 4725–4733.
2. Tettelin H, Saunders NJ, Heidelberg J et al (2000) Complete genome sequence of *Neisseria meningitidis* serogroup B strain MC58. Science 287: 1809–1815.
3. Parkhill J, Achtman M, James KD et al (2000) Complete DNA sequence of a serogroup A strain of *Neisseria meningitidis* Z2491. Nature 404: 502–506.
4. Bentley SD, Vernikos GS, Snyder LA et al (2007) Meningococcal genetic variation mechanisms viewed through comparative analysis of serogroup C strain FAM18. PLoS Genet 3: e23.
5. Hotopp JC, Grifantini R, Kumar N et al (2006) Comparative genomics of *Neisseria meningitidis*: core genome, islands of horizontal transfer and pathogen-specific genes. Microbiology 152: 3733–3749.
6. Rabilloud T (2009) Membrane proteins and proteomics: love is possible, but so difficult. Electrophoresis 30 Suppl 1: S174–S180.
7. Williams JN, Skipp PJ, Humphries HE et al (2007) Proteomic analysis of outer membranes and vesicles from wild-type serogroup B *Neisseria meningitidis* and a lipopolysaccharide-deficient mutant. Infect Immun 75: 1364–1372.
8. Vaughan TE, Skipp PJ, O'Connor CD et al (2006) Proteomic analysis of *Neisseria lactamica* and *Neisseria meningitidis* outer membrane vesicle vaccine antigens. Vaccine 24: 5277–5293.
9. Vipond C, Wheeler JX, Jones C et al (2005) Characterization of the protein content of a meningococcal outer membrane vesicle vaccine by polyacrylamide gel electrophoresis and mass spectrometry. Hum Vaccin 1: 80–84.
10. Uli L, Castellanos-Serra L, Betancourt L et al (2006) Outer membrane vesicles of the VA-MENGOC-BC vaccine against serogroup B of *Neisseria meningitidis*: Analysis of protein components by two-dimensional gel electrophoresis and mass spectrometry. Proteomics 6: 3389–3399.
11. Ferrari G, Garaguso I, Adu-Bobie J et al (2006) Outer membrane vesicles from group B *Neisseria meningitidis gna33* mutant: Proteomic and immunological comparison with detergent-derived outer membrane vesicles. Proteomics 6: 1856–1866.
12. Vipond C, Suker J, Jones C et al (2006) Proteomic analysis of a meningococcal outer membrane vesicle vaccine prepared from the group B strain NZ98/254. Proteomics 6: 4203.
13. Williams JN, Skipp PJ, O'Connor CD et al (2009) Immunoproteomic Analysis of the Development of Natural Immunity in Subjects Colonized by *Neisseria meningitidis* Reveals Potential Vaccine Candidates. Infect Immun 77: 5080–5089.
14. Mendum TA, Newcombe J, McNeilly CL et al (2009) Towards the Immunoproteome of *Neisseria meningitidis*. PLoS ONE 4: e5940.

15. Abel A, Sanchez S, Arenas J et al (2007) Bioinformatic analysis of outer membrane proteome of *Neisseria meningitidis* and *Neisseria lactamica*. Int Microbiol 10: 5–11.
16. Bernardini G, Renzone G, Comanducci M et al (2004) Proteome analysis of *Neisseria meningitidis* serogroup A. Proteomics 4: 2893–2926.
17. Pennington K, McGregor E, Beasley CL et al (2004) Optimization of the first dimension for separation by two-dimensional gel electrophoresis of basic proteins from human brain tissue. Proteomics 4: 27–30.
18. Anonymous, 2-D Electrophoresis: Principles and Methods, in: Anonymous, Handbook 80-6429-60AC,GE Healthcare, 2004, pp. 1–168.
19. Jones GR, Williams JN, Christodoulides M et al (2000) Lack of immunity in university students prior to an outbreak of serogroup C meningococcal infection. J Infect Dis 181: 1172–1175.
20. Jones GR, Christodoulides M, Brooks JL et al (1998) Dynamics of carriage of *Neisseria meningitidis* in a group of military recruits: subtype stability and specificity of the immune response following colonisation. J Infect Dis 178: 451–459.

Chapter 21

Antigen Identification Starting from the Genome: A "Reverse Vaccinology" Approach Applied to MenB

Emmanuelle Palumbo*, Luigi Fiaschi*, Brunella Brunelli, Sara Marchi, Silvana Savino, and Mariagrazia Pizza

Abstract

Most of the vaccines available today, albeit very effective, have been developed using traditional "old-style" methodologies. Technologies developed in recent years have opened up new perspectives in the field of vaccinology and novel strategies are now being used to design improved or new vaccines against infections for which preventive measures do not exist. The Reverse Vaccinology (RV) approach is one of the most powerful examples of biotechnology applied to the field of vaccinology for identifying new protein-based vaccines. RV combines the availability of genomic data, the analyzing capabilities of new bioinformatic tools, and the application of high throughput expression and purification systems combined with serological screening assays for a coordinated screening process of the entire genomic repertoire of bacterial, viral, or parasitic pathogens. The application of RV to *Neisseria meningitidis* serogroup B represents the first success of this novel approach. In this chapter, we describe how this revolutionary approach can be easily applied to any pathogen.

Key words: Vaccine, Reverse vaccinology, Genomics, *Neisseria meningitidis*, Antigen selection, Antigen expression and purification, Bactericidal assay

1. Introduction

New technologies developed in the last years have opened prospects in the field of vaccinology and strategies are now being undertaken to design improved or new vaccines against infections for which preventive measures do not exist. Most of the vaccines available today, albeit very effective, have been developed using traditional "old-style" methodologies. New technologies, based on molecular biology and genetics, have allowed cloning and large-scale purification of antigens in homologous and heterologous systems and have

* Both authors equally contributed to the work

Myron Christodoulides (ed.), *Neisseria meningitidis: Advanced Methods and Protocols*, Methods in Molecular Biology, vol. 799,
DOI 10.1007/978-1-61779-346-2_21, © Springer Science+Business Media, LLC 2012

been applied to the vaccine field. Modern examples include the hepatitis B and papilloma vaccines (1, 2), in which the capsid proteins (the main antigenic components) are expressed and purified from recombinant yeast, or the pertussis vaccine, in which the pertussis toxin has been genetically detoxified, expressed and purified from an engineered *Bordetella pertussis* strain (3).

Knowledge on the mechanisms of microbial pathogenesis and immunity has allowed the rational design of live attenuated strains in which genes identified as important in virulence virulence are deleted from the chromosome so that the mutations are not generated randomly but rationally, and are easy to characterize. An additional important revolution in the vaccine field has been the discovery that conjugation of capsular polysaccharides to a carrier protein can increase immunogenicity, in particular in infants, and induce immunological memory. Improvements in chemical technology have allowed the development of efficacious vaccines against 7, 10, and 13 serotypes of *Streptococcus pneumoniae* and four *Neisseria meningitidis* serogroups (A, C, Y, and W135) (4, 5).

The basic components of all these vaccines are antigenic determinants, which are highly expressed by the microorganisms, relatively easy to identify and characterize and, in many cases, are known virulence factors. The genomic era has seen a breakthrough in the identification of novel proteins of unknown function, which are expressed by a microorganism at different stages of its life cycle. These proteins, when localized on the surface, have the potential to be ideal targets for antibody recognition and therefore ideal vaccine antigens (6).

Screening of the genome using bioinformatic tools allows the identification of a high number of new proteins and the prediction of their localization. In this chapter, we will describe the in silico methods for selecting surface-exposed or secreted proteins. After antigen selection, great effort needs to be applied in amplifying the genes, cloning them in suitable vectors for protein expression, purifying the proteins and testing them "in vitro" for functionality, or in vivo for their ability to raise protective antibodies. Only once the antibodies are available it is possible to complete the analysis and evaluate whether each protein is expressed, has the predicted molecular weight, and most importantly, is surface-exposed (Fig. 1). Although there are a lot of critical steps in this process, going from the best "in silico" tools to the selection of the high throughput cloning and expression systems, the most challenging step is the evaluation of the functionality of the antibodies. This is mainly driven by the existence of correlates of protection for each particular disease. In the case of the meningococcus, as discussed in this chapter, it is possible to select the best antigens on the basis of their ability to induce antibodies with bactericidal activity (7). This "in vitro" correlate of protection allows the potential of each antigen as vaccine candidate to be evaluated. When such an "in vitro" assay for correlation is not available, the screening has to

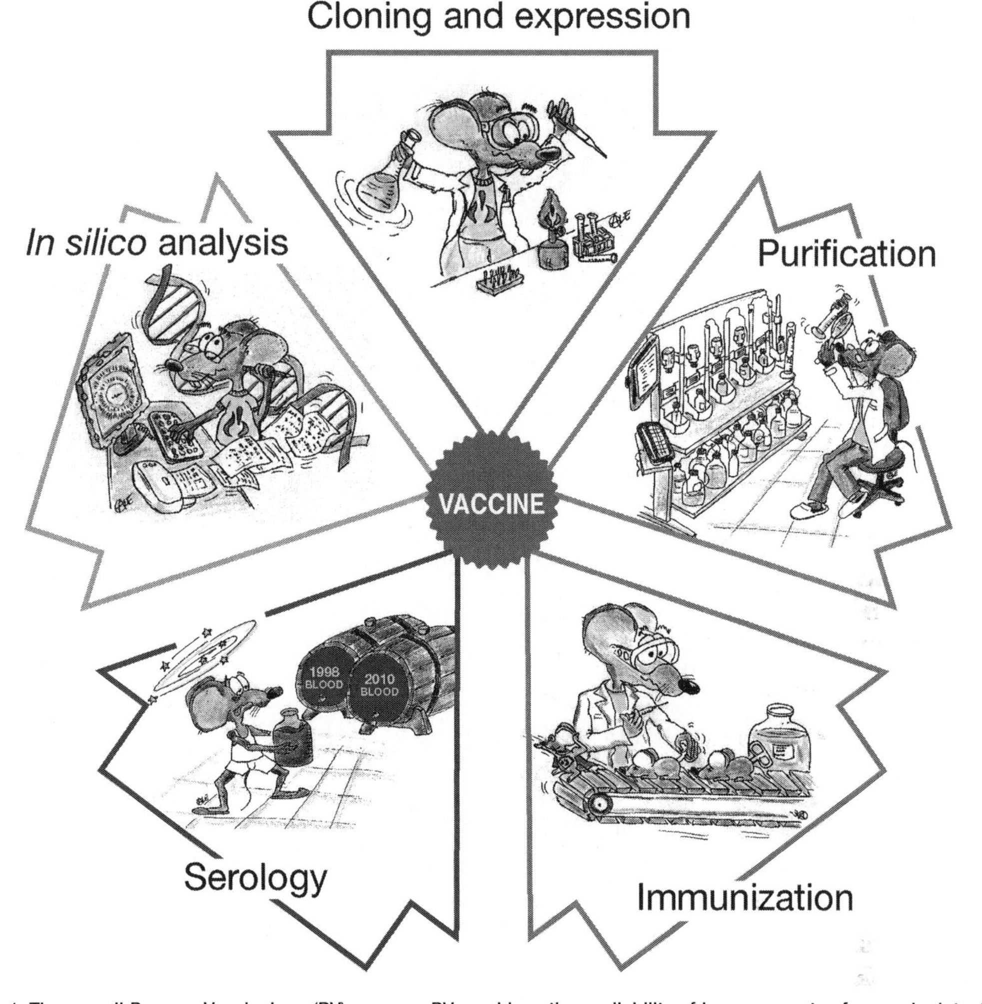

Fig. 1. The overall Reverse Vaccinology (RV) process. RV combines the availability of huge amounts of genomic data, the analyzing capabilities of new bioinformatic tools, the application of high throughput expression and purification systems combined with serological screening assays, for a coordinated screening process of the entire genomic repertoire of bacterial pathogens, viruses or parasites, for selecting the best vaccine antigens.

be performed "in vivo" and animals immunized with each of the antigens are evaluated for their survival from challenge with the virulent strains.

A further complexity of this approach is the evaluation of gene presence and conservation in different isolates to measure the potential ability of the selected antigens to induce protection against different strains. Once protective and conserved antigens are identified, the ultimate step is to ascertain their putative function in virulence and pathogenesis. To do this, the sequences undergo further bioinformatic analyses to drive the biochemical and functional characterization. Ideal vaccines are created by a combination of the most protective antigens identified through the "Reverse Vaccinology (RV)" approach (Fig. 2).

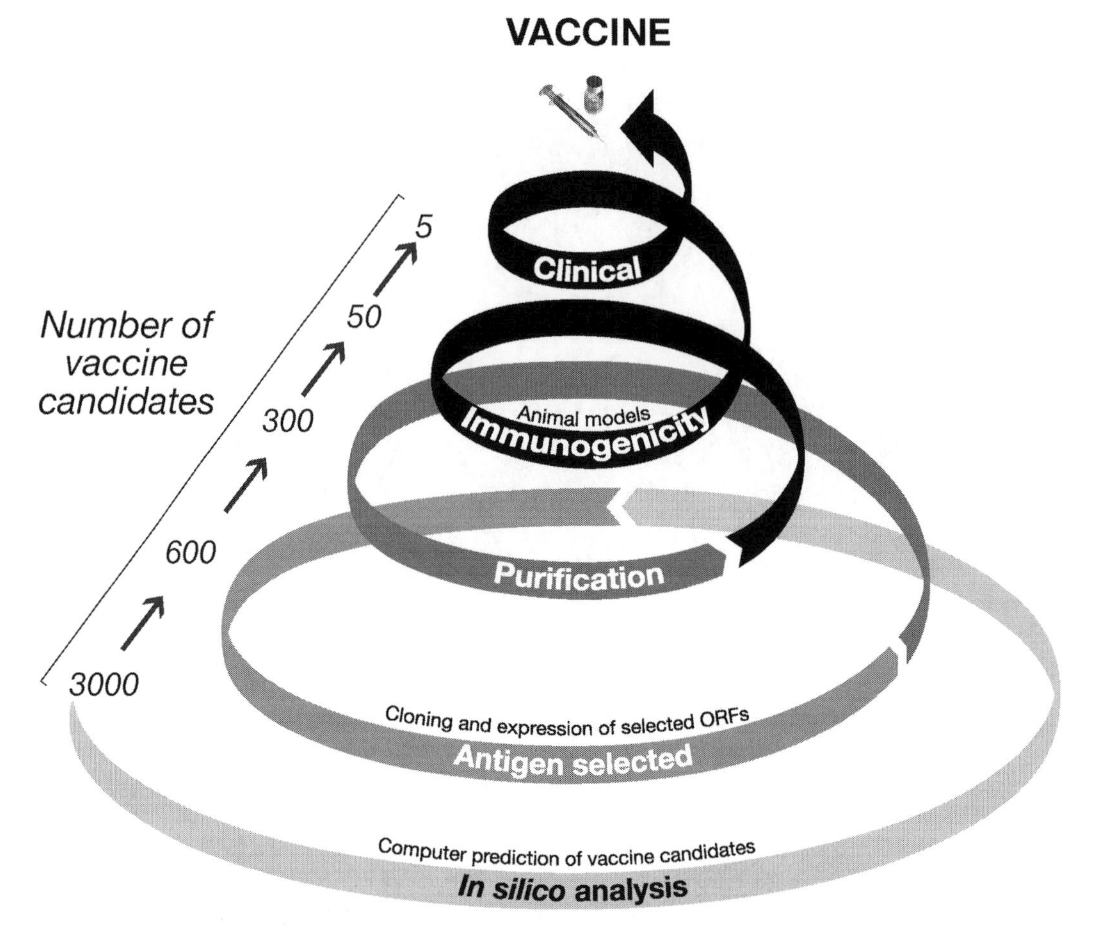

Fig. 2. Antigen screening in RV process. In the RV approach, the entire genomic repertoire encoding the vaccine candidate pool is filtered, reducing the number of antigens step after step, until a reasonable number of antigens is selected for inclusion in the final vaccine formulation to be tested in clinical trials.

2. Materials

2.1. In Silico Analyses: Gene Identification, Cellular Localization Prediction, and Functional Characterization for Vaccine Candidate Selection

2.1.1. Gene Finding

1. The National Center for Biotechnology Information (NCBI) Entrez Genome Project database is a collection of complete and incomplete (in progress) large-scale sequencing, assembly, annotation, and mapping projects for cellular organisms, which can be interrogated. Currently, this database consists of projects that have submitted data to NCBI. http://www.ncbi.nlm.nih.gov/genomes/lproks.cgi
2. Genomes Online Database is a World Wide Web resource for comprehensive access to information regarding complete and ongoing genome projects as well as metagenomes and metadata. http://www.genomesonline.org/

3. Sanger Pathogen Genomics aims to sequence the genomes of organisms relevant to human and animal health. A large portfolio of projects ranging from bacterial plasmids and phages to bacterial genomes, from protozoic parasites to helminths and insect vectors is available online. The projects themselves range from EST sequencing to comparative genomic sequencing, to fully finished genomes. http://www.sanger.ac.uk/Projects/Pathogens/
4. GLIMMER is a software package for predicting genes in microbial DNA, identifying the coding regions and distinguishing from noncoding DNA. This software is particularly useful when dealing with genomes of bacteria, archaea, and viruses. http://www.cbcb.umd.edu/software/glimmer/
5. The FASTA programs find regions of local or global similarity between Protein or DNA sequences, either by searching Protein or DNA databases, or by identifying local duplications within a sequence. Other programs provide information on the statistical significance of an alignment. Like Alignment Search Tool (BLAST), FASTA can be used to infer functional and evolutionary relationships between sequences and to identify members of gene families. The current FASTA package contains programs for Protein:Protein, DNA:DNA and Protein:translated DNA (with frameshifts) comparisons. http://www.ebi.ac.uk/Tools/fasta/

2.1.2. In Silico Localization Prediction

1. PSORT.org provides links to the PSORT family of programs for subcellular localization prediction as well as other datasets and resources relevant to localization prediction. http://www.psort.org/
2. LocateP is a genome-scale subcellular location predictor for bacterial proteins that combines many of the existing protein subcellular-location identifiers within Gram-positive bacteria: intracellular, multitransmembrane, N-terminally membrane-anchored, C-terminally membrane-anchored, lipid-anchored, LPxTG-type cell wall-anchored and secreted/released proteins. http://www.cmbi.ru.nl/locatep-db/cgi-bin/locatepdb.py

2.1.3. In Silico Functional Characterization

1. The Basic Local Alignment Search Tool (BLAST) finds regions of local similarity between sequences. The program compares nucleotide or protein sequences to sequence databases and calculates the statistical significance of matches. As in the case of FASTA, BLAST can be used to infer functional and evolutionary relationships between sequences and to help identify members of gene families. http://blast.ncbi.nlm.nih.gov/Blast.cgi
2. SMART (Simple Modular Architecture Research Tool) allows the identification and annotation of internal domains and the analysis of domain architectures. http://smart.embl-heidelberg.de/

3. PFAM is a database of protein families that includes their annotations and multiple sequence alignments generated using Hidden Markov Models. Proteins are generally composed of one or more functional regions, commonly termed domains. Different combinations of domains give rise to the diverse range of proteins found in nature. The identification of domains that occur within proteins can therefore provide insights into their function. http://pfam.sanger.ac.uk/

2.2. Cloning and Expression

2.2.1. Primer Design

Primers are designed in silico according to the nucleotidic sequences of the genes of interest and following the recommendations outlined in Subheading 3.2.1.

2.2.2. Genomic DNA (gDNA) Preparation

1. GC medium broth for growing *N. meningitidis*: Proteose Peptone n 3 15 g/L; soluble starch 1 g/L; K_2HPO_4 4 g/L; KH_2PO_4 1 g/L; NaCl 5 g/L. Sterilize by autoclaving for 15 min at 121°C and 2.68 kg/cm^2 pressure. After cooling, add 10 mL/L of Supplement I and 1 mL/L of Supplement II.
2. Supplement I: Glucose 400 g/L; L-Glutamine 10 g/L; Cocarboxylase 20 mg/L. Solubilize glucose in 800 mL of milliQ water by heating. After cooling, add L-glutamine and cocarboxylase, solubilize, adjust the volume to 1 L and aliquot with 0.22 μm filtration into 15 mL tubes. Store the stock at −20°C.
3. Supplement II: $Fe(NO_3)_3 \cdot 9H_2O$ 5 g/L, divided into 15 mL aliquots after 0.22 μm filtration. Store the stock at −20°C.
4. GC agar plates: GC agar 36.25 g/L in water; autoclave at 121°C for 15 min and 2.68 kg/cm^2 pressure. Bring to 50°C in heated bath, and add supplement I and II. Pour into Petri dishes (25 mL/plate) in sterile conditions. Plates are stored at 2–8°C.
5. Incubator, 37°C with a 5% (v/v) CO_2 atmosphere.
6. NanoVue Spectrophotometer (GE healthcare) or equivalent for measuring optical density of bacterial growth in medium.
7. Eppendorf micro-centrifuge.
8. DNeasy Blood and Tissue kit (Qiagen).

2.2.3. Amplification of the DNA Inserts by Polymerase Chain Reaction (PCR), Purification of DNA, and Agarose Gel Electrophoresis

1. GENE AMP PCR SYSTEM 9700 (Applied Biosystems).
2. Accuprime kit for DNA amplification for cloning (Invitrogen): contains Accuprime Pfx DNA Polymerase, 10× Accuprime *Pfx* reaction mix ready to use containing thermostable Accuprime proteins, 1 mM $MgSO_4$, 0.3 mM dNTPs. Alternatively, use Platinum® Taq DNA Polymerase High Fidelity (Invitrogen).
3. QIAquick PCR purification kit (Qiagen) or PureLink™ PCR purification Kit (Invitrogen).

4. Generuler 1 kb DNA Ladder, 0.5 mg/mL (MBI Fermentas)
5. DNA loading buffer: 6× DNA Loading Dye (MBI Fermentas).
6. TAE buffer for agarose gel electrophoresis: 40 mM Tris-acetate containing 2 mM EDTA prepared from a 50× solution (Biorad).
7. UltraPure Agarose melted as a 1% (w/v) solution in 1× TAE.
8. SYBR SAFE DNA gel stain 10,000× solution (Invitrogen): added to the agarose gel to visualize DNA.
9. Electrophoresis cell, power supply, and gel trays.
10. Ultraviolet (UV) transilluminator.

2.2.4. Digestion and Purification of PCR Fragments

1. Restriction enzymes and corresponding buffers.
2. QIAquick DNA Gel extraction kit (Qiagen).
3. Reagents for agarose gel electrophoresis as described in Subheading 2.2.3 (items 4–10).
4. NanoVue Spectrophotometer (GE healthcare) or equivalent.

2.2.5. Preparation and Digestion of Expression Vector

1. Luria Bertani (LB) bacterial culture medium: 10 g/L of tryptone, 5 g/L of yeast extract, 10 g/L of NaCl, pH adjusted to 7.6 with NaOH. Sterilize by autoclaving (15 min at 121°C and 2.68 kg/cm^2 pressure).
2. Qiaprep Spin minipreps for plasmid DNA isolation (Qiagen).
3. Restriction enzymes and corresponding buffers.
4. Heated water bath (37°C).
5. Ampicillin (100 μg/mL) for *E. coli* strain containing the expression vector pET21b.

2.2.6. Ligation of DNA Inserts into Expression Vector and Transformation of Cloning Strain

1. Rapid DNA ligation Kit (Roche).
2. *E. coli* strain suitable for transformation with pET vectors.
3. LB agar plates: add 15 g/L of agar *nobile* to liquid medium before sterilization, by autoclaving for 15 min at 121°C for 2.68 kg/cm^2. Bring to 55–56°C and add ampicillin (100 μg/mL) for pET21b(+) vector and pour into Petri dishes in sterile conditions. Plates can be stored at 2–8°C and used within 1 month.
4. LB bacterial culture medium supplemented with ampicillin (100 μg/mL).
5. $CaCl_2$ for preparation of competent cells.
6. Glycerol, sterilized by autoclaving (15 min at 121°C at 2.68 kg/cm^2).
7. SOC medium: Bacto-tryptone 20 g/L; bacto-yeast extract 5 g/L; NaCl 0.5 g/L; 1 M KCl 2.5 mL/L or LB medium for recovery of transformed cells. Sterilize by autoclaving.
8. Incubator, 37°C with a shaking platform.

2.2.7. Screening for Recombinant Clones

1. LB bacterial culture medium supplemented with ampicillin (100 μg/mL).
2. Qiaprep Spin minipreps (Qiagen) for plasmid DNA isolation and restriction enzymes and corresponding buffers.
3. Reagents for agarose gel electrophoresis as described in Subheading 2.2.3 (items 4–10).
4. Glycerol, sterilized by autoclaving.

2.2.8. Transformation of Expression Strain

1. *E. coli* strain containing isopropyl-β-D-1-thiogalactopyranoside (IPTG)-inducible source of T7 RNA polymerase suitable for expression of genes cloned into pET vectors.
2. LB culture medium and LB agar plates.

2.2.9. Small-Scale Protein Expression Trials

1. LB bacterial culture medium supplemented with ampicillin (100 μg/mL).
2. IPTG.
3. Ultrospec 3000 Spectrophotometer (Pharmacia Biotech), or equivalent.
4. Branson Sonifier 450 sonicator.
5. Phosphate Buffered Saline (PBS), pH 7.4.
6. Precast gel NuPAGE Novex 4–12% Bis–Tris Gel 1.0 mm, 12 well.
7. NuPAGE sample reducing agent.
8. NuPAGE LDS sample buffer 4×.
9. NuPAGE MES 20× Running buffer.
10. SEEBLUE PLUS2 molecular weight marker for protein.
11. Simply Blue™ Safe Stain.

All reagents for NuPAGE (items 6–11) are from Invitrogen.

2.3. Purification of Proteins

2.3.1. Protein Expression for Production on a Medium Scale

1. Sterile Erlenmeyer flasks for bacterial culture.
2. LB bacterial culture medium supplemented with ampicillin (100 μg/mL).
3. IPTG.

2.3.2. Bacterial Lysis and Centrifugation

1. For mechanical lysis: use a Branson sonifier 450 Sonicator and protease inhibitor cocktail (Complete EDTA-free, Roche), which is prepared by adding 1 tablet to 25 mL of equilibration buffer for soluble proteins. Equilibration buffer for soluble proteins: 50 mM NaH_2PO_4, pH 8.0 containing 300 mM NaCl.
2. Equilibration buffer for insoluble proteins: 10 mM Tris–HCl, pH 7.5 containing 100 mM NaH_2PO_4 and 6 M guanidine hydrochloride.

3. Branson Sonifier 450 sonicator.
4. For chemical Lysis: use Bacterial-Protein Extraction Reagent (B-PER, Pierce) in PBS containing 0.1 mM $MgCl_2$, 100 K units of DNAse I, and 1 mg/mL of lysozyme.
5. Beckman J2-21MIE centrifuge, or equivalent.
6. Dounce homogenizer.

2.3.3. Metal Chelating Affinity Chromatography Purification of His-Tag Proteins by Gravity Column

1. HisGraviTrap nickel (Ni^{+2}) prepacked column (GE Healthcare).
2. Nonprepacked column with resin Ni^{+2} Sepharose (Fast Flow, GE Healthcare).
3. Equilibration buffer for soluble proteins: 50 mM NaH_2PO_4, pH 8.0 containing 300 mM NaCl.
4. Washing buffer for soluble proteins: 50 mM NaH_2PO_4, pH 8.0 containing 300 mM NaCl and 30 mM imidazole.
5. Elution buffer for soluble proteins: 50 mM NaH_2PO_4, pH 8.0 containing 300 mM NaCl and 250 mM imidazole.
6. Equilibration buffer for insoluble proteins: 10 mM Tris–HCl, pH 7.5 containing 100 mM NaH_2PO_4 and 6 M guanidine hydrochloride.
7. Washing buffer for insoluble proteins; 50 mM Tris–HCl, pH 8.8 containing 8 M Urea, 1 mM tris(2-carboxyethyl)phosphine (TCEP)-HCl (Pierce), and 30 mM imidazole.
8. Elution buffer for insoluble proteins: 50 mM Tris–HCl, pH 8.8 containing 8 M Urea, 1 mM TCEP-HCl, and 250 mM imidazole.

2.3.4. Metal Chelating Affinity Chromatography Purification of His-Tag Proteins Using 96-Well Vacuum Plates

1. His MultiTrap™ HP 50 μL NiSepharose High-Performance 96-well plates (GE Healthcare).
2. Use the same gravity column buffers as described in Subheading 2.3.3.

2.3.5. Metal Chelating Affinity Chromatography Purification of His-Tag Proteins Using the AKTA x-PRESS System

1. AKTA x-PRESS system (GE Healthcare): using a Ni^{2+}-HisTrap™ FF column (1 mL) for first-step affinity chromatography with an elution buffer containing 500 mM imidazole; a HiTrap™ Desalting column (3×5 mL) and a HiTrap™ QHP ion exchange column for second-step chromatography using 50 mM Tris–HCl buffer, pH 8.0 as the equilibration buffer and the same buffer containing 1 M NaCl as the elution buffer.

2.3.6. Resolubilization of Purified Insoluble Proteins by Dialysis

1. Elution buffer: 50 mM Tris–HCl, pH 8.8 containing 8 M Urea, 1 mM TCEP-HCl, and 250 mM imidazole.
2. Dialysis buffer: 50 mM NaH_2PO_4, pH 8.8 containing 10% (v/v) glycerol, 0.5 M arginine, 5.0 mM of reduced glutathione, 0.5 mM of oxidized glutathione with 4, 2, or 0 M urea present.

2.3.7. Quality Control of His-Fusion Proteins by SDS-PAGE Analysis and Protein Assay

1. CRITERION XT Precast Gel, 26-well comb, 15 μL, 1.0 mm or 18-well comb, 30 μL, 1.0 mm (Biorad, or equivalent system).
2. MES-SDS running buffer (Invitrogen).
3. NuPAGE LDS Sample Buffer 4× (Invitrogen).
4. NuPAGE Sample Reducing Agent 10× (Invitrogen).
5. SEEBLUE PLUS2 molecular weight marker for protein (Invitrogen).
6. Simply Blue™ Safestain (Invitrogen).
7. Micro BCA Protein Assay Reagent Kit (Pierce).
8. Phoretix 1D software for densitometry analysis (TotalLab).

2.4. Immunogenicity of Meningococcal Vaccine Antigens

2.4.1. Immunization of Mice

1. CD1 mice.
2. Freund's Complete and Incomplete Adjuvants.
3. Aluminium hydroxide ($Al(OH)_3$: Superfos) adjuvant.
4. 25-gauge, 5/8′ needle and 1 mL sterile syringe (Terumo) for injections.
5. Eppendorf micro-centrifuge for serum separation.

2.4.2. Preparation of Whole Cells as Antigens for ELISA and Western Blotting

1. Appropriate microbiological safety cabinet (Biosafety level 2 safety cabinet) for working with pathogens.
2. *N. meningitidis* strains.
3. Chocolate round agar plates, store at 4°C.
4. Mueller–Hinton Broth (MHB) containing 0.25% (w/v) glucose for liquid culture.
5. 25% (w/v) glucose solution in water, filter-sterilized before use.
6. Humidified CO_2 incubator and dry 37°C incubator.
7. NanoVue Spectrophotometer and Heraeus Multifuge 3SR Plus centrifuge, or equivalent.
8. PBS containing 0.025% (v/v) para-formaldehyde.

2.4.3. ELISA on Whole Cells

1. Microtiter Plates: MaxiSorp high-binding plates (Nunc).
2. Washing Buffer (PBT): PBS, pH 7.4 + 0. 1% (v/v) Tween-20.
3. Saturation Buffer: 1% (w/v) gelatin solution in PBS.
4. Fixative solution: Saline containing 4% (w/v) polyvinylpyrrolidone and 10% (w/v) saccharose.
5. Dilution Buffer: PBS + 0.1% (v/v) Tween-20 + 1% (w/v) Bovine Serum Albumin (BSA, RIA Grade).
6. Secondary Antibody: Horseradish Peroxidase (HRP)-conjugated rabbit antimouse antibody (Dako).
7. Substrate Buffer: 25 mL of citrate buffer, pH 5 containing 10 mg of *O*-phenylenediamine and 10 μL of H_2O_2 (30% (v/v) solution).

8. Stop solution: 12.5% (v/v) H_2SO_4.
9. ELISA Readers: There are a variety of high-quality ELISA readers available commercially. The ELISA reader for different wavelengths (492–620/650 nm) must be interfaced with dedicated software for data transfer and analysis.

2.4.4. Western Blot

1. Sample buffer: 4× NuPAGE (Invitrogen).
2. Reducing Agent: 10× NuPAGE (Invitrogen).
3. 12% prepared SDS-PAGE gel (Invitrogen).
4. Nitrocellulose membrane iBlot Dry Blotting System (Invitrogen), or equivalent.
5. Blocking Solution: PBS containing 10% (w/v) skimmed milk powder and 0.01% (v/v) Triton X100.
6. Washing Buffer: PBS containing 0.01% (v/v) Triton X100.
7. Dilution Buffer: PBS containing 0.01% (v/v) Triton X100 and 3% (w/v) skimmed milk powder.
8. Secondary Antibody: HRP labeled antimouse Ig (Dako).
9. Chemiluminescent Substrate: Supersignal West Pico (Pierce).
10. Chemiluminescence films (Amersham Hyperfilm ECL).
11. Film cassette.
12. Dark room.
13. Autoradiographic process machine (AGFA), or equivalent.
14. Appropriate developing and fixing reagents (AGFA), or equivalent.

2.4.5. FACS

1. 96 U-bottom well plates.
2. Blocking and washing buffer: PBS containing 1% (w/v) BSA.
3. Goat antimouse IgG-fluorescein isothiocyanate (FITC).
4. PBS containing 0.5% (v/v) para-formaldehyde: dilute a stock solution of 4% (v/v) para-formaldehyde in PBS to 0.5% (v/v) fresh before the assay and filter-sterilize (0.22 μm pore filter).
5. PBS containing 1% (w/v) BSA. To prepare this solution, dissolve 1% (w/v) BSA in PBS, making at least 100 mL for each strain. Filter-sterilize the solution (0.22 μm filter) and prepare fresh for use.
6. *FACScan* tubes (Becton Dickson).
7. *FACScalibur* flow cytometer (Becton Dickinson).

2.4.6. Bactericidal Assay

1. *N. meningitidis* test strains.
2. Mueller–Hinton (MH) agar square plates (store at 4°C).
3. Chocolate agar round plates.
4. Mueller–Hinton Broth (MHB).

5. Milli Q H_2O.
6. Dulbecco's saline phosphate buffer (Sigma-Aldrich) containing 0.5 mM $MgCl_2$ and 0.9 mM $CaCl_2$.
7. Assay Buffer: Dulbecco's buffer containing 1% (w/v) BSA (min. 96% electrophoresis) and 0.1% (w/v) glucose, pH 7.4. Filter-sterilize (0.22 μm) the buffer and store at 4°C.
8. Baby rabbit complement (Cederlane or equivalent supplier). Store at −80°C.
9. Sera to be assayed.
10. Heat-inactivated rabbit complement: prepared by incubation at 56°C in a water bath for 30 min.
11. Serum samples or monoclonal antibodies to use as positive controls.
12. Sterile 96-well tissue culture U-bottomed plates.

2.4.7. In Vivo Passive Antibody Protection Assay

1. *N. meningitidis* test strains.
2. PBS, pH 7.4.
3. Chocolate agar plates.
4. 10% (w/v) glucose.
5. MHB containing 0.25% (w/v) glucose.
6. Wistar rat pups, 5–7 days old.
7. Animal sera or monoclonal antibodies.
8. Heparin, disodic (5,000 U).
9. Sterile L-form spreaders.

3. Methods

3.1. In Silico Analyses: Gene Identification, Cellular Localization Prediction, and Functional Characterization for Vaccine Candidate Selection

Starting from the concept that surface proteins as well as secreted proteins are more easily accessible to antibodies and therefore represent ideal vaccine candidates, selection in RV is mainly based on localization prediction (Fig. 3). All proteins fulfilling this initial criterion may be considered as vaccine candidates and selected for the following steps of the analysis, i.e., cloning and expression, purification, and in vivo immunogenicity testing.

In order to define a list of potential vaccine candidates, the initial step is performed in silico, by analyzing the genomes of the pathogen of interest. This implies: translation of the genome to produce the complete list of putative encoding genes; analysis of the protein sequences to predict their subcellular localization; evaluation of gene distribution in multiple strains for the analysis of the conservation profile; and analysis of homologues in other species according to sequence similarity to predict their putative function.

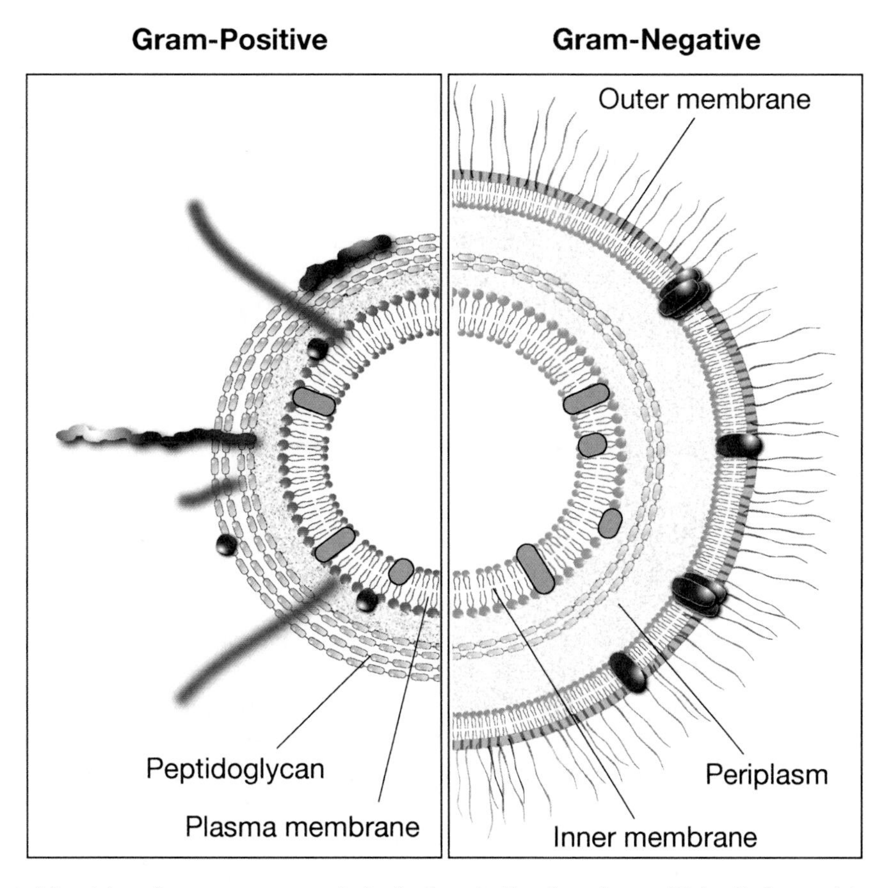

Fig. 3. Bacterial protein surface exposure as a criterion for the selection of vaccine candidates. Surface proteins as well as secreted proteins (*red* colored proteins) are more easily accessible to antibodies and therefore represent putative ideal vaccine candidates: selection in RV is based mainly on localization prediction.

The gene finding procedure allows the translation of the bacterial genome sequence into a systematic list of all the proteins that a bacterium could express at any time of its life cycle. This is accomplished by predictive algorithms, which scan the sequence in search of genomic regions that are likely to contain genes. Furthermore, since the lack of availability of genomic sequences is no longer a major constraint, comparison of sequences of different strains of the same species allows evaluating whether genes of interest are widely distributed. A single genomic sequence is not sufficient to represent the variability of any bacterial species (8) as not all genes are present in all genomes. Therefore, multiple sequences might be needed to identify multiple antigens that could be combined in a vaccine to render it effective against different strains belonging to the same species (9).

The repertoire of putative proteins identified in silico is then analyzed by dedicated software programs to deduce their putative cellular localization. Several computer programs are available to search for gene products with a subcellular localization spanning from the cytoplasm to the cell wall (in Gram-positive bacteria, Fig. 3) and to the outer membrane (in Gram-negative bacteria, Fig. 3), according to their specific structures and signatures for surface localization or secretion, e.g., secretory leader sequences, trans-membrane helices, cell wall anchor motifs for Gram-positive bacteria. Although a large majority of proteins can be associated with a specific subcellular localization, some others require further studies; sequence similarity searches represent a good complementary tool to analyze these proteins in order to predict their putative localization. Any nucleotide/amino acid sequence is not informative *per se*, but it must be analyzed by comparative methods against existing databases, i.e., sequence similarity searches, in order to build hypotheses about their possible function and hence their protein localization.

Once all data have been collected, it is possible to start the analytical process that will lead to the final selection of vaccine candidates from the total repertoire of genes/proteins. These steps generally produce a long list of putative candidates to be analyzed in all further steps, but the list can be reduced by using additional criteria based on conservation profile (10), a low number of predicted transmembrane domains, predicted molecular weight, and a match with transcriptomic and/or proteomic data where available.

The final step of the analysis is the preparation of the list of sequences for the cloning and purification steps: each gene can be cloned as full length or as a domain, in order to possibly optimize expression and solubility. It is also worth mentioning that bioinformatics can be used to design modified forms of the most promising vaccine candidates, which could be useful for reducing toxicity or for engineering multiple protective epitopes (11). This modification process is antigen-specific, and requires an extensive knowledge of its function, mode of action, and structural conformation.

3.1.1. Gene Finding

1. Retrieve from the NCBI database the list of all publicly available genomes, both complete and in-progress genome projects, of the pathogen of interest, then add to the list all other public/nonpublic genome sequences available. Publicly available sequences may be downloaded from NCBI, SANGER, or other online sources.
2. Assuming that there is more than one genome in the list choose the reference one, preferring a complete genome to an unfinished one.
3. For each genome in the list retrieve the genome sequences, gene and protein coding sequences, building a small database.
4. Draft-assembled genomes may contain DNA sequences without annotation; these sequences can be analyzed with GLIMMER for gene identification. Run GLIMMER on all draft DNA sequences and then add all the predicted genes to the database (see Note 1).

5. Once all protein sequences have been retrieved, run the FASTA algorithm comparing all the protein sequences belonging to the reference genome to all the protein sequences belonging to each of the other genomes. The FASTA algorithm is applied in order to obtain a conservation profile of each sequence across the ensemble of genomes (see Note 2).
6. Next, run the TFASTA algorithm comparing all the protein sequences belonging to the reference genome to all the DNA genome sequence(s) belonging to each one of the other genomes, in order to obtain the conservation profile of each sequence across the ensemble of genomes (see Note 3).
7. Compare the data obtained from FASTA and TFASTA analyses and update the list of all genes contained in the reference genome with their conservation profile, *i.e.*, to denote the presence/absence of the gene in each genome.
8. At this point, the complete list of all genes/proteins encoded by the reference genome will be obtained and their conservation profile, i.e., the presence and conservation of each gene across the genomes, will be established.

3.1.2. In Silico Localization Prediction

1. The PSORT program is readily available online for download to personal computers (see Note 4). Run PSORT on all the protein sequences belonging to the reference genome: this will result in the putative subcellular localization associated with each protein.
2. Download the "localization prediction of proteins" from the LocateP metadatabase (see Note 5).
3. Compare the localization predictions obtained from the different sources and assign to each protein in the list a curated localization prediction (see Note 6).
4. As result of this analysis, a complete list of all genes/proteins encoded by the reference genome with their conservation profile and their curated localization prediction will be produced and will be used for further studies.

3.1.3. In Silico Functional Characterization

Sequence similarity searches provide functional and evolutionary clues about the structure and the function of each single protein leading to a curated annotation.

1. Run PSI-BLAST on each protein sequence listed; retrieve information about putative function, putative domain presence, and conservation across the species, the genus and outside the genus (see Note 7).
2. The SMART tool provides automatic identification and annotation of domains in user-supplied protein sequences. Perform sequence analysis of domain architecture using the SMART web server, retrieving information about function and domain presence for each predicted protein sequence.

3. Retrieve all data regarding the domains associated with the listed proteins from the PFAM database, whereas domain presence has already been discovered by BLAST and SMART analysis.
4. Compare the results obtained from BLAST and SMART analyses and the PFAM database and then assign a curated annotation to each individual protein in the list.
5. Update the list of all genes/proteins contained in the reference genome with their curated annotation, adding information about the presence and conservation of each gene/protein across the genera and the species, including commensal strains, genus-related organisms and not genus-related organisms.
6. As a result of this analysis, a complete list of all genes/proteins encoded by the reference genome and their conservation profile across the species, across and outside the genus, their curated localization prediction and their curated annotation is now available. These data will be used in the analysis process that will generate the list of vaccine candidates for cloning and expression.

3.1.4. Data Analysis

1. First of all, discard all the predicted cytoplasmic, inner membrane, and periplasmic (in case of a Gram-negative bacterium) proteins from the complete list of predicted proteins associated with the reference genome (see Note 8). At this point, only surface and secreted proteins (Fig. 3) should remain in the list of protein vaccine candidates.
2. From this list of surface and secreted protein vaccine candidates, remove all the proteins that do not show conservation (at least 75% of sequence identity) across at least 50% of the strains analyzed.
3. Next, remove all the antigens present in commensal strains/species.
4. Then, remove all predicted proteins shorter than 150 amino acids and longer than 1,000 amino acids.
5. As a result of these actions, all the remaining antigens in the list are considered as promising vaccine candidates and are prepared for the next step, i.e., cloning. Each gene sequence is processed before being inserted in the vectors suitable for cloning and expression by removing secretory leader peptides and, when possible, hydrophobic membrane spanning segments, in order to optimize protein expression and solubility (12).

3.2. Cloning and Expression

Having prepared a list of vaccine candidates, the corresponding genes are cloned and expressed in the heterologous host *Escherichia coli*, in order to obtain sufficient amounts of protein to proceed with the further steps as shown in Fig. 4. An efficient expression

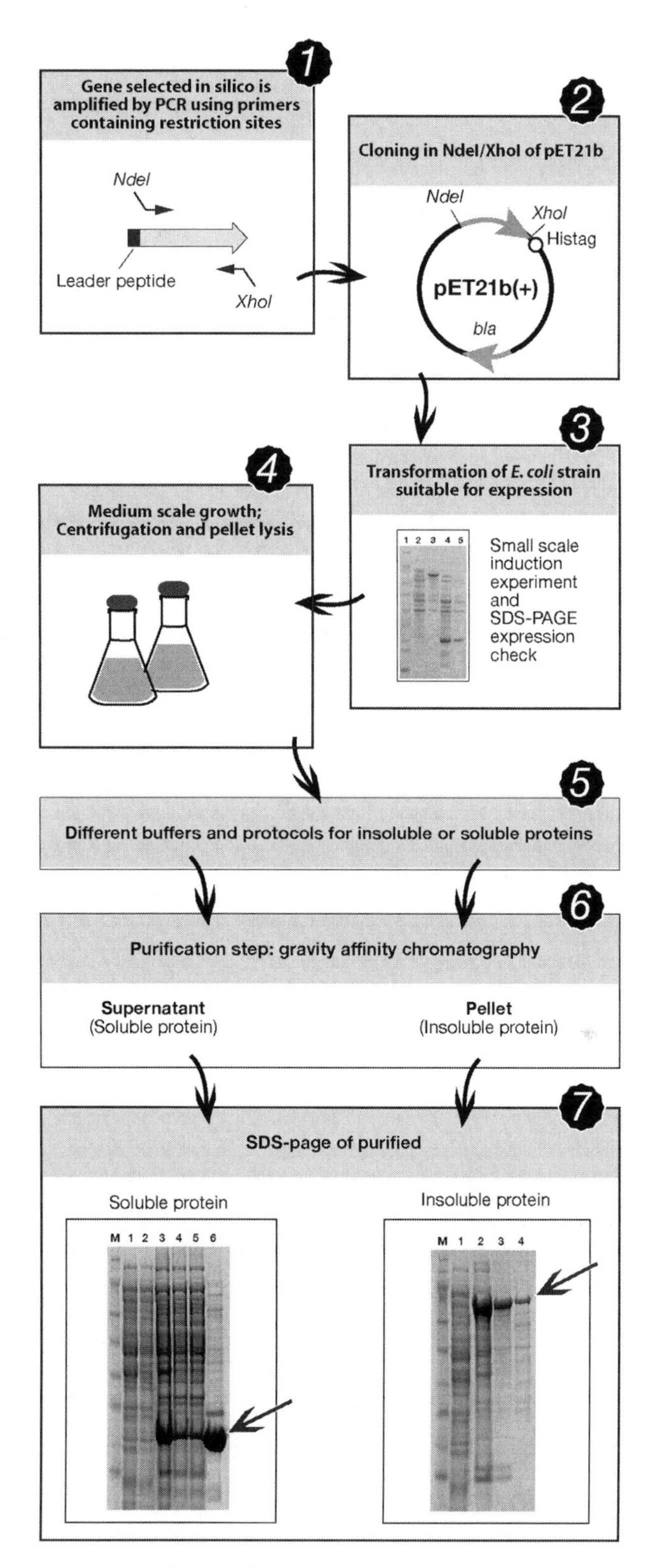

Fig. 4. Successive steps from selection to production of promising vaccine antigens. Having prepared a list of vaccine candidates (1), the corresponding genes are cloned (2) and expressed in a suitable heterologous system. Small-scale expression experiments (3) are done to evaluate the solubility of the antigens, which provides a guide to the best purification protocols to be used with medium-scale growth experiments (4–7).

system to use is the pET system (see Note 9), which allows IPTG (lactose analogue)-inducible gene expression. Target genes are cloned in pET plasmids under control of strong T7 bacteriophage transcription and translation signals so that expression is induced by providing an inducible source of T7 RNA polymerase in the host cell. An important benefit of this system is its ability to maintain target genes transcriptionally silent in the un-induced state.

To facilitate and accelerate cloning of a large number of candidates, we describe a common cloning strategy for all genes (see Note 10). One plasmid preparation is digested by two restriction enzymes and used to clone all genes amplified by Polymerase Chain Reaction (PCR) from bacterial genomic DNA (gDNA). Target genes are cloned using *E. coli* strains that do not contain the T7 RNA polymerase gene necessary for expression, thus eliminating plasmid instability due to the production of proteins potentially toxic to host cells. Colonies are screened by plasmid DNA extraction (mini-preparations) and digestion to verify the presence of the correct expression plasmid. After cloning of each insert into expression vectors, target protein expression is initiated by transferring the plasmid into an *E. coli* expression host containing a chromosomal copy of the T7 RNA polymerase gene under lacUV5 control. In order to check recombinant protein expression, an induction experiment is performed on a small volume of culture. IPTG induces the expression of T7 polymerase that consecutively starts transcription of the gene of interest. After 3 h of growth with IPTG, recombinant protein expression is examined and protein solubility is evaluated. Total bacterial lysates and soluble fractions are analyzed by SDS-PAGE to verify the presence of the recombinant protein and the amount of protein expressed.

3.2.1. Primer Design

Several factors need to be considered when designing primers for gene cloning.

1. Design synthetic oligonucleotide primers on the basis of the coding sequence of each gene previously established in silico as described in Subheading 3.1 (see Note 11).
2. Direct cloning of each amplified gene into vector pET21b(+) (or in any other vector system chosen) by including restriction enzyme sites, respectively, in the forward and reverse primers.
3. Include within the 5′ forward primers a 5′-tail with a restriction enzyme recognition site and additional nucleotides to allow cleavage by the enzyme.
4. Design the 3′ reverse primer without the stop codon, in order to allow C-terminal fusion of the protein to vector His-tag and include a tail containing the restriction site.
5. Design both primers with similar melting temperatures. In addition to the classical melting temperature (T_{m1}, see Note 12)

calculated for annealing part of the primer, a second melting temperature is calculated using the formula: T_{m2} (whole primer) = 64.9 + 0.41 (% GC) − 600/N, where N is the total number of nucleotides. T_{m2} is usually between 65 and 70°C.

3.2.2. Genomic DNA (gDNA) Preparation

Grow the bacteria according to species-specific procedures in order to obtain biomass for gDNA extraction.

1. For meningococci, plate bacteria taken from frozen bacterial culture stocks, stored at −80°C in 15% (v/v) glycerol broth, onto GC agar plates and incubate overnight at 37°C with a 5% (v/v) CO_2 atmosphere. The day after, suspend about five colonies in 7 mL of GC medium in order to start with OD at λ_{600}nm of 0.04–0.06 and incubate the culture at 37°C with 5% (v/v) CO_2 with shaking until exponential phase is reached (ODλ_{600}nm of 0.7). Harvest the bacterial cells from 1.5 mL of culture by centrifugation for 10 min at 16,000 × g and discard the supernatant.
2. Extract gDNA from the bacterial pellet (2×10^9 cells) using the DNeasy Blood and Tissue kit, following the manufacturer's instructions (see Note 13).
3. Treat the extracted DNA with RNase according to the kit manufacturer's instructions and elute gDNA in 100 μL of elution buffer. A concentration of about 50–150 ng/mL is obtained (see Note 14).

3.2.3. Amplification of the DNA Inserts by Polymerase Chain Reaction (PCR), Purification of DNA, and Agarose Gel Electrophoresis

1. The PCR is done using forward and reverse primers on bacterial gDNA as the template. Use the Accuprime kit according to the manufacturer's instructions (see Note 15).
2. Prepare the PCR amplification mix in a final volume of 100 μL in PCR tubes, containing 20 ng of gDNA as template, 10 mM of each oligonucleotide primer, 10 μL of 10× Accuprime Pfx reaction mix, 1 mM $MgSO_4$, 0.3 mM dNTPs, 2.5 U of Accuprime Pfx DNA polymerase and sterile water up to 100 μL.
3. The most commonly used PCR program is: 94°C 5 min; [94°C for 30 s, 1–5°C below T_{m1} for 30 s, 68°C for 30–60 s] for 5 cycles; [94°C for 30s, 1–5°C below T_{m2} for 30s, 68°C for 30–60 s] for 30 cycles; 68°C for 7 min; 4°C (see Note 16). Different programs are designed for pools of reactions depending on the T_m of primers and on the length of the fragment to amplify.
4. To visualize the PCR products, load 1/20th of the PCR reaction volume onto a 1% (w/v) agarose gel, stain with SYBR SAFE DNA gel stain and reveal by UV light.
5. Prepare the 1% (w/v) agarose gel by melting 1 g of agarose in 100 mL of 1× TAE buffer by heating. After cooling to ~65°C, add SYBR SAFE DNA gel stain as recommended by the manufacturer and mix gently. Then pour the solution into the gel tray.

When the gel is polymerized, put the tray in the electrophoresis cell and fill the cell with 1× TAE buffer to cover the gel.

6. Add 6× DNA loading buffer to each sample and load the samples into the agarose gel wells. Load a molecular weight marker (0.5 μg of Generuler 1 kb DNA Ladder) in parallel, as reference to compare the size of each amplified fragment. Connect the electrophoresis cell to a power supply and run at a voltage of 100 V for 1 h or until adequate separation of the bands is obtained. After the electrophoresis, photograph the gel on a UV-transilluminator to visualize the DNA bands stained by SYBR SAFE.
7. If the PCR product presents only one band compatible with the theoretical expected size, purify the DNA using a commercial PCR purification kit, according to the manufacturer's instructions.
8. If the PCR product shows an abundant band of expected size but additional nonspecific bands, extract the correct band from the gel and purify it using a commercial DNA Gel extraction kit, according to the manufacturer's instructions.
9. Regardless of the purification kit used, elute the DNA in a volume of 30 μL or 50 μL sterile water.
10. Determine the amount of purified DNA by quantitative 1% (w/v) agarose gel electrophoresis by loading 2 μL of sample in the presence of a titrated molecular weight marker. Alternatively, if the expected size is smaller than 400 bp, increase the agarose gel concentration and use 1.5 or 2% instead of 1% (w/v) (13).

3.2.4. Digestion and Purification of PCR Fragments

1. Double-digest the purified DNA corresponding to the amplified fragment with the appropriate restriction enzymes for cloning into the pET21b(+) vector. For each reaction, digest 1 μg of purified DNA (insert) with 30 U of each restriction enzyme in a final volume of 80 μL containing the appropriate buffer and BSA if needed, following the manufacturer's recommendations.
2. Incubate the mixture overnight at 37°C.
3. Analyze the PCR reaction mixture by agarose gel electrophoresis. Purify the digested fragments of expected size using the QIAquick DNA gel extraction kit, following the manufacturer's instructions and elute the fragments in a volume of 30 μL with H_2O.
4. Determine the DNA concentration by quantitative agarose gel electrophoresis as described in step 10 of Subheading 3.2.3 and by using a spectrophotometer to read optical density (OD) at λ_{260} nm (see Note 14).

3.2.5. Preparation and Digestion of Expression Vector

1. Streak out a single colony containing the expression vector selected for cloning, i.e., vector pET21b(+), onto a LB-Amp (100 μg/mL) agar plate and grow overnight at 37°C.

2. Inoculate an isolated colony from this plate into 20 mL of LB-Amp (100 μg/mL) liquid medium and grow overnight at 37°C with shaking.
3. Perform five minipreparations of the plasmid on 4 mL fractions of the 20 mL liquid culture using the Plasmid DNA isolation kit, following the manufacturer's instructions. Pool all the minipreparations and measure the plasmid concentration using a spectrophotometer at OD λ_{260} nm (see Note 14).
4. Double-digest ~5–10 μg of plasmid overnight with 80 U of two appropriate restriction enzymes in a volume of 100 μL, following the enzyme manufacturer's recommended buffers and temperatures.
5. Load 2 μL of the digestion reaction onto a 1% (w/v) agarose gel and carry out electrophoresis as described in steps 4–6 of Subheading 3.2.3. This is to check that the plasmid is actually linear and not degraded and shows a size compatible with the expected size (see Note 17).
6. If the plasmid is linear, purify the remaining reaction volume using a PCR purification kit and elute the DNA preferentially in a small volume of water (30 μL).
7. Quantify the DNA concentration by measuring ODλ_{260}nm (see Note 14) and adjust the concentration to 20 ng/μL. Use 1 μL of plasmid for each cloning procedure.

3.2.6. Ligation of DNA Inserts into Expression Vector and Transformation of Cloning Strain

1. Each insert is ligated into the expression vector pET21b(+). Both partners of the ligation, previously digested and purified, are mixed in an appropriate ratio. A molar ratio of insert/vector of 10:1 gives good results with pET21b(+) (see Note 18). For example, mix ~20 ng of pET21b(+) (5.4 kb) with ~30 ng of 0.8 kb insert.
2. Use the Rapid DNA Ligation kit according to manufacturer's instructions. Thaw the DNA ligation buffer on ice, add the DNA ligase (5 U/μL) last of all, and incubate the reaction at room temperature for 30 min.
3. For transformation by the heat-shock procedure, prepare chemically competent cells of the *E. coli* cloning strain as follows:
4. Plate *E. coli* bacteria onto LB agar and grow overnight at 37°C. On the next day, inoculate one single colony into 20 mL of LB medium and grow for 16 h at 37°C to prepare a starter culture.
5. Inoculate 1 L of LB medium with 10 mL of the starter culture and grow until the OD λ_{600}nm reaches 0.4.
6. Centrifuge the culture at 3,000 × *g* for 10 min at 4°C and suspend the pellet in 500 mL of cold 0.1 M $CaCl_2$ (freshly prepared and used within 1 week).

7. After 30 min on ice, centrifuge the cells at 3,000 × *g* for 10 min at 4°C, suspend the pellet in 50 mL of cold 0.1 M $CaCl_2$ and add sterile glycerol to a final concentration of 15% (v/v).
8. Freeze at −80°C in aliquots of 50–100 μL.
9. For transformation, thaw 50 μL of the −80°C frozen competent cells on ice (see Note 19).
10. Under sterile conditions, add the entire ligation reaction volume to an aliquot of 50 μL of competent cells. Mix by gently finger-tapping the bottom of the tube, avoiding pipetting up and down. Chill on ice for 30 min, and then heat-shock in a water bath at 42°C for exactly 30 s.
11. Under sterile conditions, add 250 μL of SOC medium to recover the cells. Incubate the recovery suspension with agitation at 37°C for 45 min to 1 h.
12. Under sterile conditions, plate 100 μL of the bacterial suspension onto selective medium according to the resistance marker encoded on the plasmid, i.e., LB-Amp (100 μg/mL) agar plates for pET21b(+) and incubate the plates overnight at 37°C.
13. Keep the remaining volume of recovery mix at 4°C for one night and store the plates at 4°C.

3.2.7. Screening for Recombinant Clones

1. Colonies resulting from the transformation are screened for the presence of the correct recombinant construct (see Note 20). Select four to eight colonies randomly for each transformation for analysis.
2. Inoculate 4 mL of LB medium with selective antibiotic (100 μg/mL ampicillin for pET21b(+)) with each single colony selected and incubate the cultures at 37°C overnight.
3. Harvest the bacterial cells by centrifugation at 16,000 × *g* for 2 min and extract the plasmid DNA using the Plasmid DNA isolation kit, following the manufacturer's instructions.
4. Elute plasmid DNA in 50 μL of sterile water (see Note 21).
5. Next, digest ~500 ng of each minipreparation of extracted DNA plasmid with 10 U of each restriction enzyme used for cloning, in order to identify the constructs containing the insert cloned into the correct restriction sites (see Note 22). Use the manufacturer's recommended digestion conditions.
6. Next, load the digestion reactions onto a 1% (w/v) agarose gel for electrophoresis (see Note 23) in parallel with a molecular weight marker and control fragments (linear digested vector and insert). Run the electrophoresis as described in steps 4–7 of Subheading 3.2.3. The correct construct has to display both linear vector and insert bands.
7. Streak out clones containing the correct plasmids onto new LB-Amp agar plates from the remaining volume of the 4 mL

culture and freeze the bacteria at –80°C in LB liquid medium with 10% (v/v) sterile glycerol.

8. Cloning of a large number of genes as performed in the RV process is not systematically followed by commercial gene sequencing unless expression problems are encountered in a further step, or verification is needed at the end of the process.

3.2.8. Transformation of Expression Strain

1. Transform 30–50 μL of chemically competent cells of *E. coli* expression strain with 1 μL of the minipreparation of each new construct by heat shock transformation, as described in Subheading 3.2.6.
2. Plate out a smaller volume (50 μL) of cell suspension onto LB-Amp selective medium agar plates and incubate at 37°C.

3.2.9. Small-Scale Protein Expression Trials

1. Inoculate 4 mL of LB-Amp (100 μg/mL) liquid medium with single recombinant colonies and incubate at 37°C overnight with shaking.
2. Use this starter culture to inoculate 20 mL of LB-Amp (100 μg/mL) liquid medium in a 100-mL flask, to start at an $OD\lambda_{600}$nm of between 0.15 and 0.2. Incubate the cultures at 37°C with shaking at 200 rpm until the $OD\lambda_{600}$nm indicates exponential growth, which is suitable for induction of expression (0.4–0.8 OD for pET vector).
3. Induce protein expression by addition of IPTG at a final concentration of 1 mM, followed by incubation at 37°C or at lower temperatures if appropriate (see Note 24).
4. After 3 h of induction, measure the $OD\lambda_{600}$nm and centrifuge a 10-mL sample at 2,000×*g* for 15 min. At this stage, the bacterial pellets can be frozen at –20°C until analysis.
5. Suspend the bacterial pellet in a volume of PBS calculated in order to standardize the concentration of bacteria for lysis and SDS-PAGE analysis, using the formula:

$$\text{PBS volume (in mL)} = (10\ \text{mL} \times OD_{600}) / 11.4.$$

6. Disrupt the cells on ice by sonication four times for 30 s each at 40 W.
7. Collect the total extract sample, i.e., 200 μL volume.
8. Centrifuge another sample of total lysate (500 μL) at 13,000×*g* for 30°min at 4°C to separate the soluble and insoluble fractions.
9. Collect a sample of supernatant containing the soluble proteins (about 200 μL).
10. At this stage, samples can be frozen at –20°C until analysis.
11. In order to define where the protein is localized, both total, soluble and insoluble samples are analyzed by SDS-PAGE,

using the NuPAGE system according to the manufacturer's instructions. The proteins that are abundant in the total extract fraction and present in low amount in the soluble fraction are considered insoluble (see Note 25).

12. Prepare the samples for NuPAGE by mixing 20 μL of sample, 12 μL of 4× sample buffer, and 16 μL of PBS. Boil this mixture for 10 min at 95°C and transfer onto ice.
13. Analyse 12 μL of each sample on the precast NuPAGE gels using MES running buffer. Load a molecular weight marker as a size reference.
14. After electrophoresis, wash the gel for 5 min in water, stain it with Simply Blue Safe Stain for 1–2 h and then de-stain in water for several hours or overnight.

3.3. Purification of Proteins

The small-scale expression trials and relative solubility tests described in Subheading 3.2.9 suggest the protocols to adopt for medium to large-scale protein production (Fig. 4). The two critical aspects for purification are expression level and solubility of the protein. Protein expression level is crucial to obtain a sufficient protein with a satisfactory level of purity for animal immunization and further antigen characterization. Soluble proteins are preferred, as solubility is usually an indication of correct folding. Moreover, solubility is an important requirement in process industrialization.

Three different chromatographic methods – gravity column, 96-well vacuum plate, and AKTA x-PRESS (14, 15) – may be used for purification. The choice of method depends on the number of samples, the target level of purity, the amount of protein(s) needed, and the time available for the purification. Both gravity column and 96-well vacuum plate methods are used when only a small/medium amount of protein is needed, regardless of purity level. For soluble proteins, when a high purity is required, then the AKTA x-PRESS system can be used for a two-step purification process consisting of metal affinity chromatography followed by ion exchange chromatography. For purified insoluble proteins, dialysis is done in buffer with decreasing urea concentrations in order to solubilize the proteins in absence of denaturing agents. Finally, quality control of the purified His-fusion proteins is done by SDS-PAGE densitometry and protein estimation.

3.3.1. Protein Expression for Production on a Medium Scale

1. Inoculate each single recombinant colonies into 30 mL of LB-Amp liquid medium in 250 mL Erlenmeyer flasks, incubate at 37°C overnight and then add 20 mL of culture into 500 mL of LB-Amp in 2 L flasks, to obtain an $OD\lambda_{600}$ nm of between 0.1 and 0.15.
2. Incubate the larger cultures at 37°C on a shaker (200 rpm) until the exponential phase is reached ($OD\lambda_{600}$ nm of between 0.4 and 0.8).

3. Induce protein expression by adding 1.0 mM IPTG (the amount of IPTG can be decreased to improve protein solubility).
4. After 3 h of incubation at 37, 30, or 25°C (based on the protein solubility evaluated on the small-scale expression trials), centrifuge the bacterial cultures at 3,000 × *g* for 45 min.
5. Discard the culture supernatant and store the pellets at −20°C.

3.3.2. Bacterial Lysis and Centrifugation

1. Lysis can be chemical or mechanical (16) depending on the number of clones to process and the amount of protein needed. The speed of the chemical lysis allows several samples to be processed simultaneously, while mechanical lysis allows a more efficient protein recovery from the soluble fraction.
2. For mechanical lysis, suspend the frozen bacterial biomass in equilibration buffer for soluble proteins containing protease inhibitor cocktail and then disrupt the bacteria by sonication on ice, applying five impulses of 30 s each at 40 W.
3. For chemical lysis, suspend the frozen bacterial biomass in B-PER lysis buffer (see Note 26) and incubate at room temperature for 40 min with constant pipetting to produce a homogeneous suspension.
4. Centrifuge the bacterial lysates at 13,000 × *g* for 30 min at 4°C and collect the supernatants for purification of soluble proteins.
5. Retain the pellets for the purification of insoluble proteins. Suspend 1 g of pellet in 10 mL of equilibration buffer (for insoluble proteins) and treat the suspension with a Dounce homogenizer. Centrifuge the homogenized material at 31,000 × *g* for 30 min at 4°C and retain the supernatant for further purification.

3.3.3. Metal Chelating Affinity Chromatography Purification of His-Tag Proteins by Gravity Column

This protocol can be used for both soluble and insoluble proteins, following the method essentially as described by Bornhorst and Falke (17).

1. Apply the starting bacterial lysates onto both HisGravTrap prepacked and nonprepacked gravity columns, which have been prewashed with water and the relevant equilibration buffer for soluble or insoluble protein. Collect the flow-through.
2. Wash the column with washing buffer at 4°C, until the $OD\lambda_{280}$nm of the flow-through falls to 0.02–0.01.
3. Elute the His-tagged protein in three fractions by the addition of 1 mL volumes of elution buffer, waiting 10 min between each elution. Pool the elution fractions that contain the recombinant His-protein.

3.3.4. Metal Chelating Affinity Chromatography Purification of His-Tag Proteins Using 96-Well Vacuum Plates

This protocol can also be used for both soluble and insoluble proteins, as follows:

1. All purification steps are done applying a vacuum with a maximum pressure of 5 mmHg. Apply the bacterial lysates onto the

His MultiTrap 96-well plate, which has been washed previously with water and the relevant equilibration buffer for soluble or insoluble protein. Collect the flow-through.

2. Wash the plate with 600 μL/well of washing buffer at room temperature for four times.
3. Elute the His-fusion proteins in two steps by adding 300 μL/well of elution buffer. Wait for 10 min and repeat the elution.
4. Pool all of the eluted fractions containing the His-protein (see Note 27).

3.3.5. Metal Chelating Affinity Chromatography Purification of His-Tag Proteins Using the AKTA x-PRESS System

1. The first purification step uses metal chelating affinity chromatography. Automatically inject the bacterial lysate containing soluble protein(s) onto a 1-mL Ni^{2+}-HisTrap™ FF column at a flow rate of 1 mL/min. Collect the flow-through and then wash the column with a variable amount of washing buffer until the ODλ_{280} nm reading reaches a constant value below 20 mAU. Elute the His-fusion protein with elution buffer.
2. Desalting follows the metal chelating affinity chromatography, incorporating a buffer exchange suitable for the subsequent ion exchange chromatography step. Automatically load the eluted His-tag fusion protein(s) onto 3×5 mL HiTrap™ Desalting columns connected in series and elute the protein at a flow rate of 5 mL/min.
3. Next, automatically load the His-fusion protein eluted from the desalting column onto a 1-mL-HiTrap™ QHP ion exchange chromatography column, at a flow rate of 1 mL/min. During the loading of the sample onto the column, the flow-through is automatically collected into a collection plate (2 mL/fraction). The elution uses two linear gradients: the first gradient is between 50 mM Tris–HCl buffer, pH 8.0 and 50 mM Tris–HCl buffer, pH 8.0 containing 500 mM NaCl, in 10 column volumes collected in fractions of 1 mL; the second gradient is between 50 mM Tris–HCl buffer, pH 8.0 containing 500 mM NaCl, and 50 mM Tris–HCl buffer containing 1.0 M NaCl (see Note 28).

3.3.6. Resolubilization of Purified Insoluble Proteins by Dialysis (18)

1. Dilute each insoluble protein to 300 μg/mL using elution buffer and dialyze against dialysis buffer containing 2 M urea, overnight at 4°C.
2. If the proteins are insoluble in dialysis buffer containing 2 M urea, dialyze against buffer containing 4 M urea, overnight at 4°C. However, if the proteins cannot be solubilized in 4 M urea buffer, then store them in the initial elution buffer that contains 8 M urea.
3. By contrast, if the proteins are soluble, dialyze overnight at 4°C against buffer that does not contain urea.
4. After dialysis, collect all His-fusion proteins and centrifuge them at 31,000 × *g* for 30 min at 4°C in order to verify precipitation.

3.3.7. Quality Control of His-Fusion Proteins by SDS-PAGE Analysis and Protein Assay

1. For soluble His-fusion proteins, load 8 μL of marker, 1 μL of total lysate (obtained after lysis, before centrifugation), 1 μL of starting material, 1 μL of flow-through, 10 μL of collected elution or eluted fraction(s) onto the CRITERION XT precast gels. Suspend all samples in 1× Sample Buffer.
2. For insoluble proteins, load the same samples as in step 1, but omit the starting material.
3. Analyze the proteins by SDS-PAGE, following the manufacturer's instructions.
4. Quantify protein using the Micro BCA Protein Assay, following the manufacturer's instructions.
5. Finally, assess the purity of the protein by loading 10 μg of the purified protein suspended in 1× sample buffer onto a SDS-PAGE gel and after electrophoresis, analyze the gel with densitometry scanning.
6. Store all purified proteins at –20°C in 40% (v/v) glycerol.

3.4. Immunogenicity of Meningococcal Vaccine Antigens

The next steps of the RV approach involve evaluating the immunogenicity of vaccine candidates obtained from the previous screening and the functionality of the antibodies raised. These protocols are applicable to any bacterial pathogen, but in this section we describe in detail the specific approach used for the serogroup B meningococcal vaccine.

For raising antibodies, mice are immunized with the recombinant proteins. The sera obtained are analyzed by enzyme-linked immunosorbent assay (ELISA), Western blot analysis on both purified proteins and total cell extracts and flow cytometry (FACS, fluorescence activated cell sorting). ELISA is used to quantify the amount of antibody induced; Western blot on purified protein is used to confirm the specificity of the antibodies, whereas Western blot on total cell extract reveals whether the protein is actually expressed by bacteria; FACS analysis on whole cells is used to determine whether the antigen is surface-exposed (Fig. 5) (6).

For evaluating antibody functionality, the main assays are divided into three categories: the serum bactericidal assay (SBA), the opsonophagocytosis assay (OPA), and the passive protection assay (19). SBA measures the amount of antibodies in serum required to kill bacteria by the classical pathway of complement activation. The SBA titer has been accepted as the correlate of protection against *N. meningitidis*; therefore, during the development of meningococcal vaccines the SBA is predominantly used for the evaluation of antibodies and is considered as the "gold standard" (19). Human serum is the preferred complement source for this assay. However, as humans are frequently carriers of meningococci, human sera usually contains endogenous antibodies to meningococcal antigens that may influence the assay. It is therefore essential to screen potential human complement sources for their lack of endogenous antibodies to the bacterial isolates that are used in the

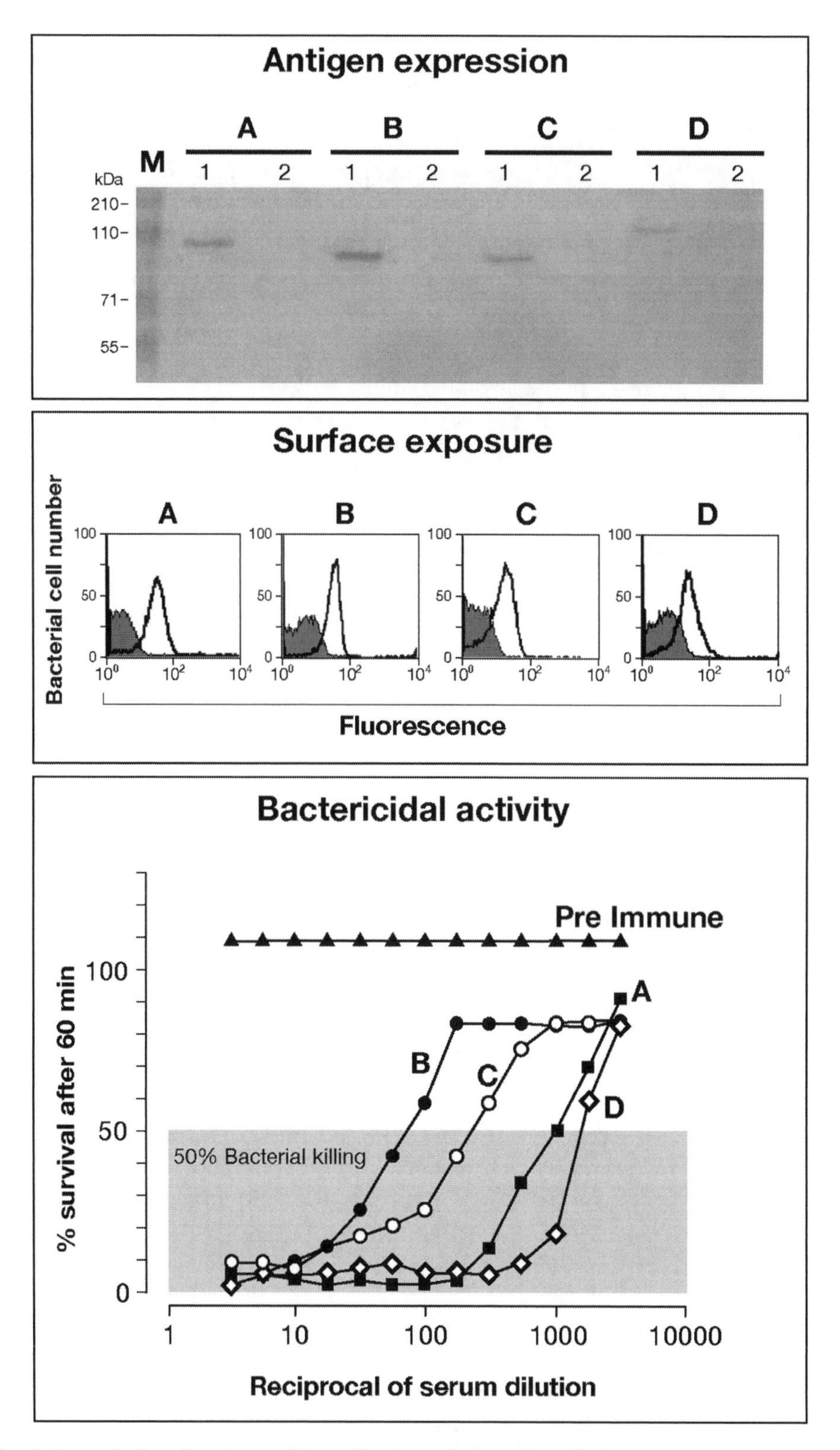

Fig. 5. Serological evaluation of vaccine candidates. After immunization with purified antigens, the generated antisera are used to check protein expression levels in total extracts of bacteria (e.g. A,B,C,D strains). The surface exposure of antigens is verified by *FACS* analysis on whole bacterial cells and sera are screened through bactericidal assays to evaluate their protective capacity.

assay. Alternatively, serum from a different species free of cross-reactive bactericidal antibodies can be used. Commercially available sources are recommended, to ensure an adequate supply of complement (20). Therefore, commercial rabbit complement is used for the high numbers of SBA needed in the case of a genome screening.

For vaccines based on novel proteins that vary in sequence and level of expression, none of the existing genetic or biochemical typing systems correlates with SBA and therefore protection can only be determined by SBA against a large panel of strains (20). In this case, it is necessary to select a panel of representative strains that ensures that the resulting SBA data can be extrapolated to estimate protection against disease-causing strains in the population. For MenB, the best vaccine candidates selected are further analyzed on a larger panel to estimate the breadth of coverage in the meningococcus population. Unfortunately, not all the proteins are able to induce a bactericidal response and in these cases it is difficult to establish whether these proteins contribute to protective immunity. Consequently, other assays that can predict protective meningococcal immunity are needed to assess the vaccine potential of such antigens; some of these candidates can be evaluated through the support of an animal model. Several animal models have been used to study the pathogenesis of meningococcal infection and the protective effect of vaccines (21). Most of them require either a large inoculum, giving little space for bacterial growth in vivo, or the use of enhancers such as mucin or iron compounds that may have multiple and poorly definable effects on both the bacteria and the host. An infant mouse model with intranasal (*i.n.*) inoculation of meningococci has been described (7), which appears to be useful for studying the early pathogenesis of infection and the role of mucosal immunity, but less suitable for assessment of protection because of the low and variable rate of invasive disease. The choice of the assay conditions needs to be evaluated carefully for sensitivity and reproducibility, taking into account all ethical considerations for the use of experimental animals (22). Protection in the model correlates with the ability of the antibodies to activate C3b deposition on the surface of live, encapsulated cells of *N. meningitidis*, which for certain strains does not proceed to bacteriolysis via the membrane attack complex. Protection by nonbactericidal antibody is conferred by opsonic activity, since this has been observed in rats deficient for complement component six, which allows opsonization but not bacteriolysis.

The best candidates selected by these screening processes will be considered as the most promising as components of a final vaccine composition.

3.4.1. Immunization of Mice

1. To prepare antiserum, 20 μg of each purified His-protein are used to immunize 6-week-old CD1 mice (see Note 29) in groups of five to eight animals (see Note 30).

2. Initially, collect preimmune serum from individual mice by tail-tipping under anesthesia, to be used as negative control sera. To prepare serum, incubate blood at 37°C for 1 h to allow clot formation and then store the sample at 4°C overnight to express the serum. Separate the serum from the blood clot by centrifugation at 3,000 × *g* for 15 min at 4°C. Remove the serum from the clot using a glass Pasteur into a clean tube. Store the serum at 4°C or at –20°C for long term.
3. Immunize mice on days 0, 21, and 35 and take sample bleeds by tail-tipping on days 0, 34, and 49. Store all sera at 4°C.
4. Inject the recombinant proteins intraperitoneally (*i.p.*), together with Freund's Complete Adjuvant (FCA) for the first dose and Freund's Incomplete Adjuvant (FIA) for the second (day 21) and third (day 35) booster doses. Add an equal volume of the adjuvant to the sample immediately before use and mix the two to form a stable emulsion before injection.
5. Inject the recombinant proteins also after adsorption to the human-compatible adjuvant aluminum hydroxide (3 mg/mL) (see Note 31), following the same immunization schedule. To prepare the antigen-$Al(OH)_3$ suspension, thaw the antigen and dilute with PBS and mix with $Al(OH)_3$ gel to obtain the appropriate concentration(s). Keep the mixture overnight at 4°C to allow antigen adsorption to the adjuvant gel. On the following morning and immediately before injecting the animals, mix again by vortex to suspend any possible precipitated complexes.
6. Take blood samples for analysis at 14 days after the third immunization (day 49) by cardiac puncture with terminal anesthesia. Collect blood in appropriate tubes and prepare serum as described in step 2 above.
7. Pool all the serum samples from individual groups of mice.

3.4.2. Preparation of Whole Cells as Antigens for ELISA and Western Blotting

1. Plate *N. meningitidis* strains from frozen stocks onto chocolate agar plates and incubate overnight at 37°C in a humidified incubator with 5% (v/v) CO_2 (see Note 32).
2. Collect the bacterial growth from the agar plates using a sterile dracon swab and inoculate into 7 mL of MHB containing 0.25% (w/v) glucose.
3. Grow the bacteria at 37°C with 5% (v/v) CO_2, starting from an $OD\lambda_{600}$nm of 0.05 until an OD value of 0.3–0.4 is reached.
4. Centrifuge the culture for 10 min at 3,500 × *g*.
5. Discard the supernatant and wash the bacteria three times with PBS.
6. Centrifuge the PBS-washed suspension for 10 min at 3,500 × *g*.

7. For preparation of the cells for ELISA, discard the supernatant and suspend the bacteria in PBS containing 0.025% (v/v) paraformaldehyde. Incubate for 1 h at 37°C and then keep overnight at 4°C with stirring, in order to fix the bacteria.
8. For preparation of cells for Western blotting, grow the meningococci in MHB broth until an ODλ_{600}nm value of 0.5–0.6 is reached, which usually takes 1.5–2.0 h. Centrifuge the culture for 10 min at 3,500×*g*, discard the supernatant and suspend the pellet in 500–600 μL of PBS, according to the final OD reached by the culture. Kill the meningococci by heating at 56°C for 1 h or by at least three consecutive cycles of rapid freeze thawing.

3.4.3. ELISA on Whole Cells

For studies of the total antibody responses, whole cells are often preferred as the antigen in the ELISA.

1. Add 100 μL of bacterial cells, prepared as described in step 7 of Subheading 3.4.2. uniformly to each well of a 96-well ELISA plate and store the plate overnight at 4°C to allow adsorption of the antigens to the plastic.
2. On the next day, wash the wells three times with PBT washing buffer.
3. Add 250 μL of Saturation Buffer to each well and incubate the plates for 2 h at 37°C.
4. After blocking, wash the plates three times with PBT washing buffer and then add 250 μL of Fixative Solution to each well. Incubate the plate for 2 h at room temperature and then remove the fixative solution from the wells.
5. To each well, add 100 μL volumes of test dilutions of mouse antisera sera prepared in Dilution Buffer. In addition to the test samples, each ELISA plate should contain a two or threefold dilution series of a standard serum for calibration. Two to three coated wells, incubated with buffer instead of antibodies, should be included as blanks and their average values subtracted from all standard curves and test values prior to further calculations.
6. Incubate the plates for 2 h at 37°C.
7. After incubation of serum samples, wash the wells three times with PBT.
8. Add 100 μL of HRP-conjugated rabbit antimouse Ig antibody, diluted 1/2,000 in Dilution Buffer, to each well (see Note 33) and incubate the plates for 90 min at 37°C (see Note 34).
9. Wash the wells three times with PBT buffer.
10. Add 100 μL of Substrate Buffer for HRP to each well and leave the plates at room temperature for 20 min.
11. Stop the reaction by addition of 100 μL of 12.5% (v/v) H_2SO_4 to each well.

12. Read the plates in the ELISA spectrophotometer at OD λ_{490}nm. The results from a whole-cell ELISA are reported as OD values using a defined serum dilution and are calculated arbitrarily as the dilution of sera which gave an OD λ_{490}nm value of 0.4 above the level of preimmune sera (see Note 35). The ELISA is considered positive when the dilutions of sera with OD_{490} of 0.4 are higher than a 1/400 dilution.

3.4.4. Western Blot

1. Suspend purified protein samples (1 μg) and whole cell lysates (15 μL of preparation described in step 8 of Subheading 3.4.2) in 1× Sample Buffer containing 1× Reducing Agent.
2. Separate the samples by SDS-PAGE (Nu-PAGE), following the manufacturer's instructions.
3. Transfer the samples onto nitrocellulose membrane using an iBlot Dry Blotting System, following the manufacturer's instructions.
4. After transfer, remove the membrane and leave it overnight at 4°C in Blocking Solution, with agitation in order to block the membrane and reduce the nonspecific background.
5. Wash the membrane twice at room temperature for about 5–10 min in Washing Buffer.
6. Incubate the membrane for 2 h at room temperature with murine sera diluted 1/200 in Washing Buffer.
7. Next, wash the membrane three times (from 5 to 10 min each) with Washing Buffer, add a 1/2,000 dilution of secondary antibody HRP-labeled antimouse Ig in Dilution Buffer to the membrane and incubate for 90 min at room temperature.
8. Wash the membrane twice with PBS and finally incubate for 5 min with the Chemiluminescent Substrate using enough volume to cover the surface of the membrane.
9. Remove the blot from the chemiluminescent substrate solution and place it in a film cassette with the protein side facing up
10. Use a darkroom and turn off all lights except those appropriate for exposure.
11. Carefully place a piece of film on top of the membrane. A recommended first exposure time is 60 s, but exposure time can be varied to achieve optimal results.
12. Develop the film using a film processor machine and appropriate developing solution and fixative.

3.4.5. FACS

This surface labeling assay allows the surface exposure of the selected antigens and the levels of expression in different strains to be examined.

1. Prepare bacterial cultures as described in step 8 of Subheading 3.4.2 and when an ODλ_{600} nm value of 0.6 is

reached (see Note 36), transfer 1 mL of culture to a sterile 1.5 mL Eppendorf tube and centrifuge at 13,000 × *g* in a micro-centrifuge for 3 min to pellet the bacteria. Discard the supernatant and suspend the pellet in 1 mL of PBS containing 1%(w/v) BSA. Finally, dilute the bacterial suspension 1/50 in PBS containing 1% (w/v) BSA.

2. Add 50 μL samples of sera diluted in Blocking Buffer (at 1/100, 1/200, and 1/400) in a 96-well plate (see Note 37). Include positive controls, such as SEAM 12, a monoclonal antibody specific for the serogroup B capsular polysaccharide.
3. Add 50 μL of bacterial cells to each well and store the plate at 4°C for 2 h.
4. Centrifuge the cells for 5 min at 3,500 × *g*, discard the supernatant and wash the cells by adding 200 μL/well of Washing Buffer.
5. Add 50 μL of a 1/100 dilution of FITC-conjugated goat antimouse Ig to each well and store the plate at 4°C for 1 h.
6. Centrifuge the cells at 3,500 × *g* for 5 min and wash the pellet with 200 μL/well of PBS.
7. Repeat the centrifugation step, discard the supernatant and add 200 μL/well of PBS containing 0.5% (v/v) para-formaldehyde, in order to fix the cells.
8. Transfer the fixed samples to individual *FACScan* tubes and analyze by flow cytometry, following the equipment manufacturer's instructions.

3.4.6. Bactericidal Assay

The assay is done in sterile 96-well plates in a final volume of 50 μL per well. Refer to Fig. 6 for a schematic of the assay.

1. Add 25 μL of Dulbecco's PBS containing 1% (w/v) BSA and 0.1% (w/v) glucose to each well in columns 1–11.
2. Prepare serial twofold dilutions of sera by adding 25 μL of prediluted sera (at least 1/16) to 25 μL of buffer previously added to the wells, mix and transfer 25 μL in the next well (repeating the procedure from column 2 to 10). Discard 25 μL from last well (column 10) (Fig. 6).
3. The wells of Column 12 contain the complement independent control, i.e., 5 μL of prediluted sera with 12.5 μL of heat-inactivated complement and 20 μL of buffer.
4. The wells of Column 11 contain the complement dependent control, i.e. 25 μL of buffer, 12.5 μL of active complement, and 12.5 μL of bacteria only.
5. Prepare bacteria as described in Subheading 3.4.2 until the $OD\lambda_{600}$ nm reaches 0.23–0.25.
6. Dilute the bacterial culture 1/10,000 by serial dilution and add 12.5 μL volumes of bacteria to each well.

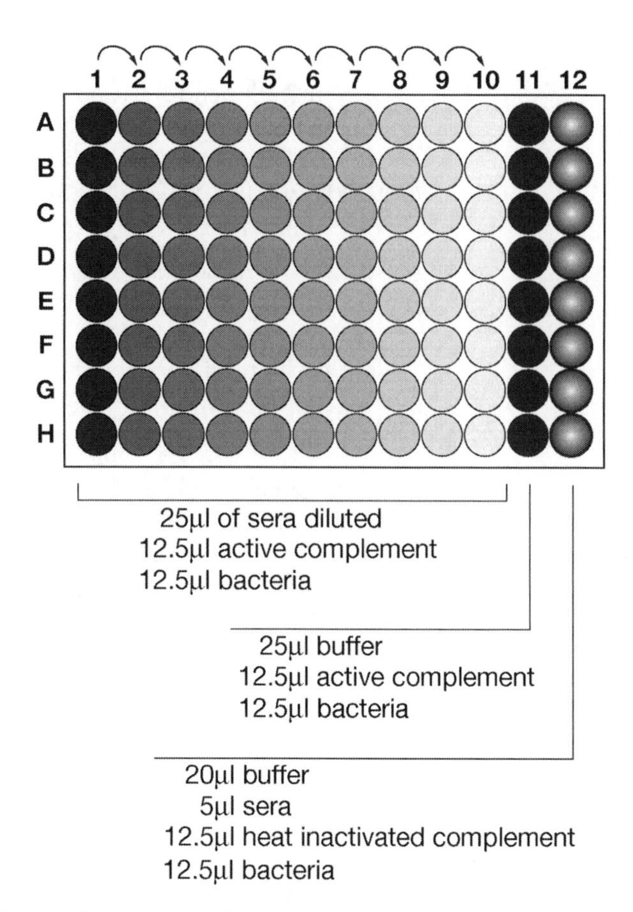

Fig. 6. The plate layout (96-well) for the serum bactericidal assay (SBA). For the preparation of serum dilutions, where the final well volume is 50 μL, initially add 25 μL of Dulbecco's buffer with 1% (w/v) BSA and 0.1% (w/v) glucose to each wells from 1 to 11. Prepare serial-step twofold dilutions by adding 25 μL of prediluted serum (at least 1/16) to an aliquot of 25 μL of buffer, mix and transfer 25 μL to the next well (repeating the procedure from column 2 to 10). Discard 25 μL from last well (column 10). Column 12 wells contain the complement independent control, i.e., 5 μL of prediluted serum with 12.5 μL of heat-inactivated complement and 20 μL of buffer. Column 11 wells contain 25 μL of buffer, 12.5 μL of active complement, and 12.5 μL of bacteria.

7. Reconstitute the lyophilized baby rabbit complement in 1 mL of ice water (see Note 38) and add 12.5 μL of active complement to the wells in Columns 1–11, according to the template shown in Fig. 6.
8. Sample the bacteria at time = 0 h by pipetting the well contents up and down three times and then placing 10 μL from the controls, column 11 and 12, to the top of a square MH agar plate. Tilt the plate so that the 10 μL volume runs down to the bottom of the plate in a straight line.
9. Cover the 96-well plate and seal with Parafilm. Incubate for 1 h at 37°C with 5% (v/v) CO_2 in a humidified atmosphere and on a soft orbital rotating shaker.

10. Sample the bacteria at t=60 min. Using a multichannel pipettor, mix the contents of each well by pipetting up and down three times, and then spot 7 μL of each well onto square MH agar plates. Sample the bacteria at t= 60 min from columns 11 and 12 and tilt the plates as described in step 8.
11. Incubate all the MH agar plates overnight at 37°C in a humidified atmosphere with 5% (v/v) CO_2.
12. On the following day, count and record the number of colony forming units (CFU) in each spot on each of the plates. All wells are plated in duplicate and duplicate counts must be recorded (see Note 39).
13. Bactericidal titers are defined as the serum dilution that gives a 50% decrease of CFU after 60 min incubation in the reaction mixture, compared with the mean number of CFU in the control reactions at time 0. The T0 average of both controls (column 11, the complement control: killing by complement alone (CDC), in the absence of antibodies, and column 12, the complement-independent control (CIC), must be between 30 and 100 CFU. Killing by serum alone in presence of inactivated complement (CIC or column 12), should be minimal. In the case of the meningococcal SBA, these values have to be less than 25% of T0 to be considered valid. (see Notes 40 and 41).

3.4.7. In Vivo Passive Antibody Protection Assay

An animal model may be used as an alternative/complementary method for testing protection conferred by vaccine candidates.

1. The day before the challenge, inoculate freshly thawed meningococci (see Note 42) onto chocolate agar plates and incubate the plates overnight at 37°C in a humidified atmosphere with 5% (v/v) CO_2.
2. Before challenge, transfer single colonies into Mueller–Hinton Broth containing 0.25% (v/v) glucose in order to start bacterial growth from an initial $OD\lambda_{600}$ nm of 0.05.
3. Grow the bacteria for 2 h with shaking (200 rpm) at 37°C in a humidified atmosphere with 5% (v/v) CO_2, until an $OD\lambda_{600}$ nm of 0.23 is reached.
4. Dilute the bacterial suspension from 10^9 to 10^4 CFU/mL in PBS (see Note 43).
5. Divide 5- to 6-day-old infant rats into groups for immunization (see Note 44) and distribute the pups equally to the nursing mothers in different cages.
6. Prepare monoclonal antibody and test antisera in PBS for inoculation and inject groups of five to six animals *i.p.* at time 0 h with 100 μL of antibody/antisera.
7. After 3 h, challenge the animals *i.p.* with 100 μL of PBS containing ~4×10^3 CFU of bacteria.

8. Eighteen hours after the bacterial challenge, obtain blood samples by puncture of rat pup cheeks using a needle and put the samples into accurately labeled 2 mL Eppendorf tubes prepared with 20 μL of disodic heparin (5,000 U).
9. Dilute the collected blood samples by a factor of 1/10 and 1/100, by transferring 100 μL into 900 μL of PBS.
10. Plate 100 μL of undiluted blood, 100 μL of blood diluted at 1/10 and 1/100, onto chocolate agar plates using sterile spreaders and incubate the plates overnight at 37°C in a humidified atmosphere with 5% (v/v) CO_2. Count the CFU on each agar plate on the following day.

In summary, this RV approach, as described in Subheading 3, led to the discovery of protein antigens that were included in new vaccine compositions against *N. meningitidis* group B infection, which are currently under development and in clinical trials (20).

4. Notes

4.1. In Silico Analysis: Gene Identification, Cellular Localization Prediction, and Functional Characterization for Vaccine Candidate Selection

1. A GLIMMER server is available at the NCBI website, so it is possible to run GLIMMER without installing it on your local computer.
2. FASTA analysis can also be performed online without installing it on your local machine. http://www.ebi.ac.uk/Tools/fasta33/nucleotide.html
3. The use of both algorithms is not redundant: since TFASTA translates the nucleotide sequences in all six frames before performing the comparison, it is helpful in finding genes missed by the gene finding procedure as well as the presence of frame-shifted sequences.
4. An alternative method is to use the PSORT meta-database containing the localization prediction of all proteins encoded in all publicly available complete genomes.
5. There are plenty of localization prediction algorithms that can be used besides those mentioned above, *e.g.*, Cello, PSLpred, Proteome Analyst, LOCtree. It is recommended that several of these programs are used and the different output data compared.
6. It may be useful to perform a direct *ad oculum* analysis on amino acid sequences, in order to recognize simple localization signals such as leader peptide sequences, LPxTG motifs and other localization signals in the case of particularly interesting proteins or dubious (low confidence) ones (23, 24).
7. An alternative method to downloading and locally installing the BLAST package is to use the BLAST web server provided by NCBI or SANGER.

8. Identification of periplasmic proteins can be a particularly challenging task, as their amino acid sequences have many signatures in common with the outer membrane-associated proteins.

4.2. Cloning and Expression

9. The pET System (Novagen/Invitrogen) is a powerful system developed for cloning and expression of recombinant proteins in *E. coli*. T7 RNA polymerase is so selective and active that when fully induced, almost all of the cell's resources are converted to target gene expression; the desired product can comprise more than 50% of the total cell protein within a few hours after induction. Although this system is extremely powerful, it is also possible to attenuate the expression level simply by lowering the concentration of inducer (IPTG). Decreasing the expression level may enhance the soluble yield of some target proteins.
10. It is not always possible to use the same cloning strategy for all ORFs because some genes can contain one or more restriction sites that were selected for cloning. In this case, alternative restriction enzymes can be used for cloning within the same vector, depending on the restriction pattern of the gene.
11. The sequence of each gene to amplify may start directly at the ATG start codon or after the signal sequence and ends just before the stop codon or before the C-terminal motif that is removed. This is defined in silico in the antigen selection section. In some cases, the presence of the leader peptide can facilitate production and purification of the protein, for instance, from the supernatant or from the periplasm (12).
12. The usual formula for calculating melting temperature

$$T_{m1}(\text{tail excluded}) = 4 \times (\text{G} + \text{C}) + 2 \times (\text{A} + \text{T}).$$

13. For gDNA preparation from pathogenic bacteria, the 56°C incubation step indicated in the protocol is performed overnight and corresponds to inactivation of the total bacterial extract. Effective inactivation is checked by plating a sample of total extract and checking for the absence of bacterial colonies after at least overnight incubation.
14. Use of a NanoVue spectrophotometer is recommended as it allows the DNA concentration to be measured directly in a small sample of DNA preparation (2 μL), thereby avoiding waste of sample and the necessity of using a quartz vial.
15. The Accuprime kit has high specificity, fidelity, and yield that makes it a good tool for PCR cloning and expression. It is provided in an antibody-bound form that is inactive at ambient temperature. The enzyme regains activity after the initial denaturation step at 94°C, providing an automatic hot start that

increases specificity, sensitivity and yield, while allowing room temperature assembly. It is easier to use the same DNA polymerase kit for all PCR reactions, but if exceptionally long genes have to be amplified, a specific polymerase will probably be necessary.

16. Elongation time in the PCR program varies according to the length of the ORF to be amplified. Usually, 1 min/1 kb is recommended. In each program, after denaturation and enzyme activation at 94°C, each sample undergoes a two-step amplification. The first step is designed by considering T_{m1} and the second step is designed considering T_{m2}. Cycles are completed with a 10-min extension step at 68°C or adequate temperature for another polymerase.
17. If the vector is only partially digested (presence of additional bands of different size compared to linear plasmid size), repeat the digestion with more restriction enzyme or in a larger volume. If the vector is degraded (presence of smear), repeat the digestion with new sterile water, new reagents, sterile tubes, or cleaner pipettes. Indeed, the quality of vector digestion is very important to facilitate all ligation reactions and accelerate the cloning procedure.
18. Usually, a molar ratio of 3:1 insert/vector is recommended, but it was observed that an excess of insert molecules enabled correct ligation to occur. Therefore, a ratio of 10:1 was used. DNA ligation buffer has to be thawed and kept on ice and freeze/thaw cycles of this buffer must be avoided. It is recommended that the buffer is stored in small aliquots at –20°C. DNA Ligase enzyme should be kept on ice and added as the last reagent. The ligation product can be stored at –20°C, but better results are generally obtained by transforming *E. coli* on the same day.
19. Chemically competent cells are very fragile and have to be gently thawed on ice. Absolutely avoid vortexing and pipetting the cells up-and-down.
20. Alternatively, screening for recombinant clones can be performed by PCR on colonies using primers that hybridize around cloning sites in the vector or specific primers for each PCR fragment. Minipreparations are then performed for a reduced number of clones that contain a construct that gave the right PCR product.
21. Water is recommended for DNA elution if the DNA sequence has to be determined in a further step.
22. If a plasmid with high copy number per cell is used, reduce the amount of minipreparation included in the digestion reaction.
23. If DNA fragments shorter than 600 bp have to be visualized, use a higher percentage of agarose, e.g., 1.5–2% (w/v) to provide better resolution.

24. IPTG is a lactose analogue that activates transcription of the T7 polymerase gene under the control of the lacUV5 promoter. For proteins expected to be insoluble (transmembrane domain), expression trials can also be done at 25°C. A bacterial culture incubated at 25°C will reach the OD indicated for induction only after about 2.5 h instead of 1 h, but the induction time with IPTG is the same (3 h).
25. There are several options to explore if no expression is observed. For example, other colonies can be analyzed, cloning can be refined by dividing the protein into several domains, and the use of other expression system plasmids and tags can be considered.

4.3. Purification of Proteins

26. During chemical lysis, pellets are suspended by pipetting to avoid excessive foam production due to the presence of detergents in B-PER. Suspension during mechanical lysis can be done directly by vortex.
27. The 96-well vacuum plate method is more flexible because it allows purification of small amounts of up to 96 different proteins or of higher amounts of selected proteins.
28. The His-fusion protein is present in the flow-through or eluted in the two-salt gradients depending on the protein's isoelectric point. Each operation is performed automatically by AKTA x-PRESS. With this system it is possible to have a two-step purification of three proteins per module. Each individual module works in parallel with the other modules and samples in the same module are processed in series. All purification steps are monitored at $OD\lambda_{280}$ nm.

4.4. Immunogenicity of Meningococcal Vaccine Antigens

29. Because the CD1 is an outbred strain of mouse, the use of such animals is probably more indicative of the variability of the immune response.
30. New Zealand rabbits are normally used for immunization to obtain larger volumes (about 50 mL or more) of antisera. This strain has half-lop ears, which make blood collection from the marginal veins a fairly straightforward procedure. For rabbit immunization, we usually use 50 μg of protein per dose.
31. Most antigens require an adjuvant to increase their immunogenicity and a number of different formulations can be used. Regulations on their use should be consulted prior to embarking on a course of immunizations. For many years Freund's Complete and Incomplete Adjuvant were the formulations of choice for all immunization work, but welfare issues have been raised over their use in recent years.
32. Multiple aliquots (0.5 mL) of the target strain(s) must be stored to prevent sub-culturing. Our recommended storage medium is Glycerol Broth (Nutrient Broth from Becton

Dickenson or equivalent) with 15% (v/v) glycerol. The stock of glycerol broth is prepared by taking a swab of an overnight meningococcal culture from a chocolate agar plate and emulsifying the bacteria in the glycerol broth to make a heavy suspension. This is then dispensed into 0.5 mL aliquots in sterile plastic vials and immediately stored frozen at −70°C.

33. Repeated freezing (−20°C) and thawing and heat inactivation of sera may influence the results in ELISA.
34. The diffusion of antibodies from aqueous solution to the solid phase may be a limiting factor for antibody binding. The effect of shaking the plates has been studied by incubating with serum with or without rapid shaking of the microtiter plates. In the absence of shaking, 4 h incubation steps are needed to obtain the same quantitative results as observed with 1 h incubation at room temperature with rapid shaking.
35. Switch on the spectrophotometer 30 min before taking a reading.
36. If bacterial culture takes 3 h or longer, growth is likely to have been compromised and the organism from that culture should not be used.
37. Eppendorf tubes can be used instead of 96-well microplates if only a few samples need to be analyzed; in this "tube procedure"
    (a) Prepare the bacteria as described in Subheading 3.4.2.
    (b) Transfer the bacterial culture to a sterile 50 mL polypropylene, high-speed conical tube (Fisher). Pour in sterile PBS containing 1% (w/v) BSA up to the 50 mL mark on the tube. Cap securely and wrap the cap with Parafilm. Centrifuge at 2,908 × *g* for 10 min in a Beckman Coulter Allegra using a specific rotor for 50 mL Falcon tubes.
    (c) Remove the tubes from the rotor in a Class II biosafety cabinet. Aspirate the supernatant and suspend the pellet from the original 7 mL of culture into 2 mL of PBS containing 1% (w/v) BSA (~3 to 5×10^8 CFU/mL).
    (d) Dispense 85 μL of bacterial suspension per test sample into 1.5 mL Eppendorf tubes. Next, add 10 μL of each test serum dilution and then incubate the mixture at either room temperature or at 37°C for 30 min.
    (e) Centrifuge the tubes at 13,400 × *g* in a microcentrifuge for 3 min to pellet the bacteria. Aspirate the supernatant and suspend each pellet in 100 μL of PBS containing 1% (w/v) BSA and R-FITC-conjugated goat antimouse Ig (final working concentration of 1/100). Incubate for 15 min at room temperature.
    (f) Pellet the bacteria by centrifugation (13,400 × *g* for 3 min), aspirate the supernatant and suspend each bacterial pellet

in 1 mL of freshly made, filter-sterilized (0.2 μm filter) PBS containing 0.5% (v/v) formaldehyde.

(g) Transfer the fixed samples to FACS tubes. The bacteria are completely killed in 10 min and safe to transport and process through the FACS machine for fluorescence measurements. However, sterility should be confirmed with pilot studies.

38. Clearly, the SBA cannot be used with bacteria that are too sensitive to complement in the absence of antibody, so this should be tested in advance and controlled in the assay. Pooled baby rabbit serum may be used for SBA. Keep at –70°C and transport on dry ice. Aliquot the stock of supplied commercial serum, which must only be defrosted for a minimum of time, into small volumes (1–3 mL) that must also be kept frozen at –70°C and defrosted only immediately prior to use. It is not recommended to freeze again any thawed complement that remains after the experiment. Using lyophilized lots of baby rabbit sera and reconstituting with ice water immediately prior to use is recommended. New lots of rabbit complement must be tested to determine the bactericidal activity of the complement itself. Positive control sera should also be used to demonstrate that new lots of complement maintain the same sensitivity as the previous lots, i.e., that lot to lot variation is minimal.

39. We recommend using an automated colony counter to facilitate counting of CFU and to standardize the SBA.

40. Decrease in CFU in Column 12. Column 12 is used to measure non-complement-mediated lysis of bacteria. The dilution of serum in this control well must always be the initial dilution that is tested and must be heat-inactivated, which is done in a water bath at 56°C for 30 min.

41. The controls are necessary to assess the growth of the bacteria, in order to detect any killing due to the diluent or complement alone and to detect any non-complement-mediated killing in the test sera. Internal controls are usually positive and negative sera or monoclonal antibodies with predetermined titers (high, low, or negative titer ranges) and they are analyzed in the same manner as the test sera. These controls are used to detect any trends in the response of the assay over time.

42. Each meningococcal strain has been passaged three times through infant rats. For new strains, it is important to first perform growth curves to assess the time required for the bacteria to reach the mid-log phase of growth.

43. The number of bacteria present in the suspension is calculated from measuring the OD at λ_{600}nm, where a reading of 0.23 is equivalent to 10^9 CFU. The calculation used to quantify the number of bacteria present in a suspension is not unique to this protocol but is a useful tool when the number of bacteria, live

or dead is required and it relies on the correlation between the amount of DNA present and the number of bacteria. The constant value given in the equation of 18 is only applicable to overnight plate grown bacteria and must be amended under differing growth conditions. For example when bacteria are grown to mid-log phase in MHB supplemented with 10% (w/v) glucose, the constant value of 22 must be substituted in order to obtain an accurate CFU quantification. The equation can also usefully be rewritten as follows when a specific number of bacteria are required.

An OD at λ_{260}nm of 18 is equivalent to 10^{10} bacteria per mL and this correlation can then be used to calculate the number of bacteria in the suspension as follows:

OD at λ_{260}nm × dilution factor × mL = number of bacteria × 10^{10}

1.8/OD at λ_{260}nm × dilution factor = the volume in mL that contains 1×10^9 CFU

These bacteria are in mid-log phase and in order to determine the concentration of the suspension it is crucial to use the constant of 2.2 in the equation, i.e.,

2.2/OD at λ_{260}nm × dilution factor = the volume in mL that contains 1×10^9 CFU

Prepare a 1×10^9-CFU/mL bacterial suspension and then dilute in PBS to the appropriate concentration(s). The total injection volume is 100 μL per pup.

44. The pups are randomly redistributed to the nursing mothers to obtain more homogeneous groups.

Acknowledgments

We are grateful to both Giorgio Corsi and Alessandro Aronica for artwork, Timothy Trevor Perkins and Jeannette Adu-Bobie for critical reading and manuscript editing, Enrico Luzzi and Francesca Ferlicca for their precious help in describing the ELISA and in vivo passive antibody protection methods and Beatrice Aricò, Maurizio Comanducci and Sara Comandi for their contribution.

References

1. Keating GM, Noble S (2003) Recombinant hepatitis B vaccine (Engerix-B (R)) - A review of its immunogenicity and protective efficacy against hepatitis B. Drugs 63: 1021–1051.
2. Cutts FT, Franceschi S, Goldie S et al (2007) Human papillomavirus and HPV vaccines: a review. Bull World Health Org 85: 719–726.
3. Pizza M, Covacci A, Bartoloni A et al (1989) Mutants of pertussis toxin suitable for vaccine development. Science 246: 497–500.
4. Stephens DS (2007) Conquering the meningococcus. Fems Micro Rev 31: 3–14.
5. Riordan A (2010) The implications of vaccines for prevention of bacterial meningitis. Curr Opin Neurol 23: 319–324.

6. Pizza M, Scarlato V, Masignani V et al (2000) Identification of vaccine candidates against serogroup B meningococcus by whole-genome sequencing. Science 287: 1816–1820.
7. Plotkin SA (2008) Correlates of vaccine-induced immunity. Clin Infect Dis 47: 401–409.
8. Medini D, Serruto D, Parkhill J et al (2008) Microbiology in the post-genomic era. Nature Rev Microbiol 6: 419–430.
9. Margarit I, Rinaudo CD, Galeotti CL et al (2009) Preventing bacterial infections with pilus-based vaccines: the Group B Streptococcus paradigm. J Infect Dis 199: 108–115.
10. Bagnoli F, Baudner B, Mishra R et al (2011) Designing the next generation of vaccines for global public health. OMICS: J Integrat Biol 17. [Epub ahead of print].
11. Scarselli M, Aricò B, Brunelli B et al (2011) Rational design of a meningococcal antigen inducing broad protective immunity. Sci Transl Med 3(91): 91–62.
12. Capecchi B, Adu-Bobie J, Di Marcello F et al (2005) *Neisseria meningitidis* NadA is a new invasin which promotes bacterial adhesion to and penetration into human epithelial cells. Mol Microbiol 55: 687–698.
13. Sambrook J, Russell DW (2001) Molecular cloning: a laboratory manual. CSHL Press,
14. Coligan JE, Dunn BM, Ploegh HL et al (2002) Curr Prot Prot Sci 1: 6.0.1–6.0.4.
15. Scopes RK (1994) Protein purification: principles and practices. Springer, New York.
16. Hopkins TR (1991) Physical and chemical cell disruption for the recovery of intracellular proteins, in: Seetharam R, Sharma SK (Eds.), Purification and analysis of recombinant proteins, Marcel Dekker, New York, pp. 57–83.
17. Bornhorst JA, Falke JJ (2000) Purification of proteins using polyhistidine affinity tags. Methods Enzymol 326: 245–254.
18. Kneusel RE, Crowe J, Wulbeck M, Ribbe J (2000) Procedures for the analysis and purification of His-tagged proteins, in: Rapley R (Ed.), The Nucleic Acid Protocols Handbook, Springer, pp. 921–934.
19. Feavers I, Walker B (2010) Functional Antibody Assays. Meth Molec Biol 626: 199–211.
20. Giuliani MM, du-Bobie J, Comanducci M et al (2006) A universal vaccine for serogroup B meningococcus. Proc Nat Acad Sci USA 103: 10834–10839.
21. Qin L, Gilbert PB, Corey L et al (2007) A framework for assessing immunological correlates of protection in vaccine trials. J Infect Dis 196: 1304–1312.
22. Welsch JA, Moe GR, Rossi R et al (2003) Antibody to genome-derived neisserial antigen 2132, a *Neisseria meningitidis* candidate vaccine, confers protection against bacteremia in the absence of complement-mediated bactericidal activity. J Infect Dis 188: 1730–1740.
23. Tjalsma H, Bolhuis A, Jongbloed JDH et al (2000) Signal peptide-dependent protein transport in *Bacillus subtilis*: a genome-based survey of the secretome. Microbiol Mol Biol Rev 64: 515–547.
24. Tjalsma H, Antelmann H, Jongbloed JDH et al (2004) Proteomics of protein secretion by *Bacillus subtilis*: Separating the "secrets" of the secretome. Microbiol Mol Biol Rev 68: 207–233.

Chapter 22

A DNA Vaccine Strategy for Effective Antibody Induction to Pathogen-Derived Antigens

Jason Rice and Myron Christodoulides

Abstract

DNA-based vaccines are currently being developed for treating a diversity of human diseases including cancers, autoimmune conditions, allergies, and microbial infections. In this chapter, we present a general protocol that can be used as a starting point for developing DNA vaccines to pathogen-derived antigens, using *Neisseria meningitidis* as an example. In addition, we describe a fusion gene-based vaccine protocol for increasing the potency of DNA vaccines that are based on poorly immunogenic antigens such as short pathogen-derived polypeptides. Finally, we provide a safe and effective protocol for delivery of DNA vaccines, based on intramuscular injection followed by electroporation.

Key words: DNA vaccine, Neutralizing humoral antibody, T cell help, Infectious disease

1. Introduction

DNA vaccines have been the subject of intense investigation over the last 2 decades and the technology provides an adaptable and powerful means for activating both innate and adaptive immune pathways and allowing multiple vaccines to be made and tested quickly and cost-effectively (1). The concept is very simple: DNA encoding the antigen of choice is inserted into a bacterial plasmid, with gene expression usually driven by a strong viral promoter. Delivery of the plasmid vaccine into muscle or skin cells leads to antigen production and presentation to the immune system, and both antibody and cell-mediated immune responses can be effectively induced. In addition, the plasmid also has excellent intrinsic adjuvant properties: the dsDNA acts as a pathogen-associated molecular pattern that can trigger a range of cellular receptors,

Myron Christodoulides (ed.), *Neisseria meningitidis: Advanced Methods and Protocols*, Methods in Molecular Biology, vol. 799, DOI 10.1007/978-1-61779-346-2_22,

leading to type I interferon production and an inflammatory response by cells of the innate immune system. Thus, plasmid DNA not only delivers antigen, but engages multiple routes to activate innate immunity.

DNA-based vaccines are in development for infectious diseases and as therapies against autoimmune diseases, allergy and cancer, and several DNA vaccines are already licensed for veterinary use (2). Importantly, DNA vaccines can induce antibody responses against bacterial pathogens where humoral immunity is believed to be essential, e.g. against peptide mimotopes of meningococcal group B and group C polysaccharides (3, 4) and porin polypeptide (5) and *Streptococcus pneumoniae* capsular polysaccharide (6) and surface antigens (7), *Borrelia burgdorferi* outer surface proteins (8, 9), *Brucella* outer membranes (OM) (10) and OM porin OprF of *Pseudomonas aeruginosa* (11).

In this chapter we present a general protocol that can be used as a starting point for developing DNA vaccines to pathogen-derived antigens, using *Neisseria meningitidis* as an example. In addition, for antigens that can be considered as poorly immunogenic, such as short pathogen-derived polypeptides (5), we also provide a protocol for adopting a fusion gene-based vaccine design similar to the hapten-carrier system to increase the potency of our DNA vaccines (2). This strategy links highly immunostimulatory sequences, such as the fragment C (FrC) sequence of tetanus toxin, to weak target antigens within the DNA vaccine format. This fusion gene vaccine strategy can induce high-affinity antibodies to the linked target antigen through several mechanisms, the most important of which is the cognate $CD4^+$ T-cell help provided by linked immunostimulatory elements. Finally, we provide a protocol for efficient delivery of DNA vaccine, based on intramuscular injection followed by electroporation, which delivers an immediate electrical current across the injection site. This causes an influx of inflammatory cells and improved cell transfection efficiency, leading to increased antigen production and presentation to the immune system. Importantly, this delivery system is safe and effective (12, 13) and appears to overcome the translational block to DNA vaccination in the clinic.

2. Materials

2.1. Bacterial Strains and Growth Conditions

1. 1% (w/v) proteose-peptone (Oxoid) and 8% (v/v) glycerol in water, for storage of *Neisseria meningitidis* under liquid nitrogen.
2. Luria Bertani (LB) medium (Oxoid) containing 10% (v/v) glycerol for storage of *Escherichia coli* strain JM109 competent cells (Promega) under liquid nitrogen. LB medium: 10 g bacto-tryptone (Sigma-Aldrich), 5 g yeast extract (Sigma-Aldrich) and

10 g NaCl per litre and autoclave (15 min at 6.8 kg per 0.4 cm^2).

3. GC agar medium for growth of meningococci (14).
4. LB agar medium for growth of *E. coli*, prepared by adding 15 g agar per litre of autoclaved LB medium.

2.2. Polymerase Chain Reaction (PCR)

1. Genomic template DNA.
2. Oligonucleotide primers. These are generally purchased as lyophilised pellets, which are resuspended in distilled, de-ionised water to prepare high concentration stocks (100 pmol/μL). Aliquots of working stocks (50 μL) are stored at –20°C to minimize freeze/thaw cycles.
3. HotStarTaq DNA polymerase (QIAGEN, supplied at 5 U/μL) and the proprietary 10× PCR buffer supplied with the enzyme. The enzyme is stored at –20°C in the supplied buffer and during use it is kept on ice in an insulated polystyrene cooler to protect it from temperature fluctuations (see Note 1).
4. 10× dNTP solution: 2 mM dATP, 2 mM dTTP, 2 mM dCTP, 2 mM dGTP in de-ionised water. We use a dNTP set supplied by Promega in which each dNTP is supplied separately at 100 mM. These are combined and diluted 1:50 using de-ionised water to produce a working stock containing all four dNTPs at 2 mM. Aliquots are stored at –20°C to minimize freeze/thaw cycles.
5. Suitable thin-walled microcentrifuge tubes (e.g. STARLAB 0.2 mL PCR tube).
6. Thermal cycler, e.g. GeneAmp PCR System 9700 (Applied Biosystems).

2.3. Agarose Gel Electrophoresis

1. Agarose (Fisher Scientific).
2. 1× TAE buffer: 40 mM Tris-acetate, 1 mM EDTA. A 50× TAE stock is made using 242 g Tris–HCl, 57.1 mL glacial acetic acid, and 200 mL 0.25 M Na_2EDTA (pH 8.0) per litre.
3. Agarose gel: 1% (w/v) agarose in 1× TAE buffer.
4. 10 mg/mL ethidium bromide (EtBr) stock solution in TE buffer: 10 mM Tris–HCl, pH 8.0, containing 0.1 mM Na_2EDTA.
5. 6× gel loading buffer: 0.25% (w/v) bromophenol blue, 0.25% (v/v) xylene cyanol, 3 mL glycerol made up to 10 mL with TE buffer.
6. DNA molecular weight markers (Bioline HyperLadder I).

2.4. Purification of PCR Products and Digested Vector

1. QIAquick Gel Extraction kit (QAIGEN).
2. Geneclean II kit (BIO 101).
3. Elution buffer: 10 mM Tris–HCl, pH 8.5.

2.5. Restriction Digestion of PCR Products and Expression Vectors

1. For the selected antigen, purchase the appropriate restriction enzymes (typically 10–20 U/μL) together with the appropriate 10× reaction buffers from the same supplier (we usually use enzymes supplied by New England Biolabs or Promega) (see Note 1).
2. Elution buffer: 10 mM Tris–HCl, pH 8.5.

2.6. Vector Ligation

1. Purchase a commercial plasmid vector that promotes gene expression in mammalian cells. We use several vectors for our vaccine construction, including pcDNA3.1 (Invitrogen) and pCI (Promega). Both vectors are designed to promote constitutive expression of cloned DNA inserts in mammalian cells. The pCI vector includes a chimaeric intron that is reported to frequently increase the level of gene expression. These vectors are generally suitable for preclinical studies in animals; however, vector choice for clinical testing will be influenced by the appropriate regulatory authorities.
2. T4 DNA ligase (Promega, supplied at 1–3 U/μL) and the proprietary 10× reaction buffer supplied with the enzyme. The enzyme is stored frozen at –20°C in aliquots to avoid freeze–thaw cycles (see Note 1).

2.7. Transformation of Vector into Expression Hosts

1. *E. coli* strain JM109 competent cells.
2. Ampicillin at 100 mg/mL stock concentration in de-ionised, sterile water.
3. LB medium (step 2, Subheading 2.1).
4. LB agar plates with ampicillin: 15 g agar per litre of LB medium; autoclave the mixture and add ampicillin when hand-cool to a final concentration of 100 μg/mL; pour the plates and allow to set.

2.8. Screening and Validation of Expression Constructs

1. QIAprep Spin Miniprep kit (QIAGEN) with supplied elution buffer: 10 mM Tris–HCl, pH 8.5.
2. Inoculation loops, 1 μL (Greiner Bio One).
3. Appropriate restriction enzymes and suitable 10× reaction buffer (e.g. New England Biolabs or Promega).
4. ABI Prism Big Dye Terminator v1.1 Cycle Sequencing Kit (Applied Biosystems).
5. Oligonucleotide sequencing primers: 1.6 pmol/μL working stock.
6. Suitable thin-walled microcentrifuge tubes (e.g. STARLAB 0.2 mL PCR tube).
7. Hi-Di formamide (Applied Biosystems).
8. 3 M sodium acetate buffer: 24.61 g in 50 mL de-ionised water, adjust pH to 5.2 with glacial acetic acid and make up to 100 mL with water.

9. 3130xl Genetic Analyzer (Applied Biosystems) for DNA sequencing.
10. MacVector (MacVector, Inc.) and Lasergene (DNASTAR) DNA sequencing software.

2.9. Large Scale DNA Vaccine Production and Storage

1. QIAfilter Plasmid Giga kit (QIAGEN).
2. Alcohols: isopropanol, 70% (v/v) ethanol.
3. 50% (v/v) glycerol in de-ionised water: aliquot 0.5 mL of the solution into sterile cryo-tubes (e.g. Cryo.s tubes, Greiner Bio-One), re-cap them and autoclave (15 min at 6.8 kg per 0.4 cm^2).
4. BioPhotometer (Eppendorf AG).

2.10. DNA Vaccine Preparation and Recommended Vaccination Schedule

1. 70% (v/v) ethanol.
2. Sterile sodium chloride 0.9% (w/v) for in vivo use. We use "NaCl 0.9% for intravenous infusion BP" (Baxter).
3. The appropriate animal model (e.g. an inbred strain of mouse, usually BALB/c) and the necessary permissions to conduct animal research.
4. Myjector 27 G× ½″ 0.5 mL syringe (Terumo).
5. Elgen pulse generator and associated software (Inovio Pharmaceuticals, Inc.) for electroporation.
6. Conductance gel: 0.57 g NaCl, 2.3 g methyl cellulose, 1 M NaOH, make up to 100 mL with de-ionized water.

3. Methods

3.1. Bacterial Strains and Growth Conditions

Grow *Neisseria meningitidis* on GC agar medium plates at 37°C for 18 h in an atmosphere of 5% v/v CO_2 (14) and *E. coli* strains on LB agar medium plates and in LB liquid medium in a dry 37°C incubator.

3.2. Polymerase Chain Reaction (PCR)

1. Genomic DNA template: single colonies of overnight growth of *N. meningitidis* are picked and suspended in 10 μL distilled, de-ionised water in a sterile tube. To extract DNA, the bacteria are lysed with the addition of 10 μL 0.25 M KOH with boiling for 5 min. To adjust the pH for PCR, 10 μL of 0.5 M Tris–HCl, pH 7.5 buffer is then added. The lysed samples are then diluted to 300 μL with water, briefly centrifuged to remove any particulate material, and stored at –20°C. A negative control is also prepared from 10 μL de-ionised water treated similarly.
2. In thin-walled PCR tubes assemble the PCR reaction components consisting of: 10–500 ng template DNA (or de-ionised

water for control reaction), 10 μL 10× reaction buffer, 10 μL dNTP solution (2 mM of each; final concentration in reaction: 200 μM each), 1 μL of each forward and reverse primer (100 pmol each), 1 μL HotStarTaq DNA polymerase (5 U) and de-ionised water to make a final volume of 100 μL (see Note 1).

3. Place the reaction tubes in a thermal cycler and subject to PCR using a typical PCR program as follows: an initial denaturation step at 95°C for 15 min (required to activate the HotStarTaq enzyme), followed by 30 cycles each at 94°C for 0.5 min (denaturation), 50–68°C for 0.5 min (annealing), 72°C for 1 min (extension). Finish the reaction with a final extension step at 72°C for 10 min (see Notes 5 and 6).
4. Several techniques can be used to incorporate a leader sequence or immunostimulatory sequence, such as FrC of tetanus toxin, into the final vaccine if required. One method involves serial restriction digestion and ligation of each PCR component into the vector backbone, building up the insert in stages, although careful planning is required to ensure that unwanted coding sequence does not remain from restriction enzyme sites. We recommend the technique of PCR SOEing (15), which links separate PCR products together in frame and amplifies the resulting product. To achieve this, perform a separate PCR reaction for each component (e.g. leader, antigen, FrC) using PCR primers that provide an overlap with the adjacent vaccine component. To link, e.g. the PCR product encoding the antigen sequence with that encoding FrC:
   (a) Visualize and purify each separate PCR fragment (Subheadings 3.3 and 3.4).
   (b) Set up a new PCR reaction (step 2, Subheading 3.2) but using 1–5 μL of each purified PCR product as template DNA and the two primers from the original PCR reactions that will anneal at the extreme 5′ and 3′ termini of the final bonded PCR product.
   (c) Subject to PCR using a typical program (step 3, Subheading 3.2), but with slight modification: after the initial denaturation step run the program for five cycles with a low annealing temperature (e.g. 5°C below the lowest that was used in the original PCR reactions) followed by 25 cycles with the annealing temperature identical to the lowest used in the original PCR reactions.
   (d) Visualize and purify the larger PCR product (Subheadings 3.3 and 3.4). Now repeat step 4, Subheading 3.2 to add a leader sequence, if required.

3.3. Agarose Gel Electrophoresis

1. Mix 5 μL aliquots of the PCR reactions with 1 μL of 6× loading buffer and load onto a 1% (w/v) agarose gel containing 0.5 μg/

mL EtBr for electrophoresis; load an additional lane with DNA molecular weight markers (see Note 4).

2. Carry out electrophoresis at ~5 V/cm (measured as the distance between the electrodes) until the dyes in the loading buffer have separated 2–4 cm. The distance travelled by the PCR product is size-dependent, so check the gel regularly to ensure the product does not travel too far in the gel.
3. View the gel on a UV light box and gauge the size of the resulting PCR product; there should be no corresponding band in the negative control lane.

3.4. Purification of PCR Products and Digested Vector

1. Following UV visualisation of the PCR product, purify the remaining PCR product using one of a variety of commercially available kits. We use the QIAquick Gel Extraction kit in two ways:
   (a) If the PCR reaction produces a single DNA band of the expected size, purify this directly from the PCR reaction liquid. Elute the purified PCR product in 50 μL elution buffer.
   (b) If the PCR reaction produces several discrete DNA products, then subject some/all of the remaining PCR reaction to electrophoresis on an agarose gel and view the DNA products using a UV lamp. Use a clean scalpel blade to cut the DNA band of the expected size from the gel and transfer it to a sterile 1.5 mL microcentrifuge tube. Take care to excise only the required band with minimal excess gel. Purify the DNA using the Gel Extraction kit and elute the PCR product in 50 μL elution buffer.

3.5. Restriction Digestion of PCR Products and Expression Vectors

Digest the PCR product and the vaccine expression vector with the same restriction enzymes to enable cloning of the product into the vector.

1. Set up a restriction digestion reaction in a 1.5-mL microcentrifuge tube containing 0.5–2 μg PCR product, 10–20 U restriction enzyme, 10× reaction buffer (final concentration 1×), and de-ionised water. It may be possible to use two restriction enzymes in one reaction if they are compatible in the same 10× reaction buffer (see Note 2). Ensure that enzyme is the final ingredient added to the reaction and that the total volume of enzyme does not exceed 10% of the reaction volume.
2. Incubate the reaction in a heating block or water bath at 37°C for 2–3 h.
3. Purify the digested PCR product using a commercial kit (Subheading 3.4) and elute the DNA in 50 μL elution buffer.
4. If digestion with a second restriction enzyme is required then repeat the above steps 1–3 (see Note 2).
5. Digest the expression vector using the method given above (steps 1–4).

6. Run the digested vector on an agarose gel as described above (Subheading 3.3).
7. Visualise the band representing the linearised vector using a UV lamp and cut the band from the gel using a scalpel blade.
8. Purify the linearised vector using a commercial kit (Subheading 3.4) and elute the DNA in 50 μL elution buffer. Perform a second digestion if required and re-purify the digested vector.

3.6. Vector Ligation

1. Set up a ligation reaction for the vaccine construct in a sterile 1.5 mL microcentrifuge tube. This should include 1 μL of linearised plasmid vector (~100 ng/μL), 5 μL of purified, restriction-digested PCR product (10–20 ng/μL), 1 μL of 10× ligase reaction buffer, 1 μL of T4 DNA ligase (1–3 U), and 2 μL of de-ionised water.
2. Set up a control reaction containing no PCR product, instead replacing it with de-ionised water. This can be used to determine background levels of vector re-ligation or incomplete linearization.
3. Keep the ligation reactions overnight at 4°C.

3.7. Transformation of Vector into Expression Hosts

1. Gently thaw the *E. coli* JM109 competent cells on ice (see Note 10).
2. Using separate, pre-chilled 1.5 mL microcentrifuge tubes mix aliquots (50 μL) of the JM109 cells with aliquots (2–5 μL) of the ligation reactions described above (Subheading 3.6); flick the tubes gently to mix the contents and keep on ice for 20 min.
3. Heat shock the cells by placing the reaction tubes in a water bath for 45–50 s at exactly 42°C, then return them to ice for 2 min.
4. Add 0.5 mL LB medium to each transformation reaction, gently invert the tubes to mix and incubate for 1 h at 37°C on a shaker (200 rpm).
5. Gently pellet the cells using a microcentrifuge (~955×*g* for 10 min), resuspend in 50–100 μL LB medium and plate them out onto LB agar containing 100 μg/mL ampicillin. Incubate the plates overnight at 37°C.
6. Following incubation (~16 h) each agar plate containing bacteria transformed with PCR product should have numerous colonies (50–500) (see Note 3). The control plate should have few or no colonies.

3.8. Screening and Validation of Expression Constructs

1. Select single colonies, using individual sterile micropipette tips or sterile inoculation loops, and inoculate each into separate 25 mL sterile plastic universal tubes containing 5 mL LB medium supplemented with 100 μg/mL ampicillin; select up

to 12 colonies per plate. Plates secured with Saran wrap can be stored upside down at 4°C for ~1 month.

2. Incubate overnight (~16 h) at 37°C on a shaker (200 rpm).
3. Purify plasmid DNA from 1.5 mL of each culture using a suitable commercial kit. Elute the purified DNA in 100 μL elution buffer supplied. Excess culture volume can be stored for the short term at 4°C for subsequent use in large-scale plasmid preparation (Subheading 3.9).
4. Run 5 μL of each purified DNA sample on a 1% (w/v) agarose gel to ensure sufficient yield, together with a DNA size marker.
5. This small-scale DNA preparation is suitable for screening by restriction digestion and DNA sequencing to identify plasmids containing the correct DNA insert.
   (a) Use the appropriate restriction enzymes to digest the purified plasmids, using 5 μL of each DNA preparation; if the two desired restriction enzymes cannot be used together then two digestion reactions should be performed as described above (Subheading 3.5).
   (b) Run the digested plasmids on a 1% (w/v) agarose gel, together with a separate molecular weight marker; the presence of a band of the correct size corresponding to the excised DNA insert provides initial confirmation that the undigested plasmid may contain the PCR product insert.
6. Using a commercially available kit, sequence the DNA insert to confirm its integrity and orientation/location within the plasmid vector, as follows (see Notes 7 and 11):
   (a) Set up sequencing reactions in 0.2 mL PCR tubes that contain 2 μL Big Dye Terminator Ready Reaction Mix (labelled A-dye, C-dye, G-dye and T-dye terminators, dNTPs, AmpliTaq DNA polymerase), 2 μL 5× sequencing buffer, 5 μL miniprep DNA, 1 μL sequencing primer (1.6 pmol/μL).
   (b) Perform sequencing reactions on a thermal cycler (step 6, Subheading 2.2) using the following conditions: 94°C for 10 s (denaturation), 50°C for 5 s (annealing), and 60°C for 4 min (extension) for 25 cycles.
   (c) Precipitate the DNA from each sequencing reaction by adding the 10 μL reaction volume to a 1.5-mL microcentrifuge tube containing 50 μL ethanol (100%) and 2 μL 3 M sodium acetate buffer, pH 5.2 and keep on ice for 30 min.
   (d) Pellet the DNA precipitate in a microcentrifuge at 20,817×g for 30 min at 4°C; remove the liquid carefully with a micropipette tip (care! the pellet will not be visible).
   (e) Wash the DNA pellet by adding 150 μL of 70% (v/v) ethanol and microcentrifuge at 20,817×g for 5 min at 4°C; remove the liquid carefully and air dry the pellet for 5–10 min.

(f) Resuspend the DNA pellet in 10 μL formamide and run the reaction products on a DNA sequencing machine.

(g) Analyze the sequencing data using the appropriate software; to achieve this, first produce a word processing file containing a template of the single-letter nucleotide sequence for the DNA insert that you are trying to create; the published DNA sequence of known proteins can be found by searching a suitable nucleotide database (e.g. http://www.ncbi.nlm.nih.gov/nuccore). Use the analysis software (according to the manufacturer's instructions) to align and compare your expected template sequence with the data generated from the sequencing reactions to determine whether your DNA inserts are correct (see Notes 8 and 9).

3.9. Large-Scale DNA Vaccine Production and Storage

1. Once the correct vaccine vector has been identified, plate out 50 μL of the corresponding bacterial culture (from step 3, Subheading 3.8 above) onto a fresh LB agar plate supplemented with 100 μg/mL ampicillin. Incubate overnight (~16 h) at 37°C.
2. Pick a single bacterial colony and inoculate into a 25-mL universal tube containing 10 mL LB plus 100 μg/mL ampicillin. Incubate this starter culture during the day (~8 h) at 37°C on a shaker (200 rpm). Culture a separate colony as insurance against failure of the first to grow sufficiently. Colony growth should be evident after 8 h with the LB medium beginning to look cloudy.
3. Use the 10 mL starter culture to inoculate 1.2 L of LB medium supplemented with 100 μg/mL ampicillin. Incubate this large-scale culture overnight at 37°C on a shaker (200 rpm).
4. Reserve 2 mL of the overnight grown culture for long-term storage of this vaccine-containing host. Add 0.5 mL of this to a sterile cryotube containing 0.5 mL 50% (v/v) glycerol and mix gently by inversion. We recommend preparing several glycerol stocks of each vaccine strain and storing them separately at –20 and –80°C. Colonies can be recovered by using a sterile loop to streak out a small volume of the stock onto a fresh LB agar plate containing 100 μg/mL ampicillin.
5. Purify the DNA vaccine construct using the suggested commercial kit, which will purify up to 10 mg plasmid DNA (see Note 11).
6. Resuspend the resulting purified DNA pellet in 5 mL of 10 mM Tris–HCl buffer, pH 8.0.
7. Determine the DNA yield and concentration by measuring the absorbance at 260 nm wavelength. The ratio of absorbance at 260 and 280 nm wavelength (A_{260}/A_{280}) should be above 1.8 to ensure minimal contamination with protein and RNA.

8. Validate DNA vaccine integrity by DNA sequencing as described above (step 6, Subheading 3.8).
9. Mix 1 mg aliquots of the DNA vaccine vector with 0.8× volume of isopropanol (100%) and 0.1× volume of 3 M sodium acetate buffer, pH 5.2 in 1.5 mL microcentrifuge tubes. Mix the contents by inversion and store the 1 mg DNA aliquots long-term as isopropanol precipitates at −20°C.

3.10. DNA Vaccine Preparation and Recommended Vaccination Schedule

1. Pellet the 1 mg DNA vaccine aliquot in a microcentifuge at 20,817×*g* for 10 min at 4°C.
2. Decant the liquid, add 1 mL 70% (v/v) ethanol and microcentifuge for a further 5 min.
3. Decant the liquid, microcentrifuge at 20,817×*g* for 5 s at 4°C and remove all remaining liquid with a pipette tip. Allow the pellet to air dry for no longer than 5 min.
4. Resuspend the pellet in 1 mL sterile 0.9% (w/v) NaCl (saline) for injections. The pellet will usually dissolve with gentle agitation at 37°C within no more than 30 min.
5. Check the DNA vaccine concentration (absorbance at 260 nm wavelength) and prepare samples for injection by dilution with sterile saline.
6. For the generation of antibody responses in murine models (see Note 12), we recommend either of the following two schedules:
   (a) Single injection into each tibialis anterior muscle in the hind limbs of 50 μL DNA vaccine (0.5 mg/mL) in sterile saline using a 27-G needle. Repeat injections on days 21 and 42.
   (b) Single injection into each rear tibialis anterior muscle as above. Give booster injections at day 21 in combination with electroporation, using a commercial pulse generator as described below.
7. To perform DNA vaccination with electroporation first anaesthetize the mice using an appropriate anaesthetic, then shave the fur overlaying the quadriceps muscle at the front of each rear limb exposing bare skin.
8. Inject the DNA vaccine directly into the quadriceps muscle of one limb and then immediately apply conductance gel to the skin overlaying this muscle and the surrounding area.
9. Immediately place the silver electrodes of the Elgen pulse generator on the skin on either side of the vaccine injection site and apply a local electrical field using the pulse generator and associated software according to the manufacturer's instructions. We use an electrical field comprising 10 trains of 1,000 square wave pulses delivered at a frequency of 1,000 Hz, with each pulse lasting 400 μs (200 μs positive and 200 μs negative).

Each train is delivered at 1 s intervals with the electric pulse kept constant at ±50 mAmp.

10. As the electrical pulses are delivered, move the electrodes slowly up and down the muscle length to ensure that the whole muscle receives electroporation. Make sure that skin contact is maintained and that conductance gel fully covers the area to be electroporated.
11. Repeat the DNA vaccine injection and electroporation procedure into the second rear quadriceps muscle and allow the mice to recover from anaesthesia; no adverse effects should be observed following electroporation. Excluding the induction and recovery from anaesthesia, the whole procedure (steps 7–11, above) takes approximately 2 min per animal.
12. Collect blood/tissues for immune screening according to your model (e.g. day 42/56) (see Notes 13, 14 and 15).

4. Notes

1. Store all enzymes at −20°C in the buffer in which they are supplied. For use, keep the enzyme(s) in an insulated bench top cooler or ice box to protect it from temperature fluctuations which could inactivate it.
2. For "double" restriction digestion reactions (i.e. when two restriction enzymes are used together in one digestion reaction) ensure that both enzymes retain an appropriate level of activity in the reaction buffer; this may not be possible with certain enzyme combinations and two separate digestion reactions may need to be performed, with the template DNA cleaned up between each reaction to remove enzyme and buffer (e.g. using the QIAquick Gel Extraction kit). Commercial suppliers will provide information on reaction buffer compatibility.
3. If difficulty is experienced in cloning the PCR product into the desired vaccine vector then we recommend the use of pGEM-T Easy Vector System I (Promega), if appropriate. PCR amplification with Taq DNA polymerase introduces a template-independent single A nucleotide overhang at the 3′ end of the DNA strand. This allows relatively straightforward cloning of the purified PCR product directly into the T vector which has a complementary 3′ single T residue overhang. Once cloned into this plasmid, the DNA insert may then be excised by restriction digestion and inserted into the vaccine vector of choice, assuming that the appropriate restriction enzyme sites occur in both vectors.
4. When preparing agarose gels, ensure that the melted gel is hand cool and no longer giving off steam before adding the EtBr

solution. EtBr is a ***mutagen*** that is stored at room temperature in the dark: ensure that suitable protective clothing and containment procedures are used and that the "spent" gel is disposed of in a suitable manner.

5. For optimal PCR specificity and product yield, temperatures and cycling times should be optimized for each new template sequence or primer set. Generally, the annealing temperature should be 5°C below the lowest melting temperature (T_m) of the primers used.
6. For PCR products larger than 1 kb, use an extension time of 1 min per kb DNA product. If using a different DNA polymerase then check the manufacturer's instructions for recommended extension times.
7. For complete DNA sequencing of vaccine inserts, design and purchase sequencing primers in the forward and reverse orientations that anneal at ~300 base intervals covering the whole of the DNA insert plus the insertional junction with the vector sequence.
8. Bacterial clones that do not contain the correct plasmid insert should be discarded: e.g. the insert may be missing, certain sequences may be deleted/inserted or DNA sequencing may reveal nucleotide mutations leading to single amino acid changes. For these cases, there are several options: (a) screen more bacterial colonies; (b) repeat the restriction enzyme digestion/ligation/cloning process using stored PCR product; and/or (c) begin the PCR construction process of the insert DNA from scratch.
9. Although we perform our own sequencing reactions, various commercial enterprises offer DNA sequencing as a service. The exact requirements may vary, but many commercial services simply require a sample of the plasmid DNA and sequencing primers; they will carry out the reactions and subsequent gel/capillary analysis. We routinely use Geneservice (http://www.geneservice.co.uk/).
10. *E. coli* competent cells: we prefer to buy ready-competent bacteria for transformations, for reasons of reproducibility and convenience. Alternatively, standard methods can be used to make *E. coli* competent for DNA uptake (16).
11. Whilst we recommend various commercial kits, reagents, services, equipment and software that we routinely use in our laboratories, we do not endorse these; alternatives are usually available from other sources and all should be used according to the manufacturer's instructions.
12. We routinely include a signal peptide/leader sequence at the N-terminus of our DNA vaccine-encoded antigen sequence. This guides the encoded antigen into the endoplasmic reticulum

where folding and initial glycosylation occur. We have found that this can promote the induction of both antibody and T-cell responses. We routinely use the leader sequence derived from the V_H gene utilized by the IgM of the murine BCL1 tumour (17). The leader sequence is usually amplified by PCR using a reverse primer that overlaps with the 5′-terminus of the antigen of choice. The leader and antigen sequences can be separately amplified by PCR and subsequently linked by PCR SOEing (15). Alternatively, the target protein may contain its own leader sequence.

13. Analysis of humoral immune response is largely dependent on the meningococcal antigens(s) of interest, but should include enzyme-linked immunosorbent assays against native protein, OM and vesicles and whole bacteria; Western immunoblotting; immuno-fluorescence assays; serum bactericidal activity and opsonophagocytosis assays. Detailed protocols for these assays can be found in a previous series volume (18).
14. If the resulting immune response is weak, there is an option to introduce an in-built immunostimulatory sequence. We recommend the FrC portion of tetanus toxin protein as a non-toxic, immunogenic fusion partner for the generation of a DNA fusion vaccine.
15. Ensuring the correct tertiary structure of the antigen encoded by a DNA vaccine is crucial for the induction of an appropriate and effective humoral antibody response. When incorporating FrC as a fusion partner for the pathogen-derived target antigen of choice we routinely include a short "linker" between the two sequences, allowing each to fold into its natural conformation following protein expression in vivo. Previously, we have used the linker sequence AAAGPGP (5).

Acknowledgments

We are indebted to Leukaemia and Lymphoma Research (grant number 08025) and the University of Southampton Strategic Development Fund for supporting this research.

References

1. Gurunathan S, Klinman DM, Seder RA (2000) DNA vaccines: Immunology, application, and optimization. Ann Rev Immunol 18: 927–974.
2. Rice J, Ottensmeier CH, Stevenson FK (2008) DNA vaccines: precision tools for activating effective immunity against cancer. Nat Rev Cancer 8: 108–120.
3. Prinz DM, Smithson SL, Kieber-Emmons T et al (2003) Induction of a protective capsular polysaccharide antibody response to a multi-epitope DNA vaccine encoding a peptide mimic of meningococcal serogroup C capsular polysaccharide. Immunol 110: 242–249.
4. Beninati C, Midiri A, Mancuso G et al (2006) Antiidiotypic DNA vaccination induces serum

bactericidal activity and protection against group B meningococci. J Exp Med 203: 111–118.

5. Zhu D, Williams JN, Rice J et al (2008) A DNA fusion vaccine induces bactericidal antibodies to a peptide epitope from the PorA porin of *Neisseria meningitidis*. Infect Immun 76: 334–338.
6. Lesinski GB, Smithson SL, Srivastava N et al (2001) A DNA vaccine encoding a peptide mimic of *Streptococcus pneumoniae* serotype 4 capsular polysaccharide induces specific anti-carbohydrate antibodies in Balb/c mice. Vaccine 19: 1717–1726.
7. Miyaji EN, Dias WO, Gamberini M et al (2001) PsaA (pneumococcal surface adhesin A) and PspA (pneumococcal surface protein A) DNA vaccines induce humoral and cellular immune responses against *Streptococcus pneumoniae*. Vaccine 20: 805–812.
8. Wallich R, Siebers A, Jahraus O et al (2001) DNA vaccines expressing a fusion product of outer surface proteins A and C from *Borrelia burgdorferi* induce protective antibodies suitable for prophylaxis but not for resolution of Lyme disease. Infect Immun 69: 2130–2136.
9. Scheiblhofer S, Weiss R, Durnberger H et al (2003) A DNA vaccine encoding the outer surface protein C from *Borrelia burgdorferi* is able to induce protective immune responses. Microb Infect 5: 939–946.
10. Cassataro J, Velikovsky CA, de la Barrera S et al (2005) A DNA vaccine coding for the *Brucella* outer membrane protein 31 confers protection against *B.melitensis* and *B.ovis* infection by eliciting a specific cytotoxic response. Infect Immun 73: 6537–6546.
11. Price BM, Galloway DR, Baker NR et al (2001) Protection against *Pseudomonas aeruginosa* chronic lung infection in mice by genetic immunization against outer membrane protein F (OprF) of *P-aeruginosa*. Infect Immun 69: 3510–3515.
12. Buchan S, Gronevik E, Mathiesen I et al (2005) Electroporation as a "prime/boost" strategy for naked DNA vaccination against a tumor antigen. J Immunol 174: 6292–6298.
13. Low L, Mander A, McCann K et al (2009) DNA vaccination with electroporation induces increased antibody responses in patients with prostate cancer. Human Gene Ther 20: 1269–1278.
14. Tinsley CR, Heckels JE (1986) Variation in the expression of pili and outer membrane protein by *Neisseria meningitidis* during the course of meningococcal infection. J Gen Microbiol 132: 2483–2490.
15. Horton R M (1995) PCR-mediated recombination and mutagenesis. SOEing together tailor-made genes. Mol Biotechnol 3: 93-99.
16. Sambrook J, Russell D W (2001) Molecular Cloning: a laboratory manual. Cold Spring Harbor Laboratory Press, New York.
17. Rice J, King CA, Spellerberg MB et al (1999) Manipulation of pathogen-derived genes to influence antigen presentation via DNA vaccines. Vaccine 17: 3030–3038.
18. Pollard A J, Maiden M C J. (2001) Meningococcal vaccines: methods and protocols. Humana Press, Totowa, New Jersey.

Index

A

B

C

Myron Christodoulides (ed.), *Neisseria meningitidis: Advanced Methods and Protocols*, Methods in Molecular Biology, vol. 799, DOI 10.1007/978-1-61779-346-2, © Springer Science+Business Media, LLC 2012

G

H

I

K

L

M

R

S